1805437162
WITHDRAWN FROM STOCK
CPC UCW
LLYFRGELL
LIBRARY
ABERYSTWYTH
AF744633

CONTROL OF VIRUS DISEASES

SYMPOSIA OF THE SOCIETY FOR GENERAL MICROBIOLOGY*

1 THE NATURE OF THE BACTERIAL SURFACE
2 THE NATURE OF VIRUS MULTIPLICATION
3 ADAPTATION IN MICRO-ORGANISMS
4 AUTOTROPHIC MICRO-ORGANISMS
5 MECHANISMS OF MICROBIAL PATHOGENICITY
6 BACTERIAL ANATOMY
7 MICROBIAL ECOLOGY
8 THE STRATEGY OF CHEMOTHERAPY
9 VIRUS GROWTH AND VARIATION
10 MICROBIAL GENETICS
11 MICROBIAL REACTION TO ENVIRONMENT
12 MICROBIAL CLASSIFICATION
13 SYMBIOTIC ASSOCIATIONS
14 MICROBIAL BEHAVIOUR 'IN VIVO' AND 'IN VITRO'
15 FUNCTION AND STRUCTURE IN MICRO-ORGANISMS
16 BIOCHEMICAL STUDIES OF ANTIMICROBIAL DRUGS
17 AIRBORNE MICROBES
18 THE MOLECULAR BIOLOGY OF VIRUSES
19 MICROBIAL GROWTH
20 ORGANIZATION AND CONTROL IN PROKARYOTIC AND EUKARYOTIC CELLS
21 MICROBES AND BIOLOGICAL PRODUCTIVITY
22 MICROBIAL PATHOGENICITY IN MAN AND ANIMALS
23 MICROBIAL DIFFERENTIATION
24 EVOLUTION IN THE MICROBIAL WORLD
25 CONTROL PROCESSES IN VIRUS MULTIPLICATION
26 THE SURVIVAL OF VEGETATIVE MICROBES
27 MICROBIAL ENERGETICS
28 RELATIONS BETWEEN STRUCTURE AND FUNCTION IN THE PROKARYOTIC CELL
29 MICROBIAL TECHNOLOGY: CURRENT STATE, FUTURE PROSPECTS
30 THE EUKARYOTIC MICROBIAL CELL
31 GENETICS AS A TOOL IN MICROBIOLOGY
32 MOLECULAR AND CELLULAR ASPECTS OF MICROBIAL EVOLUTION
33 VIRUS PERSISTENCE
34 MICROBES IN THEIR NATURAL ENVIRONMENTS
35 INTERFERONS: FROM MOLECULAR BIOLOGY TO CLINICAL APPLICATIONS
36 THE MICROBE 1984 PART I: VIRUSES
PART II: PROKARYOTES AND EUKARYOTES
37 VIRUSES AND CANCER
38 THE SCIENTIFIC BASIS OF ANTIMICROBIAL CHEMOTHERAPY
39 REGULATION OF GENE EXPRESSION – 25 YEARS ON
40 MOLECULAR BASIS OF VIRUSES DISEASE
41 ECOLOGY OF MICROBIAL COMMUNITIES
42 THE NITROGEN AND SULPHUR CYCLES
43 TRANSPOSITION
44 MICROBIAL PRODUCTS: NEW APPROACHES

* Published by the Cambridge University Press, except for the first Symposium, which was published by Blackwell's Scientific Publications Limited.

CONTROL OF VIRUS DISEASES

EDITED BY

N. J. DIMMOCK, P. D. GRIFFITHS

AND C. R. MADELEY

FORTY-FIFTH SYMPOSIUM OF

THE SOCIETY FOR GENERAL MICROBIOLOGY

HELD AT

THE UNIVERSITY OF WARWICK

APRIL 1990

The right of the University of Cambridge to print and sell all manner of books was granted by Henry VIII in 1534. The University has printed and published continuously since 1584.

Published for the Society for General Microbiology

CAMBRIDGE UNIVERSITY PRESS

CAMBRIDGE

NEW YORK PORT CHESTER

MELBOURNE SYDNEY

Published by the Press Syndicate of the University of Cambridge
The Pitt Building, Trumpington Street, Cambridge CB2 1RP
40 West 20th Street, New York NY 10011, USA
10 Stamford Road, Oakleigh, Melbourne 3166, Australia

© The Society for General Microbiology Limited 1990

First published 1990

Printed in Great Britain at The Bath Press, Avon

British Library cataloguing in publication data

Society for General Microbiology, *Symposium: 45th: 1990*:
University of Warwick
1. Man. Virus diseases.
I. Title: Dimmock, N. J. III. Griffiths, P. D., Madeley, C. R.
IIII. Series
616.925

Library of Congress cataloguing in publication data

Society for General Microbiology. Symposium (45th: 1990:
University of Warwick)
Control of virus diseases: Forty-fifth Symposium of the Society
for General Microbiology, held at the University of Warwick,
April 1990/N. J. Dimmock, P. D. Griffiths, and C. R. Madeley.
p. cm.
Published for the Society for General Microbiology
ISBN 0 521 38562 8
1. Virus diseases—Immunotherapy—Congresses. 2. Virus diseases—Chemotherapy—Congresses. 3. Viral vaccines—Congresses.
I. Dimmock, N. J. II. Griffiths, P. D. Madeley, C. R.
IV. Title.
RC114.5.S63 1990
616.9′2506–dc20 90–31047 CIP

ISBN 0 521 38562 8

Coleg Prifysgol Cymru
The University College Of Wales
Aberystwyth
Acc. No. 8077083395
Class No. QR 1 S6 n.45

BS

CONTRIBUTORS

BEDFORD, H. E., Department of Paediatric Epidemiology, Institute of Child Health, 30 Guilford Street, London WC1N 1EH, UK

BOMFORD, R., Department of Experimental Immunobiology, Wellcome Biotech, Langley Court, Beckenham, Kent BR3 3BS, UK

CAMERON, J. M., Virology Department, Glaxo Group Research Ltd, Greenford Road, Greenford, Middlesex UB6 0HE, UK

DARBY, G. K., Department of Molecular Sciences, Wellcome Biotech, Langley Court, Beckenham, Kent BR3 3BS, UK

DIMMOCK, N. J., Department of Biological Sciences, University of Warwick, Coventry CV4 7AL, UK

GNANN, J., Departments of Medicine, Pediatrics and Microbiology, University of Alabama, Birmingham, Alabama 35294, USA

GOLDSMITH, S., Departments of Medicine, Pediatrics and Microbiology, University of Alabama, Birmingham, Alabama 35294, USA

GRIFFITHS, P. D., Department of Virology, Royal Free Hospital School of Medicine, Pond Street, Hampstead, London NW3 2QG, UK

KING, L. A., School of Biological and Molecular Sciences, Oxford Polytechnic, Gipsy Lane, Oxford OX3 0BP, UK

MCINTOSH, K., Division of Infectious Diseases, Children's Hospital, and Department of Pediatrics, Harvard Medical School, 300 Longwood Avenue, Boston, MA02115, USA

MADELEY, C. R., Department of Virology, Royal Victoria Infirmary, University of Newcastle, Newcastle upon Tyne NE1 4LP, UK

MELNICK, J. L., Department of Virology and Epidemiology, Baylor College of Medicine, Texas Medical Center, Houston, Texas 77030, USA

OXFORD, J., Department of Medical Microbiology, The London Hospital Medical College, Turner Street, London E1 2AD, UK

PECKHAM, C. S., Department of Paediatric Epidemiology, Institute of Child Health, 30 Guilford Street, London WC1N 1EH, UK

POSSEE, R. D., NERC Institute of Virology, Mansfield Road, Oxford OX1 3SR, UK

RANDALL, R. E., Department of Biochemistry and Microbiology, Irvine Building, University of St Andrews, St Andrews, Fife, KY16 9AL, UK

RICHMAN, D. D., Department of Infectious Diseases, San Diego VA Medical Center, San Diego, CA 92161, USA

ROWLANDS, D. J. Department of Virology, Wellcome Biotech, Langley Court, Beckenham, Kent BR3 3BS, UK

SMITH, G. L., Sir William Dunn School of Pathology, University of Oxford, South Parks Road, Oxford OX1 3RE, UK

SOUBERBIELLE, B. E., Department of Biochemistry and Microbiology, University of St Andrews, Irvine Building, North Street, St Andrews, Fife KY16 9AL, UK

THOMAS, H. C., Department of Medicine, Queen Elizabeth the Queen Mother Wing, St Mary's Hospital Medical School, London W2 1NY, UK

WEYER, U., NERC Institute of Virology and Environmental Microbiology, Mansfield Road, Oxford, Oxford OX1 3SR, UK

WHITLEY, R. J., Departments of Medicine, Pediatrics and Microbiology, The University of Alabama, Birmingham, Alabama 35294, USA

CONTENTS

EDITORS' PREFACE

The current AIDS pandemic has brought home only too clearly the realization that past successes in vaccination have more to do with serendipity than a real understanding of the basis of immune responses to virus infections. While we know that protection against infection requires activation of a particular part or parts of the immune system, each infection is unique and each must be understood before rational intervention is possible. Without such understanding, introduction of new vaccines is often a leap in the dark with sometimes unfortunate consequences of which the inactivated measles and respiratory syncytial virus vaccines are obvious examples. Study of infection is not a simple matter: animal model systems can be used but close parallels with the human infection are not always available. In the end, we must have detailed knowledge of the human immune responses, and this is only obtained with immense difficulty. Finally, even supposing that we have arrived at this advanced state of knowledge, we do not yet know how to stimulate specifically each part of the immune system. Much is being done to fill these gaps in our knowledge and to provide vaccines which are urgently needed.

The following chapters are based on a Symposium organized to address these formidable problems. It begins with an assessment of the effectiveness of current conventional vaccines and continues with a forward look at the ways in which the problems of immunization are being tackled. Presentation of antigen, new adjuvants, factors influencing the uptake of vaccine by the man or woman in the street, advantages and drawbacks of modern molecular technologies to immunization are amongst the topics covered by leading exponents in the field.

The second part of the book deals with chemotherapy, so successful against bacterial infections, but only rarely effective against viruses. Using empirical screening procedures, industry has been searching diligently for the past three decades for antivirals, yet today there is only one unequivocally effective compound (Acyclovir) on the market. Knowledge of viruses at the molecular level has been steadily increasing over this period and we now know that viruses are recalcitrant to chemotherapy because they have few gene products, and most of these are too similar to those of the cell to provide differential toxicity. However, unique viral functions, potential targets for chemotherapy, have been discovered and the recent explosion of knowledge about virus structure and virus–cell interactions at the molecular level, based primarily on recombinant DNA technology, has reinforced hopes of rational progress in chemotherapy. Again, the AIDS pandemic has brought unprecedented finance, manpower and determination to bear on solving the problems associated with HIV and these efforts will have far-reaching beneficial consequences for all virus infections. This section of the volume starts with a discussion of the current philosophy and practice of chemotherapy and then examines a number of individual virus systems where progress is being made.

This book, the parent Symposium and discussions during the Meeting in April 1990 at the University of Warwick focus attention on a vitally important area of human disease. In particular, we hope it will be useful to the new generation of virologists and if it encourages them to help solve the problems of controlling virus infections we shall count it a success.

N. J. Dimmock
P. D. Griffiths
C. R. Madeley
April 1990

PART I: IMMUNISATION

CONVENTIONAL VIRAL VACCINES AND THEIR INFLUENCE ON THE EPIDEMIOLOGY OF DISEASE

JOSEPH L. MELNICK

Division of Molecular Virology, Baylor College of Medicine, Houston, Texas, USA

There are some general principles that apply to most virus vaccines used in the prevention of human disease. Vaccination (or natural infection) does not always result in total immunity against a subsequent exposure to the wild virus. Even when vaccine-induced resistance does not totally block infection, but only serves to limit the multiplication of wild virus within the individual, it prevents spread to the target organs where the severe pathological damage occurs. Thus, wild poliovirus may replicate in the gut of a vaccinated person, but it will be prevented from invading the brain and spinal cord.

In many instances a number of susceptibles in a population may fail to be covered by vaccine, yet the herd immunity based on proper immunisation of a significant number of other susceptibles may serve to break the chain of transmission to a degree that wild virus no longer is able to sustain endemic circulation in the population.

Killed-virus vaccines prepared from whole virions generally elicit circulating antibody against the coat proteins of the virus. For some diseases, killed-virus vaccines or viral subunit vaccines are currently the only ones licensed.

Attenuated live-virus vaccines have the advantage of behaving like the natural infection in their effect on immunity. They multiply in the host and not only stimulate production of long-lasting humoral antibody but also induce cellular immunity and resistance at the portal of entry. However, some disadvantages, particularly that of genetic mutability, are associated with live attenuated vaccines.

Until recent years, virus strains suitable for live-virus vaccines were developed chiefly by selecting naturally attenuated strains or by cultivating the virus serially in various hosts and cell cultures in the hope of deriving an attenuated strain. The search for such strains is now being approached by laboratory manipulations aimed at specific genetic alterations in the virus. Newer vaccines and vaccine candidates (derived from deletion or other mutants, from reassortants, or from genetic recombinants) none the less must go through most of the same processes of development and demonstration of efficacy and safety – including examination of their epidemiological effects – that were requisite to the development of conventional viral vaccines.

One fact cannot be overemphasised. An effective vaccine does not protect against disease until it is administered at the proper time and in the proper dosage and potency to the proper target population. With the introduction of a new vaccine, there often follows not only a marked decrease in incidence of the target disease, but also a significant change in its epidemiology, brought about by the new age-distribution of susceptibles in the population. Some examples are described in this article, which is concerned with the conventional vaccines that have been used throughout the world.

HISTORY

As with history in general, the history of vaccines needs to be re-examined and updated. My task is to look back to see what has been successful. The present vaccines consist either of infectious, living, attenuated viruses or of non-infectious killed viruses or subviral antigens. When we look at the record, it is the live vaccines with which the greatest successes have been achieved in controlling diseases around the world. Examples are smallpox, yellow fever, poliomyelitis, measles, mumps, and rubella.

The chronological course of the development of vaccines for virus diseases of humans is shown in Table 1. One of the greatest triumphs of mankind has been the purposeful eradication of smallpox from the world's population. Almost 200 years ago, Jenner noted that cowpox conferred immunity against smallpox. This observation led him to inoculate material from the cowpox lesions as a means of protecting persons against smallpox, and the first successful vaccine came into use. More recent dramatic results of widespread vaccination can be seen in the rapid decreases, in the United States, of cases of four viral diseases that had previously been common in childhood (Fig. 1).

CURRENT ERA

The viral vaccines now in use are listed in Table 2. Those in the top section of the table are recommended for the general public, whereas those in the lower section are recommended only for special populations. However, there are some general principles that apply to most virus vaccines used in the prevention of human disease. Vaccination (or recovery from natural infection) does not always result in total immunity against a subsequent exposure to the wild virus. This situation holds true for diseases for which successful control measures are available, including polio, smallpox, influenza, rubella, measles, mumps, and adenovirus infections. But the replication of the wild virus, if it occurs, is markedly restricted. The restriction

Table 1. *Chronology of virus vaccines*

Year	Type of vaccine
1721	Variolation
1798	Smallpox attenuated
1885	Rabies attenuated and inactivated
1936	Yellow fever attenuated
1940s	Influenza inactivated (and later subviral)
1955	Poliomyelitis inactivated
1960	Poliomyelitis attenuated
1960s	Measles inactivated (no longer used)
	Measles attenuated
	Mumps attenuated
	Rubella attenuated
1970s	Japanese encephalitis inactivated
1980s	Japanese encephalitis attenuated
	Varicella attenuated
1982	Hepatitis B subviral particles
1986	Hepatitis B recombinant

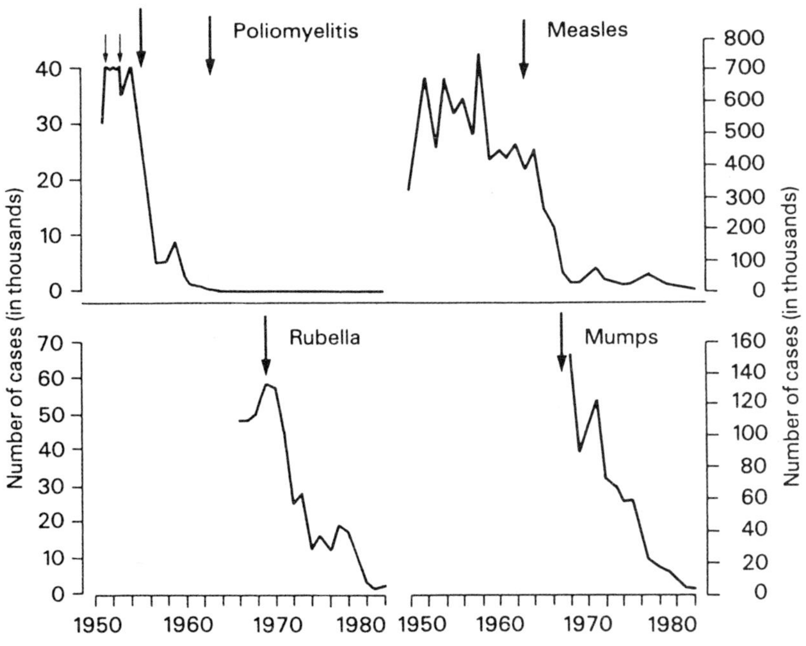

Fig. 1. Record of the decrease in cases of poliomyelitis, rubella, measles, and mumps in the USA following widespread administration of vaccines. Numbers of cases are indicated in thousands.

of replication thus not only can prevent spread of virus within the community but also can prevent, within the infected individual, spread of virus to target organs where the severe pathologic damage is done. (For example,

Table 2. *Principal vaccines used in prevention of virus diseases*

Disease	Source of vaccine	Vaccine virus	Route
Immunization recommended for everyone			
Poliomyelitis	Monkey kidney cells, human diploid cells	Live attenuated	Oral
		Killed	Subcutaneous
Measles	Chick embryo cells	Live attenuated	Subcutaneous
Mumps	Chick embryo cells	Live attenuated	Subcutaneous
Rubella	Human diploid cells	Live attenuated	Subcutaneous
Hepatitis type B	Purified HBsAg from human plasma	Killed, subunit	Subcutaneous
	Recombinant HBsAg from yeast	Subunit	Subcutaneous
Immunization recommended only under certain conditions (epidemics, exposure, travel, military)			
Smallpox	Lymph from calf or sheep	Live vaccinia	Intradermal
Yellow fever	Chick embryo	Live attenuated	Subcutaneous
Influenza	Chick embryo	Killed, subunit	Subcutaneous
Rabies	Human diploid cells	Killed	Subcutaneous
Adenovirus	Human diploid cells	Live attenuated	Oral
Japanese B encephalitis	Hamster kidney cells	Killed	Subcutaneous
Varicella	Human diploid cells	Live attenuated	Subcutaneous

polio and measles viruses may be prevented from invading the brain and spinal cord; rubella virus may be prevented from infecting the embryo.)

Killed-virus vaccines

Those prepared from whole virions generally stimulate the development of circulating antibody against the coat proteins of the virus, conferring some degree of resistance. For some diseases, killed-virus or subunit vaccines are currently the only ones licensed. However, some caveats are associated with them:

1. Since virulent strains often have been the ones chosen as source material for vaccine, extreme care is required in the manufacture of these vaccines to make certain that no residual live virulent virus is present in the vaccine.
2. The immunity conferred may be relatively short-lived, and therefore booster doses may be necessary. The use of killed-virus vaccines not only involves the logistical problem of repeatedly reaching the persons in need of immunisation but also has caused concern about the possible induction of hypersensitivity.
3. Sometimes protection conferred by parenteral administration of killed-virus vaccine is limited. In these instances, even when the vaccine stimulates circulating antibody (IgM, IgG) to satisfactory levels, local resistance (IgA) is not induced adequately at the natural portal of entry or the primary site of multiplication of the wild virus – for example, the nasopharynx for respiratory viruses, and the alimentary tract for poliovirus.

Attenuated live-virus vaccines

These have the advantage of behaving like the natural virus. They multiply in the host and tend to stimulate production of longer-lasting humoral antibody and also to induce cellular immunity and resistance at the portal of entry. However, some problems are associated with live attenuated vaccines:

1. There may be a risk that the vaccine virus reverts to greater virulence during its multiplication in the vaccinee. Although reversion of conventional vaccines has not proved to be a major problem in practice, its potential exists. Long-term monitoring of vaccine programmes should be required.
2. Unrecognised adventitious agents latently infecting the culture substrate (eggs, primary cell cultures) may enter the vaccine stocks. Viruses found in vaccines have included avian leukosis virus, simian papovavirus SV40, and simian cytomegalovirus. The problem of adventitious contaminants

has been circumvented through the use of pretested normal cells serially propagated in culture as substrates for cultivation of vaccine viruses. Vaccines prepared in such cultures have been in use for years and have been safely administered to many millions of persons.

3. The storage constraints and limited shelf-life of live attenuated vaccines present problems, but these can be overcome in some cases by extraordinary efforts to maintain the 'cold chain' of refrigeration even under difficult field conditions, and also by the use of viral stabilisers (e.g. molar $MgCl_2$ for polio vaccine).

SMALLPOX

Control of smallpox by deliberate infection with material from lesions of mild cases of the disease was practised for centuries. This process, called variolation, was dangerous but it decreased the disastrous effects of major epidemics, in some instances variolation reduced the case-fatality rate from 25% to 1%. Then Jenner introduced the first live attenuated vaccine in 1798. Almost two centuries later, WHO introduced a worldwide campaign to eradicate smallpox. Epidemiological features of the disease made it feasible to attempt total eradication. At that time, in 1967, there were 33 countries with endemic smallpox, 10 to 15 million cases per year. The programme has been a huge success, with the last naturally infected victim diagnosed in Somalia in 1977. A new concept of disease eradication by a co-ordinated global vaccination strategy came into being. Furthermore, with the eradication of the disease, it is no longer necessary to spend large sums and effort on a vaccination programme.

The following epidemiological features made smallpox amenable to total eradication, there was no known non-human reservoir; there was one stable serotype; there was an effective vaccine; since there were no subclinical cases, asymptomatic carriage of the virus did not occur; the patient's contacts could be readily identified, and thus specifically targeted control measures could be instituted quickly to interrupt the cycle of transmission.

In view of the current investigations of vaccinia virus as a vector for introducing immunising genes of other viruses into susceptible persons, a word of caution has been raised. Smallpox vaccination is associated with a definite, measurable risk. In the USA, the risk of death from all complications was 1 per million for primary vaccinees and 0.1 per million for revaccinees. For children under 1 year of age, the risk of death was 5 per million primary vaccinations. Among primary vaccinees, the combined incidence of post-vaccinal encephalitis and vaccinia necrosum was 3.8 per million in persons of all ages. Even in revaccinees, these two complications occurred, at a rate of 0.7 per million.

Since strict and effective barriers had been maintained against imported smallpox cases for many years, routine vaccination of children was discontinued in the USA in 1971, several years before global eradication of the virus. For this age-group, the risk of complications from the vaccination proved to be greater than that of contracting smallpox. Nevertheless, more than 4 million doses of smallpox vaccine were administered in the USA as recently as 1978. Severe complications of vaccination occurred in conjunction with immunodeficiency, immunosuppression, hematological or other malignancies, and pregnancy.

POLIOMYELITIS

The vaccine era in poliomyelitis began after 1955 when inactivated poliovirus vaccine was introduced, and was moved forward after 1960 when attenuated oral poliovaccine became available on a large scale. It is rare that a serious disease has been controlled so quickly and dramatically as was poliomyelitis in the developed countries of the world. In 1955, a total of more than 76,000 cases of poliomyelitis were reported from the Soviet Union, 23 other European countries, the United States, Canada, Australia, and New Zealand. By 1967, 1013 cases were recorded in these same countries, a reduction of almost 99%. Numbers of poliomyelitis cases have continued to be reduced; in industrialised countries, poliomyelitis is now a rare disease.

In the United States, before the inactivated virus vaccine became available, there were 10,000 to 21,000 paralytic cases per year (that is, rates between 5 and 10 per 100,000 population). Cases were far fewer after the inactivated vaccine came into use; in some years incidence rates as low as 0.5 per 100,000 population were achieved. Nevertheless, this meant that significant numbers of cases were still occurring; in 1960 there were more than 2500. And some of these cases were in fully vaccinated individuals. In a study of several thousand paralytic cases, 17% were in children who had received three injections of the inactivated vaccine. Some of the disappointing results were due to potency problems which have since been corrected in the few small north European countries that have used inactivated virus vaccines solely through most of the vaccine era.

After live attenuated vaccine was introduced and came into widespread use in the United States, the numbers of cases were reduced precipitously. Thus far in the 1980s, the annual number of cases has averaged less than 10. This translates into case rates consistently ranging from 0.02 down to 0.001 per 100,000 population.

For many years, the United States has relied almost completely on live poliovirus vaccine. It appears that a true break in the chain of infection has been achieved, and that there is no longer any endogenous reservoir of wild polioviruses within the country. Wild strains continue to be introduced, particularly from Mexico, but even such imported cases are sporadic and almost never result in secondary cases. The use of live poliovirus vaccine has achieved this result by establishing intestinal resistance extensively within the population, thereby reducing the pool of susceptible individuals below the level required for perpetuation of wild polioviruses. Many other countries with extensive and continuing live vaccine programmes are also reporting virtually no cases and few, if any, poliovirus isolates other than vaccine-like strains.

In Sweden, Finland, and The Netherlands, where the inactivated poliovaccines have been used almost exclusively, good results also have been obtained. These are countries of relatively small populations (a total of 28 million persons), that are culturally homogeneous and socially advanced. They have excellent public health services, and inactivated virus vaccine has been administered in intensive and regularly maintained immunisation programmes, achieving vaccine coverage among children that has approached 100%.

But there can be important deficiencies in protection, even in well-vaccinated countries. This vulnerability was demonstrated in 1978 and 1979 by outbreaks of poliomyelitis among members of closely knit, interconnected religious groups who refused vaccine on religious grounds. Imported into The Netherlands from the Middle East, a virulent type 1 strain spread to related religious groups in Canada and the United States. In none of the countries involved did paralytic cases occur beyond the unvaccinated members of these interconnected religious communities. However, infections with the epidemic virus did occur in some vaccinated children: in nursery and primary schools in some of the affected Netherlands communities not only were 71% of the non-vaccinated children excreting the wild poliovirus but also 24% of the IPV-vaccinated children were found to be excreting the virus.

In Finland, where the use of standard IPV had produced 20 years of freedom from poliomyelitis, ten cases of polio occurred between August 1984 and January 1985. On the basis of virus isolations from healthy individuals and from sewage, it was estimated that at least 100,000 persons in the general population were infected. The 1984 epidemic strain was a wild-type 3 variant that differed in both immunological and molecular properties from the type 3 vaccine strains. Analysis of sera collected before the outbreak indicated that neutralising antibodies against new type 3 isolates were much less prevalent in the population than antibodies to the type 3 strain (Saukett) used in the killed vaccine.

The precipitating factor in the outbreak was judged to be the appearance of the wild strain of poliovirus type 3 that was sufficiently aberrant to break through the type 3 immunity of the population. However, immunisation with the new enhanced IPV or with standard OPV induced high serum antibody titres against all known type 3 strains, including the Finnish epidemic type 3 strain. Thus, although differences among strains from various parts of the world have indeed been demonstrated for many years, the major neutralising antigen of each poliovirus serotype has proved to be remarkably stable.

The data from the 1988 outbreak in Israel also raise serious concerns with the use of IPV alone, as was done for several years in children in one part of the country. In the 1988 outbreak not a single case occurred among the IPV-vaccinated children, but it appears that they played a role in the circulation of wild virus in that year. This outbreak was striking in the number of paralytic cases in adolescents and adults. It has been demonstrated repeatedly that, while IPV vaccine is highly effective in protecting individuals from the paralytic disease, such persons subsequently can become infected subclinically and may transmit virus to their susceptible contacts. As mentioned, OPV has been highly effective but there are some concerns that its use may cause paralysis in one person for every million immunised. A combined schedule of IPV and OPV should theoretically give maximum benefit of both efficacy and safety.

Such a combined regimen was instituted in 1978 in Gaza and the West Bank, where cases, particularly in infants, were occurring year in, year out, despite extensive campaigns of immunisation with OPV. The combined schedule has almost eradicated paralytic polio from the areas. The schedule includes administration of OPV (type 1 monovalent) during the first month of an infant's life; then at two months and again at four months of age trivalent OPV is given orally and at the same time IPV is injected subcutaneously together with DPT. This is followed by OPV at six months, and again at 12 months. OPV is needed to provide long-lasting protection by inducing both circulating humoral antibodies and intestinal immunity, and IPV is needed to provide, in the critical first months of life, an immediate immunogenic stimulus that is not subject to the interfering or inhibiting factors described above. In addition, early administration of IPV should provide protection against the rare occurrence of vaccine-associated disease.

Another type of combined programme has been used for many years in Denmark where, despite extensive coverage with IPV alone, epidemics still occurred and antibody prevalence was less than desired. The combined programme in that country consists of three doses of IPV at 5, 6, and 15 months, followed by OPV at 2, 3, and 4 years of age. As a result, antibody levels in the population are close to 100% for all three types; and polio no longer occurs in Denmark.

The outbreaks mentioned above were unfortunate reminders that virulent polioviruses do indeed still exist and can readily circulate if given the right opportunity. Against the threat from virulent wild polioviruses, no nation can be secure at this time except by maintaining high levels of immunity through vaccination.

While cases have almost vanished from the industrialised world, polio continues to be an urgent problem, particularly in developing nations in tropical and subtropical zones. In the 1980s, well over 300,000 cases of poliomyelitis are occurring each year. In India alone, more than 100,000 cases were reported in 1985 and again in 1986. Of these cases, 98% were in children under 5 years of age; 42% of the patients were infants in the first year of life. As has been shown by surveys of lameness indicative of past paralytic poliomyelitis, in many countries the reported cases may represent less than 10% of the actual number of cases that occurred. However, with the advent of the WHO Expanded Programme on Immunisation there has been a marked improvement in live-virus vaccine coverage in the developing world, and indeed the World Health Assembly adopted a resolution in 1988 to eradicate poliomyelitis from the world by the year 2000. This programme is heading towards success in many regions.

Since it has been proven in several regions that paralytic disease caused by wild polioviruses can be eliminated, we can outline the features that should make this possible worldwide:

(a) Virus is spread by human-to-human contact. No animal reservoirs exist.
(b) Only three immunological types are known. No additional types have emerged in the twentieth century either naturally or under the immunological pressures of the poliovaccines.
(c) Two excellent vaccines are available. Each can prevent epidemics of paralytic disease, but for eradication of polio in some regions of the world a combined schedule of IPV and OPV seems to be required.
(d) Surveillance programmes will be necessary, particularly in view of the high ratio of unapparent infections to clinical disease. This will require monitoring for maintenance of immunity as well as for absence of wild poliovirus.

MEASLES

The key epidemiological features of measles are as follows: the virus is highly contagious; there is a single serotype; there is no animal reservoir; unapparent infections are rare, and infection confers lifelong immunity. Prevalence and age incidence of measles have long been related to population density, economic and environmental factors, and, more recently, to use of an effective live-virus vaccine.

Transmission occurs predominantly via the respiratory route. A continuous supply of susceptible individuals is required for the virus to persist in a community. A population size approaching 500,000 is necessary to sustain measles as an endemic disease. In smaller communities, the virus disappears until it is reintroduced from the outside after a critical number of non-immune persons accumulates.

When the disease is introduced into isolated communities where it has not been endemic, an epidemic builds rapidly and attack rates are almost

100%. All age-groups develop clinical measles. A classic example of this phenomenon occurred in 1946 when measles was introduced into the Faroe Islands; only people over age 60 years, who had been alive during the last epidemic, escaped the disease. In places where the disease strikes rarely, its consequences are often disastrous and the mortality rate may be as high as 25%.

At one time, prevention was limited to use of gamma globulin, which protects for about 4 weeks, and can modify or prevent the disease. Both killed-virus and live-virus vaccines have been developed. Killed-virus vaccine is no longer used, as some vaccinees became sensitised and developed local reactions when revaccinated with live attenuated virus, or a severe illness upon contracting natural measles. Live attenuated virus vaccine, which had been recommended for children at 15 months of age when maternal antibodies had disappeared, in most instances effectively prevents measles from occurring. Prior to the introduction of the vaccine, over 500,000 cases of measles occurred annually in the United States. With emphasis on immunisation, the number of cases declined to a low of 1700 in 1982. However, there has been a recent upsurge, with over 9000 cases in 1989.

Measles virus is responsible for subacute sclerosing panencephalitis (SSPE, Dawson's inclusion body encephalitis), a rare chronic degenerative brain disorder. The disease develops a number of years after the initial measles infection. With the widespread use of measles vaccine SSPE has almost disappeared as a clinical entity.

Because early lots of measles vaccine proved to be unstable, people vaccinated before 1980 may not be immune. This explains why some measles outbreaks have occurred in the USA during the late 1980s in high schools and colleges. When a measles case occurs among school children, it is now the policy to revaccinate all students at the school who were initially vaccinated before 1980.

Proper use of vaccine should lead to the eradication of measles. This goal has been thwarted by continuing outbreaks among unvaccinated inner-city children, many of them younger than the age of 15 months generally recommended for routine vaccination. The age for measles vaccination in this group should be lowered from 15 months to 9 months of age, but children who receive single-antigen measles vaccine before their first birthday should be revaccinated at 15 months of age, with trivalent measles–mumps–rubella (MMR) vaccine. For routine immunisation, children should continue to receive the trivalent MMR vaccine at 15 months, but a second dose of the trivalent MMR vaccine is now recommended upon entrance to middle or junior high school, at age 10 or 12 years.

A new recommendation has been made in the USA to prevent future outbreaks in those who were born in the measles vaccine era and had not been exposed and naturally immunised by the highly contagious wild virus. The aim is to have all 125 million Americans born since 1957 receive a second measles vaccination. It is estimated that the single vaccination that these persons received in childhood failed to protect about 5% of them. Since the susceptible fraction cannot be identified, a second dose is recommended for all. Obviously this is an enormous programme which will have to be carried out gradually.

MUMPS

Outbreaks of mumps occur where crowding favours dissemination of the virus. Mumps is primarily an infection of children, but epidemics may occur in adolescents and young adults.

An effective live attenuated vaccine is recommended for young children and for non-vaccinated adolescents and adults who have not had mumps parotitis. Vaccine is available in monovalent form (mumps only) but is usually administered in combination with measles and rubella (MMR) live vaccines. Combination live-virus vaccines induce antibodies to each of the component viruses in about 95% of vaccinees.

In 1967, the year mumps vaccine was licensed, there were about 200,000 mumps cases (plus 900 patients with mumps virus encephalitis) in the USA. By 1985, the annual number of mumps cases had decreased to less than 3000, with fewer than 20 cases of encephalitis. However, mumps cases increased severalfold during the 1986–1988 period, particularly among persons over 14 years old, who in 1987 accounted for 38% of the cases (compared to 8% in the prevaccine period). Although mumps increased in all age-groups in this period, the most dramatic increases were among those 10–14 years old (a sevenfold increase) and in those 15–19 years old (more than an eightfold increase). Nevertheless, the risk of disease in adolescents and young adults is still much less than that in the prevaccine era. The recent shift in risk of disease to adolescents and young adults does not appear to be due to waning immunity in vaccinated persons, but is probably attributable to the relative under-immunisation of the cohort of individuals born between 1967 and 1977. The shift in risk to older persons through 1988 is limited to states without school laws that include comprehensive mumps immunisation; the relative resurgence of mumps in the USA is not due to vaccine failure but to a failure to vaccinate. With the new emphasis on controlling measles by increased use of trivalent MMR vaccine, a decrease in mumps cases should also occur.

RUBELLA

Rubella occurs throughout the year, with a peak incidence in the spring. Epidemics have occurred every 6–10 years, with explosive pandemics every 20–25 years, increasing the risk of congenital rubella with its tragic consequences of severe abnormalities at birth. Live attenuated rubella vaccines have been available since 1969. The vaccine is now grown in human diploid cells. It induces high antibody titres and an enduring and solid immunity; secretory IgA antibody in the respiratory tract interferes with establishment of infection by wild virus. This vaccine is available as a single antigen or combined with measles and mumps vaccines (MMR vaccine).

The vaccine virus multiplies in the body and is shed in small amounts, but it does not spread to contacts. Vaccinated children pose no threat to mothers who are susceptible and pregnant. In contrast, non-immunised children can bring home wild virus and spread it to susceptible family contacts. The vaccine induces immunity in at least 95% of recipients, and that immunity endures for at least ten years.

In the USA, control of rubella is being attempted by routine vaccination of children aged 1–12 years and selective immunisation of adolescents and women of childbearing age. Before vaccine became available in 1969, about 70,000 cases were being reported annually. Vaccination decreased the incidence of rubella to only 550 cases in 1986 – a decrease of 99%. However, the decrease occurred primarily in children; in persons 15 years of age and older, only a small decrease in incidence occurred.

Scattered outbreaks of rubella still occur, chiefly among non-vaccinated adolescents in high school and college. The change in age incidence of rubella since introduction of vaccine is similar to that observed with measles.

Because immunity may wane in individuals vaccinated as children, pregnant women may be at risk of infection and of giving birth to abnormal infants. Therefore, vaccination of prepubertal girls and of women in the postpartum period has been proposed.

INFLUENZA

Despite intensive efforts on vaccine development and use, influenza viruses continue to be a major cause of morbidity and mortality worldwide. The high frequency of genetic reassortment characteristic of orthomyxoviruses, and the resultant antigenic changes in viral surface glycoproteins, make influenza viruses formidable challenges for control efforts. Influenza type A is highly variable antigenically and is responsible for most cases of epidemic influenza. Influenza type B may exhibit antigenic changes and may also cause epidemics. Influenza type C is antigenically stable; it causes only mild illness, at most.

Periodic outbreaks appear because of antigenic changes in one or both surface glycoproteins of the virus, to which the population is again susceptible. When the number of susceptible persons in a population reaches a sufficient level, the new strain of virus causes an epidemic. The change may be gradual (hence the term 'antigenic drift') owing to point mutations reflected in alterations at major antigenic sites on the glycoprotein, or drastic and abrupt (hence the term 'antigenic shift') owing to genetic reassortment during co-infection with an unrelated strain.

Inactivated viral vaccines are the primary means for preventing influenza in the USA. However, the characteristics of influenza viruses mentioned above make prevention and control of the disease by immunisation especially difficult. Existing vaccines are continually being rendered obsolete as the viruses undergo antigenic drift and shift.

WHO and national health agencies make recommendations each year on which strains should be included in the vaccine. The vaccine usually contains both a type A and a type B virus of the strains isolated in the previous winter's outbreaks. Two A subtypes are often included (in addition to the B type), particularly when both A strains circulated widely the previous winter and no one is able to predict with confidence which strain might be most important in the succeeding influenza season. Such trivalent vaccines have been in recent use. Made from the chief strains circulating in 1988–1989, the vaccine for use in the 1989–1990 season contains two type A strains (A/H1N1/Taiwan/86) (A/H3N2/Shanghai/87), and one type B strain (B/Yamagata/88).

Selected seed strains are grown in embryonated eggs, the substrate used for vaccine production. Sometimes the natural isolates grow too poorly in eggs to permit vaccine production, in which case a reassortant virus is made in the laboratory. The reassortant virus, which carries the HA and NA genes for the surface antigens of the newly circulating virus along with the replication genes from an egg-adapted laboratory virus, is then used in vaccine production. Virus is harvested from the egg allantoic fluid, purified, concentrated, and inactivated with formalin. Each dose of vaccine contains the equivalent of about 10^{10} virus particles, or their split products.

Protection is at best about 70% for a year following immunisation and may be much lower. Inactivated vaccines do not generate local IgA or cell-mediated immune responses. Vaccination is also complicated by the phenomenon of original antigenic stimulus (i.e. annual immunisations may predominantly boost antibody levels directed against irrelevant strains to which the person had once been exposed).

During the seven influenza epidemics that occurred in the USA from 1972 to 1988, influenza resulted in an average of 20,000 excess deaths per year. More than 80% of these deaths in the epidemic years were among persons aged 65 years or older. Despite continuing mortality caused by influenza among older adults, most do not accept annual immunisation.

Modern technologies offer new strategies for producing inactivated vaccines against influenza. Cloned DNA copies of the HA gene can be used to express the neutralising protein in bacteria or yeast. Secondly, oligopeptides that correspond to the sequence of crucial functional domains on the HA molecule are being synthesised *in vitro* and used as immunogens. These approaches have not yet resulted in practical influenza vaccines.

No live influenza virus vaccine is currently licensed for use in the USA, but research efforts to develop a reliable live-virus vaccine continue. In view of the constantly changing face of influenza viruses in nature and the long-term laboratory efforts required to attenuate a virulent virus, attempts are being made by genetic reassortment to transfer defined attenuating genes from an attenuated master donor virus to each new epidemic or pandemic isolate. Much remains to be done before influenza is brought under control.

HEPATITIS

In the USA viral hepatitis is reportable, according to the virus type causing the disease. About 70% of cases were categorised as hepatitis A until 1983, when cases of hepatitis B exceeded those of hepatitis A. In 1985, of 60,000 reported hepatitis cases, 39% were hepatitis A, 45% hepatitis B, 7% NANB (which now includes hepatitis C), and 9% unspecified hepatitis. The actual incidence is undoubtedly much higher, because many persons contract so mild a form of hepatitis that they do not seek treatment, and physicians often report only 10–20% of the cases they see.

HBV infection is now recognised to be one of the leading causes of death among adults. Deaths occur from liver cirrhosis and primary liver cancer resulting from chronic HBV infection. Worldwide, about 300 million persons are chronically infected, with at least 300,000 deaths annually.

In the USA about 300,000 new HBV infections occur each year. Of these patients, over 300 die of acute hepatitis. However, in about 60% of children and 8% of adults, the HBV infection becomes persistent. Of the one million chronic carriers in the USA, more than 5000 die each year from cirrhosis and over 1000 from primary liver cancer directly attributable to the chronic HBV infection.

A vaccine for hepatitis B is available. Vaccine is prepared by purifying HBsAg associated with the 22-nm particles from healthy HBsAg-positive carriers, and treating the particles with virus-inactivating agents (formalin, urea, heat). Protection is conferred by antibody to the *a* antigen, an antigen common to all subtypes. Preparations containing intact 22-nm particles have proven highly effective in reducing HBV infection when used in haemodialysis patients and staff, in homosexuals, and in newborns born to HBV-infected mothers.

Other sources of immunogen have been developed: 1) HBsAg produced in continuous mammalian cell lines or by a recombinant DNA in yeast cells in which a plasmid containing the gene for HBsAg has been incorporated; 2) polypeptides derived from the 22-nm particles; and 3) a recombinant of live vaccinia virus and the HBsAg gene. Other vectors, including attenuated adenovirus, have also been developed.

A product made from yeast is in use as the second-generation vaccine;

it has proven to be as effective as the plasma-derived vaccine. The HBsAg is expressed in yeast as particles 15–30 nm in diameter, with the morphological characteristics of free surface antigen in plasma.

The plasma-derived hepatitis B vaccine became available in 1982 in the USA, and by mid-1989 more than two million persons have been vaccinated. However, there has been no decrease in the incidence of acute hepatitis B. With the exception of health-care workers – who comprised about 5% of hepatitis B cases before vaccine was available – members of a number of the high-risk groups are not taking the vaccine; these include homosexual males, heterosexuals with multiple sexual partners, intravenous drug abusers, prisoners, residents and staff members of mental institutions, and immigrants from areas of high endemicity.

Although in the USA perinatal transmission of HBV is relatively uncommon, when it does occur there follow the same highly associated morbidity and mortality, including chronic hepatitis, cirrhosis, and liver cancer, that occur in other parts of the world. Routine screening of pregnant women for HBsAg and anti-HBc is now recommended in the USA, and infants born at risk should be immunised, beginning in the perinatal period, with either the plasma-derived or the recombinant vaccine. It is becoming evident that for effective control of hepatitis B, there must be widespread immunisation of infants, not only in hyperepidemic countries like China but also in countries like the USA as well. Now under consideration for the USA is a programme combining both infant and adolescent immunisation.

In many regions another viral genome, that of the hepatitis delta virus (HDV), is associated with HBV infections. HDV is found throughout the world, but its highest prevalence has been reported in Italy, the Middle East, Africa, and South America. Persons who have received multiple transfusions, intravenous drug abusers, and their close contacts are at high risk. The primary routes of transmission are believed to be similar to those of HBV. Infection is dependent on HBV replication with synthesis of HBsAg. HBV provides a rescue function for the delta agent. Delta hepatitis can be prevented by vaccinating HBV-susceptible persons with hepatitis B vaccine. However, for persons who are already hepatitis B carriers, hepatitis B vaccination does not prevent superinfection by hepatitis D virus.

BIBLIOGRAPHY

Evans, A. S. (Editor) (1985). *Viral Infections of Humans: Epidemiology and Control*, 3rd edn. New York, Plenum Publishing Corporation.

Fields, B. N., Knipe, D. M., Chanock, R. M., Hirsch, M., Melnick, J. L., Monath, T. & Roizman, B. (Editors) (1989). *Virology*, 2nd edn. New York, Raven Press.

Melnick, J. L. (1989) Virus vaccines: principles and prospects, *Bulletin of the World Health Organization*, **67**, 507–12.

Oxford, J. S. & Öberg, G. (1985). *Conquest of Viral Diseases: A Topical Review of Drugs and Vaccines*. Amsterdam, Elsevier.
Plotkin, S. A. & Mortimer, E. A., Jr (eds) (1988). *Vaccines*, Philadelphia, W. B. Saunders.

PRESENTATION OF VIRUS ANTIGENS FOR THE INDUCTION OF PROTECTIVE IMMUNITY

R. E. RANDALL AND B. E. SOUBERBIELLE

Department of Biochemistry and Microbiology, University of St Andrews, Irvine Building, North Street, St Andrews, Fife KY16 9AL, Scotland, UK

The proof of the pudding is in the eating
Don Quixote de la Mancha, Miguel de Cervantes (1615)

INTRODUCTION

Successful virus vaccines may need to induce three types of immune responses, namely 1) secretory (neutralising) antibody, 2) serum antibody and 3) specific T cell responses. The relative importance of each of these immune responses may vary in different virus infections. Any immunisation schedule should also result in long-term immunity with minimal side-effects. Most vaccines currently available utilise preparations containing either killed or live attenuated viruses as the immunising antigens (Melnick, 1986; Plotkin & Mortimer, 1988). However, for some virus diseases these approaches have not so far been successful or may not be practical. A great deal of research is therefore being undertaken to develop alternative vaccination strategies against these diseases and to improve the efficacy and safety of existing vaccines. This research includes schemes such as the construction of expression vectors or virus chimaeras, and the incorporation of proteins into immunogenic structures such as ISCOMS and solid matrix–antibody–antigen (SMAA) complexes. Immunogenic peptides are also being manipulated to improve their antigenicity, e.g. by coupling T helper cell determinants onto B cell determinants. In this chapter some of the advantages and disadvantages of these various immunisation strategies will be discussed in the light of our current knowledge of how the immune response is orchestrated.

IMMUNITY AND PROTECTION

The outcome of virus infections, in terms of the severity of disease invoked, is dependent upon (amongst other things) the speed of induction of a protective immune response when compared to the replication of virus. The purpose of vaccinating against a specific virus is to tilt the balance in favour of the immune response, thereby minimising the replication of the virus and hence preventing or reducing the severity of the disease. It is thus important to note that while effective vaccines prevent the pathological symptoms of disease they may not prevent the initial virus infection. Consequently, when designing vaccines against specific viruses it is important to consider how protective immune responses are generated. Protective immunity following a natural infection is invoked through the induction and interaction of specific antibody and cell mediated immune responses. Thus there is a wide spectrum of both B and T cell activation with the expansion of cell clones that recognise many different target antigens on various virus proteins. The antibody responses are primarily concerned with the inactivation, or neutralisation, of free virus while cytotoxic T cells (CTL) recognise and kill virally infected cells, thus reducing or preventing the production of progeny virus (Fig. 1). A further class of T cells, T helper (Th) cells, are involved in controlling antibody production (Stevens, *et al.*, 1988) and CTL immune responses (Abbas, 1988; Cantor *et al.*, 1976).

INDUCTION OF IMMUNE RESPONSES AND RECOGNITION OF ANTIGENS BY LYMPHOCYTES

The immune response depends upon the interaction of different lymphoid cell types. In the absence of foreign antigens, lymphocytes are small resting cells which circulate through the body ready to react specifically with their corresponding antigen (Nossal, 1987). Through vaccination a naive animal is primed (sensitised) to a virus so that the animal will mount a more efficient (in duration and in strength) secondary immune response following a subsequent encounter with virus. This is achieved by the clonal expansion of specific lymphocytes (Burnet, 1959) following their stimulation by the interaction of their antigen specific cell surface receptors (immunoglobulins for B cells and T cell receptors (TCR) for T cells) with virus antigens (Ashman, 1984). While B cell surface immunoglobulin binds directly to antigen, T cells can only recognise an antigen when it is associated with major histocompatibility complex (MHC) molecules expressed on the surface of antigen presenting cells (APC) (Acuto & Reinherz, 1985). Stimulation of lymphocytes with antigen results in the generation of numerous circulating memory cells, thus increasing the chance of specific antigen recognition during subsequent contact with the same virus. Along with these memory cells, specific effector B and CTL cells are produced (Nossal, 1987). The majority of B and CTL responses are T helper cell dependent (Cantor *et al.*, 1976). Consequently, to achieve clonal expansion of B lymphocytes and CTLs following antigen stimulation, as well as interacting with their appropriate antigen, lymphocytes usually have to be activated with soluble factors (cytokines) secreted

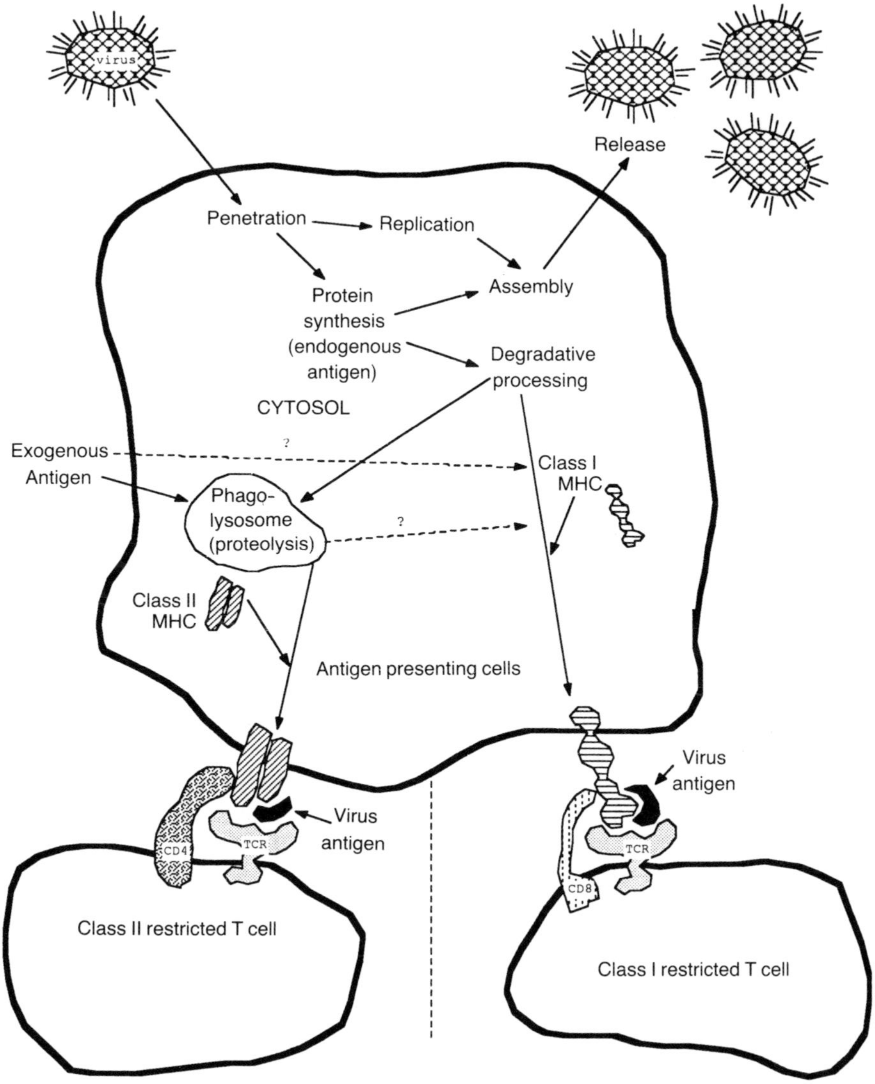

Fig. 1. Diagrammatic representation of intracellular processing pathways for endogenous and exogenous virus proteins in antigen presenting cells. Following processing of the proteins certain peptides become associated with either class I or class II MHC antigens. The MHC–peptide complexes are transported to the cell surface where they are specifically recognised by T cell receptors (TCR) present on either class I or class II restricted T cells.

by antigen-activated T helper lymphocytes (Stevens *et al.*, 1988; Hamblin, 1988; Balkwill *et al.*, 1989). However, T cell independent B cell stimulation can occasionally occur, but this usually involves the binding of multiple surface immunoglobulins present on the surface of a B cell to repetitive epitopes on an antigen (Chused *et al.*, 1976). More recently it has also been reported that viral CTL responses can sometimes be generated following depletion of T helper cells (Buller *et al.*, 1987).

ACCESSORY CELLS AND ANTIGEN PRESENTING CELLS

To generate a T cell response most antigens have first to be processed for presentation to the T cells in conjunction with MHC antigens (Grey *et al.*, 1985) (below). There are certain specialised cells within the body, such as dendritic cells and macrophages, whose role is to capture foreign antigens and present them to T cells (Grey, *et al.*, 1985; Austyn, 1989). However, to do this these cells have first to bind the foreign antigen, and, at the outset of an immune response, this is performed in the absence of antigen specific receptors on their cell surface. For antigens to be recognised by these accessory cells they must therefore have certain properties. For example, soluble antigens may not be efficiently recognised by these cells and thus fail to elicit vigorous immune responses. To use soluble antigens as immunogens it is therefore necessary to use adjuvants (below) which modify the antigen and stimulate the immune response. Particulate antigens are more immunogenic than soluble antigens possibly because they may be more efficient at triggering various non specific immune recognition signals that result in their capture by accessory cells. Once the immune response has been initiated then the whole process becomes more efficient as specific antibodies bound to foreign antigens help accessory cells capture the antigens (Celis *et al.*, 1984). This may be because the antigen–antibody complexes activate the complement system (Arvieux *et al.*, 1988; Klaus, 1988) or because the complexes themselves bind directly to the accessory cells via Fc receptors (Chang, 1985; McCullough *et al.*, 1988). In addition, B cells can act as antigen presenting cells, so that once antigen specific B cell clones have been established they act as efficient antigen presenting cells for T cells (Lanzavecchia 1985; Manca *et al.*, 1988).

Antigen presentation following a virus infection is somewhat more complex than that outlined above. This is because, as viruses infect and replicate in cells, infected cells in their own right may act as antigen presenting cells to CTLs. However, it is still controversial as to whether any cell can initiate a T cell response, or whether this function is restricted *in vivo* to a few cell types such as dendritic cells and macrophages. In other words, activation of resting T cells (immunostimulation) may be a different process from antigen presentation to sensitised T cells. While the former may require specialised cells, any cell may act as an antigen presenting cell for sensitised T cells (Austyn, 1989). Moreover, unprimed CTLs need viable APC to be activated whereas sensitised CTLs can be activated by non-viable APC. It is thus thought that additional factors, over and above those required for the activation of sensitised CTLs, may be necessary for the initial priming of CTL responses, e.g. soluble factors secreted by APC, or direct contact with viable APC (Sprent *et al.*, 1989).

PROCESSING OF ANTIGEN FOR T CELL RECOGNITION

Both Th cells and CTLs recognise target antigens consisting of short peptides (Babbit *et al.*, 1985; Townsend *et al.*, 1986; Maryanski *et al.*, 1986; Buss *et al.*, 1987; Reddehase *et al.*, 1989), bound within a groove on the ectodomain of MHC molecules (Bjorkman *et al.*, 1987*a*, *b*) that are located on the surface of the APC. Normally, for T cell recognition to occur, viral proteins must therefore first be processed into short peptides in the APC (Allen, 1987). Some, but not all, of the peptides generated by processing of a protein may have affinities for particular MHC antigens and thus act as target antigens for T cells (McMichael *et al.*, 1986; Bodmer *et al.*, 1989).

There are two different classes of MHC antigens termed class I and class II. While class I MHC antigens are found on the majority of cells, class II MHC antigens are present in greatest concentrations on haematopoietic cells (Cresswell, 1987). It is nevertheless known that many non-haematopoietic cells begin to express class II MHC antigen following the release of lymphokines as a result of the immune response to antigenic stimulation (Botazzo *et al.*, 1983; Wong *et al.*, 1984). With few exceptions (Aparicio *et al.*, 1987), a T cell will recognise a foreign antigen associated with either class I or class II antigens but not both. T cell subtypes can be distinguished by cell surface antigens (Knapp *et al.*, 1989), those recognising class I MHC antigens in humans having a $CD8^+$, $CD4^-$ phenotype (Norment *et al.*, 1988) and those recognising class II MHC antigens; a $CD4^+$, $CD8^-$ phenotype (Gay *et al.*, 1987). These phenotypes correlate to some extent with their functions. T helper cells have a $CD4^+$, $CD8^-$ phenotype as they recognise virus antigens in association with class II MHC antigens. In the majority of virus infections the CTL response is mainly class I restricted (Zinkernagel & Doherty, 1979; Biddison *et al.*, 1981; Meuer *et al.*, 1983; Enssle *et al.*, 1987; Borysiewicz *et al.*, 1983), although there can be class II restricted CTL responses, which may be important in controlling certain viral infections (Long & Jacobson, 1989).

Since peptides to be presented to T cells are selected in part by their affinity for MHC antigens, only some parts of the protein (called immunodominant epitopes) are able to initiate a T cell response. The general rules of 'immunodominancy' for peptides are not completely understood though it appears to be influenced by the primary structure of the peptide (Berzofsky *et al.*, 1987). Two computer programs have so far been generated to predict T cell epitopes on a protein. One originated from the study of different helper antigenic determinants and is based on the fact that immunodominant peptides are likely to adopt the conformation of amphipathic alpha helices (the hydrophobic side being in contact with the MHC molecule and the hydrophilic part of the peptide facing the T cell receptor (Margalit *et al.*, 1987). The other program predicts that T cell epitopes are likely to show a consensus amino acid (aa) sequence (Rothbard & Taylor, 1988). Although these programs have proved useful in identifying many T cell epitopes on a protein they have also often failed to identify other important ones. This is probably because an important characteristic which influences the immunodominancy of a peptide is the MHC antigen itself. MHC antigens are extremely polymorphic (Strachan, 1987) and different MHC molecules may bind different target antigens (Babbit *et al.*, 1985; Buss *et al.*, 1986; McMichael *et al.*, 1986; Gotch *et al.*, 1987). Consequently, as any one individual has only a limited number of different MHC antigens, in a heterozygous population different T cell target antigens may be recognised by different individuals depending on their MHC status. Thus to ensure that all individuals respond appropriately to a particular vaccine it may be necessary to incorporate multiple T cell epitopes into the vaccine. However it should be noted that recent studies on the target antigens of class II restricted T cell clones generated following influenza virus infections of mice have shown that there is extensive diversity in the memory repertoire of these cells. Indeed there can be multiple T cell epitopes on a single virus protein, indicating that a particular MHC molecule may possess a greater capacity to bind structually dissimilar peptides than has been generally assumed (Burt *et al.*, 1989; Barnett *et al.*, 1989; Graham *et al.*, 1989). The class I restricted repertoire on the other hand appears to be more restricted (Townsend *et al.*, 1986; Bastin *et al.*, 1987; Taylor *et al.*, 1987; Gotch *et al.*, 1986; McMichael *et al.*, 1986). This may be because antigen is processed in different ways for presentation to class I and class II restricted cells (below).

In addition, in influenza studies commonality between B and T cell repertoires was observed in that the majority of epitopes recognised by class II restricted T cell clones mapped to the same region of the HA molecule as neutralising antibodies (Barnett *et al.*, 1989; Graham *et al.*, 1989; Burt *et al.*, 1989). While the mechanism that leads to this observed commonality is speculative it suggests that the B cell repertoire can influence the selection of the T cell repertoire. Such a situation could arise, for example, following antigen specific capture by B cells, e.g. antibody bound to antigen could influence the processing of the antigen, by protecting specific epitopes from proteolysis and thus preserving them for MHC capture (Barnett *et al.*, 1989; Manca *et al.*, 1988). However, the importance of these findings in terms of vaccine design has still to be evaluated.

EXOGENOUS vs ENDOGENOUS ANTIGEN FOR THE INDUCTION OF CLASS I AND CLASS II RESTRICTED T CELLS

The processing pathways of protein degradation that result in the association of peptides with either class I or class II MHC antigens are different

(Germain, 1986; Mills, 1986; Morrison *et al.*, 1986; Braciale *et al.*, 1987; Delovitch *et al.*, 1988; Sekaly *et al.*, 1988; Long, 1989). For class I antigen presentation it appears that the antigen at some point must be located in the cytosol of the cell (Moore *et al.*, 1988; Yewdell *et al.*, 1988) from where it is processed and the appropriate peptide(s) becomes associated with newly synthesised MHC class I antigens (Townsend *et al.*, 1989). Conversely, for class II presentation the antigen is first processed in endosomes before the resulting peptides become associated with class II MHC antigens (Fig. 1). Antigens for class II presentation can either be exogenous or endogenous antigens, although processing of exogenous antigen for class II presentation is particularly efficient.

There is a body of data which suggests that exogenous antigens (e.g. non-infectious virus antigens) are not normally processed for class I presentation (Germain, 1986). However, this is a contentious point. Indeed, it has been reported that class I restricted T cells can be generated *in vivo* after immunisation with non-infectious virus antigens, e.g. recombinant hepatitis B vaccine (Jin *et al.*, 1988) the NP protein of influenza virus (Wraith & Askonas, 1985) and a hybrid influenza HA protein (Kuwano *et al.*, 1989). Class I restricted CTL induction has also been described using albumin as the immunising antigen (Staerz *et al.*, 1987). Both class I and class II restricted CTLs also appear to be important in controlling infections with intracellular bacteria, protozoa and fungi, organisms that do not synthesise their proteins in the cytosol of the infected cell (Kaufmann, 1988). Our own studies on presenting non infectious virus antigens as solid matrix–antibody–antigen (SMAA) complexes also indicated that class I restricted CTLs are generated by this immunisation procedure *in vivo*. Lymphocytes isolated from mice immunised with simian virus 5 (SV5) proteins in this manner are capable of lysing virally infected P815 target cells, a cell line that does not express class II MHC antigens (Randall & Young, 1988). In addition, it is possible to protect immunodeficient x-irradiated mice from infection with SV5 by transferring lymphocytes isolated from mice immunised with SV5 specific SMAA complexes. Removal of class I restricted T cells from this population abolishes the protection observed (Young & Randall unpublished observations).

Bevan (1987) pointed out that if exogenous antigen cannot give rise to class I restricted CTLs then sensitisation *in vivo* would have to occur peripherally for viruses that do not productively infect cells in lymphoid organs. This concept is at odds with the old dogma that immune sensitisation occurs centrally in draining lymph nodes or the spleen. It has therefore been suggested that there may be specialised cells *in vivo* whose function is to process exogenous antigen to class I restricted T cells (Bevan, 1987; Staerz *et al.*, 1987). Bone marrow macrophages have been suggested as a possible example of such a cell type. These cells have good phagocytic activity and normally express class I but not class II MHC antigens (Kaufmann, 1988). In terms of vaccine design, it is therefore important to unravel the mystery of how exogenous antigen can be presented *in vivo* to induce class I restricted CTLs and more generally to understand the CTL priming mechanism (Sprent *et al.*, 1989; Widmer *et al.*, 1987).

To complicate the matter still further, a new class of T lymphocytes has recently been described, characterised by the TCR structure. These lymphocytes possess two polypeptide chains named gamma and delta unlike the TCRs of the usual peripheral mature T lymphocytes (Th or CTL), which consist of two polypeptide chains designated alpha and beta (Janeway, 1988). These gamma-delta lymphocytes form less than 5% of the peripheral lymphocytes but are prominent in some parts of the body, particularly the gut mucosal system. It has been hypothesised that this class of lymphocyte may play a role in immunosurveillance and may thus be particularly important in controlling some viral infections (Janeway, 1988; Augustin *et al.*, 1989).

CLASS I vs CLASS II RESTRICTED CTLs

Although in the majority of virus infections class I restricted CTL responses are the predominant responses, for some viruses it appears that class II restricted CTLs are also important. Thus, for example, while following influenza infections over 90% of the CTLs are class I restricted (Bourgault *et al.*, 1989), it appears that class II restricted CTLs form a major part of the CTL response after measles (Jacobson *et al.*, 1984; Jacobson *et al.*, 1989) and herpes simplex virus (Yasukawa & Zarling, 1984; Schmid, 1988) infections. Differences in the types of cells infected by these viruses and differences in the way in which these viruses infect cells (e.g. by fusion at the plasma membrane or entry through endosomes) have been proposed as possible explanations for why they primarily induce either class I or class II restricted CTLs (Long & Jacobson, 1989). Consequently in terms of vaccine design, it may be important to induce the most appropriate type of CTL depending on the virus concerned. Thus for measles virus immunisation, schemes that give rise to class II restricted CTLs may be more appropriate than those that give rise to class I CTLs, while for influenza the converse would be true.

B CELL RECOGNITION OF ANTIGEN

So far we have mainly considered the immune response in terms of the T cell. However, when designing vaccines it is equally important to consider how to induce an appropriate antibody response. Because of the differences in the ways in which antibodies and T cell receptors recognise and bind to their respective target antigens, it may be necessary to present antigens in a different form to induce antibody responses as opposed to T cell responses. Although antibodies recognise small antigenic determinants (usually a sequence of between 5–10 amino acids on proteins; Novotny *et al.*, 1987) denaturation and proteolysis of the antigen is usually harmful as it can destroy important epitopes. There are two main types of epitopes on a protein that are recognised by antibodies, termed continuous and discontinuous (Barlow *et al.*, 1986). Continuous epitopes are composed of amino acids in continuous peptide bond linkage and vary in their conformational sensitivity. Continuous epitopes present on globular proteins, such as most viral surface glycoproteins, are usually conformationally sensitive. Discontinuous epitopes are always conformationally sensitive as the amino acids which participate in the formation of these epitopes are not in continuous peptide bond linkage. It is now widely believed that the majority of epitopes on a protein are discontinuous (Barlow *et al.*, 1986; van Regenmortel, 1989*a*, *b*). Consequently, linear peptides rarely represent the exact structure of conformationally sensitive epitopes (van Regenmortel, 1987). Furthermore, any processing or denaturation of the antigen, either *in vivo* or in preparing the antigen as an immunogen, may destroy important antigenic determinants. Nevertheless, there are a number of ways in which it is possible to predict areas of a protein that will be antigenic (Novotny *et al.*, 1987; Blundell *et al.*, 1987; van Regenmortel, 1989*a*, *b*). These methods often attempt to predict those regions of a protein that are hydrophobic (and hence buried within the protein) and those regions which are hydrophilic and potentially antigenic. However, as with programs for predicting T cell target epitopes, whilst these programs are very good at identifying some potentially important antigenic sites they fail to identify others.

MUCOSAL IMMUNITY

The induction of mucosal immunity, and in particular secretory antibodies, may be particularly important in clearing virus infections of mucosal membranes (Liew *et al.*, 1984). However, although there is a common mucosal system whereby the induction of a secretory immune response at one body surface may result in the presence of secretory antibody at other mucosal surfaces, this system seems to be largely independent of humoral immune responses (Challacombe, 1987; Mestecky & McGhee, 1987). Consequently, immunisation schedules that give rise to efficient humoral immune responses may not necessarily result in the induction of mucosal immunity. In fact, antigens capable of inducing vigorous immune responses upon systemic injection, may induce mucosal tolerance when administered orally (Mowat, 1987; Challacombe & Tomasi, 1987). However, it has been shown that the problem of orally induced tolerance can potentially be overcome since the co-administration of cholera toxin can prevent the induction of tolerance to an unrelated antigen (Elson & Ealding, 1984; Lycke *et al.*, 1989). Nevertheless, even if efficient methods are developed for the induction of secretory antibody, there may be problems associated with the development of long-term protective immunity. Indeed, even after natural infections of mucosal surfaces, the production of secretory antibody tends to be short lived and can be detected for only about 3–6 months post-infection (Challacombe, 1987). The lack of long-lived mucosal immunity may be one of the reasons why individuals can be repeatedly infected with viruses such as the parainfluenza viruses and respiratory syncytial virus (Knight, 1973; Wright, 1984; Belshe *et al.*, 1984; Randall & Russell, 1990). However, upon secondary infection with these viruses disease symptoms are often reduced. This is presumably because of the speed of induction of secondary immune responses and gives rise to the belief that useful vaccines can be produced to viruses which infect mucosal membranes.

LONG-LIVED IMMUNITY

A desirable property of all vaccines is their ability to induce long term immunological memory. Consequently vaccines that induce a broad immune response (thereby increasing the speed of a secondary immune response) may be beneficial in situations where the initial level of antibodies (secretory or humoral) and/or cytotoxic T cells in a patient at the time of the infection may be insufficient to prevent disease but where a rapid secondary immune response can. Thus immunisation schemes that prime a B cell response to only one antigenic determinant on a virus, without inducing memory T helper cells to the same virus, may not prime the immune system for a rapid secondary immune response. In this respect it is pertinent to note that in a number of viral systems T helper cells which recognise target antigens on internal virus proteins may enhance B cell responses to surface antigens present on the same virus particle through the mechanism of B cell antigen presentation (Milich *et al.*, 1987; Russell & Liew, 1979; Scherle & Gerhard, 1988).

IDENTIFICATION OF VIRUS ANTIGENS FOR INCORPORATION INTO VACCINES

Because of the mechanism of intracellular antigen processing the important point to note in terms of vaccine design is that potentially any virus protein, including internal structural proteins and even non-structural proteins, may contain important target antigens for T cells (Townsend *et al.*, 1984; Yewdell *et al.*, 1985; Yewdell *et al.*, 1988; Taylor *et al.*, 1986; Wraith *et al.*, 1987; Dietzschold *et al.*, 1987; Jacobson *et al.*, 1988; Nixon *et al.*, 1988; Walker *et al.*, 1988; Randall *et al.*, 1988). Thus when considering which virus proteins are likely to contain the most important target antigens for inducing T cell responses, it is also important to consider aspects of the biology of the virus. For some viruses the most important target antigens for CTLs may be those present on virus proteins made early during the infectious cycle or those made in persistently or latently infected cells. Such proteins are usually distinct from those found on the surface of the virus particle which display target antigens for neutralising antibody. For example, it has been shown that CTLs that recognise the immediate early protein of murine cytomegalovirus protect mice from lethal infections with this virus (Jonjic *et al.*, 1988). Similarly CTLs that recognise the reverse transcriptase and gag proteins of the human immunodeficiency virus have been described (Nixon *et al.*, 1988; Walker *et al.*, 1988). Other criteria concerning the biology of the virus for which a vaccine is to be designed should also be taken into account. For viruses that undergo antigenic changes (e.g HIV; Starcick, *et al.*, 1986; Modrow *et al.*, 1987; Castro *et al.*, 1988; Fisher *et al.*, 1988; Saag *et al.*, 1988) it may be important to include multiple antigens from a variety of strains to induce protection against all the circulating strains of virus. This principle already applies to the manufacture of influenza virus vaccines (Kilbourne, 1988; Maassab *et al.*, 1988).

It should also be noted that in some viral systems (e.g. HIV) 'suppressor' peptides, which can inhibit lymphocyte proliferation *in vitro*, have been identified (Ruegg *et al.*, 1989*a*, *b*, Klasse *et al.*, 1988). Like immunostimulation, immunosuppression by some viral proteins is strongly influenced by the MHC phenotype of the immunised animal (Morrison *et al.*, 1987). Although the full molecular basis of these mechanisms is unknown, it may be important to avoid the incorporation of antigens which contain such epitopes into subunit vaccines unless, of course, an antibody response to the suppressor peptide itself is beneficial (Klasse *et al.*, 1988).

PROTECTIVE vs DELETERIOUS IMMUNE RESPONSES

While it may be possible and even beneficial to induce protective immunity to some viral diseases by inducing high levels of neutralising antibody to a restricted number of antigenic determinants on a virus (e.g. foot-and-mouth disease; Francis *et al.*, 1988), for many other diseases it may be necessary to induce both appropriate antibody and cell-mediated immune responses. It is clear from studies on immunodeficient patients that the relative importance and balance of humoral and cell-mediated immune responses may vary from one virus infection to another. For example, children with thymic aplasia who do not develop normal cell-mediated immune responses become fatally infected with measles virus but children with agammaglobulinaemia can contract and survive measles in a relatively normal fashion (Burnet, 1968; Ewan & Lachmann, 1982). Yet it is recognised that passive transfer of immune globulins to non immune individuals after exposure to measles is effective in modifying or preventing subsequent disease (Modlin, 1984). While there is a very effective attenuated virus vaccine to measles, inactivated vaccines failed to induce long lived protective immunity in children who had normal immune responses. Moreover subsequent infection of some of these children with wild type measles resulted in a more serious atypical measles infection (Rauh & Schmidt, 1965). An explanation for the atypical immunopathology caused by the use of these inactivated vaccines has been proposed on the basis that the vaccines failed to induce an appropriate antibody response to the fusion glycoprotein of the virus (Norrby *et al.*, 1975; Norrby, 1987). However, as we will discuss later, inactivated vaccines may also induce aberrant cell-mediated immune responses. Similar findings occurred when an inactivated respiratory syncytial virus (RSV) vaccine was used in that some individuals developed a more serious disease than non-immune people upon subsequent infection with virulent virus (Kim *et al.*, 1969).

Useful information about deleterious immune response can also be obtained experimentally. For example, immunisation of Cotton rats with formalin-inactivated RSV enhanced the subsequent pathology upon infection with virus (Prince *et al.*, 1986). While CTLs probably play an essential role in clearing RSV infections (Wells *et al.*, 1983; Isaacs *et al.*, 1987) it may also be that inappropriate CTL responses contribute to the increased pathology observed in individuals immunized with inactivated virus (Kim *et al.*, 1976). This idea is supported by the finding that transfer of CTLs specific for RSV to immunodeficient mice can increase the pathology upon subsequent infection with RSV (Cannon *et al.*, 1988). On the other hand, passive transfer of monoclonal antibodies has not been shown to enhance pathology in RSV-infected mice, and some antibodies protect against pulmonary disease (Taylor *et al.*, 1984; Prince *et al.*, 1987).

In other systems an antibody response alone may have deleterious consequences. Antibody enhancement of viral infectivity has been recorded in a number of virus systems (Mims & White, 1984) and with flaviviruses, such as dengue, can lead to disease exacerbation *in vivo* (Halstead, 1979). More recently it has been suggested that antibodies may enhance the infection of cells with HIV (Bolognesi, 1989). It has also been suggested that antibody-induced antigenic modulation may be an important factor in the establishment of persistent infections *in vivo* (Fujinami & Oldstone, 1979; Fujinami & Oldstone, 1980; Oldstone & Fujinami, 1982; Randall & Russell, 1990). There is some experimental evidence for such a phenomenon occurring *in vivo*. For example, inoculation of either polyvalent antibodies or monoclonal antibodies to the measles virus haemagglutinin led to the establishment of a chronic infection in mice, whereas the inoculation of antibody to measles virus nucleocapsid did not alter the course of the acute infection (Rammohan *et al.*, 1981). Furthermore, Albrecht *et al.*, (1977) could only establish a persistent measles virus infection in monkeys with pre-existing antibodies to measles virus and not in non-immune monkeys.

A fine balance may thus exist between the induction of protective immunity and the induction of immunity that will lead to a detrimental response upon subsequent infection with virulent virus. Indeed for some infections it may be harmful to induce a restricted immune response by inappropriate vaccination. On the other hand, it has been suggested that some viruses may induce autoimmunity and hence lead to disease (Mims & White, 1984). Consequently, if at all possible, virus antigens which may lead to the induction of such responses should not be incorporated into vaccines (Ada, 1989).

VACCINES AND VACCINE DESIGN

Due to the inherent difficulties in making attenuated virus or whole killed virus vaccines (below) to some diseases much recent research has been expended on developing alternative immunisation strategies for the production of virus vaccines. Many of these approaches have taken a reductionist strategy following multiple observations that immunisation of experimental animals with a restricted number of virus antigens, or passive transfer of monoclonal antibodies, can result in the induction of protective immunity. The use of peptides and virus chimaeras represent the extremes of such an approach. Whilst a reductionist approach is fraught with difficulties, it does have the potential advantage that if all the conditions required to achieve long-lasting immunity can be met then it may be possible to prevent some of the unnecessary and potentially harmful immune reactions to unimportant or contaminating antigens that may be present in other vaccines.

Much data relevant to vaccine design and identification of important target antigens to be incorporated into subunit vaccines has been generated using animal model systems. However, a number of general points should be borne in mind when extrapolating data generated using these model systems to the actual design of vaccines. For example, while it may be demonstrated that immunising animals with particular antigens in a given way may induce protective immunity to the appropriate virus, it should not be forgotten that the route of infection and type of disease induced by that virus may be quite different from that seen in the natural infection. In addition, through the use of animal model systems it may not be possible to identify important human T cell target epitopes because different MHC antigens will have different affinities for different peptides. Also often little is learnt about the induction of long term immunity in these systems. This is not to say that animal model systems do not have an important role to play in identifying the type of virus antigens that should be incorporated into vaccines but it is important to be aware of some of the drawbacks of such systems.

Multiple factors can influence the efficacy of vaccines namely: the type(s), number and association of antigens used, their shape and conformation, the timing of their administration, the addition or absence of adjuvants, and the immune status of the vaccinee (below). It is also important to stress that the route and dose of immunising antigen is important in the induction of T and B cell responses. Small amounts of antigen can fail to induce a response but too large a quantity can lead to immune tolerance to the antigen (Mitchison, 1964). In the light of these principles we will now briefly consider some of the advantages and disadvantages of various immunisation procedures that have been used to make vaccines, or are being proposed as methods of making novel vaccines.

Attenuated virus vaccines

Attenuated viruses are ones which replicate *in vivo* but do not induce disease. There are a number of very successful live attenuated virus vaccines, e.g. to mumps, measles, polio and yellow fever (Melnick, 1986; Plotkin & Mortimer, 1988). However, the way in which the attenuated viruses were first isolated has largely been empirical in that the virulent virus was passaged a number of times in tissue culture cells and/or experimental animals to generate the attenuated virus. Consequently the molecular basis of attenuation is often unknown and this therefore is one of the potential weaknesses of this general approach. However, the use of live attenuated viruses as immunogens from a hypothetical point of view is one of the most attractive ways of producing vaccines. This is because as the virus replicates *in vivo* a broad spectrum immune response should be generated, with the induction of both antibody (potentially both humoral and secretory) and cell-mediated immune responses. Furthermore, any virus protein, including structural and non-structural proteins, could potentially give rise to T cell responses with individuals selecting the appropriate T cell epitopes depending on their MHC status. Also the most appropriate class of CTL should be generated as presumably the attenuated virus will infect the same types of cell as the virulent virus (Long & Jacobson, 1989). Despite these obvious advantages in using attenuated viruses as vaccines there are a number of practical problems associated with their use. The most obvious problem is the potential for the virus to revert to a virulent from, e.g. immunisation with attenuated polio virus can very occasionally lead to vaccine related poliomyelitis (Nomoto & Wimmer, 1987). In addition the attenuated virus itself may occasionally produce serious disease symptoms, e.g. immunisation with vaccinia virus did sometimes lead to serious complications (Henderson, 1988). Furthermore, for a variety of reasons, it may not be possible to produce attenuated virus vaccines to some diseases. With hepatitis B virus, for example, at present there is no good tissue culture system to produce large amounts of virus so even if an attenuated strain could be isolated it could not be produced industrially. By contrast, it may not be appropriate to use certain attenuated viruses because of their theoretical potential to induce disease, e.g. attenuated retrovirus vaccines could potentially be oncogenic.

Killed virus vaccines

Another successful approach to the production of vaccines has been to use inactivated virulent virus as the immunogen. Usually viruses are inactivated with heat and by treatment with protein modifying agents such as formaldehyde, a chemical that is also capable of cross-linking proteins. The principal advantage that killed virus vaccines have over attenuated virus vaccines is that the problem of reversion to virulent virus does not occur. However, there are also certain other advantages and disadvantages. Thus greater antigenic mass may be required for a killed vaccine to be effective and, in order to induce long-term protective immunity, booster inoculations are required. Nevertheless it is easier to quantitate the actual amount of immunogen necessary to induce protective immunity when using killed vaccines and there is not the problem of vaccine 'take' (replication *in vivo*). Killed virus vaccines also tend to be more stable to storage than live attenuated virus. There are, however, a number of other problems associated with the use of killed virus vaccines. Thus, for example, non-structural proteins are not presented to the immune system and therefore cannot induce T cell responses. Also the processes by which the virus is inactivated may lead to the destruction of critical antigenic sites and the induction of deleterious immunity (Rauh & Schmidt, 1965; Kim *et al.*, 1969). In this regard it is important to realise that treatments which crosslink proteins may affect both B and T cell responses. For example, not only can cross linking virus proteins result in the denaturation of certain critical antigenic sites recognised by antibodies it may also result in the incorrect processing of the protein for the induction of T cell responses. Furthermore, it may be that antigens present in killed virus vaccines will preferentially be presented to class II restricted T cells due to the lack of endogenous synthesis of viral antigens.

SUBUNIT VACCINES

(a) Peptides

The reductionist approach of producing virus vaccines is best exemplified by considering the advantages and disadvantages of using synthetic peptides. In this approach it is first necessary to identify critical antigenic sites present in the native virus particle and then to construct short peptides which represent the appropriate epitopes in the expectation that immunisation with these synthetic peptides will induce protective immunity. Indeed, it has been shown in a number of virus systems that immunisation of experimental animals with synthetic peptides can give rise to virus neutralising antibodies (e.g. foot-and-mouth (Bittle *et al.*, 1982), polio (Emini *et al.*, 1983; Ferguson *el al.*, 1985), herpes simplex virus (Cohen *et al.*, 1984) and hepatitis B virus (Neurath *et al.*, 1985)). However, while the use of synthetic peptides has a number of potential advantages over other methods of antigen production (e.g. they are chemically defined by synthesis, stable, non-infectious and potentially may be modified to stimulate different immune responses; Brown, 1987), for synthetic peptides to be used generally in vaccines a lot more research on their use as immunogens will have to be undertaken. This is because there are a number of fundamental immunological problems in using peptides as immunogens (Schwartz, 1986). For example, many critical antigenic sites on viruses that induce neutralising antibodies are conformationally sensitive. Consequently the appropriate chemistry will have to be developed to produce synthetic peptides that will take up the appropriate epitope conformation found on the native virus particle (Satterthwait *et al.*, 1989). Furthermore the whole problem of how to present peptides (with or without adjuvants–below) in such a way to induce a vigorous immune response remains. For example, to induce antibody responses it is also usually necessary to induce T helper cell responses and this may not be possible as not all synthetic peptides contain target antigens for T cells. Attempts have been made to overcome this problem by fusing predicted T helper cell epitopes onto B cell epitopes (Francis *et al.*, 1987; Good *et al.*, 1987), but there may still be the problem of inducing long-term immunological memory with such immunogens. Long-term protective immunity may be dependent upon the presence of both virus specific memory B cells and memory T helper cells. Immunisation with specific B cell peptides coupled to unrelated T helper cell target antigens would not result in the induction of T helper memory cells that would be induced by infection with wild-type virus. This potential problem may, however, be partially overcome if B memory cells so generated act as efficient antigen presenting cells for appropriate T cell responses following wild-type infection.

When attempting to induce any form of T cell response with synthetic peptides there is also the fundamental problem of MHC polymorphism to be considered. Thus, even if delivery systems can be developed so that peptides will induce T cell responses, in a heterogeneous population a single peptide will only give rise to T cell responses in individuals with appropriate MHC antigens. Consequently for such an approach to be successful multiple T cell determinants will have to be identified and incorporated into any potential vaccine (Rothbard, 1987).

The innate problems of using peptides as vaccines does not mean that peptides will not be successfully used in vaccines. Clearly, when presented in the correct manner, peptides can induce protective immunity and in certain situations it may be that the use of synthetic peptides as immunogens may be particularly effective.

(b) Virus chimaeras

Many of the drawbacks of using peptides as vaccines also apply to the potential use of virus chimaeras. In this approach a small antigenic determinant on the surface of a virus particle is replaced with a foreign antigenic determinant against which an immune response is to be generated (Evans *et al.*, 1989). Thus a chimaera will only generate a restricted immune response to a particular antigenic determinant on the virus of interest. While it is clear that the chimaera itself should induce a vigorous T cell response this will not in the main be directed against the virus for which the vaccine is being developed. Nevertheless virus chimaeras may have a role to play in the development of certain vaccines. For example, because potentially they are infectious it may be that they can be used in situations where the induction of mucosal immunity is required. However, the use of virus chimaeras raises the whole question of vaccine safety. This is because one of the ways in which viruses may evolve is through changes in their receptor binding specificity a property that may be specifically altered in the construction of the chimaera. Consequently, before virus chimaeras could be used as vaccines either rigorous safety testing will have to be carried out or they would first have to be inactivated, a process which would abolish many of the potential advantages of using the chimaeras in the first place. However, the potential difficulties that may arise from the use of virus chimaeras, as with the use of peptides, do not mean that they may not have a valuable role to play in the design of certain vaccines. Indeed it has been shown that polio chimaeras which express a conserved epitope on HIV virus can give rise to neutralising antibody (Evans *et al.*, 1989).

(c) Infectious expression vectors

There are a number of different expression vectors which have been proposed as potential vehicles for constructing live infectious recombinant vaccines. Since vaccinia virus was used so successfully as a vaccine in the world-wide eradication of smallpox, the use of vaccinia virus expression vectors as potential candidates for recombinant vaccines has received particularly strong backing (Zarling *et al.*, 1986; Zagury *et al.*, 1988; Moss *et al.*, 1988). The use of any expression vector as a vaccine has the potential advantage that the appropriate antigens will be endogenously synthesised *in vivo* and should thus be efficiently presented to the immune system. Furthermore complete foreign genes can be cloned into expression vectors and thus the expressed protein, which may contain multiple antigenic sites, should be folded and processed (e.g. glycosylated) correctly. However, there is a limit to the number of genes that can be cloned into any one expression vector. In addition when attempting to produce CTL responses it may be that the expression vector will induce an inappropriate class of CTL (Long & Jacobson, 1989). However, whether this is just a perceived problem remains to be established.

A further potential problem in using live virus expression vectors (or virus chimaeras) as vaccines is how the immune response to the virus vector itself will affect the response to the antigen of interest. Thus it may be difficult to successfully immunise individuals with previously acquired immunity to the virus vector or to reimmunise with the same vector (Rooney *et al.*, 1988). As with the use of any live virus there is also the potential problem of vaccine related disease induced by the replication of the expression vector *in vivo*. While it may be possible to molecularly engineer viruses such as vaccinia virus to totally abolish virulence (Moss *et al.*, 1988) it is not clear whether any reduction in virulence will always be accompanied by a reduction in the replication of the virus *in vivo* and a concomitant lessening of the immune response to the antigen of interest.

(d) Purified proteins as potential vaccines

The use of purified proteins as antigens for possible incorporation into subunit vaccines has a number of potential advantages and disadvantages over and above those discussed for other methods of antigen presentation. For example, unlike peptides, proteins may contain multiple antigenic determinants including both B cell and T cell putative target epitopes. Also unlike the use of killed virus vaccines it is potentially possible to incorporate structural and non-structural proteins into multivalent single virus vaccines. Such a strategy may be particularly desirable when designing vaccines against certain viruses, such as herpes viruses, with temporal control of gene expression.

The main problem of using purified proteins as immunogens is how to present non-infectious exogenous antigens in such a way as to elicit vigorous antibody and CTL responses. As soluble antigens are not very immunogenic they have to be presented to the immune system in conjunction with adjuvants. However, as mentioned earlier, particulate antigens tend to be more antigenic than soluble proteins. Thus with regard to certain viral proteins which self-aggregate into small particles, such as the hepatitis B surface antigen (Valenzuela *et al.*, 1982; McAleer *et al.*, 1984), the problem of adjuvants is not so acute as the particles alone are immunogenic. For this reason and as a variation on using synthetic peptides as immunogens, genetic material encoding small antigenic epitopes have been cloned onto the end of HBsAg or HBcAg in order to create novel immunogens that will self aggregate (Rutgers *et al.*, 1988; Clarke *et al*, 1987). The same principle applies for the formation of recombinant Ty particles (Kingsman & Kingsman, 1988), the latter approach being particularly promising because of the relatively large amount of foreign antigen that can be incorporated.

For proteins that do not self aggregate the problem of adjuvants remains and is discussed in greater detail in Bomford (this volume). The older adjuvants, such as Freund's complete adjuvant which has been extensively used in animals, are often too toxic to be used in man. In fact to date the only adjuvants licensed for use in man in most countries are aluminium hydroxide and aluminium phosphate. Here the adjuvant probably works in part by forming artificial aggregates of the protein. However, when compared with other more toxic adjuvants such as Freund's complete adjuvant, alum is often less effective in inducing antibody and cell mediated responses in experimental animals (Allison, 1987). Consequently there is a lot of effort going into the design of novel non-toxic adjuvants for use in man (e.g. Adams, 1985; Adams *et al.*, 1987; Allison, 1987; Bomford, this volume). Another particularly promising approach has been to incorporate antigens into what has been termed immunostimulating complexes or ISCOMS. These particles have a cage like structure with a mean diameter of 35nm and consist of the antigen of interest together with the adjuvant Quil A. While they are particularly potent in inducing antibody responses, they appear to induce good cell mediated responses as well (Morein *et al.*, 1987; Morein, 1988).

Given that to produce vaccines against certain viruses it may be desirable to incorporate multiple virus antigens into the vaccine we have also been developing an alternative immunisation strategy. The basis of this scheme is to use monoclonal antibodies, linked to a solid matrix, to first purify and then use the solid matrix–antibody–antigen (SMAA) complexes as immunogens (Fig. 2). SMAA complexes themselves are particulate and can induce both vigorous antibody responses as well as class I restricted T cell responses (Randall & Young, 1988; Randall *et al.*, 1988; Young & Randall, unpublished observations). The potential advantages and disadvantages of such a method of antigen presentation have been discussed in greater detail elsewhere (Randall, 1989). Principally the major advantage of such a system is that the complexes are extremely easy to construct and it is relatively simple to incorporate multiple virus antigens, including structual and non-structural proteins, into the SMAA complexes (Randall & Young, 1988; Randall & Young 1989). Also antigens derived from virtually any source, be it expression vectors or virally infected tissue culture cells, can simply be incorporated into SMAA complexes.

As alternatives to conventional adjuvants it has also been suggested that it may be possible to induce an immune response to antigens by targeting the binding of antigens directly to particular host cells (Randall, 1989). This could be achieved by coupling the antigens to monoclonal antibodies that recognise cell surface determinants, such as immunoglobulins (Kawamura & Berzofsky, 1986) or MHC class I and class II antigens (Snider & Segal, 1987; Carayanniotis & Barber, 1987). Furthermore it may be possible to enhance particular arms of the immune response by coinjecting specific cytokines with the antigen (Stevens *et al.*, 1988). In this respect

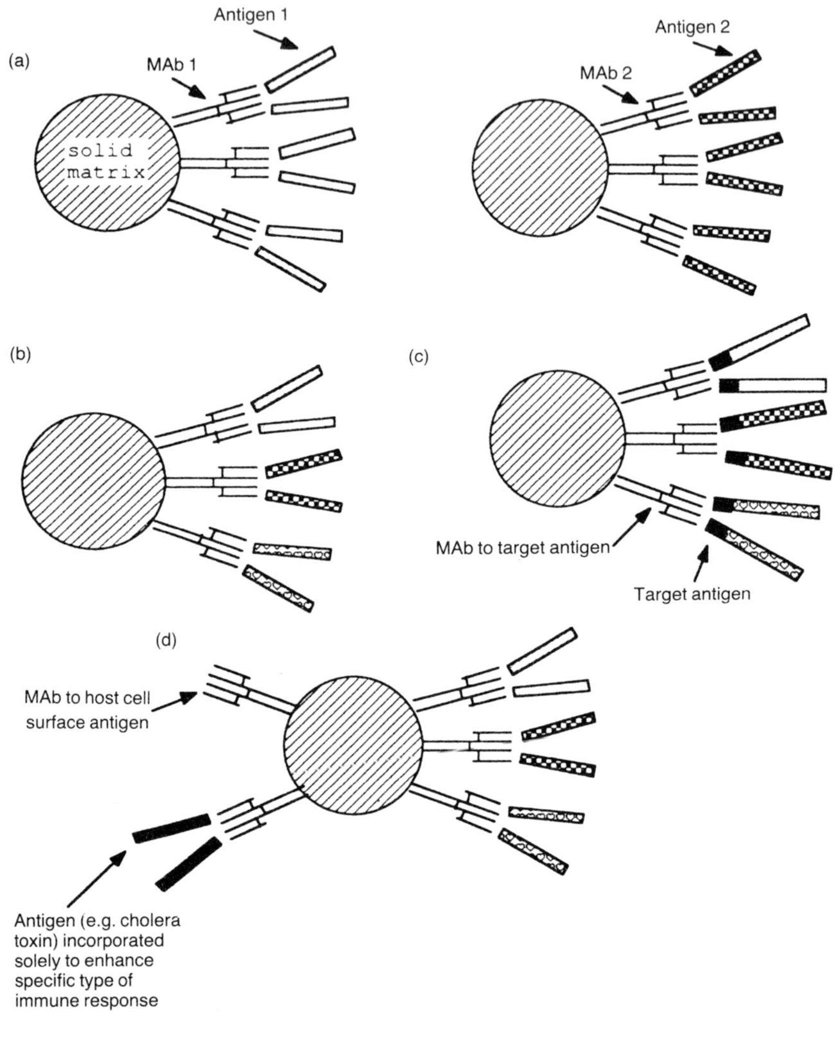

Fig. 2. Schematic illustrations of how SMAA complexes may be constructed to form multivalent immunogens. Panel (a), different MAbs are attached to independent solid matrices prior to the binding of antigens. The resulting SMAA complexes are mixed together before being used as immunogens. Panel (b), different MAbs are attached to the same solid matrix prior to the formation of the SMAA complexes. Panel (c), a single MAb that recognises a target antigen, which may be attached to different microbial antigens, is used in the construction of SMAA complexes. Panel (d), the same as panel (b) except that additional MAbs (e.g. to host cell surface antigens) and proteins (e.g. cholera toxin) are also attached to the solid matrix specifically to enhance particular types of immune responses. (Modified from Randall, 1989 and with permission from *Immunology Today*, Elsevier Science Publishers Ltd.)

it is of interest to note that Meuer *et al.*, (1989) reported that injection of interleukin-2 induced systemic immune responses against HBsAg in immunodeficient patients who were non-responders to hepatitis B vaccination.

CONCLUSIONS

From the foregoing discussion it is clear that there are many alternative strategies for making virus vaccines. More comparative analyses of these alternative strategies will have to be carried out to determine their effectiveness in inducing different immune responses. For example, if there are specialised antigen presenting cells *in vivo* capable of processing and presenting exogenous antigen to class I restricted T cells it may be that different adjuvants and/or immunisation strategies will affect the ability of these cells to capture antigen and hence to induce class I restricted CTLs. It may even be found that the best way of inducing long term protective immunity in some instances will be to use more than one method of antigen presentation, e.g. the most effective way to use virus chimaeras may be in conjunction with other immunisation protocols capable of inducing broader virus specific T cell responses. However, while intellectual arguments for and against different strategies can be made based on our current understanding of how the immune system works and while many of us may have our own personal windmills to charge, ultimately for each virus vaccine, as Don Quixote commented, 'The proof of the pudding will be in the eating'.

ACKNOWLEDGEMENTS

RER is a recipient of a Wellcome Trust University Award and BES is supported by a Wellcome Trust Medical Training Fellowship.

REFERENCES

Abbas, A. K. (1988). A reassessment of the mechanisms of antigen-specific T cell-dependent B-cell activation. *Immunology Today*, **9**, 89–94.

Acuto, O. & Reinherz, E. L. (1985). The human T-cell receptor. Structure and Function. *New England Journal of Medicine* **312**, 1100–11.

Ada, G. (1989). Prospect for a vaccine against HIV. *Nature, London*, **339**, 331–2.

Adams, A. (1985). Synthetic adjuvants. In *Modern Concepts in Immunology* vol. 1. New York: John Wiley & Sons.

Adams, S. E., Dawson, K. M. Gull, K. Kingsman, S. M. & Kingsman, A. J. (1987). The expression of hybrid HIV: Ty virus-like particles in yeast. *Nature, London*, **329** 68–73.

Albrecht, P., Burnstein, T., Klutch, M. J., Hicks, J. J. & Ennis, F. A. (1977). Subacute sclerosing panencephalitis: Experimental infection in primates. *Science*, **195**, 64–6.

Allen, P. M. (1987). Antigen processing at the molecular level. *Immunology Today*, **8**, 270–3.

Allison, A. C. & Byars, N. E. (1987). Vaccine Technology: Adjuvants for increased efficacy. *Biotechnology*, **5**, 1041–5.

Aparicio, P., Jaraquemada, D. & Lopez de Castro, J. A. (1987). Alloreactive cytolytic T cell clones with dual recognition of HLA-B27 and HLA-DR2 antigens. *Journal of Experimental Medicine*, **165**, 428–43.

Arvieux, J., Yssel, H. & Colomb, M. G. (1988). Antigen-bound C3b and C4b enhance antigen-presenting cell function in activation of human T-cell clones. *Immunology*, **65**, 229–35.

Ashman, R. F. (1984). Lymphocyte activation, In '*Fundamental Immunology*', W. E. Paul, ed. pp 267–300. Raven Press, New York.

Augustin, A., Kubo, R. T. & Sim, G-K (1989). Resident pulmonary lymphocytes expressing the γ/δ T-cell receptor. *Nature, London*, **340**, 239–41.

Austyn, J. M. (1989). *Antigen-presenting cells*. IRL Press, Oxford.

Babbit, B. P., Allen, P. M., Matsueda, G., Haber, E. & Unanue, E. R. (1985). Binding of immunogenic peptides to 1a histocompatibility molecules. *Nature, London*, **317**, 359–61.

Balkwill, F. R. & Burke, F. (1989). The cytokine network. *Immunology Today*, **10**, 299–304.

Barlow, D. J., Edwards, M. S. & Thornton, J. M. (1986). Continuous and discontinuous protein antigenic determinants. *Nature, London*, **322**, 747–8.

Barnett, B. C., Graham, C. M., Burt, D. S., Skehel, J. J. & Thomas, D. B. (1989). The immune response of BALB/c mice to influenza hemagglutinin: commonality of the B and T cell repertoires and their relevance to antigenic drift. *European Journal of Immunology*, **19**, 515–21.

Bastin, J., Rothbard, J., Davey, J., Jones, I & Townsend, A. R. M. (1987). Use of synthetic peptides of influenza nucleoprotein to define epitopes recognised by class 1 restricted cytotoxic T lymphocytes. *Journal of Experimental Medicine*, **165**, 1508–23.

Belshe, R. B., Bernstein, J. M. & Dansby, K. N. (1984). Respiratory Syncytial Virus. In *Textbook of Human Virology*, R. B. Belshe ed., pp. 361–84, PSG Publishing Company.

Berzofsky, J. A., Cease, K. B., Cornette, J. L., Spouge, J. L., Margalit, H., Berkower, I. J., Good, M. F., Miller, L. H. & DeLisi, C. (1987). Protein antigenic structures recognised by T cells: Potential applications to vaccine design. *Immunological Reviews*, **98**, 9–52.

Bevan, M. J. (1987). Class discrimination in the world of immunology. *Nature, London*, **325**, 192–4.

Biddison, W. E., Sharrow, S. O. & Shearer, G. M. (1981). T cell subpopulations required for the human cytotoxic T lymphocyte response to influenza virus; evidence for T cell help. *Journal of Immunology*, **127**, 487–91.

Bittle, J. L., Houghten, R. A., Alexander, H., Shinnick, T. M., Sutcliffe, J. G., Lerner, R. A., Rowlands, D. J. and Brown, F. (1982). Protection against foot-and-mouth disease by immunization with a chemically synthesized peptide predicted from the viral nucleotide sequence. *Nature, London*, **298**, 30–3.

Bjorkman, P. J., Saper, M. A., Samraoui, Bennett, W. S., Strominger, J. L. & Wiley, D. C. (1987a). Structure of the human class 1 histocompatibility antigen, HLA-A2. *Nature, London*, **329**, 506–12.

Bjorkman, P. J., Saper, M. A., Samraoui, B., Bennett, W. S., Strominger, J. L. & Wiley, D. C. (1987b). The foreign antigen binding site and T cell recognition of class 1 histocompatibility antigens. *Nature, London*, **329**, 512–18.

Blundell, T. L., Sibanda, B. L., Sternberg, M. J. E. & Thornton, J. M. (1987).

Knowledge-based prediction of protein structures and the design of novel molecules. *Nature, London*, **326**, 347–52.

Bodmer, H. C., Gotch, F. M. & McMichael, A. J. (1989). Class 1 cross-restricted T cells reveal low responder allele due to processing of viral antigen. *Nature, London*, **337**, 653–5.

Bolognesi, D. P. (1989). Do antibodies enhance the infection of cells by HIV? *Nature, London*, **340**, 431–2.

Borysiewicz, L. K., Morris, S., Page, J. D. & Sissons, J. G. D. (1983). Human cytomegalovirus-specific cytotoxic T lymphocytes: Requirements for in vitro generation and specificity. *European Journal of Immunology*, **13**, 804–9.

Botazzo, G. F., Pujol-Borrell, R., Hanafusa, T. & Feldmann, M. (1983). Role of aberrant HLA-DR expression and antigen presentation induction of endocrine autoimmunity. *Lancet*, **ii**, 1115–19.

Bourgault, I., Gomez, A., Gomrad, E., Picard, F. & Levy, J. P. (1989). A virus-specific $CD4^+$ cell-mediated cytolytic activity revealed by $CD8^+$ cell elimination regularly develops in uncloned human antiviral cell lines. *Journal of Immunology*, **142**, 252–6.

Braciale, T. J., Morrison, L. A., Sweetser, M. T., Sambrook, J., Gething, M.-J. & Braciale, V. L. (1987). Antigen presentation pathways to Class I and Class II MHC-restricted T lymphocytes. *Immunological Reviews*, **98**, 95–114.

Brown, F. (1987). From Jenner to Genes – The next generation of virus vaccines. Biochemical Society Symposium 53, 75–84.

Buller, R. M. L., Holmes, K. L., Hugin, A, Frederickson, T. N. & Morse III, H. C. (1987). Induction of cytotoxic T-cell responses *in vivo* in the absence of CD4 helper cells. *Nature, London*, **328**, 77–9.

Burnet, F. M. (1968). Measles as an index of immunological function. *Lancet* **ii**, 610–13.

Burnet, M. (1959). *The Clonal Selection Theory of Acquired Immunity*. Cambridge University Press, Cambridge.

Burt, D. S., Mills, K. H. G., Skehel, J. J. & Thomas, D. B. (1989). Diversity of the class II (1-A^k/1-E^k)-restricted T cell repertoire for influenza hemagglutinin and antigenic drift. Six nonoverlapping epitopes on the HA1 subunit are defined by synthetic peptides. *Journal of Experimental Medicine*, **170**, 383–97.

Buss, S., Colon, S., Smith, C., Freed, J. H., Miles, C. & Grey, H. M. (1986). Interaction between a 'processed' ovalbumin peptide and la molecules. *Proceedings of the National Academy of Sciences, USA*, **83**, 3968–71.

Buss, S., Sette, A., Colon, S. M., Miles, C. & Grey, H. M. (1987). The relation between major histocompatibility complex (MHC) restriction and the capacity of la to bind immunogenic peptides. *Science*, **235**, 1353–8.

Cannon, M. J., Openshaw, P. J. M. & Askonas, B. A. (1988). Cytotoxic T cells clear virus but augment lung pathology in mice infected with respiratory syncytial virus. *Journal of Experimental Medicine*, **168**, 1163–8.

Cantor, H. & Boyse, E. A. (1976). Regulation of cellular and humoral immune responses by T-cell subclasses. *Cold Spring Harbor Symposia on Quantitative Biology*, **41**, 23–32.

Carayanniotis, G. & Barber, B. H. (1987). Adjuvant-free IgG responses induced with antigen coupled to antibodies against class II MHC. *Nature, London*, **327**, 59–61.

Castro, B. A., Cheng-Mager, C., Evans, L. A. and Levy, J. A. (1988). HIV heterogeneity and viral pathogenesis. *AIDS (suppl. 1)*, S17-S27.

Celis, E. & Chang, T. W. (1984). Antibodies to hepatitis B surface antigen potentiate the response of human T lymphocyte clones to the same antigen. *Science*,

224, 297–9.
Challacombe, S. J. (1987). The induction of secretory IgA responses. In *Food Allergy and Intolerance*, J. Brostoff & S. J. Challacombe, eds, pp. 269–85. Bailliere Tindall.
Challacombe, S. J. & Tomasi, T. B. (1987). Model systems of antigen handling. In *Food Allergy and Intolerance*, J. Brostoff & S. J. Challacombe eds, pp. 255–68. Bailliere Tindall.
Chang, T. W. (1985). Regulation of immune response by antibodies: the importance of antibody and monocyte Fc receptor interaction in T-cell activation. *Immunology Today*, **6**, 245–9.
Chused, T. M., Kassan, S. S. & Mosier, D. E. (1976). Macrophage requirement for the in vitro response to TNP Ficoll: A thymic independent antigen. *Journal of Immunology*, **116**, 1579–81.
Clarke, B. E., Newton, S. E., Carroll, A. R., Francis, M. J., Appleyard, G., Syred, A. D., Highfield, P. E., Rowlands, D. J. & Brown, F. (1987). Improved immunogenicity of a peptide epitope after fusion to hepatitis B core protein. *Nature, London*, **330**, 381–4.
Cohen, G. H., Dietzschold, B., Ponce de Leon, M., Long, D., Golub, E., Varrichio, A, Pereira, L. & Eisenberg, R. J. (1984). Localization and synthesis of an antigenic determinant of herpes simplex virus glycoprotein D that stimulates the production of neutralizing antibody. *Journal of Virology*, **49**, 102–8.
Cresswell, P. (1987). Regulation of HLA Class I and Class II antigen expression. *British Medical Bulletin*, **43**, 66–80.
Delovitch, T. L., Semple, J. W. & Phillips, M. L. (1988). Influence of antigen processing on immune responsiveness. *Immunology Today*, **9**, 216–17.
Dietzschold, B., Wang, H., Rupprecht, C. E., Celis, E., Tollis, M., Ertl, H., Heber-Katz, E. & Koprowski, H. (1987). Induction of protective immunity against rabies by immunization with rabies virus ribonucleoprotein. *Proceedings of the National Academy of Sciences, USA*, **84**, 9165–9.
Elson, C. O. & Ealding, W. (1984). Cholera toxin feeding did not induce oral tolerance in mice and abrogated oral tolerance to an unrelated protein antigen. *Journal of Immunology*, **133**, 2892–7.
Emini, E. A., Jameson, B. A. & Wimmer, E. (1983). Priming for and induction of anti-polio neutralizing antibodies by synthetic peptides. *Nature, London*, **304**, 699–703.
Enssle, K. H., Wagner, H. & Fleischer, B. (1987). Human mumps virus-specific cytotoxic T lymphocytes: quantitative analysis of HLA restriction. *Human Immunology*, **18**, 135–40.
Evans, D. J., McKeating, J., Meredith, J. M., Burke, K. L., Katrak, K., John, A., Ferguson, M., Minor, P. D., Weiss, R. A. & Almond, J. W. (1989). An engineered poliovirus chimaera elicits broadly reactive HIV-1 neutralizing antibodies. *Nature, London*, **339**, 385–8.
Ewan, P. W. & Lachman, P. J. (1982). In *Clinical Aspects of Immunology*, vol. 2, pp. 1482–1514, Blackwell Scientific Publications.
Ferguson, M., Evans, D. M. A., McGrath, D. I., Minor, P. D., Almond, J. W. & Schild, G. C. (1985). Induction by synthetic peptides of broadly reactive, type specific neutralizing antibody to poliovirus type 3. *Virology*, **143**, 505–15.
Fisher, A. G., Ensoli, B., Looney, D., Rose, A., Gallo, R. C., Saag, M. S., Shaw, G. M., Hahn, B. H. & Wong-Staal, F. (1988). Biologically diverse molecular variants within a single HIV-1 isolate. *Nature, London*, **334**, 444–7.
Francis, M. J., Hastings, G. Z., Syred, A. D., McGigg, B., Brown, F. & Rowlands, D. J. (1987). Non-responsiveness to a foot-and-mouth disease virus peptide over-

come by addition of foreign helper T-cell determinants. *Nature, London*, **330**, 168–70.

Francis, M. J., Fry, C. M., Rowlands, D. J. & Brown, F. (1988). Qualitative and quantitative differences in the immune responses to foot-and-mouth disease virus antigens and synthetic peptides. *Journal of General Virology*, **69**, 2483–91.

Fujinami, R. S. & Oldstone, M. B. A. (1979). Antiviral antibody reacting on the plasma membrane alters measles virus expression inside the cell. *Nature, London*, **279**, 529–30.

Fujinami, R. S. & Oldstone, M. B. A. (1980). Alterations in expression of measles virus polypeptides by antibody: molecular events in antibody-induced antigenic modulation. *Journal of Immunology*, **125**, 78–85.

Gay, D., Maddon, P., Sekaly, R., Talle, M. A., Godfrey, M., Long, E., Goldstein, G., Chess, L., Axel, R., Kappler, J. & Marrack, P. (1987). Functional interaction between human T-cell protein CD4 and the major histocompatibility complex HLA-DR antigen. *Nature, London*, **328**, 626–9.

Germain, R. N. (1986). The ins and outs of antigen processing and presentation. *Nature, London*, **322**, 687–9.

Good, M. F., Maloy, W. L., Lunde, M. N., Margalit, H., Cornette, J. L., Smith, G. L., Moss, B., Miller, L. H. & Berzofsky, J. A. (1987). Construction of a synthetic immunogen: use of a new T-helper epitope on malaria circumsporozoite protein. *Science*, **235**, 1059–62.

Gotch, F., Rothbard, J., Howland, K, Townsend, A & McMichael, A. (1987). Cytotoxic T lymphocytes recognize a fragment of influenza virus matrix protein in association with HLA-A2. *Nature, London*, **326**, 881–2.

Graham, C. M., Barnett, B. C., Hartlmayr, I., Burt, D. S., Faulkes, R., Skehel, J. J. & Thomas, D. B. (1989). The structural requirements for class II (I-A^k)-restricted T cell recognition of influenza hemagglutinin: B cell epitopes define T cell epitopes. *European Journal of Immunology*, **19**, 523–8.

Grey, H. M. & Chesnut, R. (1985). Antigen processing and presentation to T cells. *Immunology Today*, **6**, 101–6.

Halstead, S. B. (1979). In vivo enhancement of dengue virus infection in rhesus monkeys by passively transferred antibodies. *Journal of Infectious Diseases*, **140**, 527–33.

Hamblin, A. S. (1988). *Lymphokines*, IRL Press, Oxford.

Henderson, D. A. (1988). Smallpox and Vaccinia. In *Vaccines*, S. A. Plotkin & E. A. Mortimer eds, pp. 8–30. Saunders Company.

Issacs, D., Bangham, C. R. M. & McMichael, A. J. (1987). Cell-mediated cytotoxic response to respiratory syncytial virus in infants with bronchiolitis. *Lancet*, **ii**, 769–71.

Jacobson, S., Richert, J. R., Biddison, W. E., Satinsky, A., Hartzman, R. J. & McFarland, H. F. (1984). Measles virus-specific $T4^+$ human cytotoxic T cell clones are restricted by class II HLA antigens. *Journal of Immunology*, **133**, 754–7.

Jacobson, S., Rose, J. W., Flerlage, M. L., McFarlin, D. E. & McFarland, H. F. (1988). Induction of measles virus-specific human cytotoxic T cells by purified measles virus nucleocapsid and hemagglutinin polypeptides. *Viral Immunology*, **1**, 153–62.

Jacobson, S., Sekaly, R. P., Jacobson, C. L., McFarland, H. F. & Long, E. O. (1989). HLA Class II-restricted presentation of cytoplasmic measles virus antigens to cytotoxic T cells. *Journal of Virology*, **63**, 1756–62.

Janeway, C. A. (1988). Frontiers of the immune system. *Nature, London*, **333**, 804–6.

Jin, Y., Shih, J. W.-K. & Berkower, I. (1988). Human T cell response to the surface antigen of hepatitis B virus (HBsAg). Endosomal and nonendosomal processing pathways are accessible to both endogenous and exogenous antigen. *Journal of Experimental Medicine*, **168**, 293–306.

Jonjic, S., del Val, M., Keil, G. M., Reddehase, M. J. & Koszinowski, U. H. (1988). A nonstructural viral protein expressed by a recombinant vaccinia virus protects against lethal cytomegalovirus infection. *Journal of Virology*, **62**, 1653–8.

Kaufmann, S. H. E. (1988). $CD8^+$ T lymphocytes in intracellular microbial infections. *Immunology Today*, **9**, 168–74.

Kawamura, H. & Berzofsky, J. A. (1986). Enhancement of antigenic potency in vitro and immunogenicity in vivo by coupling the antigen to anti-immunoglobulin. *Journal of Immunology*, **136**, 58–65.

Kilbourne, E. D. (1988). Inactivated Influenza Vaccines. In *Vaccines* S. A. Plotkin & E. A. Mortimer eds. pp. 420–34. W. B. Saunders Company.

Kim, H. W., Canchola, J. G., Brand, C. D., Pyles, G., Chanock, R. M., Jensen, K & Parrot, R. M. (1969). Respiratory syncytial virus disease in infants despite prior administration of antigenic inactivated vaccine. *American Journal of Epidemiology*, **89**, 422–34.

Kim, H. W., Keikin, S. L. & Arrobio, J. O. (1976). Cell-mediated immunity to respiratory syncytial virus induced by inactivated vaccine or by infection. *Pediatric Research*, **10**, 74–8.

Kingsman, S. M. & Kingsman, A. J. (1988). Polyvalent recombinant antigens: a new vaccine strategy. *Vaccine*, **6**, 304–6.

Klasse, P. J., Pipkorn, R. & Blomberg, J. (1988). Presence of antibodies to a putatively immunosuppressive part of human immunodeficiency virus (HIV) envelope glycoprotein gp41 is strongly associated with health among HIV-positive subjects. *Proceedings of the National Academy of Sciences, USA*, **85**, 5225–9.

Klaus, G. G. B. (1988). Role of complement in the induction of antibody responses. In *The Complement System*, K. Rother & G. O. Till eds, Wien & New York: Springer-Heidelberger.

Knapp, W., Rieber, P., Dorken, B., Schmidt, R. E., Stein, H. & Borne, A. E. G. Kr, V. D. (1989). Towards a better definition of human leucocyte surface molecules. *Immunology Today*, **10**, 253–8.

Knight, V. (1973). In *Viral and Mycoplasmal Infections of the Respiratory Tract*, V. Knight, ed., pp. 131–40, Lea and Febiger, Philadelphia, Pennsylvania.

Kuwano, K., Scott, M., Young, J. F. & Ennis, F. A. (1989). Active immunization against virus infections due to antigenic drift by induction of crossreactive cytotoxic T lymphocytes. *Journal of Experimental Medicine*, **169**, 1361–71.

Lanzavecchia, A. (1985). Antigen specific interaction between T and B cells. *Nature, London*, **314**, 537–9.

Liew, F. Y., Russell, S. M., Appleyard, G., Brand, C. M. & Beale, J. (1984). Cross-protection in mice infected with influenza A virus by the respiratory route is correlated with local IgA antibody rather than serum antibody or cytotoxic T cell reactivity. *European Journal of Immunology*, **14**, 350–6.

Long, E. O. (1989). Intracellular traffic of antigen processing. *Immunology Today*, **10**, 232–4.

Long, E. O. & Jacobson, S. (1989). Pathways of viral antigen processing and presentation to CTL: defined by the mode of virus entry? *Immunology Today*, **10**, 45–8.

Lycke, N., Bromander, A. K., Ekman, L., Karlsson, U. & Holmgren, J. (1989). Cellular basis of immunomodulation by cholera toxin *in vitro* with possible association to the adjuvant function *in vivo*. *Journal of Immunology*, **142**, 20–7.

McAleer, W. J., Buynak, E. B., Maigetter, R. Z., Wampler, D. E., Miller, W. J. & Hilleman, M. R. (1984). Human hepatitis B vaccine from recombinant yeast. *Nature, London*, **307**, 178–80.
McCullough, K. C., Parkinson, D. & Crowther, J. R. (1988). Opsonization-enhanced phagocytosis of foot-and-mouth disease virus. *Immunology*, **65**, 187–91.
McMichael, A. J., Gotch, F. M. & Rothbard, J. (1986). HLA B37 determines an influenza A virus nucleoprotein epitope recognised by cytotoxic T lymphocytes. *Journal of Experimental Medicine*, **164**, 1397–1406.
Maassab, H. F., LaMontagne, J. R. & Deborde, D. C. (1988). Live Influenza Virus Vaccine. In *Vaccines*, S. A. Plonkin & E. A. Mortimer, eds: pp. 435–57. W. B. Saunders Company.
Manca, F., Fenoglio, D., Kunki, A., Cambiaggi, C., Li Para, G. & Celada, F. (1988). B cells on a podium: regulatory roles of surface and secreted immunoglobulins. *Immunology Today*, **9**, 300–3.
Margalit, H., Spouge, J. L., Cornette, J. L., Cease, K. B., Delisi, C. & Berzofsky, J. A. (1987). Prediction of immunodominant helper T cell antigenic sites from the primary sequence. *Journal of Immunology*, **138**, 2213–29.
Maryanski, J. L., Pala, P., Corrandin, G., Jordan, B. R. & Cerottini, J.-C. (1986). H-2-restricted cytolytic cells specific for HLA can recognise a synthetic HLA peptide. *Nature, London*, **324**, 578–9.
Melnick, J. L. (1986). Virus Vaccines: 1986 Update. *Progress in Medical Virology*, **33**, 134–70.
Mestecky, J. & McGhee, J. R. (1987). Immunoglobulin A (lgA): Molecular and cellular interactions involved in IgA biosynthesis and immune response. *Advances in Immunology*, **40**, 153–245.
Meuer, S. C., Hodgdon, J. C., Cooper, D. A., Hussey, R. E., Fitzgerald, K. A., Schlossman, S. F. & Reinherz, E. L. (1983). Human cytotoxic T cell clones directed at autologous virus-transformed targets: further evidence for linkage of genetic restriction to T4 and T8 surface glycoproteins. *Journal of Immunology*, **131**, 186–90.
Meuer, S. C., Dumann, H., Meyer zum Buschenfelde, K-H. & Kohler, H. (1989). Low-dose interleukin-2 induces systemic immune responses against HBsAg in immunodeficient non-responders to hepatitis B vaccination. *Lancet*, **i**, 15–17.
Milich, D. R., McLachlan, A., Thornton, G. B. & Hughes, J. L. (1987). Antibody production to the nucleocapsid and envelope of hepatitis B virus primed by a single synthetic T cell site. *Nature, London*, **329**, 547–9.
Mills, K. H. G. (1986). Processing of viral antigens and presentation to class II-restricted T cells. *Immunology Today*, **7**, 260–6.
Mims, C. A. & White, D. O. (1984). In *Viral Pathogenesis and Immunology*, Blackwell Scientific Publications.
Mitchison, N. A. (1964). Induction of immunological paralysis in two zones of dosage. *Proceedings of the Royal Society*, series B, **161**, 275–92.
Modlin, J. F. (1984). Measles Virus. In *Textbook of Human Virology*, R. B. Belshe, ed. pp. 333–360. PSG Publishing Company.
Modrow, S., Hahn, B. D., Shaw, G. M., Gallo, R. C., Wong-Staal, F. & Wolf, H. (1987). Computer-assisted analysis of envelope protein sequences of seven human immunodeficiency virus isolates: prediction of antigenic epitopes in conserved and variable regions. *Journal of Virology*, **61**, 570–8.
Moore, M. W., Carbone, F. R. & Bevan, M. J. (1988). Introduction of soluble protein into the Class I pathway of antigen processing and presentation. *Cell*, **54**, 777–85.
Morein, B. (1988). The ISCOM antigen-presenting system. *Nature, London*, **332**,

287–8.
Morein, B., Lovgren, K., Hoglund, S. & Sundquist, B. (1987). The ISCOM: an immunostimulating complex. *Immunology Today*, **8**, 333–8.
Morrison, L. A., Lukacher, A. E., Braciale, V. L., Fan, D. P. & Braciale, T. J. (1986). Differences in antigen presentation to MHC class I- and class II-restricted influenza virus-specific cytolytic T lymphocyte clones. *Journal of Experimental Medicine*, **163**, 903–21.
Morrison, R. P., Earl, P. L., Nishio, J., Lodmell, D. L., Moss, B. & Chesebro, B. (1987). Different H-2 subregions influence immunization against retroviruses and immunosuppression. *Nature, London*, **329**, 729–32.
Moss, B., Fuerst, T. R., Flexner, C. & Hugin, A. (1988). Roles of vaccinia virus in the development of new vaccines. *Vaccine*, **6**, 161–3.
Mowat, A. M. (1987). The regulation of immune responses to dietary protein antigens. *Immunology Today*, **8**, 93–8.
Neurath, A. R., Strich, N. & Keub, S. B. M. (1985). In *Vaccines 85*, R. A. Lerner, R. M. Chanock & F. Brown, eds., pp 185–189, Cold Spring Harbor Laboratory, New York.
Nixon, D. F., Townsend, A. R. M., Elvin, J. G., Rizza, C. R., Gallwey, J. & McMichael, A. J. (1988). HIV-1 gag-specific cytotoxic T lymphocytes defined with recombinant vaccinia virus and synthetic peptides. *Nature, London*, **336**, 484–7.
Nomoto, A. & Wimmer, E. (1987). Genetic studies of the antigenicity and the attenuation phenotype of poliovirus. In *Molecular Basis of Virus Disease*. SGM symposium 40, W. C. Russell & J. W. Almond, eds, pp. 107–34. Cambridge University Press.
Norment, A. M., Salter, R. D., Parham, P., Engelhard, V. H. & Littman, D. R. (1988). Cell-cell adhesion mediated by CD8 and MHC class I molecules. *Nature, London*, **336**, 79–81.
Norrby, E. (1987). Towards new viral vaccines for man. *Advances in Virus Research*, **32**, 1–34.
Norrby, E., Enders-Ruckle, G. & ter Meulen, V. (1975). Differences in the appearance of antibodies to structural components of measles virus after immunization with inactivated and live virus. *Journal of Infectious Diseases*, **132**, 262–9.
Nossal, G. J. V. (1987). The basic components of the immune system. *New England Journal of Medicine*, **316**, 1320–6.
Novotny, J., Handschumacher, M. & Bruccoleri, R. E. (1987). Protein antigenicity: a static surface property. *Immunology Today*, **8**, 26–31.
Oldstone, M. B. A. & Fujinami, R. S. (1982). Virus persistence and avoidance of immune surveillance: How measles can be induced to persist in cells, escape immune assault and injure tissues. In *Virus Persistence*, B. W. J. Mahy, A. C. Minson & G. K. Darby, eds, Cambridge University Press.
Plotkin, S. A. & Mortimer, E. A. (1988). *Vaccines*, W. B. Saunders Company.
Prince, G. A., Jenson, A. B., Hemming, V. G., Murphy, B. R., Walsh, E. E., Horswood, R. L. & Chanock, R. M. (1986). Enhancement of respiratory syncytial virus pulmonary pathology in cotton rats by prior intramuscular inoculation of formalin-inactivated virus. *Journal of Virology*, **57**, 721–8.
Prince, G. A., Hemming, V. G., Horswood, R. L., Baron, P. A. & Chanock, R. M. (1987). Effectiveness of topically administered neutralizing antibodies in experimental immunotherapy of respiratory syncytial virus infection in cotton rats. *Journal of Virology*, **61**, 1851–4.
Rammohan, K. W., McFarland, H. F. & McFarlin, D. E. (1981). Induction of subacute murine measles encephalitis by monoclonal antibody to virus haemag-

glutinin. *Nature, London*, **290**, 588–9.
Randall, R. E. (1989). Solid matrix antibody antigen (SMAA) complexes for constructing multivalent subunit vaccines. *Immunology Today*, **10**, 336–9.
Randall, R. E. & Russell, W. C. (1990). Paramyxovirus persistence, consequences for host and virus. In *The Paramyxoviruses*, D. W. Kingsbury, ed. Plenum.
Randall, R. E. & Young, D. F. (1988). Humoral and cytotoxic T cell immune responses to internal and external structural proteins of simian virus 5 induced by immunization with solid matrix–antibody–antigen complexes. *Journal of General Virology*, **69**, 2505–16.
Randall, R. E. & Young, D. F. (1989). Immunization against multiple viruses by using solid matrix–antibody–antigen complexes. *Journal of Virology*, **63**, 1808–10.
Randall, R. E., Young, D. F. & Southern, J. A. (1988). Immunization with solid matrix–antibody–antigen complexes containing surface or internal virus structural proteins protects mice from infection with the paramyxovirus, Simian Virus 5. *Journal of General Virology*, **69**, 2517–26.
Rauh, L. W. & Schmidt, R. (1965). Measles immunization with killed virus vaccine. *American Journal of Diseases of Children*, **109**, 232–7.
Reddehase, M. J. Rothbard, J. B. & Koszinowski, U. H. (1989). A pentapeptide as minimal antigenic determinant for MHC class I-restricted T lymphocytes. *Nature, London*, **337**, 651–3.
Rooney, J., Wohlenberg, C., Cremer, K., Moss, B. & Notkins, A. (1988). Immunization with a vaccinia virus recombinant expressing herpes simplex virus type 1 glycoprotein D: Long-term protection and effect of revaccination. *Journal of Virology*, **62**, 1530–4.
Rothbard, J. B. (1987). Synthetic peptides as vaccines. *Nature, London*, **330**, 106–7.
Rothbard, J. B. & Taylor, W. R. (1988). A sequence pattern common to T cell epitopes. *The EMBO Journal*, **7**, 93–100.
Ruegg, C. L., Monnell, C. R. & Strand, M. (1989a). Identification, using synthetic peptides, of the minimum amino acid sequence from retroviral transmembrane protein p15E required for inhibition of lymphoproliferation and its similarity to gp21 of human T-lymphotrophic virus types I and II. *Journal of Virology*, **63**, 3250–6.
Ruegg, C. L., Monell, C. R. & Strand, M. (1989b). Inhibition of lymphoproliferation by a synthetic peptide with sequence identity to gp41 of human immunodeficiency virus type 1. *Journal of Virology*, **63**, 3257–60.
Russell, S. M. & Liew, F. M. (1979). T cells primed by influenza virion internal components can cooperate in the antibody response to haemagglutinin. *Nature, London*, **280**, 147–8.
Rutgers, T., Gordon, D., Gathoye, A. M., Hollingdale, M., Hockmeyer, W., Rosenberg, M. & de Wilde, M. (1988). Hepatitis B surface antigen as carrier matrix for the repetitive epitope of the circumsporozoite protein of *Plasmodium falciparum. Biotechnology*, **6**, 1065–70.
Saag, M. S., Hahn, B. H., Gibbons, J., Li, Y., Parks, E. S., Parks, W. P. and Shaw, G. M. (1988). Extensive variation in human immunodeficiency virus type-1 *in vivo. Nature, London*, **334**, 440–4.
Satterthwait, A. C., Arrhenius, T., Hagopian, R. A., Zavala, F., Nussenzweig, V. and Lerner, R. A. (1989). The conformational restriction of synthetic peptides, including a malaria peptide, for use as immunogens. *Philosophical Transactions of the Royal Society of London series B*, **323**, 565–72.
Scherle, P. A. & Gerhard, W. (1988). Differential ability of B cells specific for

external vs. internal influenza virus proteins to respond to help from influenza virus-specific T-cell clones *in vivo*. *Proceedings of the National Academy of Sciences, USA*, **85**, 4446–50.

Schmid, D. S. (1988). The human MHC-restricted cellular response to herpes simplex virus type 1 is mediated by $CD4^+$, $CD8^-$ T cells and is restricted to the DR region of the MHC complex. *Journal of Immunology*, **140**, 3610–16.

Schwartz, R. H. (1986). The value of synthetic peptides as vaccines for eliciting T-cell immunity. *Current Topics in Microbiology and Immunology*, **130**, 79–85.

Sekaly, R. P., Jacobson, S., Richert, J. R., Tonnelle, C., McFarland, H. F. & Long, E. O. (1988). Antigen presentation to HLA class II-restricted measles virus-specific T-cell clones can occur in the absence of the invariant chain. *Proceedings of the National Academy of Sciences, USA* **85**, 1209–12.

Snider, D. P. & Segal, D. M. (1987). Target antigen presentation using crosslinked antibody heteroaggregates. *Journal of Immunology*, **139**, 1609–16.

Sprent, J. & Schaefer, M. (1989). Antigen-presenting cells for unprimed T cells. *Immunology Today*, **10**, 17–23.

Staerz, U. D. Karasuyama, H. & Garner, A. M. (1987). Cytotoxic T lymphocytes against a soluble protein. *Nature, London*, **329**, 449–51.

Starcick, B. R., Hahn, B. H., Shaw, G. M., McNeely, P. D., Modrow, S., Wolf, H., Park, E. S., Josephs, S. F., Gallo, R. & Wong-Staal, F. (1986). Identification and characterisation of conserved and variable regions in the envelope gene of HTLV-III/LAV, the retrovirus of AIDS. *Cell*, **45**, 637–48.

Stevens, T. L., Bossie, A., Sanders, V. M., Fernandez-Botran, R., Coffman, R. L., Mosmann, T. R. & Vitetta, E. S. (1988). Regulation of antibody isotype secretion by subsets of antigen-specific helper T cells. *Nature, London*, **334**, 255–8.

Strachan, T. (1987). Molecular genetics and polymorphism of class 1 HLA antigens. *British Medical Bulletin*, **43**, 1–14.

Taylor, G., Stott, E. J., Bew, M., Fernie, B. F., Cote, P., Collins, A. P., Hughes, M. & Jebbett, J. (1984). Monoclonal antibodies protect against respiratory syncytial virus infection in mice. *Immunology*, **52**, 137–42.

Taylor, P. M. & Askonas, B. A. (1986). Influenza nucleoprotein-specific cytotoxic T-cell clones are protective *in vivo*. *Immunology* **58**, 417–20.

Taylor, P. M., Davey, J., Holland, K., Rothbard, J. B. & Askonas, B. A. (1987). Class I MHC molecules rather than other mouse genes dictate influenza epitope recognition by cytotoxic T cells. *Immunogenetics*, **26**, 267–72.

Townsend, A. R. M., McMichael, A. J., Carter, N. P., Huddleston, J. A. & Brownlee, G. G. (1984). Cytotoxic T cell recognition of influenza nucleoprotein and hemagglutinin expressed in transfected mouse L cells. *Cell*, **39**, 13–25.

Townsend, A. R. M., Rothbard, J., Gotch, F. M., Bahadur, G., Wraith, D. & McMichael, A. J. (1986). The epitopes of influenza nucleoprotein recognized by cytotoxic T lymphocytes can be defined with short synthetic peptides. *Cell*, **44**, 959–68.

Townsend, A., Ohlen, C., Bastin, J., Ljunggren, H.-G., Foster, L. & Karre, K. (1989). Association of class I major histocompatibility heavy and light chains induced by viral peptides. *Nature, London*, **340**, 443–8.

Valenzuela, P., Medina, A., Rutter, W. J., Ammerer, G. & Hall, B. D. (1982). Synthesis and assembly of hepatitis B virus surface antigen particles in yeast. *Nature, London*, **298**, 347–50.

van Regenmortel, M. H. V. (1987). Antigenic cross-reactivity between proteins and peptides: new insights and applications. *Trends in Biochemical Science*, **12**, 237–40.

van Regenmortel, M. H. V. (1989a). Structural and functional approaches to the

study of protein antigenicity. *Immunology Today*, **10**., 266–72.
van Regenmortel, M. H. V. (1989b). The concept and operational definition of protein epitopes. *Philosophical Transactions of the Royal Society of London, series B*, **323**, 1–16.
Walker, B. D., Flexner, C., Paradis, T. J., Fuller, T. C., Hirsch, M. S., Schooley, R. T. & Moss, B. (1988). HIV-1 reverse transcriptase is a target for cytotoxic T lymphocytes in infected individuals. *Science*, **240**, 64–6.
Wells, M. A., Daniel, S., Djeu, J. Y., Kiley, S. C. & Ennis, F. A. (1983). Recovery from viral respiratory tract infection. IV. Specificity of protection by cytotoxic lymphocytes. *Journal of Immunology*, **130**, 2908–14.
Widmer, M. B. & Grabstein, K. H. (1987). Regulation of cytolytic T-lymphocyte generation by B-cell stimulatory factor. *Nature, London*, **326**, 795–8.
Wong, G. H. W., Bartlett, P. F., Clark-Lewis, I., Battye, F. & Schrader, J. W. (1984). Inducible expression of H-2 and Ia antigens on brain cells. *Nature, London*, **310**, 688–91.
Wraith, D. C. & Askonas, B. A. (1985). Induction of influenza A virus cross-reactive cytotoxic T cells by a nucleoprotein/haemagglutinin preparation. *Journal of General Virology*, **66**, 1327–31.
Wraith, D. C., Vessey, A. E. & Askonas, B. A. (1987). Purified influenza virus nucleoprotein protects mice from lethal infection. *Journal of General Virology*, **68**, 433–40.
Wright, P. F. (1984). Parainfluenza viruses. In *Textbook of Human Virology*, R. B. Belshe ed., pp. 299–309, PSG Publishing Company.
Yasukawa, M. & Zarling, J. M. (1984). Human cytotoxic T cell clones directed against herpes simplex virus-infected cells. I Lysis restricted by HLA class II MB and DR antigens. *Journal of Immunology*, **133**, 422–39.
Yewdell, J. W., Bennink, J. R., Smith, G. L. & Moss, B. (1985). Influenza A virus nucleoprotein is a major target antigen for cross-reactive anti-influenza A virus cytotoxic T lymphocytes. *Proceedings of the National Academy of Sciences, USA*, **82**, 1785–9.
Yewdell, J. W., Bennink, J. R., Mackett, M., Lefrancois, L., Lyles, D. S. & Moss, B. (1986). Recognition of cloned vesicular stomatitis virus internal and external gene products by cytotoxic T lymphocytes. *Journal of Experimental Medicine*, **163**, 1529–38.
Yewdell, J. W., Bennink, J. R., & Hosaka, Y. (1988). Cells process exogenous proteins for recognition by cytotoxic T lymphocytes. *Science*, **239**, 637–40.
Zagury, D., Bernard, J., Cheynier, R., Desportes, I., Leonard, R., Fouchard, M., Reveil, B., Ittele, D., Lurhuma, Z., Mbayo, K., Wane, J., Salaum, J. J., Goussard, B., Dechazal, L., Burny, A., Nara, P. & Gallo, R. C. (1988). A group specific anamnestic immune reaction against HIV-1 induced by a candidate vaccine against AIDS. *Nature, London*, **332**, 728–31.
Zarling, J. M., Morton, W., Moran, P. A., McClure, J., Kosowski, S. G. & Hu, S-L. (1986). T cell responses to human AIDS virus in macaques immunized with recombinant vaccinia virus. *Nature, London*, **323**, 344–6.
Zinkernagel, R. M. & Doherty, P. C. (1979). MHC-restricted cytotoxic T cells: studies on the biological role of polymorphic major transplant antigens determining T cell restriction-specificity, function, and responsiveness. *Advances in Immunology*, **27**, 51–177.

RECOMBINANT ANTIGEN PRODUCTION USING BACULOVIRUS EXPRESSION VECTORS

ROBERT D. POSSEE, ULRIKE WEYER & LINDA A. KING[1]

NERC Institute of Virology and Environmental Microbiology, Mansfield Road, Oxford OX1 3SR, UK

[1]*School of Biological and Molecular Sciences, Oxford Polytechnic, Gipsy Lane, Oxford OX3 0BP, UK*

The advent of techniques for gene manipulation has opened up exciting possibilities for new approaches to old problems in many areas of biology. Not least of these is the production of synthetic vaccines to combat the ever-present threat of infectious diseases. This chapter will consider the application of an invertebrate (insect) virus expression vector system to the synthesis of mammalian virus antigens and their potential use as sub-unit vaccines.

TRADITIONAL VIRUS VACCINES VERSUS RECOMBINANT PRODUCTS

Historically, virus vaccines have been derived through two approaches. In the first, a virus pathogen is attenuated by repeated passage in cell culture until it accumulates sufficient random mutations within its genome to produce a strain that no longer causes severe disease, but can still induce a protective immune response in animals. An example of this is the live, attenuated, poliomyelitis Sabin vaccine for all three poliovirus serotypes. Alternatively a vaccine may be produced by inactivating the virulent virus and using this as an immunogen to induce a protective response in the host. This method was used to produce the Salk inactivated poliovirus vaccine. However, both of these approaches rely on the production of large amounts of virus in cell culture, or other systems such as the embryonated egg in the case of influenza viruses. Also virulent viruses must be totally inactivated. Not all virus pathogens can be readily propagated *in vitro*. For example, because hepatitis B virus (HBV) did not replicate in cell culture, it was necessary to purify the surface (S) antigen from the serum of virus carriers to produce a vaccine. This procedure carries an inherent risk of contamination with other agents.

A further complication is that as the success of an immunisation programme increases a point is inevitably reached when the risk of side-effects of a particular vaccine outweigh its benefits. This is because, in a highly vaccinated population, the chances of catching the disease are remote. The dangers from continued vaccination, minimal though they are, can stem from reversion to a virulent virus in the case of attenuated vaccines, or failure to inactivate the virus in a killed preparation. It presents a moral dilemma as to whether one should continue to vaccinate and thus create more problems through use of the vaccine than are encountered because of the virus disease, or abandon the programme and risk a new outbreak.

Recombinant vaccines can potentially circumvent some of these problems. Their very nature means that they do not incorporate the infectious agent responsible for causing the disease. The derivation of such a vaccine requires a basic knowledge of the parts of the virus structure responsible for inducing preferably both the antibody and cell-mediated immune responses. Such an understanding can only be gained after meticulous analyses of the molecular genetics and immunology of the virus, with the result that cloned fragments of the genome are isolated encoding the required antigens. Using this method recombinant vaccines can be produced for viruses that do not grow in cell culture. Two approaches to the production of recombinant vaccines can be envisaged. First, an innocuous carrier virus can be used to deliver the desired antigen(s) to the host. The most attractive, though not always innocuous, candidate for this role is the vaccinia virus (see p. 77) which was used to eradicate smallpox and now enjoys considerable success as an expression vector of foreign genes. Poliovirus has also been considered for this role. The Sabin type 1 poliovirus has been used to carry an epitope from the glycoprotein (gp) 41 of HIV-1 which, on inoculation into rabbits producedan immune response (Evans *et al.*, 1989). The second approach can be likened to the previously described method for producing a killed virus vaccine. However, instead of using a whole virus particle, the portion of the virus responsible for inducing an immune response is produced in an expression vector, isolated to a high degree of purity, and then used as a sub-unit vaccine.

It is not the object of this paper to compare the respective merits of different methods of making new virus vaccines by genetic engineering. Undoubtedly each system will have advantages in particular areas. Rather, our task is to review, and also to look ahead, and consider how expression systems may be used to derive non-infectious, sub-unit virus vaccines by making use of high level expression vectors propagated using cell culture and other methods.

PROKARYOTIC AND EUKARYOTIC EXPRESSION SYSTEMS

Several prokaryotic and eukaryotic expression systems are now available for the synthesis of proteins, including virus antigens, to high concentrations. The main advantage of prokaryotic expression vectors is the ease with which their production may be scaled up to large volumes and their host cells cultured continuously in relatively cheap medium. However, since the viruses causing the diseases that we wish to prevent replicate in eukaryotic cells, the expression system of choice must be able to perform the prerequisite co-translational and post-translational modifications that result in the synthesis of immunogenic material suitable for use as a vaccine.

In general, prokaryotic cells are unable to carry out processes such as glycosylation and proteolytic cleavages. Therefore, it is necessary to choose a eukaryotic expression system that can produce the recombinant antigens in the correct form and be scaled up to volumes which are economically viable. Yeast cells are the obvious choice and have been used to produce an HBV recombinant vaccine by expression of the virus S antigen to derive sub-viral particles (Valenzuela *et al.*, 1982; Miyanohara *et al.*, 1983). Other workers are also investigating the use of yeast cells to produce an HIV sub-unit vaccine as part of a virus-like particle found in these cells (Adams *et al.*, 1987). However, yeast cells cannot fully process higher eukaryotic proteins as evidenced by the lack of glycosylation of the HBV S antigen (Valenzuela *et al.*, 1982). Therefore, attention has turned to using higher eukaryotic cells themselves to express the gene products. These cells are more difficult to culture but, fortunately, with the introduction of monoclonal antibody techniques, the growth of vertebrate cells in large volumes has become a more routine procedure.

It is also essential that the expression system can produce large amounts of the recombinant antigen. In this paper we shall describe the use of such an expression system based on the insect baculoviruses. Its particular attraction is that the latter have probably the most productive eukaryotic promoters that can be harnessed to synthesise foreign gene products. Furthermore, insect cells can be cultured in large volumes. We will also consider the processing events required for the formation of biologically active virus antigens and how the expression system may be developed further to produce better sub-unit vaccines.

BACULOVIRUSES AS EXPRESSION VECTORS

Baculoviruses can only replicate in invertebrate hosts: principally Lepidoptera (butterflies and moths), Hymenoptera (bees and wasps), Diptera (flies), Coleoptera (beetles), Neuroptera (lacewings) and Crustacea (prawns and shrimps). There is no evidence that they can grow in mammalian or plant cells, or in other invertebrates such as earthworms. Their safety is well accepted and they have been used on frequent occasions as biological pesticides to control insect species in agriculture. Thus, when considering their use as expression vectors very few special safety precautions need be taken in the average laboratory. Several excellent reviews describing the biology and molecular genetics of baculoviruses have been published (Doerfler & Bohm, 1986; Granados & Federici, 1986) and we will refrain from covering these areas in depth.

The two baculoviruses which have, for the most part, been used as expression vectors to infect insect larvae are the *Autographa californica* (Ac) or alfalfa looper nuclear polyhedrosis virus (NPV) (Smith *et al.*, 1983*b*; Pennock *et al.*, 1984) and the *Bombyx mori* (Bm) or silkworm NPV (Maeda *et al.*, 1985). Each has a genome of double-stranded, covalently closed, circular DNA 130,000 nucleotides in length which is packaged into rod-shaped nucleocapsid structures comprising several proteins. One of these is a highly basic, arginine-rich polypeptide thought to associate closely with the virus DNA (Bud & Kelly, 1980; Tweeten *et al.*, 1980; Burley *et al.*, 1982; Kelly *et al.*, 1983; Wilson *et al.*, 1987). The nucleocapsids are further enclosed within a lipoprotein envelope to form the virus particles and several nucleocapsids may be packaged within a single envelope. The virus particles may also be occluded within proteinaceous crystal structures called polyhedra. The polyhedron is a very resistant body that protects the virus in the natural environment between encounters with susceptible insect larvae, the main target for the virus. Polyhedra are largely composed of many copies of polyhedrin, a polypeptide of 29,000 M_r.

The replication cycle of AcNPV has been studied in the greatest detail. Polyhedra are consumed as the insect larvae feed on contaminated diet (leaves or artificial material in the laboratory). In the alkaline environment of the midgut polyhedra dissolve and release virus particles which can then invade the gut wall and initiate infection. In cell culture, because of the slightly acidic nature of the medium, the virus particles must be released from the polyhedra by prior treatment with alkali before adding them to the cells and establishing the infection. Subsequently, the events at the molecular level are believed to proceed in the same fashion in insect larvae or cell culture.

The virus DNA localises within the nucleus of the cell and there is an initial burst of transcriptional activity which is unaffected by treating the cells with cycloheximide. Thereafter and prior to DNA synthesis an early class of genes is expressed. After DNA replication begins there is a late phase of gene expression during which the structural proteins required for virus assembly are produced, for example the virus basic protein (see above). Virus particles are then released by the budding of nucleocapsids from the plasma membrane. In this process the virions acquire a 67 K glycoprotein (gp67) (Whitford *et al.*, 1989).

However, at about 15–18 h post-infection (p.i.), a fourth phase of virus gene expression begins during which time virus particles within the nuclei of infected cells are occluded into polyhedra. This phase may continue until 72 h p.i. and is characterised by the expression of two gene products in particular, the polyhedrin and p10 (10 K) proteins, which are produced in very large quantities. Their promoters have been characterised (Matsuura *et al.*, 1987; Possee & Howard, 1987; Rankin *et al.*, 1988; Weyer & Possee,

1988; 1989; Qin *et al.*, 1989) and in more recent experiments the relative activities of the two promoters have been compared. These data show that the amount of the two proteins expressed is very similar (Weyer & Possee, unpublished data) and this is mirrored by identical levels of steady state polyhedrin and p10 RNAs in infected cells (M-K. Min and D. H. L. Bishop, personal communication). The role of the polyhedrin in the baculovirus replication cycle is well documented and has been described above. The p10 protein has yet to be assigned a definitive function, but recent evidence suggests that it has a role in polyhedron formation (Vlak *et al.*, 1988; Williams *et al.*, 1989). What is certain is that neither gene product is needed for virus replication (Smith *et al.*, 1983*a*; Vlak *et al.*, 1988; Williams *et al.*, 1989; Weyer *et al.*, unpublished data). In the absence of polyhedrin protein only non-occluded virus is produced and this is infectious in cells and insects (Maeda *et al.*, 1985; Overton *et al.*, 1987). Viruses deficient in p10 synthesis continue to produce polyhedra which exhibit some abnormalities, but can still infect insect larvae (Vlak *et al.*, 1988). Therefore, it is a comparatively simple matter to produce replacement expression vectors where the virus polyhedrin and p10 coding sequences are substituted with foreign genes (Smith *et al.*, 1983*b*; Pennock *et al.*, 1984; Vlak *et al.*, 1988; Williams *et al.*, 1989). The methodology for deriving a recombinant virus is outlined in Figure 1. The baculovirus DNA is too large to permit the easy insertion of foreign DNA directly into the genome, therefore, a portion of the baculovirus DNA is first manipulated in a bacterial plasmid vector. Polyhedrin-based expression vectors are produced by inserting the foreign gene under the control of the polyhedrin promoter and transcription termination signals within a plasmid to produce a transfer vector. This is then coprecipitated with infectious virus DNA and used to cotransfect insect cells (Smith *et al.*, 1983*a*). Alternatively, virus and plasmid DNA may be introduced into insect cells via electroporation (Mann & King, 1989). Recombination occurs between the virus DNA flanking the foreign gene and the homologous sequences in the virus, with the result that the polyhedrin gene is replaced to derive a polyhedrin-minus virus. Recombinant virus is isolated by selecting under a microscope plaques which lack the easily visible polyhedra. Figure 2 shows the contrast between cells infected with wild type polyhedrin-positive virus and the recombinant polyhedrin-negative phenotype.

Expression vectors based on the p10 promoter are produced in a similar fashion, but with the complication that p10 does not give a visible phenotype in a plaque assay. Recombinants must be selected by using nucleic acid hybridisation probes or detecting with antibody. Alternatively, infectious DNA from a virus expressing the *Escherichia coli lac* Z gene under the control of the p10 promoter is used in the cotransfection with the result that recombinant virus appears as a white plaque against a background of blue plaques.

A particular advantage of using these two very late promoters for the

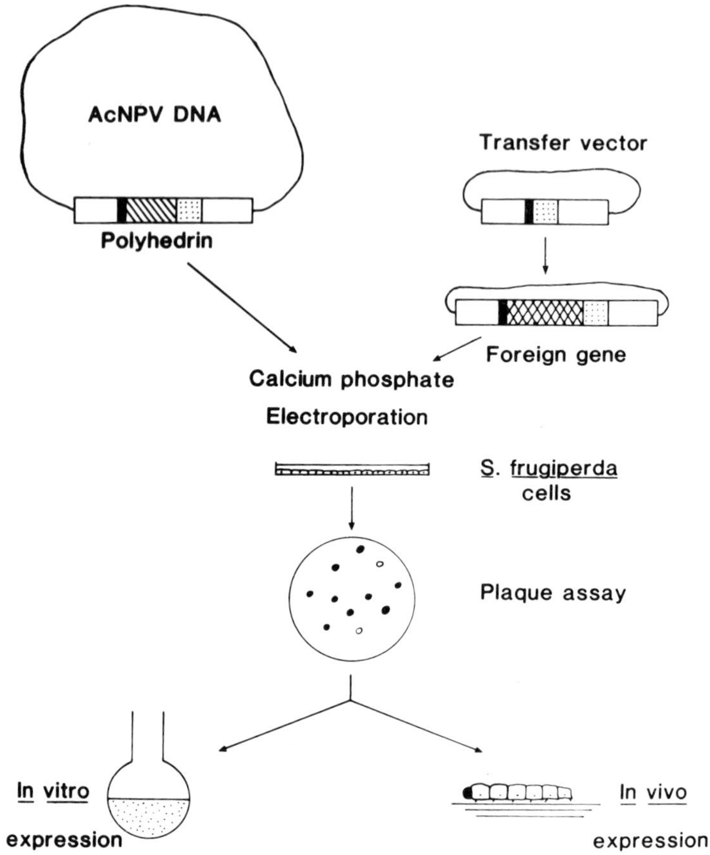

Fig. 1. Schematic representation of the production of a recombinant baculovirus. The plasmid transfer vector contains the polyhedrin promoter (solid box) and transcription termination signals (stippled box) flanked by AcNPV sequences (open box). A foreign gene (cross-hatched box) is inserted under the control of the polyhedrin promoter and then mixed with infectious AcNPV DNA. Cells are cotransfected via calcium phosphate or electroporation and the progeny virus screened for polyhedrin–negative viruses where the polyhedrin coding sequences (hatched box) are replaced with the foreign DNA. The selection is facilitated by the recombinant plaques appearing clear against a background of dark, wild-type virus plaques. Recombinant antigen may be produced in cell culture or insect larvae.

expression of foreign genes is that they are active after infectious virus particles have been assembled and released from the cell. Therefore, if a foreign gene product has a toxic effect on the cell, it is still likely to be expressed to a reasonable level because it will not accumulate to a harmful concentration until after the virus has ensured its continued replication. At very late times after infection the insect cell may be effectively turned into a factory dedicated to the synthesis of foreign gene products. This is not to say that virus gene promoters active at an earlier time in the infection cycle cannot be used in expression vectors. Although the

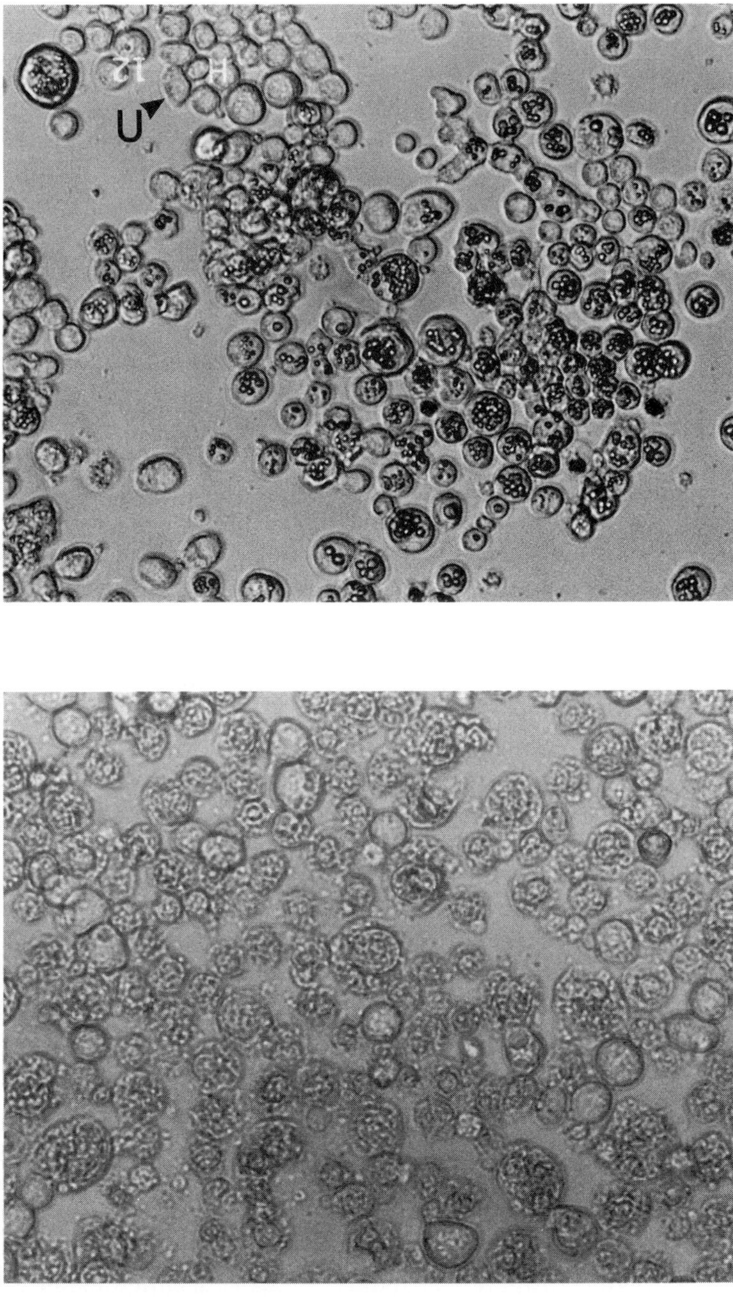

Fig. 2. *Spodoptera frugiperda* cells three days after infection with polyhedrin positive cells containing dense inclusions (upper panel; note also uninfected cells (U) or polyhedrin negative virus (lower panel).

virus gene products are likely to be non-dispensable it is possible to duplicate early and late promoters within the virus genome to permit the expression of foreign proteins while still producing the normal virus polypeptide. An example of this is the virus-encoded basic (arginine-rich) protein which is normally expressed during the late phase of virus replication (Wilson *et al.*, 1987). The promoter for this gene has been duplicated in lieu of the polyhedrin gene to facilitate foreign gene expression at an earlier stage (8–18 h p.i.) of the infection cycle (Hill-Perkins & Possee, 1990) while still retaining the synthesis of the basic protein. The significance of early and late promoter-based vectors will be discussed below.

EXPRESSION OF VIRUS ANTIGENS USING BACULOVIRUS VECTORS

There are many examples in the literature of baculovirus vectors expressing foreign genes from a wide variety of species. These have been summarised (Luckow & Summers, 1988; Miller, 1988; Fraser, 1989; Maeda, 1989; Atkinson *et al.*, 1990; Bishop & Possee, 1990) and will not be considered here; rather, those examples concerning virus antigens will be discussed.

A list of the virus proteins which have been expressed at the time of writing is presented in Table 1. Also indicated are the antigens which have been used in tests to determine the efficacy of these preparations in eliciting an immune response, which can protect against subsequent challenge with the virus pathogen. A fuller account of this topic is presented in the next section. For the present it is pertinent to consider the ability of the insect cell to perform the various cotranslational and post translational modifications which are a feature of many virus specified proteins, particularly those associated with inducing an immune response in the host.

Drawing upon the experience with the complete range of foreign genes which have been expressed in the baculovirus system it is apparent that insect cells are able to perform such processes as glycosylation, fatty acid acylation (palmitylation and myristylation), nuclear transport, phosphorylation, c-terminus amidation, disulphide bond formation, signal peptide together with other proteolytic cleavages and the formation of tertiary structures. These events have been reviewed in the literature cited above. Perhaps the processes of most concern to virologists anxious to produce a recombinant antigen for vaccine purposes are glycosylation, proteolytic cleavages, intracellular transport and the formation of secondary and tertiary structures.

Glycosylation

The role of protein glycosylation remains unclear. It may play some part in the stabilisation of proteins and also determine the nature of the immune response to virus antigens. Therefore, it is of some concern that the process of N-linked glycosylation in insect cells differs from the corresponding events in mammalian cells (Kornfield & Kornfield, 1985).

Table 1.

Virus	Antigen	Reference
Human parainfluenza virus type 3	*fusion glycoprotein	Ray *et al.* (1989)
	*HN glycoprotein	van Wyke-Coelingh *et al.* (1987)
Human respiratory syncytial	*F glycoprotein	Wathen *et al.* (1989)
Influenza virus (A/PR/8/34 H1N1)	HA glycoprotein	Possee (1986)
Influenza virus A/FPV/Rostock/34 H7N1)	*HA glycoprotein	Kuroda *et al.* (1986; 1989)
Influenza virus (A/PR/8/34)	PB1, PB2, PA	St Angelo *et al.* (1987)
Snowshoe hare virus	N and NSs	Urakawa *et al.* (1989)
Maguari bunyavirus	N	Elliot & McGregor (1989)
Rift Valley Fever virus	*G1 and *G2	Schmaljohn *et al.* (1989)
Hantaan virus	nucleocapsid protein	Schmaljohn *et al.* (1988)
Lymphocytic choriomeningitis virus	N, GPC	Matsuura *et al.* (1986; 1987)
Punta Toro phlebovirus	N, NSs	Overton *et al.* (1987)
Sindbis virus	26S cDNA	Oker-Blom & Summers (1989)
Bluetongue virus (BTV) type 17	VP3	Inumaru *et al.* (1987)
BTV type 10	VP2	Inumaru *et al.* (1987)
	NS1	Urakawa & Roy (1988)
	NS3, NS3A	French *et al.* (1989)
Simian rota virus	VP6	Estes *et al.* (1987)
Bovine rota virus	VP1	Cohen *et al.* (1989)
Human immunodeficiency virus	gp160	Hu *et al.* (1987)
	gp160	Rusche *et al.* (1987)
	gag	Madisen *et al.* (1987)
	gag	Overton *et al.* (1989)
	protease	Overton *et al.* (1989)
HTLV-1	$p40^x$	Jeang *et al.* (1987*b*)
Poliovirus type 3 (P3/Leon/37)	whole genome (less 5′ non-coding)	Urakawa *et al.* (1989)
Hepatitis A virus	VP1	Harmon *et al.* (1988)
Dengue virus type 4	core protein	Makino *et al.* (1989)
Vesicular stomatitis virus	G	Bailey *et al.* (1989)
Bovine papilloma virus type 1	E2	Tada *et al.* (1988)
Human papilloma virus type 6b	E2	Tada *et al.* (1988)
Hepatitis B virus	S	Kang *et al.* (1987)
	S	Takehara *et al.* (1988)
	S	Lanford *et al.* (1989)
	C	Takehara *et al.* (1988)
Herpes simplex virus	*Glycoprotein D	Krishna *et al.* (1989)
Polyomavirus	large T	Rice *et al.* (1987)
SV40	small T, large T	Jeang *et al.* (1987*a*)
SV40	large T	O'Reilly & Miller (1988)

* Used as immunogen in animals.

In mosquito cells glycosylation does not involve the conversion to complex N-linked oligosaccharides, furthermore the structure of the oligosaccharides is simple (Hsieh & Robbins, 1984). The glycosylation process appears to involve the addition of high-mannose oligosaccharides (Kornfield & Kornfield, 1985). Sialic acid, which is usually found at the end of oligosaccharides in mammalian cells is absent in mosquito cells (Butters & Hughes, 1981). Mammalian cells extensively modify the core oligosaccharide in terminal glycosylation events involving the transfer of glucosamine-galactose and sialic acid residues to form complex oligosaccharides (Butters & Hughes, 1981; Butters *et al.*, 1981).

Few analyses have been made of the glycosylation of foreign proteins produced using baculovirus expression vectors. A survey of some of the glycoproteins which have been expressed in this system reveals that in general their sizes are smaller than the corresponding molecule synthesized in mammalian cells (influenza virus haemagglutinin (HA), Kuroda *et al.*, 1986; 1989; Possee, 1986; lymphocytic choriomeningitis virus glycoprotein precursor, Matsuura *et al.*, 1986; 1987; vesicular stomatitis virus (VSV) glycoprotein (G), Bailey *et al.*, 1989). This is attributed to the difference in glycosylation pathways. Interestingly, in the study with VSV G (Bailey *et al.*, 1989) a non-glycosylated form of the protein was still able to fuse insect cells and produce syncytia at low pH; although it was produced in lesser quantities than the fully glycosylated version. However, other examples have shown that the sizes of the polypeptides are almost identical (human parainfluenza virus type 3 (PIV3) haemagglutination-neuraminidase (HN) glycoprotein, van Wyke-Coelingh *et al.*, 1987; PIV3 fusion (F) glycoprotein, Ray *et al.*, 1989; respiratory syncytial virus (RSV) F glycoprotein, Wathen *et al.*, 1989; Rift Valley fever virus (RVFV) glycoprotein 1 (G1) and glycoprotein 2 (G2), Schmaljohn *et al.*, 1989; HIV gp169, Hu *et al.*, 1987; Rusche *et al.*, 1987; HBV S antigen, Lanford *et al.*, 1989). The only analysis of the glycans associated with a virus glycoprotein was cited by Kuroda *et al.*, (1989) as *in press* data. They reported that the oligosaccharide side chains from influenza fowl plague virus HA expressed by AcNPV in *Spodoptera frugiperda* cells and *Heliothis virescens* larvae had lower electrophoretic mobility than HA from vertebrate cells. These authors concluded that the side chains had been trimmed but not elongated to the complex type. This problem deserves additional experimentation to resolve the issue.

Proteolytic cleavage and intracellular transport

Proteolytic cleavage is also a potential problem for those wishing to express virus proteins in insect cells. Many of these are membrane-bound structures such as HA, neuraminidase and HN glycoproteins found spanning the lipid envelope of viruses such as influenza and parainfluenza. In the course of virus replication these proteins are translated on membrane-bound polysomes, translocated via an amino-terminal signal sequence into the lumen of the endoplasmic reticulum and then transported through the Golgi apparatus to the plasma membrane where they await assembly into mature virus particles. During this process the signal peptide is cleaved from the amino terminal end of the protein.

In the baculovirus system correct signal peptide cleavage does take place since such proteins are inserted into the plasma membrane and, in the case of influenza HA proteins (Kuroda *et al.*, 1986; Possee, 1986) and PIV3 HN (van Wyke *et al.*, 1989), can bind erythrocytes in haemadsorption assays. It has also been demonstrated that the cell surface expression of influenza HA in *H. virescens* larvae was confined to the apical surface of tracheal cells (Kuroda *et al.*, 1989); mimicking the polarised expression observed in MDCK cells where it is also expressed on the apical, but not basolateral surface (Simons & Fuller, 1985; Stephens *et al.*, 1986). In addition cell surface immunofluoresence tests show that the F glycoprotein of PIV3 (Ray *et al.*, 1989) and the VSV G protein (Bailey *et al.*, 1989) are located in the plasma membrane.

In other studies, peptide analyses of some non-virus proteins has confirmed that the signal peptides of human alpha-interferon (Maeda *et al.*, 1985), human interleukin (Il) 2 (Smith *et al.*, 1985) and mouse Il-3 (Miyajima *et al.*, 1987) are cleaved. Therefore, it is reasonable to assume that membrane-bound virus glycoproteins are processed in a similar fashion.

Signal peptide cleavage is not the only proteolytic processing event undergone by virus polypeptides in virus-infected cells. Membrane-bound glycoproteins such as influenza virus HA and PIV F are usually cleaved into two subunits (HA1 and HA2, and F1 and F2, respectively) in the normal course of virus replication in the vertebrate host cell. This event is essential if the proteins are to have the fusion activity, through which the virus infects its vertebrate host cell. Although such processing is not necessarily required for the expression of these proteins in insect cells it is of interest to know whether or not insect cells can cleave proteins in the appropriate way.

Fowl plague influenza virus HA was cleaved into HA1 and HA2 in insect cells in culture (Kuroda *et al.*, 1986) and in insect larvae (Kuroda *et al.*, 1989). Interestingly, the cleavage event was more efficient in insect larvae than in cell culture. However, the HA of the PR8 strain of influenza was not cleaved in this fashion in *S. frugiperda* cells (Possee, 1986). The F glycoprotein of PIV3 was not cleaved into F1 and F2 (Ray *et al.*, 1989) and the F glycoprotein of RSV only poorly cleaved (Wathen *et al.*, 1989). HIV gp160 was only partially cleaved to gp120 and gp41 in one study (Hu *et al.*, 1987) and not at all in another (Rusche *et al.*, 1987). This discrepancy may have been a consequence of non-specific protease cleavage in the insect cells in the first report.

Some studies have indicated the direction of more sophisticated applications of the baculovirus expression systems. For example, the cDNA for the Sindbis virus 26S RNA was inserted into an AcNPV vector and expressed in insect cells (Oker-Blom and Summers, 1989). This RNA encodes the 30,000 M_r capsid (C) protein, pE2, the 62,000 M_r precursor of E3 and E2, a 6,000 M_r peptide and the 56,000 M_r E1 protein. Translation is initiated at a single site in this molecule in the course of the normal virus infection. In the *S. frugiperda* cells used in this study the cDNA insert was expressed, translated and processed to Sindbis virus C protein and the envelope glycoproteins E1 and E2. In its life cycle, Sindbis virus normally infects both vertebrate and insect (mosquito) cells so this result might have been predicted.

The experimental data reviewed above indicate that insect cells are able to process virus polypeptides in a number of complex ways. Ideally, for the production of synthetic vaccines containing multiple antigens, we would wish to derive sub-unit virus structures which faithfully reflect those found in the normal virus. Therefore, we must consider whether insect cells are able to produce such structures.

Formation of tertiary and quarternary structures

The membrane-bound influenza HA glycoprotein is assembled as a trimer. The insect cells appear able to perform this operation as evidenced by the isolation of HA trimers using gradient centrifugation (Kuroda *et al.*, 1989). However, in comparison with influenza virus-infected MDCK cells there was a predominance of the monomeric form in insect cells from larvae. This might explain why the functional fraction of total haemagglutinin as assessed by haemolytic activity was about 50% lower in insect larvae compared with the HA produced in MDCK cells. Trimer formation in cultured *S. frugiperda* cells was not affected. VSV G was identifiable as a 180 K protein in non-denaturing gels and HPLC analysis; this is consistent with the formation of trimers (Bailey *et al.*, 1989).

The expression of the HBV S antigen in insect cells was characterised by the formation of 22nm subviral particles similar to those observed in carriers of the virus. In one study (Kang *et al.*, 1987) 90% of these particles were secreted into the medium; but only when the cell cultures were agitated or maintained as spinners. In another study (Lanford *et al.*, 1989), they remained predominantly intracellular. It may be that the secretion observed in the earlier study was a consequence of cell lysis.

Although they are not regarded as important in the stimulation of a protective immune response the influenza virus polymerase genes PB1, PB2 and PA provide additional evidence that complex structures may be formed in insect cells after expression by a baculovirus (St Angelo *et al.*, 1987). When three recombinant viruses, each expressing one of the P genes, were used to co-infect cells a complex was formed between the PB1 and PB2 gene products. This partially mimics the process that occurs in influenza virus-infected cells where all three P proteins form a complex (Detjen *et al.*, 1987). The failure to detect the third P protein (PA) in the complex from insect cells may have been due to difficulties in maintaining stable structures during extraction.

Complexes between the SV40 large T antigen and mouse p53 were also seen in insect cells co-infected with two recombinant baculoviruses expressing one or other of these gene products (O'Reilly & Miller, 1988). Infection of cells with the individual viruses also saw the formation of oligomeric structures.

Perhaps the most impressive example of processing achieved within the insect cell concerns the synthesis of poliovirus particles by a baculovirus vector (Urakawa *et al.*, 1989). During poliovirus replication in vertebrate cells the positive strand RNA is translated as a single polyprotein. This is processed via a number of intermediates into the various poliovirus proteins through the action of at least two virus-encoded proteases (Hanecak *et al.*, 1982). Infectious virions contain the capsid proteins, VP1, VP2, VP3 and VP4. An intermediate protein, VP0, is a precursor to VP2 and VP4 (Holland & Kiehn, 1968; Jacobson & Baltimore, 1968; Jacobson *et al.*, 1970) and is believed to be cleaved to VP2 and VP4 when the poliovirus RNA is encapsidated.

A DNA copy of the P3/Leon/37 strain of poliovirus type 3 was inserted into AcNPV under the control of the polyhedrin promoter (Urakawa *et al.*, 1989). In insect cells the cDNA, lacking the 5′ non-coding region, was transcribed, translated and processed to form poliovirus particles (Urakawa *et al.*, 1989). Further analyses showed that these contained VP0, VP1 and VP3. Expression of the capsid region coding sequences alone in insect cells resulted in the production of an unprocessed precursor of the structural proteins. Therefore, it was concluded that the processing observed when the complete virus genome was expressed was a consequence of the action of the virus-encoded proteases. It should be added that the virus particles produced were non-infectious and appeared not to contain any RNA, possibly because of the lack of the 5′ non-coding region. The particles were, however, immunogenic and could stimulate the production of neutralising antibodies in mice.

These studies suggest that complex structures can be formed by recombinant antigens in the baculovirus-infected cell. Consequently, it is feasible not only to produce single polypeptides in insect cells, but also to form tertiary and quarternary structures after co-infection with two or more virus recombinants carrying unique genes. However, this process is inconvenient since it requires the virus titres of the recombinants to be accurately known and relies on chance for all viruses to be in the cells at the same time. Therefore, a much more useful approach would be to co-express virus antigens in insect cells from a single virus vector, thus ensuring closer regulation of the levels of expression. This exciting new approach is discussed more fully in a later section (see below).

BACULOVIRUS-EXPRESSED VIRUS ANTIGENS AS IMMUNOGENS

The data reported in the literature and reviewed above, suggest that recombinant virus proteins expressed in baculovirus-infected cells closely mimic those normally synthesised by the virus in vertebrate cells. The only proviso is that the pattern of glycosylation may not be identical. The crucial test of the worth of the baculovirus-expressed products is whether they can induce an immune response capable of protecting the host. Several studies have addressed this problem and the results are very encouraging.

The first report concerning the use of the baculovirus-expressed proteins as vaccines involved the fowl plague influenza virus (Kuroda *et al.*, 1986; 1989). In the earlier study HA protein from infected insect cell cultures was used to immunise chickens and in the latter, material from infected *H. virescens* caterpillars was utilised. In both experiments the vaccinated animals survived a lethal dose of fowl plague virus.

Cotton rats immunised with PIV3 HN glycoprotein produced by a baculovirus vector (van Wyke-Coelingh *et al.*, 1987) produced antibodies which inhibited HA activity of PIV3 and neutralised infectivity *in vitro*. Furthermore, in immunised animals challenged intranasally with PIV3, virus replication in the upper respiratory tract was reduced 100-fold compared to controls and in the lungs was almost completely abolished. More recently the F glycoprotein of PIV3 was expressed in insect cells and used in hamster protection experiments (Ray *et al.*, 1989). After challenge with PIV3 there was a seven-fold reduction in the virus titre in the lungs and trachea when compared with controls. Immunoglobulin from vaccinated hamsters inhibited the fusion activity and neutralised infectious virus.

In a similar study a truncated form of the F glycoprotein of RSV lacking the carboxyl terminal membrane anchor region was expressed using baculovirus vectors and the purifed, secreted protein used to immunise cotton rats (Wathen *et al.*, 1989). The animals produced neutralising antibody and low levels of fusion-inhibiting antibody. Approximately 30% of the vaccinated animals had no detectable virus in their lungs after challenge with RSV and in the remaining animals the amount of virus recovered from their lungs was reduced. The vaccinated cotton rats were also protected against both serogroup A (Long strain) or serogroup B (18537 strain) RSV. However, caution must be exercised when designing and using an RSV vaccine. In the past when a formalin-inactivated, parenterally injected product was assessed as a vaccine in children very unwelcome results were obtained. After subsequent natural exposure to the wild type virus 40% of the children immunised required hospital treatment compared to 1% of unvaccinated individuals (Kim *et al.*, 1969). Experiments in cotton rats later suggested that the formalin-inactivated product may have preferably induced the synthesis of non-neutralising antibodies which permitted virus replication to proceed and cause an Arthus-type reaction (Kim *et al.*, 1973).

The M genome segment of Rift Valley fever virus (RVFV) encodes G1 and G2. After insertion of this coding sequence into AcNPV G1 and G2 were expressed in insect cells (Schmaljohn *et al.*, 1989). When this material was used in a single dose to immunise mice an antibody response was generated which protected them from challenge with RVFV (10^3 LD50 or 10^3 pfu) 33 days after vaccination. Immunised mice had antibodies to G1 and G2 and these were boosted after challenge with virus. Recombinant viruses expressing only the G2 glycoprotein also produced material which, after two inoculations, protected mice from challenge with RVFV. The use of this vaccine would allow discrimination between naturally infected individuals and vaccinees since the nucleoprotein would be absent from the vaccinees.

One of the seven glycoproteins (gD) identified in HSV-1 was expressed using baculovirus vectors (Krishna *et al.*, 1989). Mice immunised either intra-peritoneally or subcutaneously were protected from subsequent challenge with 5×10^4 pfu of HSV-1.

Perhaps the most emotive issue facing virology today concerns the clinical syndrome produced after infection with HIV. The use of a live, attenuated vaccine does not appear desirable. Moreover, attempts to provide a vaccine are bedevilled by the antigenic instability of the virus. However, many groups worldwide are endeavouring to produce a vaccine using expression vector technology. Such trials of candidate vaccines are complicated because the only animal model, the chimpanzee, does not develop the immunodeficiency syndrome. The first clinical trial using expressed protein as an immunogen was announced in 1987 (Barnes, 1987). The vaccine under test consisted of gp160 expressed using the baculovirus system by MicroGenSys of West Haven, Connecticut. It remains to be seen whether it has any effect on preventing human infection with HIV.

A more optimistic note was sounded in other studies when goats were immunised with another HIV gp160 preparation expressed in baculoviruses. The resulting antiserum was efficient in neutralising the homologous HIV variant, as assessed by virus gene expression or virus-induced cell death (Rusche *et al.*, 1987). However, other HIV variants treated with the antiserum were only prevented from causing cell cytolysis, emphasising the difficulties in preparing a vaccine which can protect against all serotypes. More encouragingly, since HIV infection may be spread via cell–cell fusion, antiserum raised to the baculovirus-expressed gp160 prevented this activity when HIV-infected cells were mixed *in vitro* with susceptible T cells. Considerable work remains to be done in the quest for an effective HIV vaccine.

FUTURE DIRECTIONS IN BACULOVIRUS EXPRESSION OF VIRUS ANTIGENS

Expression of multiple genes

It has been a general approach to the problem of producing any recombinant protein that only a single gene is inserted into the vector, whether it is

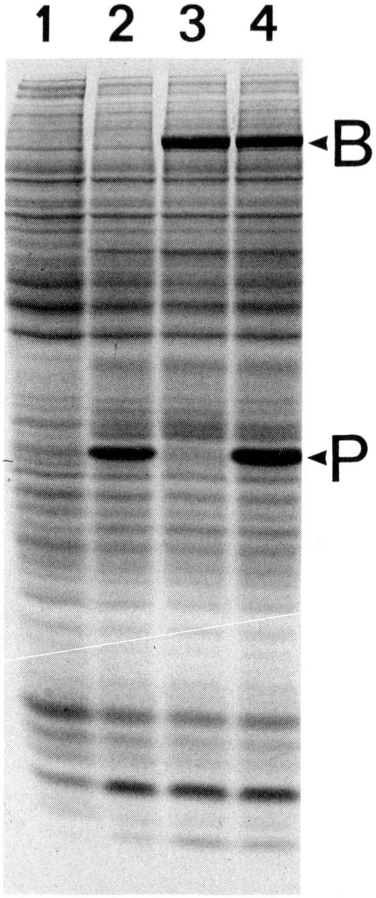

Fig. 3. Coexpression of a foreign gene with AcNPV polyhedrin protein. Lane 1, uninfected *S. frugiperda* cells; lane 2, AcNPV-infected cells; lane 3, AcRP23.*lac*Z-infected cells (Possee & Howard, 1987); lane 4, AcNPV.*lac*Z-infected cells. B, β-galactosidase; P, polyhedrin. All cells were harvested 30 hours after infection.

a bacterial plasmid or eukaryotic virus vector. Perkus *et al.*, (1985) departed from this convention by inserting the genes encoding the HBV S antigen, HSV-1 glycoprotein D and the influenza virus HA genes into a single vaccinia vector. Inoculation of rabbits resulted in the production of antibodies to all three foreign antigens. While this approach is feasible with the baculovirus system it is unlikely to be adopted since the different antigens would almost certainly require different purification protocols and overall yields would probably suffer. One possible avenue yet to be explored concerns the development of pseudotypes containing copies of foreign antigens in the baculovirus envelope for use as vaccines. It is not known whether foreign virus antigens would be incorporated into the virus particle if they were present in the plasma membrane of infected cells.

However, what is more attractive is the synthesis of multiple components of a virus to derive sub-viral particles which may conceivably be an improved immunogen. We have already seen how the insect cell permits the formation of tertiary structures such as HBV sub-virus particles and poliovirus capsid structures. Individual components of a virus usually induce a poor immune response. It may be that if sub-viral particles in insect cells can be assembled and readily purified they may be more efficacious as vaccines. This approach was first investigated by Emery & Bishop (1987) who duplicated the polyhedrin promoter of AcNPV and thus expressed LCM nucleoprotein at the same time as the polyhedrin protein. Figure 3 shows another example where the *E. coli* β-galactosidase gene was coexpressed with the AcNPV polyhedrin protein. The foreign gene with its duplicated promoters was inserted approximately 100 nucleotides upstream from the normal polyhedrin gene. The levels of protein synthesis in the dual expression vector clearly match those achieved in the respective wild type AcNPV and β-galactosidase recombinant (AcRP23.*lac*Z). The first example of true coexpression of virus antigens was the simultaneous production of HBV S and C (Takehara *et al.*, 1988). However, it was not determined whether or not these two proteins interacted in the insect cell. The VP3 and VP7 genes of bluetongue virus have also been co-expressed in insect cells using a polyhedrin-based dual expression vector (T. J. French & P. Roy, personal communication). These proteins formed core-like particles suggesting that they interacted. In other studies we have inserted an extra copy of the p10 promoter together with SV40 transcription termination signals, upstream of the AcNPV polyhedrin gene and thus produced a vector capable of expressing a foreign gene from each promoter (Weyer & Possee, unpublished data). This approach should reduce any tendency for the virus to delete the foreign genes through recombination between duplicated polyhedrin promoters.

The baculovirus genome is large (*c.*130 kbp) so it is likely that it can accommodate at least 10% extra foreign DNA. This is undoubtedly an underestimate since the rod-shaped nucleocapsid suggests that there is no physical constraint on the increase in size of the genome. The use of strong promoters active in different phases of the baculovirus replication cycle also presents an opportunity to increase the number of genes which can be expressed. This work is in its early stages, but in addition to the polyhedrin and p10 promoters from the very late phase of gene expression we have utilised the basic protein promoter (late) (Hill-Perkins & Possee, 1989, in press) and 39 K promoters (delayed early) (Pullen & Possee, unpublished data) for foreign gene expression. Other virus promoters such as that from the gp67 (late) virus membrane protein gene (Whitford *et al.*, 1989) are attractive since the protein has a signal peptide responsible for intracellular transport of the virus protein.

In vitro *production of virus antigens*

The economic production of virus antigens using the insect baculovirus system depends on the feasibility of scaling up the production from the normal laboratory requirements to cultures of at least tens, or hundreds of litres. Previously it had been reported that insect cells were very sensitive to shear forces (Weiss *et al.*, 1980; Hink, 1982) and have a requirement for high levels of oxygen (Weiss *et al.*, 1982). This latter problem was compounded by the fact that the cells were also damaged by bubbling gas through the medium to supply the necessary oxygen (Miltenburger & David, 1980; Tramper *et al.*, 1986). Moreover, the absolute requirement for foetal calf serum or some other kind of protein supplement increased the costs to unacceptable levels, while at the same time complicating the purification process for the expressed protein. Fortunately some of these problems have now been resolved by recent studies (Maiorella *et al.*, 1988) in which the scale of insect cell culture has been increased using air lift fermenters. The problems of serum requirement have been solved by the development of serum-free media.

While most foreign gene expression with AcNPV to date has been performed using the *S. frugiperda* cell line recent studies have shown that it may be worth considering alternative species. For example, production of influenza virus HA in a *Mamestra brassicae* cell line gave a considerable increase in expression and in biological activity, as assessed by haemagglutination (King *et al.*, unpublished data). It may be that different species of insect cells are better equipped to perform the necessary post-translational modifications.

In vivo *production of virus antigens*

The natural host of the insect baculoviruses, namely the humble caterpillar, may yet become one of the work horses of modern biotechnology. It is quite feasible to propagate recombinant viruses in these larvae and then recover the virus antigen after their death. Silkworms have been advocated as a suitable species for this purpose (Maeda *et al.*, 1985). They have been used by humans in silk production for many thousands of years and are largely domesticated. They grow rapidly and can yield several milligrams of recombinant protein per larvae after infection with BmNPV. Other insects such as *Trichoplusia ni* (Overton *et al.*, 1987) can be used with the AcNPV expression system. These insects are much smaller but are still easy to rear. The decision to use insects rather than cell culture for the large scale production of virus antigens may require a considerable leap of the imagination for most scientists. However, it should be remembered that the capital cost of rearing a few caterpillars is considerably less than establishing a fermentation system which may be obsolete in a few years.

CONCLUDING REMARKS

The baculovirus expression vector system undoubtedly presents virologists with an excellent opportunity to produce large quantities of vertebrate virus antigens for use as synthetic vaccines. The foreign genes are faithfully transcribed and translated in the insect cells and appropriately modified in post-translational processes. Glycosylation appears to be the only event which significantly differs in insect cells. In the few studies where the recombinant virus antigens have been assessed they have been effective in inducing an immune response capable of protecting the animal host from challenge with the vertebrate virus. Vectors are being developed for producing multi-component virus vaccines using baculoviruses and should offer exciting opportunities for producing synthetic, complex virus antigens. Furthermore, the technology for scaling up the production of the recombinant antigens is well advanced, either based on cell culture or insect larvae.

ACKNOWLEDGEMENTS

We thank Steve Howard for providing photographs of virus-infected cells.

REFERENCES

Adams, S. E., Dawson, K. M., Gull, K., Kingsman S. M. & Kingsman, A. J., (1987). The expression of hybrid HIV: Ty virus-like particles in yeast. *Nature*, London, **329**, 68–70.

Atkinson, A., Weitzman, M. D., Obosi, L., Beadle, D. J. & King, L. A. (1990). Expression of foreign genes in insect cells using baculovirus vectors. *Pesticide Science* (in press).

Bailey, M. J., McLeod, D. A., Kang, C-L. & Bishop, D. H. L. (1989). Glycosylation is not required for the fusion activity of the G protein of vesicular stomatitis virus in insect cells. *Virology*, **169**, 323–31.

Barnes, D. M. (1987). AIDS vaccine trial OKed. *Science*, **237**, 973.

Bishop, D. H. L. & Possee, R. D. (1990). Baculovirus expression vectors. *Advances in Gene Technology*, (in press).

Bud, H. M. & Kelly, D. C. (1980). An electron microscope study of partially lysed baculovirus nucleocapsids: The intranucleocapsid packaging of viral DNA. *Journal of Ultrastructural Research*, **73**, 361–8.

Burley, S. K., Miller, A., Harrap, K. A. & Kelly, D. C. (1982). Structure of the baculovirus nucleocapsid. *Virology*, **120**, 433–40.

Butters, T. D. & Hughes, R. C. (1981). Isolation and characterisation of mosquito cell membrane glycoproteins. *Biochimica et Biophysica Acta*, **640**, 655–71.

Butters, T. D. & Hughes, R. C. & Visher, R. C. (1981). Steps in the biosynthesis of mosquito cell membrane glycoproteins and the effects of tunicamycin. *Biochimica et Biophysica Acta*, **640**, 672–86.

Cohen, J., Charpilienne, A., Chilmonczyk, S. & Estes, M. K. (1989). Nucleotide sequence of bovine rotavirus gene 1 and expression of the gene product in baculovirus. *Virology*, **171**, 131–40.

Detjen, B., St Angelo, C., Katze, M. G. and Krug, R. M. (1987). The three influenza virus polymerase proteins not associated with viral nucleocapsids in the infected cell are in the form of a complex. *Journal of Virology*, **61**, 16–22.

Doerfler, W. & Bohm, P. (1986). The molecular biology of baculoviruses. *Current Topics in Microbiology and Immunology*, **131**, Springer-Verlag. Berlin.

Elliot, R. M. & McGregor, A. (1989). Nucleotide sequence and expression of the small (S) NA segment of Maguari bunyavirus. *Virology*, **171**, 516–24.

Emery, V. C. & Bishop, D. H. L. (1987). The development of multiple expression vectors for the high level synthesis of eukaryotic proteins: expression of LCMV-N and AcNPV polyhedrin protein by a recombinant baculovirus. *Protein Engineering*, **1**, 359–66.

Estes, M. K., Crawford, S. E., Penaranda, M. E., Petrie, B. L., Burns, J. W., Chan, W-K., Ericson, B., Smith, G. E. & Summers, M. D. (1987). Synthesis and immunogenicity of the rotavirus major capsid antigen using a baculovirus expression system. *Journal of Virology*, **61**, 1488–94.

Evans, D. J., McKeating, J., Meredith, J. M., Burke, K. L., Katrak, K., John, A., Ferguson M., Minor, P. D., Weiss, R. A. & Almond, J. W. (1989). An engineered poliovirus chimaera elicits broadly reactive HIV-1 neutralising antibodies. *Nature, London*, **339**, 385–8.

Fraser, M. J. (1989). Expression of eukaryotic genes in insect cell cultures. *In Vitro Cellular & Developmental Biology*, **25**, 225–35.

French, T. J., Inumaru, S. & Roy, P. (1989). Expression of two related nonstructural proteins of bluetongue virus (BTV) type 10 in insect cells by a recombinant baculovirus: production of polyclonal ascitic fluid and characterisation of the gene product in BTV-infected BHK cells. *Journal of Virology*, **63**, 3270–8.

Granados, R. R. & Federici, B. A. (eds.) (1986). *The Biology of Baculoviruses*. CRC Press, Boca Raton. F. L.

Hanecak, R., Semler, B.L., Anderson, C.W. & Wimmer, E. (1982). Proteolytic processing of poliovirus polypeptides: antibodies to polypeptide P3–7c inhibit cleavage at glutamine–glycine pairs. *Proceedings of the National Academy of Sciences, USA*, **79**, 3973–7.

Harmon, S. A., Johnston, J. M., Ziegelhoffer, T., Richards, O. C., Summers, D. F. & Ehrenfeld, E. (1988). Expression of hepatitis A virus capsid sequences in insect cells. *Virus Research*, **10**, 273–80.

Hill-Perkins, M. S. & Possee, R. D. (1990). A baculovirus expression vector derived from the basic protein promoter of Autographa californica nuclear polyhedrosis virus. *Journal of General Virology* (in press).

Hink, W. F. (1982). Production of *Autographa californica* nuclear polyhedrosis virus in cells from large-scale suspension cultures, pp. 493–506. In *Microbial and Viral pesticides* (E. Kurstak ed.) Marcel Dekker, Basel.

Holland, J. J. & Kiehn, E. D. (1968). Specific cleavage of viral proteins as steps in the synthesis and maturation of enteroviruses. *Proceedings of the National Academy of Sciences, USA*, **60**, 1015–22.

Hsieh, P. & Robbins, P. W. (1984). Regulation of asparagine-linked oligosaccharide processing: oligosaccharide processing in *Aedes albopictus* cells. *Journal of Biological Chemistry*, **259**, 2375–82.

Hu, S.-L., Kosowski, S. G. & Schaaf, K. F. (1987). Expression of envelope glycoproteins of human immunodeficiency virus by an insect virus vector. *Journal of Virology*, **61**, 3617–20.

Inumaru, S. & Roy, P. (1987). Production and characterisation of the neutralisation antigen VP2 of bluetongue virus serotype 10 using a baculovirus expression vector. *Virology*, **157**, 472–9.

Inumaru, S., Ghiasi, H. & Roy, P. (1987). Expression of bluetongue virus group-specific antigen VP3 in insect cells by a baculovirus vector: Its use for the detection of bluetongue virus antibodies. *Journal of General Virology*, **68**, 1627–35.

Jacobson, M. F., & Baltimore, D. (1968). Polypeptide cleavages in the formation of poliovirus proteins. *Proceedings of the National Academy of Sciences, USA*, **61**, 77–84.

Jacobson, M. F., Asso, J. & Baltimore, D. (1970). Further evidence on the formation of poliovirus proteins. *Journal of Molecular Biology*, **49**, 657–69.

Jeang, K-T., Giam, C-Z., Nerenberg, M., & Khoury, G. (1987*a*). Abundant synthesis of functional human T-cell leukaemia virus type I $p40^x$ protein in eukaryotic cells by using a baculovirus expression vector. *Journal of Virology*, **61**, 708–13.

Jeang, K-T., Holmgren-Konig, M. & Khoury, G. (1987*b*). A baculovirus vector can express intron-containing genes. *Journal of General Virology*, **61**, 1761–4.

Kang, C-Y., Bishop, D. H. L., Seo, J-S., Matsuura, Y. & Choe, M. (1987). Secretion of particles of hepatitis B surface antigen from insect cells using a baculovirus vector. *Journal of General Virology*, **68**, 2607–13.

Kelly, D. C., Brown, D. A., Ayres, M. D., Allen, C. J., & Walker, I. O. (1983). Properties of the major nucleocapsid protein of *Heliothis zea* nuclear polyhedrosis virus. *Journal of General Virology*, **64**, 399–408.

Kim, H. W., Canchola, J. G., Brandt, C. D., Pyles, G., Channock, R. M., Jensen, K. & Parrott, R. H. (1969). Respiratory syncytial virus disease in infants despite prior administration of antigenic inactivated vaccine. *American Journal of Epidemiology*, **89**, 422–34.

Kim, H. W., Arrobio, J., Brandt, C. D., Wright, P., Hodes, D., Channock, R. M. & Parrott, R. H. (1973). Safety and antigenicity of temperature sensitive (*ts*) mutant respiratory syncytial virus (RSV) in infants and children. *Pediatrics*, **52**, 56–63.

Kornfield, R. & Kornfield, S. (1985). Assembly of asparagine-linked oligosaccharides. *Annual Review of Biochemistry*, **54**, 631–64.

Krishna, S., Blacklaws, B. A., Overton, H. A., Bishop, D. H. L. & Nash, A. A. (1989). Expression of glycoprotein D of Herpes Simplex Virus type 1 in a recombinant baculovirus: protective responses and T cell recognition of the recombinant-infected cell extracts. *Journal of General Virology*, **70**, 1805–14.

Kuroda, K., Hauser, C., Rott, R., Klenk, H.-D. & Doerfler, W. (1986). Expression of the influenza virus haemagglutinin in insect cells by a baculovirus vector. *The European Molecular Biology Organisation Journal*, **5**, 1359–65.

Kuroda, K., Groner, A., Frese, K., Drenckhahn, D., Hauser, C., Rott, R., Doerfler, W. & Klenk, H.-D. (1989). Synthesis of biologically active influenza virus haemagglutinin in insect cells. *Journal of Virology*, **63**, 1677–85.

Lanford, R. E. (1988). Expression of Simian Virus 40 T antigen in insect cells using a baculovirus expression vector. *Virology*, **167**, 72–81.

Lanford, R. E., Luckow, V., Kennedy, R. C., Dreesman, G. R., Notvall, L. & Summers, M. D. (1989). Expression and characterisation of hepatitis B virus surface antigen polypeptides in insect cells with a baculovirus expression system. *Journal of Virology*, **63**, 1549–57.

Luckow, V.A. & Summers, M.D. (1988). Trends in the development of baculovirus expression vectors. *Bio/Technology*, **6**, 47–55.

Madison, L., Travis, B., Hu, S-L. & Purchio, A. F. (1987). Expression of the human immunodeficiency virus *gag* gene in insect cells. *Virology*, **158**, 248–50.

Maeda, S. (1989). Expression of foreign genes in insects using baculovirus vectors. *Annual Review of Entomology*, **34**, 351–72.

Maeda, S., Kawai, T., Obinata, M., Fujiwara, H., Horiuchi, T., Saeki, Y., Sato, Y. & Furasawa, M. (1985). Production of human α-interferon in silkworm using a baculovirus vector. *Nature, London*, **315**, 592–4.

Maiorella, B., Inlow, D., Shauger, A. & Harano, D. (1988). Large scale insect cell-culture for recombinant protein production. *Bio/Technology,* **6**, 1406–18.

Makino, Y., Tadano, M., Anzai, T., Ma, S-P., Yasuda, S. & Fukunaga, T. (1989). Detection of dengue 4 virus core protein in the nucleus. II. Antibody against dengue 4 core protein produced by a recombinant baculovirus reacts with the antigen in the nucleus. *Journal of General Virology*, **70**, 1417–25.

Mann, S. G. & King, L. A. (1989). Efficient transfection of insect cells with baculovirus DNA using electroporation. *Journal of General Virology* **70**, 3501–5.

Matsuura, Y., Possee, R. D. & Bishop, D. H. L. (1986). Expression of the S-coded genes of lypmphocytic choriomeningitis arenavirus using a baculovirus vector. *Journal of General Virology*, **67**, 1515–29.

Matsuura, Y., Possee, R. D., Overton, H. A. & Bishop, D. H. L. (1987). Baculovirus expression vectors: the requirements for high level expression of proteins, including glycoproteins. *Journal of General Virology*, **68**, 1233–50.

Miller, L. K. (1988). Baculoviruses as gene expression vectors. *Annual Reviews in Microbiology*, **42**, 177–99.

Miltenburger, H. G. & David, P. (1980). Mass production of insect cells in suspension, *Developments in Biological Standardization*, **46**, 83–186.

Miyajima, A., Schreurs, J., Otsu, K., Kondo, A., Arai, K & Maeda, S. (1987). Use of the silkworm, *Bombyx mori*, and insect baculovirus vector for high level expression and secretion of biologically active mouse interleukin-3. *Gene*, **58**, 273–81.

Miyanohara, A., Toh-E. A., Nozaki, C., Hamada, F., Ohtomo, N. & Matsubara, K. (1983). Expression of hepatitis B surface antigen in yeast. *Proceedings of the National Academy of Sciences, USA*, **80**, 1–5.

Oker-Blom, C. & Summers, M. D. (1989). Expression of Sindbis virus 26S cDNA in *Spodoptera frugiperda* cells, using a baculovirus expression vector. *Journal of Virology*, **63**, 1256–64.

O'Reilly, D. R. & Miller, L. K. (1988). Expression and complex formation of simian virus 40 large T antigen and mouse p53 in insect cells. *Journal of Virology*, **62**, 3109–19.

Overton, H. A., Ihara, T. & Bishop, D. H. L. (1987). Identification of the N and NS_s proteins coded by the ambisense S RNA of Punta Toro phlebovirus using monospecific antisera raised to baculovirus expressed N and NS_s proteins. *Virology*, **157**, 338–50.

Overton, H. A., Fujii, Y., Price, I. R. & Jones, I. M. (1989). The protease and *gag* gene products of the human immunodeficiency virus: authentic cleavage and post-translational modification in an insect cell expression system. *Virology*, **170**, 107–16.

Pennock, G. D., Shoemaker, C. & Miller, L. K. (1984). Strong and regulated expression of *Escherichia coli* β-galactosidase in insect cells using a baculovirus vector. *Molecular and Cellular Biology*, **4**, 399–406.

Perkus, M. E., Piccini, A., Lipinskas, B. R. & Paoletti, E. (1985). Recombinant vaccinia virus: immunization against multiple pathogens. *Science*, **299**, 981–4.

Possee, R. D. (1986). Cell-surface expression of influenza virus haemagglutinin in insect cells using a baculovirus vector. *Virus Research*, **5**, 43–59.

Possee, R. D. & Howard, S. C. (1987). Analysis of the polyhedrin gene promoter of the *Autographa californica* nuclear polyhedrosis virus. *Nucleic Acids Research*, **15**, 10233–48.

Qin, J., Liu, A. & Weaver, R. F. (1989). Studies on the control region of the p10 gene of the *Autographa californica* nuclear polyhedrosis virus. *Journal of General Virology*, **70**, 1273–9.

Rankin, C., Ooi, B. G. & Miller, L. K. (1988). Eight base pairs encompassing the transcriptional start point are the major determinant for baculovirus gene expression. *Gene*, **70**, 39–49.

Ray, R., Galinski, M. S. & Compans, R. W. (1989). Expression of the fusion glycoprotein of human parainfluenza type 3 virus in insect cells by a recombinant baculovirus and analysis of its immunogenic property. *Virus Research*, **12**, 169–80.

Rice, W. C., Lorimer, H. E., Prives, C. & Miller, L. K. (1987). Expression of polyomavirus large T antigen by using a baculovirus vector. *Journal of Virology*, **61**, 1712–16.

Rusche, J. R., Lynn, D. L., Robert-Guroff, M., Langlois, A. J., Lyerly, H. K., Carson, H., Krohn, K., Ranki, A., Gallo, R. C., Bolognesi, D. P., Putney, S. D. & Matthews, T. J. (1987). Humoral immune response to the entire human immunodeficiency virus envelope glycoprotein made in insect cells. *Proceedings of the National Academy of Sciences, USA*, **84**, 6924–8.

Schmaljohn, C. S., Parker, M. D., Ennis, W. H., Dalrymple, J. M., Collet, M.S., Suzich, J. A. & Schmaljohn, A. L. (1989). Baculovirus expression of the M genome segment of Rift Valley fever virus and examination of antigenic and immunogenic properties of the expressed proteins. *Virology*, **170**, 184–92.

Schmaljohn, C. S., Sugiyama, K., Schmaljohn, A. L. & Bishop, D. H. L. (1988). Baculovirus expression of the small genome segment of Hantaan virus and potential use of the expressed nucleocapsid protein as a diagnostic antigen. *Journal of General Virology*, **69**, 777–86.

Simons, K. & Fuller, S. D. (1985). Cell surface polarity in epithelia. *Annual Review of Cell Biology*, **1**, 295–340.

Smith, G. E., Fraser, M. J. & Summers, M. D. (1983*a*). Molecular engineering of the *Autographa californica* nuclear polyhedrosis virus genome: deletion mutations within the polyhedrin gene. *Journal of Virology*, **46**, 584–93.

Smith, G. E., Summers, M. D. & Fraser, M. J. (1983*b*). Production of human β-interferon in insect cells infected with a baculovirus expression vector. *Molecular and Cellular Biology*, **3**, 2156–65.

Smith G. E., Ju, G., Ericson, B. L., Moschera, J., Lahm, H. W., Chizzonite, R. & Summers, M. D. (1985). Modification and secretion of human interleukin 2 produced in insect cells by a baculovirus expression vector. *Proceedings of the National Academy of Sciences, USA*, **82**, 8404–8.

St Angelo, C., Smith, G. E., Summers, M. D. & Krug, R. M. (1987). Two of the three influenza viral polymerase proteins expressed by using baculovirus vectors form a complex in insect cells. *Journal of General Virology*, **61**, 361–5.

Stephens, E. B., Compans, R. W., Earl, P. & Moss, B. (1986). Surface expression of viral glycoproteins is polarised in epthelial cells infected with recombinant vaccinia vectors. *The European Molecular Biology Organisation Journal*, **5**, 237–45.

Tada, A., Fuse, A., Sekine, H., Simizu, B., Kondo, A. & Maeda, S. (1988). Expression of the E2 open reading frame of papillomaviruses BPV1 and HPV6b in silkworm by a baculovirus vector. *Virus Research*, **9**, 357–67.

Takehara, K., Ireland, D. & Bishop, D. H. L. (1988). Co-expression of the hepatitis B surface and core antigens using baculovirus expression vectors. *Journal of General Virology*, **69**, 2763–77.

Tramper, J., Williams, J. B. & Joustra, D. (1986). Shear sensitivity of insect cells in suspension. *Enzyme Microbial Technology*, **8**, 33–6.

Tweeton, K. A., Bulla, L. A. & Consigli, R. A. (1980). Characterisation of an extremely basic protein derived from granulosis virus nucleocapsids. *Journal of Virology*, **33**, 866–76.

Urakawa, T. & Roy, P. (1988). Bluetongue virus tubules made in insect cells by recombinant baculoviruses: Expression of the NS1 gene of bluetongue virus serotype 10. *Journal of Virology*, **62**, 3919–27.

Urakawa, T., Small, D. A. & Bishop, D. H. L. (1988). Expression of snowshoe hare bunyavirus S RNA coding proteins by recombinant baculoviruses. *Virus Research*, **11**, 303–17.

Urakawa, T., Ferguson, M., Minor, P. D., Cooper, J., Sullivan, M., Almond, J. W. & Bishop, D. H. L. (1989). Synthesis of immunogenic, but non-infectious, poliovirus particles in insect cells by a baculovirus vector. *Journal of General Virology*, **70**, 1453–63.

Valenzuela, P., Medina, A., Rutter, W. J., Ammerer, G. & Hall, B. D. (1982). Synthesis and assembly of hepatitis B virus surface antigen particles in yeast. *Nature, London*, **298**, 347–50.

Van Wyke-Coelingh, K. L., Murphy, B. R., Collins, P. L., Lebacq-Verheyden, A-M., & Battey, J. F. (1987). Expression of biologically and antigenically active authentic parainfluenza type 3 virus haemagglutinin-neuraminidase glycoprotein by a recombinant baculovirus. *Virology*, **160**, 465–72.

Vlak, J. M., Klinkenberg, F. A., Zaal, K. J. M., Usmany, M., Klinge-Roode, E. C., Geervliet, J. B. F., Roosien, J. & Van Lent, J. W. M. (1988). Functional studies on the p10 gene of *Autographa californica* nuclear polyhedrosis virus using a recombinant expressing a p10–beta-galactosidase fusion gene. *Journal of General Virology*, **69**, 765–76.

Wathen, M. W., Brideau, R. J. & Thomson, D. R. (1989). Immunisation of cotton rats with the human respiratory syncytial virus F glycoprotein produced using a baculovirus vector. *The Journal of Infectious Diseases*, **159**, 255–64.

Weiss, S. A., Kalter, S. S., Vaughn, J. L. & Dougherty, E. M. (1980). Effect of nutritional and biophysical parameters on insect cell culture of large scale production. *In vitro*, **16**, 222–3

Weiss, S. A., Orr, T., Smith, G. C. Kalter, S. S., Vaughn, J. L. & Dougherty, E. (1982). Quantitative measurement of oxygen consumption in insect cell culture infected with polyhedrosis virus. *Biotechnology and Bioengineering*, **24**, 1145–54.

Weyer, U. & Possee, R. D. (1988). Functional analysis of the p10 gene 5′ leader sequence of the *Autographa californica* nuclear polyhedrosis virus. *Nucleic Acids Research*, **16**, 3635–53.

Weyer, U. & Possee, R. D. (1989). Analysis of the promoter of the *Autographa californica* nuclear polyhedrosis virus p10 gene. *Journal of General Virology*, **70**, 203–8.

Whitford, M., Stewart, S., Kuzio, J. & Faulkner, P. (1989). Identification and sequence analysis of a gene encoding gp67, an abundant envelope glycoprotein of the baculovirus *Autographa californica* nuclear polyhedrosis virus. *Journal of Virology*, **63**, 1393–9.

Williams, G. V., Rohel, D. Z. & Faulkner, P. (1989). A cytopathological investigation of *Autographa californica* nuclear polyhedrosis virus p10 gene function using insertion/deletion mutants. *Journal of General Virology*, **70**, 187–202.

Wilson, M. E., Mainprize, T. H., Friesen, P. D. & Miller, L. K. (1987). Location, transcription and sequence of a baculovirus gene encoding a small arginine-rich polypeptide. *Journal of Virology*, **61**, 661–66.

VACCINIA: VIRUS, VECTOR AND VACCINE

GEOFFREY L. SMITH

Sir William Dunn School of Pathology, University of Oxford, South Parks Road, Oxford OX1 3RE, UK

INTRODUCTION

Smallpox, the 'most dreaded scourge of the human species', has been eradicated. This triumph of preventive medicine was achieved in 1977 and resulted from the global use of vaccinia virus as a live vaccine against smallpox. Jenner introduced vaccination in 1798 using a poxvirus taken from cows (probably cowpox) to immunise humans (Jenner, 1798). His belief in the value of this practice stemmed from observations made by him and others, that milkmaids who acquired dermal poxvirus infections from cows escaped death or disfigurement during smallpox epidemics. His conviction persuaded him to deliberately inoculate children with cowpox virus and subsequently to challenge them with infectious smallpox material. The vaccinees did not succumb to smallpox following this challenge nor during subsequent epidemics. One hundred and seventy nine years later the practice of vaccination eradicated smallpox as Jenner had predicted it would.

The exact nature of the virus which Jenner used for immunisation against smallpox is, and will remain, uncertain (Baxby, 1981). The vaccine used in the modern era is vaccinia virus (of which there are numerous strains) and this has been shown to be quite distinct from cowpox virus (Downie, 1939; Mackett & Archard, 1979). If cowpox was the virus which Jenner used for smallpox immunoprophylaxis, the time at which this was replaced by vaccinia is unclear. However, both these viruses and variola virus, the etiological agent of smallpox, are closely related orthopoxviruses and possess sufficient cross-reactive antigens to confer immunological cross-protection.

After the elimination of smallpox it may have been expected that research into vaccinia virus and other poxviruses would decline. On the contrary, research with these large DNA viruses is now more extensive than ever. This is in part due to the remarkable nature of poxvirus replication which occurs in the cytoplasm, unlike most other DNA viruses. Consequentially these viruses cannot directly utilise nuclear enzymes of the host cell. Instead, poxviruses encode many of their own enzymes that are necessary for synthesis of virus mRNA and for DNA replication and, therefore, provide good models for studying the processes of transcription, DNA replication and DNA recombination in a eukaryotic cell.

The other major factor contributing to the explosion of interest in poxviruses has been the development of techniques which permit the insertion of DNA into, or deletion of DNA from, poxvirus genomes without destroying virus infectivity. Thus it is possible to express, in vaccinia virus, genes derived from other pathogenic organisms so that the foreign antigens encoded by these genes are expressed during the replication of the recombinant virus. The retention of infectivity permits such recombinant viruses to be used as new live vaccines for veterinary or medical application. In addition, recombinant vaccinia viruses are finding increasing use in the research laboratory especially in the field of immunology. Although genetically modified vaccinia viruses were the first recombinant viruses to be used as live immunogens (Smith *et al.*, 1983*a, b*; Panicali *et al.*, 1983), several other viruses have now been utilised in a similar manner. These are not discussed in this chapter and the reader should refer to papers on recombinant herpes viruses (Roizman & Arsenakis, 1985), adenoviruses (McDermott *et al.*, 1989), and polioviruses (Burke *et al.*, 1988) for further details. A theme common to all these viruses is the retention of virus infectivity so that the new candidate vaccines are self-replicating, infectious agents, cheap to manufacture and administer and which induce good T cell and humoral immunity. This chapter describes vaccinia virus structure and replication, reviews the properties of recombinant vaccinia viruses and their utilisation as new vaccines and concludes with a discussion of progress made towards the attenuation of vaccinia virus to increase vaccine safety.

VIRUS STRUCTURE AND REPLICATION

Vaccinia virus is a member of the orthopoxvirus genus of the Poxviridae. Some other members of this group are variola, cowpox, monkeypox and ectromelia (mousepox). All these viruses and those of other poxvirus genera have a common replicative strategy. The description below refers to vaccinia virus, since this is the most intensively studied orthopoxvirus, but many features will probably be true for other poxviruses.

Virion structure

The virus particle is large and complex with a brick-shaped or oval appearance and overall dimensions of roughly 200 nm by 300 nm (Fig. 1). The surface of the particle is covered by irregularly arranged tubules, composed of a 58 kD protein, that are associated with the outer membrane. Antibodies directed against this protein and a 14 kD surface protein neutralise the infectivity of intracellular virus (Stern & Dales, 1976; Rodriguez *et al.*, 1985). Inside the membrane lies the biconcave virus core which is associated with two lateral bodies of unknown function. The core contains the virus genome and many virus-coded enzymes (Table 1), a characteristic feature of poxviruses. The above description refers to virus particles that

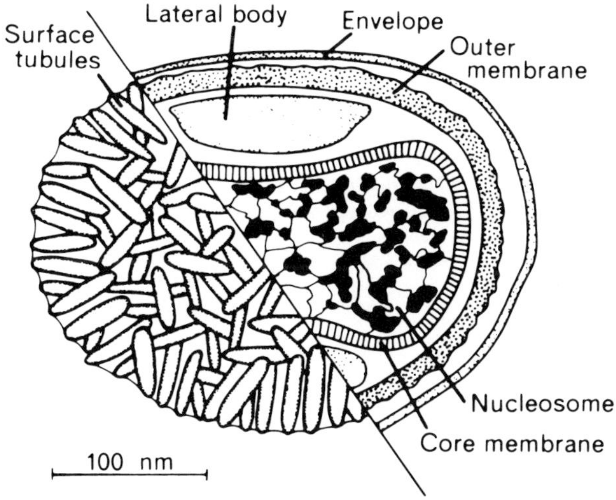

Fig. 1. Schematic representation of vaccinia virus. From Fenner, Wittek & Dumbell, *The Orthopoxviruses* 1989, Academic Press.

are mechanically released from infected cells and which may be routinely purified by sucrose density gradient centrifugation from cytoplasmic extracts (Joklik, 1962). A second type of infectious particle is naturally released from infected cells and contains an envelope derived from Golgi membrane (Hiller & Weber, 1985). This envelope contains eight proteins, seven of which are glycosylated (Payne, 1978), and which endow the 'extracellular enveloped' virus with different immunological and biological properties from the 'intracellular naked' virus. A major non-glycosylated 37 kD protein and the virus haemagglutinin (HA) are two of these proteins (Payne, 1979; Hirt *et al.,* 1986). Although the envelope is not required for infectivity, 'extracellular enveloped virus' has been implicated in the enhanced dissemination of virus through the multicellular host (Boulter & Appleyard, 1973; Payne, 1980) and antibodies directed against the envelope proteins are important components of anti-virus immunological defence.

Genome

The vaccinia genome (strain WR) is a linear double stranded molecule of 185,000 base pairs (bp) (Fig. 2). The termini have several unusual features. First, the two DNA strands are covalently linked by a covalently closed hairpin into one polynucleotide chain. This can be shown experimentally by heat denaturation of the genome into a single stranded circle and by rapid reannealing, on cooling, into a linear duplex (Geshelin & Berns, 1974). The terminal 104 nucleotides form an extremely A:T rich, incompletely base-paired hairpin which can exist in two isomeric forms. Direct

Table 1. *Enzymes present in vaccinia virions*

	Refs*
DNA-dependent RNA polymerase	1–8
RNA triphosphatase	9, 10
RNA guanyltransferase	11–15
RNA (guanine-7)methyltransferase	12, 13
RNA (nucleoside-2′)methyltransferase	16, 17
Poly (A) polymerase	18
5′-phosphate polynucleotide kinase	19
ATPase	20–24
NTPase	23, 24
Endoribonuclease	25
Deoxyribonuclease pH 4.4	26–29
Deoxyribonuclease pH 7.8	26, 27
DNA nicking-joining enzyme	30, 31
DNA topoisomerase	32–34
Protein kinase	35–37
Alkaline protease	38

* 1. Kates & McAuslan, 1967. 2. Munyon *et al.*, 1967. 3. Baroudy & Moss, 1980. 4. Spencer *et al.*, 1980. 5. Morrison *et al.*, 1985. 6. Broyles & Moss, 1986. 7. Jones *et al.*, 1987. 8. Patel & Pickup, 1989. 9. Tutas & Paoletti, 1977. 10. Venkatesen *et al.*, 1980. 11. Martin *et al.*, 1975. 12. Martin & Moss, 1975. 13. Martin & Moss, 1976. 14. Shuman & Hurwitz, 1981. 15. Shuman *et al.*, 1980. 16. Barbosa & Moss, 1978*a*. 17. Barbosa & Moss, 1978*b*. 18. Moss *et al.*, 1975. 19. Spencer *et al.*, 1978. 20. Gold & Dales, 1968. 21. Paoletti & Moss, 1974. 22. Paoletti *et al.*, 1974. 23. Rodriguez *et al.*, 1987. 24. Broyles & Moss, 1987. 25. Paoletti & Lipinskas, 1978. 26. Pogo & Dales, 1969. 27. Pogo & O'Shea, 1977. 28. Rosemond-Hornbeak & Moss, 1974. 29. Rosemond-Hornbeak *et al.*, 1974. 30. Lakritz *et al.*, 1985. 31. Reddy & Bauer, 1989. 32. Bauer *et al.*, 1977. 33. Shaffer & Traktman, 1987. 34. Shuman & Moss, 1987. 35. Kleiman & Moss, 1975*a*. 36. Kleiman & Moss, 1975*b*. 37. Howard & Smith, 1989. 38. Arzoglou *et al.*, 1978.

nucleotide sequencing of these structures showed that they are inverted and complementary copies of each other termed 'flip and flop' (Baroudy *et al.*, 1982). A second feature of the genomic termini are inverted repeats which in vaccinia virus (strain WR) are 10 kilobase pairs (kbp) long. The size of the inverted terminal repeats (ITR) is variable among orthopoxviruses. Variola virus either completely lacks ITRs or these are less than 1 kb, while terminal transposition mutants of cowpox virus contain ITRs of up to 50 kb (Pickup *et al.*, 1984; Archard *et al.*, 1984). Even within a single virus preparation there is heterogeneity in the ITRs so that the terminal restriction fragments may appear as diffuse bands on electrophoresis (Wittek *et al.*, 1978). This seems to result from recombination between blocks of tandemly repeated sequences residing near the terminal hairpin (Wittek & Moss, 1980; Moss *et al.*, 1981 *a*).

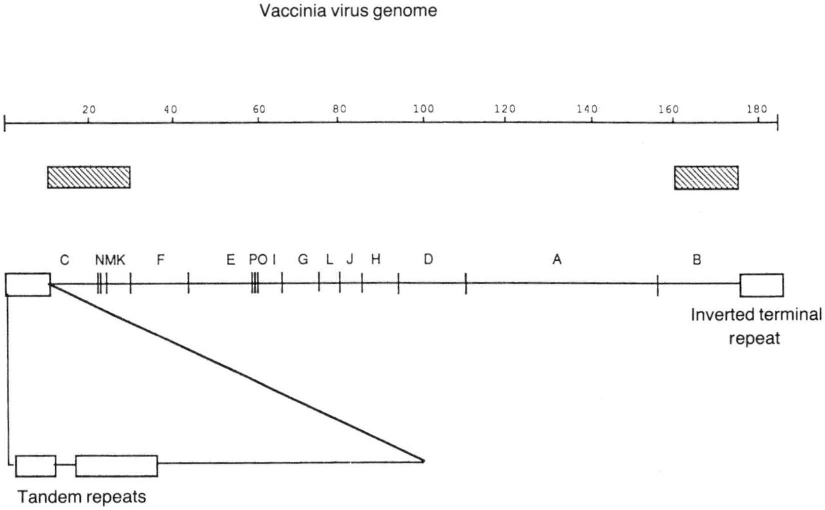

Fig. 2. *Hind* III restriction map of the vaccinia virus genome. One inverted terminal repeat is expanded to show blocks of tandem repeats adjacent to the terminal hairpin. The shaded boxes represent regions that may undergo deletion or transposition.

The regions adjacent to the ITR are also variable and may undergo deletion or transposition to the opposite end of the genome (Lake & Cooper, 1980; Moyer & Rothe, 1980; Moyer *et al.*, 1980; Moss *et al.*, 1981*b*; Panicali *et al.*, 1981; Drillien *et al.*, 1981; Perkus *et al.*, 1986). By implication, the genes residing here are not essential for virus replication in most cell culture systems. Instead they encode proteins which influence pock phenotype, host range, persistence in the presence of interferon and virulence (Fenner & Sambrook, 1966; Moyer & Rothe, 1980; Lake & Cooper, 1980; Paez & Esteban, 1985; Gillard *et al.*, 1985, 1986; Spehner *et al.*, 1988; Kotwal & Moss, 1988*a*, *b*). Relatively large amounts of DNA (20–30 kb) are either lost from deletion mutants or duplicated by transposition to the opposite end of the genome, without impairing incorporation of the genome in virions. This plasticity in size of packageable genome contrasts with some viruses that possess icosahedral symmetry, e.g. SV40 (Rigby, 1983), and permits large amounts (25 kb) of exogenous DNA to be inserted into the virus genome without a compensating deletion (Smith & Moss, 1983). The utility of the vaccinia vector system is enhanced by this property and has enabled construction of multivalent candidate vaccines (Perkus *et al.*, 1985).

In contrast to the variable 20–30 kb at the ends of the genome, the central 120 kb is conserved among orthopoxviruses. This was clearly shown by the common patterns of restriction enzyme sites of different viruses (Mackett & Archard, 1979; Esposito & Knight, 1985). In addition, conditional lethal mutations (Drillien & Spehner, 1983; Condit *et al.*, 1983) and many important enzymes and virus structural proteins have been mapped to this region.

Large sections of the vaccinia genome have been sequenced (Venkatesan *et al.*, 1982; Weir & Moss, 1983, 1984; Hruby *et al.*, 1983; Pluicienniczak *et al.*, 1985; Broyles & Moss, 1986; Rosel *et al.*, 1986; Rosel & Moss, 1985; Niles *et al.*, 1986; Weinrich & Hruby, 1986; Shida, 1986; Hirt *et al.*, 1986; Earl *et al.*, 1986*a*; Gillard *et al.*, 1986; Schmitt and Stunnenberg, 1988; Rodriguez & Esteban, 1987; Tengelsen *et al.*, 1988; Tamin *et al.*, 1988; Boursnell *et al.*, 1988; Kotwal and Moss, 1988*b*, 1989; Slabaugh *et al.*, 1988; Smith *et al.*, 1989 *a, b, c*; Howard & Smith, 1989; Slabaugh & Roseman, 1989). Analyses of the potential open reading frames deduced from nucleotide sequence show that the genes are closely packed with little or no intervening non-coding sequences and are distributed on both DNA strands. Interestingly, adjacent to both inverted terminal repeats, most genes are transcribed outwards towards the genome ends (Kotwal & Moss, 1988*b*; Smith *et al.*, unpublished data). Virus mRNAs are not spliced and protein coding sequences are contiguous in genomic DNA. This has greatly simplified the identification of protein coding sequences from raw nucleotide sequence data, a process that is further aided by the high (67%) A:T content of the genome, so that translational termination codons are frequent except in coding regions, and by comparisons of the codon usage of established vaccinia genes. It is anticipated that within the next year or two the sequence of the vaccinia genome will be complete. An increasing number of vaccinia proteins are being identified which possess homology to proteins of cells or other viruses. Possibly vaccinia virus has acquired these genes as cDNA copies of cellular mRNAs since some of the cellular genes have introns and spliced mRNAs.

Virus enzymes

Replication of a DNA virus in the cytoplasm is unusual and prevents direct use of the host cell's DNA and RNA polymerases located in the nucleus. Vaccinia virus and other cytoplasmically replicating DNA viruses, such as African Swine Fever Virus, have overcome this difficulty by encoding many enzymes necessary for their replication (Tables 1 and 2). In addition some nuclear enzymes may still be utilised by poxviruses as illustrated by the recruitment of a large subunit of the host RNA polymerase II from the nucleus to the cytoplasmic virus factories during infection (Morrison & Moyer, 1986). Vaccinia virus has played a pioneering role in the identification and study of virus enzymes. The discovery of a DNA-dependent RNA polymerase in the vaccinia virion was a revolutionary finding (Kates & McAuslan, 1967; Munyon *et al.*, 1967); it stimulated searches that led to the discovery of RNA polymerases in other viruses (e.g. reovirus and

Table 2. *Virus enzymes in infected cells*

	Refs*
DNA-dependent DNA polymerase	1–5
DNA ligase	6–9
Thymidine kinase	10–15
Thymidylate kinase	16
Ribonucleotide reductase	17–21

* 1. Challberg & Englund, 1979. 2. Jones & Moss, 1984. 3. Jones & Moss, 1985. 4. Traktman *et al.*, 1984. 5. Earl *et al.*, 1986*a*. 6. Sambrook & Shatkin, 1969. 7. Spadari, 1976. 8. Smith *et al.*, 1989*b*. 9. Kerr & Smith, 1989. 10. Dubbs & Kitt, 1964. 11. Weir *et al.*, 1982. 12. Weir & Moss, 1983. 13. Hruby & Ball, 1982. 14. Bajszar *et al.*, 1983. 15. Hruby *et al.*, 1983. 16. Smith *et al.*, 1989*a*. 17. Slabaugh *et al.*, 1984. 18. Slabaugh & Mathews, 1986. 19. Tengelsen *et al.*, 1988. 20. Schmitt & Stunnenberg, 1988. 21. Slabaugh *et al.*, 1988.

influenza virus) and culminated in the identification of reverse transcriptase in retroviruses. Subsequently many other enzymes were found in the vaccinia virion and the genes encoding several of these have now been mapped onto the virus genome (Table 1). Additional virus-induced enzymatic activities have been found in infected cells and the encoding virus genes sequenced (Table 2). Several of these are enzymes involved in synthesis of deoxynucleoside triphosphate precursors, for instance ribonucleotide reductase, thymidine kinase (TK) and thymidylate kinase (Tmpk). Others such as DNA ligase and DNA polymerase are involved in DNA replication and/or recombination. The presence of a complete enzyme system for transcription within the virus particle and enzymes for DNA replication (and recombination) are characteristics of poxviruses and have made poxviruses excellent systems for studying these processes in the eukaryotic cell. Some of the purified enzymes have been valuable tools for the molecular biologist. The TK gene has been an especially useful nonessential locus into which foreign genes may be inserted, for TK^- recombinant viruses are readily selectable by plaque assay on TK^- cells in the presence of bromodeoxyuridine (Mackett *et al.*, 1982, 1984). Moreover, TK is a factor affecting virus virulence (Buller *et al.*, 1985).

Gene expression

The virion-associated RNA polymerase and associated RNA modifying enzymes can transcribe early virus genes and produce functional mRNAs directly after penetration of the virion core into the cytoplasm. This process is independent of host RNA or protein synthesis and early mRNAs are detectable within minutes of infection. Early mRNAs are of a discrete length, possess a methylated 5′ cap structure (Wei & Moss, 1975) with

A or G as the penultimate nucleotide (Boone & Moss, 1977) and have a 3′ poly(A) tail of up to 200 nucleotides (Kates & Beeson, 1970). Correct temporal and spatial initiation of transcription requires DNA sequences of only approximately 30 nucleotides upstream from the RNA start site (Cochran *et al.*, 1985*a*; Coupar *et al.*, 1987; Weir and Moss, 1987 *a*). These nucleotide sequences are very A:T rich but do not possess the consensus sequences identified as essential for transcription by eukaryotic RNA polymerase II or by prokaryotic RNA polymerases. The promoters of eaıly (and late genes) are therefore functionally distinct from the cellular counterparts and are recognised only by the virus RNA polymerase (Puckett and Moss, 1983). Similarly, the vaccinia RNA polymerase will not initiate transcription from host cell RNA polymerase II promoters.

Early mRNA synthesis requires transcription factors in addition to the multi-subunit RNA polymerase. One such factor is a 130 kD virion protein that binds to early virus promoters. The binding of this transcription factor to an early promoter and the functional activity of the promoter are obviated by specific nucleotide substitutions (Yuen *et al.*, 1987). Another early transcription factor is composed of a complex of 87 kD and 77 kD polypeptides which possess DNA-dependent ATPase activity (Broyles *et al.*, 1988; Broyles & Moss, 1988). Termination of early transcription occurs roughly 50 nucleotides downstream of the sequence TTTTTNT (Yuen & Moss, 1986, 1987; Rohrmann *et al.*, 1986). Surprisingly, the factor responsible for transcriptional termination copurifies with the virus capping enzyme complex (Shuman *et al.*, 1987). Possibly these enzymes remain associated with the RNA polymerase complex as it transcribes the DNA template. Another interesting finding was the discovery that it is the RNA form of the termination sequence (UUUUUNU) that is recognised. This suggests that the termination factors may be located distal to the active centre of the enzyme complex involved in ribonucleotide polymerisation.

Early genes are located throughout the genome on both DNA strands and hybridisation studies have indicated that approximately half the genome is transcribed into early mRNA (Oda & Joklik, 1967). Several early genes encode enzymes involved in producing substrates for DNA replication (e.g. TK, TmpK and ribonucleotide reductase) or enzymes required for DNA replication (DNA polymerase and DNA ligase). Another early gene encodes a secretory polypeptide with homology to epidermal growth factor and which stimulates cell proliferation surrounding foci of infection (Stroobant *et al.*, 1985; Twardzik *et al.*, 1985; Buller *et al.*, 1988*b*). After these early enzymes are expressed and following a second phase of uncoating of the virus core, involving early virus gene function (Pedley & Cooper, 1987), DNA replication commences. During a high multiplicity infection in a permissive cell line, DNA synthesis has started by 2 hours post-infection and this is accompanied by a change in the pattern of transcription. Early genes are shut off and transcription of late genes begins.

Late mRNAs are structurally distinct from early transcripts in two respects. First, they are not of a defined length and are on average of greater size than early mRNAs. This size heterogeneity stems from failure to terminate transcription at specific sites, the early signal TTTTTNT no longer being recognised. Consequently, an examination of late RNAs by northern blotting reveals only a smear for individual genes. However, there is one exception to this rule. The major protein component of cowpox virus A type inclusion bodies is an abundant, late 160 kD polypeptide (Patel *et al.*, 1986; Funahashi *et al.*, 1988) and the late RNA transcribed from the encoding gene is of homogeneous length (Patel & Pickup, 1987). Presumably a specific signal resides downstream of this but not other late genes, and this signal is recognised by the RNA polymerase complex active during the late phase of infection. The second unusual feature of late mRNA is the presence of a 5′ poly (A) sequence of up to 35 nucleotides which lies immediately downstream of the cap structure and before the AUG codon for translational initiation (Bertholet *et al.*, 1987; Schwer *et al.*, 1987; Patel & Pickup, 1987; Wright & Moss, 1987; Schwer & Stunnenberg, 1988; DeMagistris and Stunnenberg, 1988; Ahn & Moss, 1989). This poly (A) sequence is not encoded in viral DNA and is proposed to be synthesised by transcriptional slippage through a TAAAT(G) motif found at the transcriptional start site of late genes (Rosel *et al.*, 1986; Bertholet *et al.*, 1986; Hänggi *et al.*, 1986; Patel & Pickup, 1987). A similar process had been proposed for the synthesis of 5′ poly (A) of certain late mRNAs from bacteriophage T4 and is supported by experimental evidence (Kassavetis *et al.*, 1986). The function of the 5′ poly (A) tract is unknown. A plausible possibility is that it confers selective translation of late mRNAs late during infection. This would augment the switch for early to late gene expression which appears to be principally controlled at the transcriptional level. Another possibility is that 5′ poly (A) influences RNA stability.

Late promoters are even shorter than early promoters and are also very A:T rich (Weir & Moss, 1987*b* ; Bertholet *et al.*, 1986; Kent, 1988). The TAAAT(G) motif is found in most, but not all, late promoters and in many cases the last three nucleotides of this motif constitute the translational initiation codon of the open reading frame. This motif is essential for function of several late gene promoters (Bertholet *et al.*, 1986; Hänggi *et al.*, 1986; Kent, 1988): however, alone it does not constitute a late promoter.

Many of the virion major structural proteins are encoded by late genes. Examples are the 28 kD protein (Weir & Moss, 1984; Yang *et al.*, 1988), the basic 11 kD protein (Wittek *et al.*, 1984*a*), the haemagglutinin (HA) (Shida, 1986), and the 4a and 4b polypeptides which are derived by proteolytic cleavage from higher molecular weight precursors (Moss & Rosenblum, 1973; Wittek *et al.*, 1984*b*; Rosel & Moss, 1985).

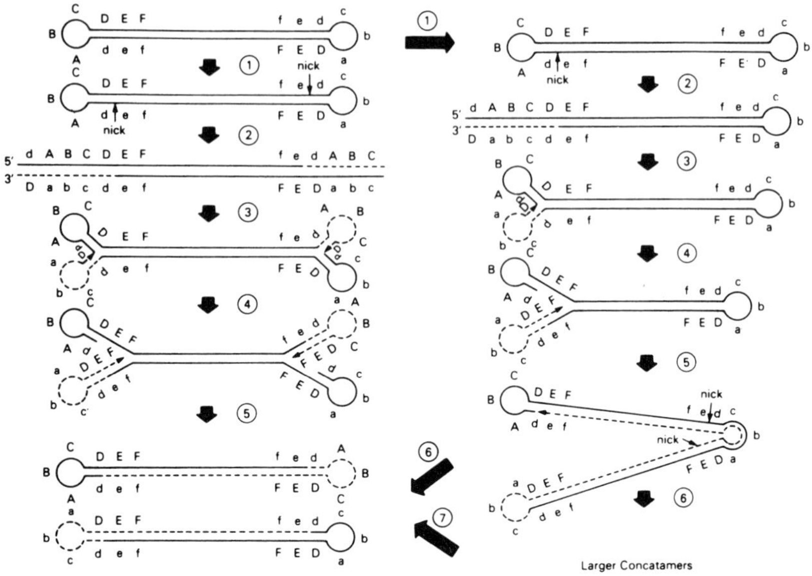

Fig. 3. Self-priming model for replication of vaccinia virus genome. Nicks are introduced into genome at both ends (left panel) or one end (right panel). From Baroudy *et al.*, 1983.

DNA replication

Vaccinia DNA replication occurs in the cytoplasm in discrete foci termed virus factories and is maximal between 2 to 3 hours post-infection. The process can occur in enucleated cells, albeit less efficiently (Pennington and Follett, 1974), suggesting that the virus encodes all necessary factors. Details of the replication mechanism remain incomplete but it is known that shortly after infection nicks are introduced into one DNA strand adjacent to one, or both, terminal hairpins (Pogo, 1977). This provides a 3′ OH group from which polymerisation may initiate around the terminal hairpin and down the opposite strand by a self-priming, semiconservative, strand displacement mechanism (Fig. 3) (Esteban & Holowczak, 1977*a*; Esteban *et al.*, 1977; Pogo, 1980; Pogo *et al.*, 1981; Moyer & Graves, 1981; Baroudy *et al.*, 1982, 1983). Some evidence for discontinuous lagging strand synthesis has been reported (Esteban & Holowczak, 1977*b*; Pogo & O'Shea, 1978). Concatemeric DNA genomes are synthesised (Moyer & Graves, 1981; Baroudy *et al.*, 1983) and resolved into unit length monomers (DeLange *et al.*, 1986; Merchlinsky & Moss, 1986; 1989). The latter process requires late gene expression and is probably performed by a virion-associated enzyme termed 'nicking-joining' (Lakritz *et al.*, 1985; Reddy & Bauer, 1989) which can introduce site-specific, single stranded nicks and subsequently crosslink the two DNA strands. This enzyme is distinct from vaccinia DNA topoisomerase (Shaffer & Traktman, 1987) and DNA ligase

(Kerr & Smith, 1989), and the encoding gene may soon be mapped using a *ts* mutant defective in concatemer resolution (DeLange, 1989). Other virus proteins required for DNA replication are DNA polymerase (Earl *et al.*, 1986*a*) and the products of genes D5R and B1R to which *ts* mutations which confer a DNA negative phenotype map (Evans & Traktman, 1987, Roseman & Hruby, 1987; P. Traktman, personal communication). The latter gene encodes a polypeptide with amino acid homology to protein kinases of serine/threonine specificity (Howard & Smith, 1989) and may constitute the virion-associated protein kinase (Kleiman and Moss, 1975*a,b*).

The processes of virus DNA recombination and DNA replication are probably linked since conditional lethal mutants are defective in both or neither process (Merchlinsky, 1989). In addition, the formation of recombinant virus genomes is dependent upon DNA replication (Ball, 1987; Spyropolous *et al.*, 1988). The proteins involved in DNA recombination are unknown although DNA ligase (Smith *et al.*, 1989*b*) and DNA topoisomerase (Shuman, 1989) are candidates.

Virus assembly

Vaccinia virus morphogenesis is an unusual and complex process. Electron microscopy has shown that crescent-shaped membranes are synthesised *de novo* in the cytoplasm and are not acquired from existing cellular membranes. The crescents are associated with virus spicules and their formation can be reversibly inhibited by the drug rifampicin (Moss *et al.*, 1969; Nagayama *et al.*, 1970). Mutations in a specific virus gene confer resistance to this drug (Tartaglia & Paoletti, 1985; Tartaglia *et al.*, 1986; Baldick & Moss, 1987). During morphogenesis the virus nucleocapsid becomes enclosed with the maturing crescent membranes and subsequently condenses to form the biconcave core and lateral bodies. A small proportion of intracellular virus particles become wrapped in double layers of Golgi membrane and migrate to the cell surface. Here they are released from the cell with loss of the outermost Golgi membrane (Dales & Pogo, 1981). The majority of infectious progeny remains intracellular.

RECOMBINANT VACCINIA VIRUSES

Shortly after the eradication of smallpox the techniques of marker rescue and transfection were applied to the insertion of exogenous DNA into vaccinia virus (Sam & Dumbell, 1981; Nakano *et al.*, 1982; Weir *et al.*, 1982; Panicali & Paoletti, 1982; Mackett *et al.*, 1982). This procedure has proved useful for the mapping of specific genes by marker rescue (see below), but has also enabled the construction of vaccinia virus recombinants which contain genes from other organisms. Expression of the foreign gene was obtained by engineering its protein coding sequences in the correct orientation downstream of a defined vaccinia promoter sequence (Mackett *et al.*, 1982; Smith *et al.*, 1983*a, b*, 1984; Mackett *et al.*, 1984), or by insertion of the foreign gene into nonessential loci of the genome close to an endogenous promoter (Panicali & Paoletti, 1982; Panicali *et al.*, 1983; Paoletti *et al.*, 1984; Perkus *et al.*, 1985, 1986). Using either method, the foreign antigen is expressed during the normal replication of the virus and, importantly, infectivity is retained. The former method has the advantage that any vaccinia promoter, or promoters from other poxviruses, may be used. With the latter method, the available promoters are restricted to those of nonessential genes. Inoculation of vaccinia recombinants into experimental animals results in immune responses to the foreign protein (Smith *et al.*, 1983*a, b*; Panicali *et al.*, 1983). The great potential of such viruses as new live vaccines against a broad range of infectious diseases, particularly in developing countries, was recognised by the World Health Organization which sponsored meetings in 1984 and 1985 specifically to discuss this field (Quinnan, 1985; Ada *et al.*, 1985; World Health Organization, 1985). Numerous vaccinia recombinants expressing foreign antigens have now been described.

Construction of recombinant viruses

The large size of the vaccinia genome, and its lack of infectivity when stripped of associated enzymes, make it necessary for recombinant viruses to be constructed in a two-stage process (Fig. 4). First, the foreign gene is cloned into a plasmid vector downstream of a vaccinia promoter and flanked by vaccinia DNA derived from a non-essential locus. Second, this plasmid is transfected into vaccinia infected cells. Within these cells recombinant genomes are produced by homologous recombination between the vaccinia sequences flanking the foreign gene in the plasmid and the corresponding genomic region of the infecting virus. The progeny virus of such transfections is mostly of parental type (99%) and consequently methods are needed to identify the recombinant viruses. Genetic selection by insertion of the foreign gene into the vaccinia TK gene and selection of TK^- viruses by plaque assay on TK^- cells in the presence of BUdR, has been the most popular method (Mackett *et al.*, 1982). Alternatively TK^+ recombinants were isolated after insertion of herpes simplex virus (HSV) TK into vaccinia TK^- mutants (Panicali & Paoletti, 1982; Mackett *et al.*, 1982; Vassef *et al.*, 1985). With the former approach not all TK^- isolates are recombinants due to the production of spontaneous TK^- mutants under BUdR selection. Consequently, DNA hybridisation with a probe specific for the foreign gene was used to identify the recombinants. The need for this secondary screening was removed by inclusion of the *E. coli* β-galactosi-

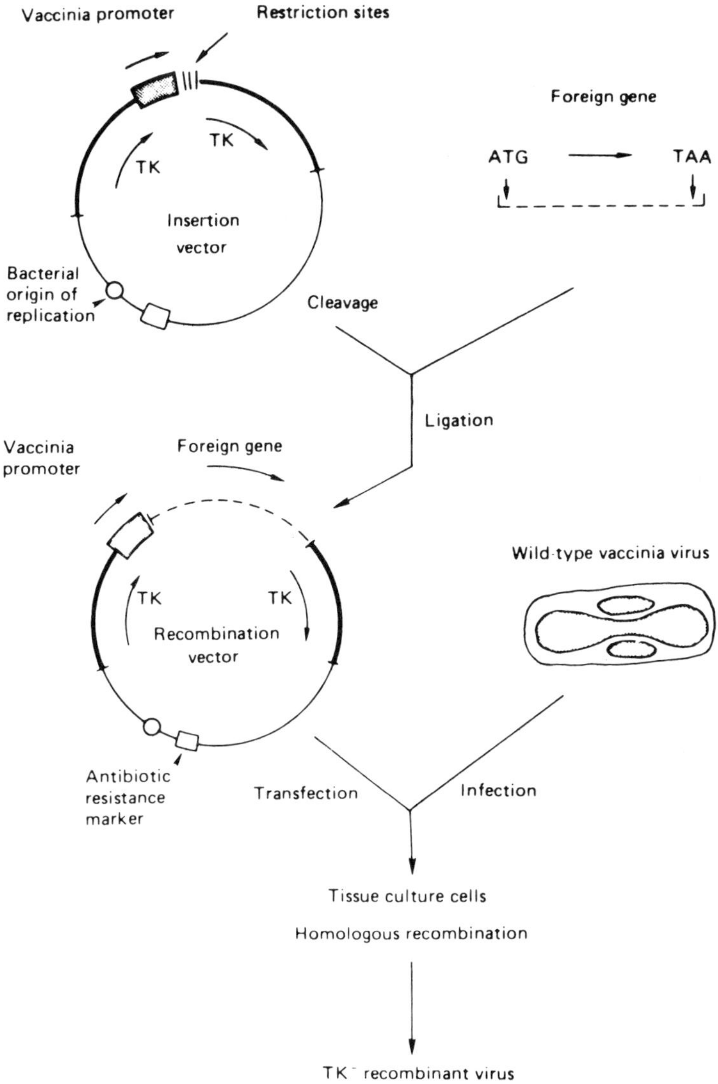

Fig. 4. Formation of vaccinia virus recombinants by insertion into the TK gene.

dase gene under control of a second vaccinia promoter (Chakrabarti *et al.*, 1985; Panicali *et al.*, 1986). Incorporation of the chromogenic substrate for this enzyme in the agarose overlay permits visual identification of blue, recombinant plaques. Other dominant selectable markers that have been expressed are a genes conferring resistance to neomycin (Franke *et al.*, 1985) and *E. coli* xanthine guanine phosphoribosyltransferase (Ecogpt) (Boyle & Coupar, 1988; Falkner & Moss, 1988). *In situ* plaque hybridisation

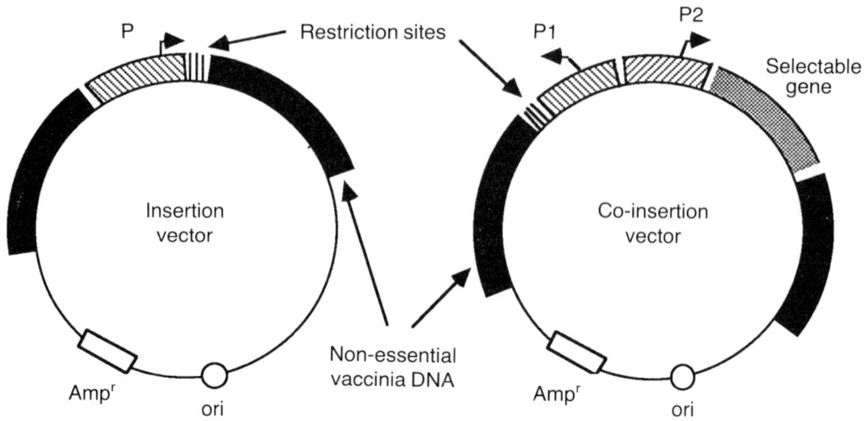

Fig. 5. Plasmid vectors for construction of vaccinia virus recombinants. P, vaccinia promoter. Amp^r, ampicillin resistance gene. Ori, plasmid origin of DNA replication.

(Panicali *et al.*, 1983), immunological screening of plaques for foreign antigen (Mackett *et al.*, 1985*a*), selection of plaque size variants (Rodriguez & Esteban, 1989), use of conditional lethal mutants under non-permissive conditions to reduce the background of parental virus (Fathi *et al.*, 1986), expression of luciferase (Rodriguez *et al.*, 1988) and selection of HA^- insertion mutants (Shida *et al.*, 1987) are all methods that have been used to identify recombinant viruses.

These selective techniques and the availability of plasmid vectors designed for the construction of recombinant vaccinia viruses expressing foreign genes (Mackett *et al.*, 1984; Chakrabarti *et al.*, 1985; Boyle *et al.*, 1985; Wilson *et al.*, 1986; Wilson & Hruby, 1987; Falkner & Moss, 1988; Boyle & Coupar, 1988) (Fig. 5) have made the production of recombinant vaccinia viruses a routine and widely used laboratory technique (Mackett *et al.*, 1985*b*; Mackett & Smith, 1986; Moss & Flexner, 1987). The major requirements for expression of foreign genes are that (1) the protein coding sequences are contiguous (where necessary cDNA copies should be used), (2) the gene should not contain the early transcriptional terminator TTTTTNT upstream of, or within the coding sequences if it is to be expressed from an early promoter, and (3) any additional ATG codons between the transcriptional start site of the vaccinia promoter and the true translational initiation site of the foreign gene should be removed.

Several properties of vaccinia virus make this vector system attractive. Large amounts of foreign DNA may be incorporated (Smith & Moss, 1983), there are many non-essential sites into which foreign DNA may be inserted without destroying virus infectivity (Mackett *et al.*, 1982; Panicali & Paoletti, 1982; Perkus *et al.*, 1985, 1986) and the virus has a broad host range. These properties have endowed vaccinia recombinants with a wide range of applications in the research laboratory in addition to their potential as new vaccines. Some of these are discussed before considering vaccines.

Research applications

(1) *Analysis of vaccinia transcriptional signals*

The introduction of vaccinia transcriptional regulatory sequences joined to reporter genes such as chloramphenicol acetyltransferase (CAT), β-galactosidase or luciferase into the vaccinia genome gives expression of the reporter gene at a level and with a temporal regulation dictated by the promoter. The consequence of specific mutation of the promoter upon this expression may be studied. Experiments of this type have provided a detailed understanding of sequences required for early or late poxvirus transcriptional initiation (Cochran *et al.*, 1985*a*; Weir & Moss, 1987*a,b*; Bertholet *et al.*, 1985, 1986; Yuen *et al.*, 1987; Coupar *et al.*, 1987; Kent, 1988) and termination (Yuen & Moss, 1986; Rohrmann *et al.*, 1986). The development of a transient expression system in vaccinia virus-infected cells (Cochran *et al.*, 1985*b*) and of *in vitro* systems based on extracts of virus-infected cells (Puckett & Moss, 1983; Foglesong, 1985; Wright & Moss, 1987) or virions (Rohrmann & Moss, 1985; Golini & Kates, 1985) have simplified and complemented the more laborious construction of recombinant viruses.

(2) *Mapping and study of poxvirus gene functions*

Poxvirus genes that encode selectable enzymes or which confer distinct phenotypes have been mapped by using marker rescue techniques to construct recombinant vaccinia viruses. For example, a vaccinia virus deletion mutant was isolated that could not replicate on human cells (Drillien *et al.*, 1981). The gene encoding this host range function was identified by recombining fragments of DNA (derived from wild type virus) into the TK gene of the deletion mutant and studying the host range phenotype of the TK^- recombinants (Gillard *et al.*, 1985). By using successively smaller DNA fragments the gene responsible was accurately mapped and subsequently sequenced (Gillard *et al.*, 1986). A similar approach was employed to determine the gene permitting cowpox virus to replicate on Chinese hamster ovary cells, a cell line nonpermissive for vaccinia virus (Spehner *et al.*, 1988). In another study, the fowlpox virus TK gene was mapped by insertion of fowlpox DNA fragments into TK^- vaccinia virus and selecting TK^+ viruses (Boyle & Coupar, 1986).

Although the majority of the vaccinia genome has been sequenced, few genes have been assigned a specific function. Moreover, unless conditional lethal mutants have been mapped to a specific gene, or viable virus deletion mutants exist, it is unknown if most genes are essential or not. Targeted insertional mutagenesis of specific genes is an application of recombinant vaccinia viruses designed to establish gene essentiality/non-essentiality and to study gene function. The coding sequences of the chosen gene cloned in a plasmid vector are interrupted by insertion of a selectable marker gene. The plasmid is then transfected into virus-infected cells and recombinant viruses expressing the marker gene selected. If the target gene is nonessential, recombinants contain the marker gene inserted into the coding region by double homologous recombination (Panicali & Paoletti, 1982; Mackett *et al.*, 1982; Kotwal *et al.*, 1989). Alternatively, if the gene codes for an essential function, recombinant viruses always retain the wild type target gene plus the entire transfected plasmid which becomes integrated via a single recombination event (Ball, 1987; Spyropolous *et al.*, 1988, Shuman *et al.*, 1989). Removal of genetic selection enables these latter, unstable genomes containing direct repeats to resolve via gene conversion to either wild type virus or to mutant virus with the target gene interrupted by the selectable marker. Careful analysis of the genomic structure of a number of recombinant plaques isolated in the presence of genetic selection may enable assignment of a gene to the essential or nonessential group. On a cautionary note, it should be remembered that insertional mutagenesis of known nonessential genes, such as TK, may also proceed via single crossover events which subsequently resolve (Spyropolous *et al.*, 1988) and, therefore, detection of the products of single crossover events may alone be insufficient to assign a gene to the essential class.

The study of non-essential gene function can be aided by construction of deletion mutants. However, a study of essential genes is more difficult, except where a conditional lethal mutant has been isolated. The development of an inducible promoter system to drive expression of selected genes in vaccinia recombinants overcomes this difficulty (Fuerst *et al.*, 1989; Rodriguez & Smith, in preparation). This system uses the bacteriophage lambda *lac* operator and repressor. The repressor protein is constitutively expressed from the early/late vaccinia 7.5K promoter, while the inducible gene is placed downstream of a second vaccinia promoter but separated from it by the *lac* operator region. The repressor protein binds to the operator and prevents expression of the target gene except when IPTG is added. Thus, the target gene may be turned on by IPTG at any time during infection and the consequences studied.

(3) *Protein modification and transport*

Post-translational modifications of viral and mammalian proteins expressed by vaccinia vectors such as glycosylation, proteolytic cleavage and carboxylation occur normally (Smith *et al.*, 1983*a*; Mackett & Arrand, 1985; Sullivan and Smith, 1987; Rice *et al.*, 1985; Chakrabarti *et al.*, 1986; de la Salle *et al.*, 1985). Protein transport within and from the cell is also normal. For example, hepatitis B virus surface antigen (HBsAg) is excreted from infected cells as characteristic 22 nm lipoprotein particles (Paoletti *et al.*, 1984); influenza virus HA and vesicular stomatitis virus (VSV) glycoprotein migrate to the cell surface apical or basolateral membranes, respectively (Stephens *et al.*, 1986), and human papilloma virus L1 protein (Browne *et al.*, 1988), influenza virus nucleoprotein (NP) (Smith *et al.*, 1987) and polyoma virus VP1 and VP2 (Stamatos *et al.*, 1987) are transported to the nucleus. Thus vaccinia recombinants have been used to study the processing and transport of individual proteins (Rice *et al.*, 1985; Smith *et al.*, 1987; Thomas *et al.*, 1986; Falkner *et al.*, 1988). Additionally, vaccinia vectors were used to identify an amino acid signal for transport to the nucleolus (Siomi *et al.*, 1988). One exception to the normal processing and transport in recombinant vaccinia infected cells is glycoprotein H of human cytomegalovirus (HCMV) and HSV (Cranage *et al.*, 1988; Sullivan, 1988). In the latter case the expressed protein also has altered antigenicity and it is proposed that other herpes gene products may participate in correct processing of this glycoprotein (Sullivan, 1988; A. Forrester and A. C. Minson, personal communication; Gompels & Minson, 1989).

(4) *Identification of vaccine antigens*

Antigens from a wide range of pathogens have been expressed in vaccinia virus as a means of evaluating their potential as components of future vaccines and of assessing their relative roles in inducing immunity (e.g. Spriggs *et al.*, 1987; Paterson *et al.*, 1987; Olmsted *et al.*, 1986). The versatility of this expression system has enabled identification of antigens that are recognised by neutralising antibody and by cytotoxic T cells (CTL) in association with class I major histocompatibility (MHC) antigens.

The identification of targets of neutralising antibodies is well illustrated by genes taken from herpes viruses. Large-scale nucleotide sequencing of HSV-1, Epstein–Barr virus (EBV), varicella zoster virus (VZV) and HCMV have identified many open reading frames potentially encoding surface glycoproteins. To characterise the encoded protein and to determine if it is a target for neutralising antibody, the gene has been expressed in vaccinia virus and the recombinant virus used to raise antibody in experimental animals. This antiserum may be tested for virus-neutralising activity or the animals may be directly challenged if an appropriate model exists. HSV glycoproteins B, D, G and I induced neutralising antibody via expression in vaccinia (Paoletti *et al.*, 1984; Cremer *et al.*, 1985; Cantin *et al.*, 1987; Sullivan and Smith, 1987, 1988). Similarly, HCMV gB and gH (Cranage *et al.*, 1986, 1988) and EBV gp340 (Mackett & Arrand, 1985; Morgan *et al.*, 1988) are targets for neutralising antibodies. Antigens encoded by genes from bovine herpes virus 1 (van Drunen Littel-van den Hurk *et al.*, 1989) and pseudorabies (Marchioli *et al.*, 1987; Kost *et al.*, 1989) have also been characterised in this way.

The cellular immune response is an important component of host defence against viruses in many cases (Mims & White, 1984). Although cellular immunity does not usually prevent infection, it is a major mechanism for destruction of virus-infected cells, and serves to control and subsequently clear virus infections. The identification of virus antigens that are recognised by this component of the immune system is, therefore, essential to the design of vaccines required to stimulate cellular immunity. Vaccinia virus recombinants have been widely used to identify virus antigens that are recognised by CTL. The value of vaccinia recombinants for this purpose stems from the observations that (1) autologous infected cells are good targets for CTL (Yewdell *et al.*, 1985) and (2) vaccinia recombinants can stimulate formation of antigen-specific CTL in experimental animals (Bennink *et al.*, 1984; Wiktor *et al.*, 1984). A collection of vaccinia recombinants individually expressing each influenza virus protein (Smith *et al.*, 1987) is enabling a thorough examination of CTL recognition of influenza antigens in murine models and in man (Bennink *et al.*, 1984, 1986, 1987; Yewdell *et al.*, 1985; McMichael *et al.*, 1986; Andrew *et al.*, 1986, 1987; Gotch *et al.*, 1987). These studies have helped to establish that 'internal' virus proteins can be effectively recognised by CTL and can be the immunodominant antigens (Yewdell *et al.*, 1985). Additionally, it was shown that antigen recognition may vary with MHC haplotype (Bennink *et al.*, 1987). CTL recognition of antigens from a wide range of other viruses has been similarly analysed (Yewdell *et al.*, 1986; Zarling *et al.*, 1986*a*, *b*; Bangham *et al.*, 1986; Pemberton *et al.*, 1987; Martin *et al.*, 1987; Borysiewicz *et al.*, 1988; McLaughlin *et al.*, 1988; Jonjiic *et al.*, 1988; Whitton *et al.*, 1988*a*, *b*; Walker *et al.*, 1988).

Most of the vaccinia recombinants employed for these CTL studies used the constitutively active 7.5K vaccinia promoter to express the foreign antigen. This was the first and best characterised vaccinia promoter, and vectors utilising it were available, but the choice was fortunate since expression of some foreign antigens (e.g. influenza HA) from exclusively late promoters resulted in a failure of presentation to CTL (Coupar *et al.*, 1986). This blockage in presentation was not due to the nature of the antigen, since good presentation could occur from an early promoter, nor due to the level of expression, since this was as high or higher than early during infection. Further investigation showed the phenomenon to be neither universal nor absolute, but to vary with different antigens and even different epitopes within the same protein (Townsend *et al.*, 1988). Moreover, the blockage could be overcome if the antigen was expressed in an unstable form (Townsend *et al.*, 1988) or by expressing peptide epitopes alone (Gould *et al.*, 1989). These data suggest that the blockage occurs at the level of peptide formation. A possible explanation for this blockage is the presence in vaccinia virus of three serine proteinase inhibitor (serpin) genes (Boursnell *et al.*, 1988; Kotwal & Moss, 1989; Smith *et al.*, 1989*c*). Possibly the encoded serpins may inhibit the unidentified cellular proteinases involved in antigen degradation. If true, this would constitute a novel mechanism by which virus-infected cells could escape lysis by host CTL.

VACCINES

A large part of the interest in recombinant vaccinia viruses arises from the potential of using these genetically modified viruses as new live vaccines. This in turn is partially due to the historical and successful use of vaccinia to prevent and finally eradicate smallpox. The reasons for vaccinia virus being so effective against smallpox are worth recalling. The vaccine was cheap to manufacture and stable without refrigeration when freeze-dried (Collier, 1955); it was easy to administer to large populations under field conditions by relatively unskilled personnel using a bifurcated needle or jet gun; the immunity induced by a single vaccine dose persisted for long periods and involved antibody and cellular responses. Moreover, variola virus induced an acute, easily recognisable disease that did not establish latent or persistent infections, and there was no animal reservoir for the virus. These latter points are major problems for the control or eradication of other virus diseases such as hepatitis B virus (HBV) or yellow fever, despite the existence of effective vaccines for these diseases. The heat stability, potency and low cost of vaccinia should be retained as useful properties of recombinant viruses. In addition, the large capacity for foreign DNA and broad host range should enable the construction of polyvalent vaccines (Perkus *et al.*, 1985) for medical or veterinary application.

There are many examples of protective immunisation of experimental animals with vaccinia virus recombinants (Table 3). There are many further examples of immune responses in animals which did not induce protection or this was not tested (Panicali *et al.*, 1983; Cheng *et al.*, 1986; Cheng & Moss, 1987; Smith *et al.*, 1984, 1986; Langford *et al.*, 1986; Chakrabarti *et al.*, 1986; Hu *et al.*, 1986, 1987; Andrew *et al.*, 1987*b*; Tomley *et al.*, 1988; Li *et al.*, 1988; Deubel *et al.*, 1988; Pensiero *et al.*, 1988; Katach *et al.*, 1988; Shida *et al.*, 1988). Although no extensive trials have been performed, in a few instances infection of humans with vaccinia recombinants has resulted in immune responses to the expressed antigen (Jones *et al.*, 1986; Zagury *et al.*, 1987, 1988).

The type and magnitude of the immune responses obtained have been variable and seem dependent upon multiple factors. Generally, surface glycoproteins have induced good primary antibody responses, but there are exceptions (Gilbert *et al.*, 1987; Moss *et al.*, 1984). Some antigens appear intrinsically more immunogenic than others, for instance, the rabies virus glycoprotein has proved an exceptionally potent immunogen inducing high levels of neutralising antibodies and CTL (Wiktor *et al.*, 1984; Kieny *et al.*, 1984). The route of vaccine administration may influence the type of antibody response. Dermal scarification or intradermal injection induce systemic IgG, but local secretory IgA may be produced following inoculation of the vaccine onto mucous membranes (Small *et al.*, 1985; Olmsted *et al.*, 1986). Revaccinations can boost antibody titres and, importantly, induce antibody responses to new antigens despite existing immunity to

Table 3. *Pathogens against which vaccinia virus recombinants have induced protective immunity*

	Refs*
Influenza virus	1–6
Hepatitis B virus	7
Herpes simplex virus	8–13
Rabies virus	14–19
Vesicular stomatitis virus	20
Measles virus	21
Respiratory syncytial virus	22–26
Parainfluenza virus	27, 28
Pseudorabies virus	29
Polyoma virus	30
Murine cytomegalovirus	31
Dengue virus	32
Venezuelan equine encephalitis virus	33
Friend murine leukaemia virus	34
Lassa fever virus	35, 36
Rinderpest virus	37–39
Streptococcus group A	40
Epstein–Barr virus	41
Human T lymphotropic virus I	42
Simian virus 5	43

* 1. Smith *et al.*, 1983*b*. 2. Small *et al.*, 1985. 3. Rota *et al.*, 1987. 4. Rota *et al.*, 1989. 5. Chambers *et al.*, 1988. 6. Jakeman *et al.*, 1989. 7. Moss *et al.*, 1984. 8. Paoletti *et al.*, 1984. 9. Cremer *et al.*, 1985. 10. Cantin *et al.*, 1987. 11. Rooney *et al.*, 1988. 12. Martin *et al.*, 1987. 13. Martin *et al.*, 1989. 14. Keiny *et al.*, 1984. 15. Wiktor *et al.*, 1984. 16. Blancou *et al.*, 1986. 17. Esposito *et al.*, 1987. 18. Esposito *et al.*, 1988. 19. Brochier *et al.*, 1989. 20. Mackett *et al.*, 1985*a*. 21. Drillien *et al.*, 1988. 22. Olmsted *et al.*, 1986. 23. Elango *et al.*, 1986. 24. Wertz *et al.*, 1987. 25. King *et al.*, 1987. 26. Stott *et al.*, 1987. 27. Spriggs *et al.*, 1987. 28. Spriggs *et al.*, 1988. 29. Marchioli *et al.*, 1987. 30. Lathe *et al.*, 1987. 31. Jonjiic *et al.*, 1988. 32. Bray *et al.*, 1989. 33. Kinney *et al.*, 1988. 34. Earl *et al.*, 1986*b*. 35. Morrison *et al.*, 1989. 36. Clegg & Lloyd, 1987. 37. Yilma *et al.*, 1988. 38. Barrett *et al.*, 1989; 39. Belsham *et al.*, 1989. 40. Fischetti *et al.*, 1988; 41. Morgan *et al.*, 1988. 42. Shida *et al.*, 1987. 43. Paterson *et al.*, 1987.

vaccinia virus or the foreign antigen (Perkus *et al.*, 1985; Flexner *et al.*, 1988). Naturally, the levels of antibody to new antigens presented during revaccination are low compared with primary vaccination, but these data demonstrate that revaccinations can be effective. This has occurred in humans also (Jones *et al.*, 1986; Zagury *et al.*, 1987). The practicality of constructing vaccinia recombinants expressing multiple foreign antigens and using such viruses as polyvalent vaccines has also been demonstrated (Perkus *et al.*, 1985; Flexner *et al.*, 1988).

Another potential use of vaccinia recombinants is for immunotherapy.

Immunisation with vaccinia recombinants expressing the external domain of rat *neu* oncogene prevented the formation of tumours by cells transformed by this gene (Bernards *et al.*, 1987). In another study, polyoma virus-induced tumours in rodents were prevented or cleared by immunisation with vaccinia viruses expressing polyoma large or middle T antigen (Lathe *et al.*, 1987). Similarly, malignant melanoma was prevented by immunisation with a vaccinia virus expressing melanoma-associated antigen p97 (Hu *et al.*, 1988; Estin *et al.*, 1988).

Although these experimental data indicate the great potential of vaccinia recombinants for immunoprophylaxis (or immunotherapy) this potential has not yet been realised. There are three specific problems which need to be overcome first. These are to increase antigen expression, to increase the immunogenicity of expressed antigens, and to attenuate the virus to improve vaccine safety.

Increase in gene expression

Higher levels of gene expression may be obtained by increasing the number of mRNA molecules available for translation or by increasing the efficiency of translation of those mRNAs. The mRNA levels reflect transcriptional efficiency and mRNA stability. Since very little is known about factors influencing mRNA stability in vaccinia virus infected cells, except in the case of the RNA produced by T7 RNA polymerase where a 5′ untranslated hairpin stabilises RNA (Fuerst & Moss, 1989), most attention has focused on increasing the levels of transcription by using stronger promoters. The genes encoding the major vaccinia structural proteins have been mapped and sequenced and their promoters utilised. However, even the late 11K and 4b promoters, which are among the strongest naturally occurring vaccinia promoters, have given levels of expression only approximately five-fold higher than obtained using the constitutive 7.5K promoter. This was encouraging but not the very large increase hoped for. Three other approaches are being tried. The first involves construction of synthetic promoters based upon information obtained from mutational studies or by random nucleotide alterations. The short length of vaccinia promoters makes this practicable. The second utilises a hybrid system using bacteriophage T7 RNA polymerase (Fuerst *et al.*, 1986, 1987; Fuerst & Moss, 1989). Two recombinant viruses are constructed and used for co-infections. One virus contains the T7 RNA polymerase gene constitutively expressed from the vaccinia 7.5K promoter, while the second virus contains a strong T7 promoter driving expression of the target gene. Coinfections produce very large amounts of mRNA but only a small fraction of this is capped and translated (Fuerst & Moss, 1989). Nonetheless, levels of protein expression are greater than from conventional vaccinia promoters. The third approach is to use promoters from other poxviruses. At least one member of each poxvirus genus produces cytoplasmic A type inclusion bodies composed predominantly of a single polypeptide. For cowpox virus the gene encoding this abundant late protein has been mapped and sequenced (Patel *et al.*, 1986; Patel & Pickup, 1987; Funahashi *et al.*, 1988). This promoter is reported to give higher levels of expression than from endogenous vaccinia promoters (Funahashi *et al.*, 1988).

Factors influencing the efficiency of vaccinia mRNA translation, other than the presence of a 5′ methylated cap structure (Muthukrishnan *et al.*, 1978) have not been rigorously investigated. From a comparison of early mRNA sequences around the AUG codon for translation initiation it is apparent that many of these sequences fit the consensus sequence for eukaryotic mRNAs (Kozak, 1983). However, in several early mRNAs the AUG and 5′ cap are extremely close, so that the Kozak consensus sequence becomes less relevant. The influence of 5′ poly (A) on late mRNA translation remains unknown.

Increase in immunogenicity

The immunogenicity of individual proteins varies considerably and it is not possible to predict which antigen will be the most immunogenic. None the less, factors which influence immunogenicity are being identified and include the location of an antigen within a cell, whether it is soluble or forms particles, the presence of T helper cell and T cytotoxic cell epitopes, the MHC haplotype of the host and the route of vaccine administration. In one study with a malarial antigen the immune response was greatly increased by engineering a naturally secreted molecule onto the cell surface (Langford *et al.*, 1986).

Increases in immunogenicity are also being sought by expression of lymphokines by recombinant viruses. Interleukin-2 expression resulted in virus attenuation, without dramatically increasing immune response (Ramshaw *et al.*, 1987; Flexner *et al.*, 1987). Bovine gamma-interferon expression has also been reported (Yilma *et al.*, 1987).

Virus attenuation

During the widespread use of vaccinia virus for smallpox immunisation, four types of complications of vaccination were recognised (Lane *et al.*, 1969). Progressive vaccinia was a grave condition caused by congenital immunodeficiency or suppression. Eczema vaccinatum occurred in patients with eczema or other skin disorders, often in contacts of vaccinees. Generalised vaccinia reflected a systemic infection and usually had a benign outcome. Most troublesome and least predictable were neurological complications of encephalopathy (in children less than 2 years old) and encephalitis (in older children and adults). These could be fatal and patients who recovered sometimes had neurological damage. Neither condition had an effective treatment. Cases of progressive vaccinia and eczema vaccination were minimised by screening out potential vaccinees possessing immunodeficiency or eczema, and treatment with immune globulin. Neurological complications were minimised by choice of safer vaccine strains, for instance, replacement of the Bern strain by the Lister strain resulted in a decline in the frequency of neurological complications. While the low frequency of these conditions was acceptable in the face of life-threatening smallpox, they were unacceptable in countries free from smallpox prior to global eradication and after eradication. Smallpox eradication was welcomed as the end of a greatly feared disease and the end of the need for vaccination with vaccinia. If vaccinia is to be reintroduced as a live vaccine in human medicine, safer attenuated strains are needed.

The attenuation of vaccinia virus is being considered in several ways. A recent approach is to express lymphokines in recombinant viruses. Interleukin-2 expression resulted in virus attenuation so that infections of immunosuppressed animals, which are fatal with wild type virus, were tolerated (Ramshaw *et al.*, 1987; Flexner *et al.*, 1987). It was proposed that these candidate vaccines might be used in immunosuppressed populations, such as those infected with HIV, so as to prevent generalised vaccinia infections that might otherwise occur in these vaccinees (Redfield *et al.*, 1987). Expression of gamma interferon is also being tested (Yilma *et al.*, 1987).

More conventional methods to obtain attenuated viruses, namely successive virus passage in tissue culture or in an unnatural host, have also been used. A strain produced by passage in CV-1 cells was shown to be safe in eczematous children (Kempe *et al.*, 1968) but the diminished immunogenicity prevented widespread use. A more encouraging strain, LC16m8, was derived from the Lister strain in Japan (Hashizume *et al.*, 1985). This virus has decreased neurovirulence, forms smaller plaques and possesses a genomic alteration near the right hand ITR (Sugimoto *et al.*, 1985). It has been used for primary smallpox vaccination in Japan since 1975 (>100,000 vaccinees) without serious complication and may provide a suitable parent strain for construction of recombinants (Morita *et al.*, 1987). Recombinants formed between the Lister parent and LC16m8 have intermediate virulence (Taakahashi-Nishamaki *et al.*, 1987). The basis of the attenuation needs determination, particularly with respect to the stability of the attenuated phenotype and the possibility of reversion. Attenuated derivatives of the laboratory vaccinia strain (WR) have been isolated which possess large deletions near the left ITR and in some cases contain mutations in structural proteins (Moss *et al.*, 1981*a*; Panicali *et al.*, 1981; Paez *et al.*, 1987; Rodriguez *et al.*, 1989). Although the WR strain is relatively neurovirulent and therefore unsuitable for use as a human vaccine, the analysis of these attenuated WR mutants is providing useful information about the genes influencing virulence.

An alternative approach to achieve stable attenuation is the deletion of specific genes. A growing number of genes are now being identified which influence virus pathogenicity in the multicellular host but which are non-essential for replication *in vitro*. These genes may be grouped according to the mode of action of their encoded proteins: 1) genes aiding virus replication within cells; 2) genes stimulating cell proliferation and thereby providing the virus with host cells of greater metabolic activity; 3) genes aiding virus dissemination; and 4) genes helping the virus to evade the host immune system. The vaccinia TK is an example of a virus protein aiding replication within quiescent cells by increasing the intracellular dTMP pool size. Deletion of this gene causes virus attenuation (Buller *et al.*, 1985). It is likely that deletion of the recently identified TmpK gene (Smith *et al.*, 1989*a*) may also result in virus attenuation since it performs the subsequent step in the same biochemical pathway of dTTP synthesis. These genes may be especially important in reducing infection of neural tissue since these cells have low metabolic activity and nucleotide pools.

The vaccinia growth factor (VGF) is a secreted protein which causes proliferation of surrounding cells (Buller *et al.*, 1988*b*) and is responsible for the dermal hyperplasia characteristic of poxvirus infections. Virus replication is enhanced in actively replicating cells and deletion of this gene, which is diploid in strain WR, results in virus attenuation (Buller *et al.*, 1988*a*). Virus spread is influenced primarily by the virion surface glycoproteins, especially of the envelope. So far the only gene encoding a glycoprotein of the extracellular virus envelope that is mapped is the HA (Shida, 1986). This gene is nonessential and its deletion causes attenuation (Shida *et al.*, 1987; Flexner *et al.*, 1987). Since naked intracellular virus is infectious, the other extracellular virion glycoproteins are non-essential for infectivity and it is probable that deletion of the encoding genes may diminish virus virulence.

The fourth group of genes influencing virus pathogenicity are those which enable the virus or virus-infected cells to evade the host's immune system. The presence of such genes is an indication of how highly adapted vaccinia virus, and presumably other orthopoxviruses, are to replication in the mammalian host. One gene encodes a secretory protein homologous to complement control proteins (Kotwal & Moss, 1988*a*). The protein probably interferes with activation of the complement cascade by the classical pathway and the availability of viable deletion mutants lacking this gene allows this to be tested experimentally. The vaccinia serpin genes also may aid immune evasion by virus-infected cells. If the serpins are responsible for the blockage in presentation of foreign antigens expressed by vaccinia recombinants to class I MHC-restricted CTL (Townsend *et al.*, 1988; Smith *et al.*, 1989*c*), then it is most probable that they may also 'hide' vaccinia antigens. These virus serpins may constitute a novel mechanism for virus immune evasion, and deletion of them may cause virus attenuation. Alternatively, the serpins may interfere in blood coagulation as has been shown for one of the cowpox serpins (Pickup *et al.*, 1986). It is noteworthy that one of the most virulent smallpox strains was highly haemorrhagic and death was often attributable to overwhelming haemorrhage of internal organs. It seems likely, in view of the genes already found, that there are other unidentified virus genes aiding immune evasion.

For widespread application of vaccinia recombinants as veterinary vaccines, the potential spread of vaccinia virus to humans must be prevented. This is particularly true for oral immunisation of wild animals (mostly foxes) against rabies by using virus-contaminated bait (Blancou *et al.*, 1986). The problem may now be overcome by using deletion mutants of vaccinia virus that are unable to replicate on human cells (Drillien *et al.*, 1981). Alternatively, if these mutants prove to have too greatly diminished immunogenicity, other naturally occurring poxviruses of wild animals, such as raccoonpox, may be used (Esposito *et al.*, 1988).

The relative degrees of attenuation obtained by deletion of the genes outlined above remains to be determined. In these studies it will also be important to evaluate the immunogenicity of the deletion mutants. The desired vaccine is one which combines safety, by attenuation, with efficiency, by immunogenicity. Achieving the right balance is a challenging problem. However, the discovery of genes such as the serpins provides a basis for optimism. Deletion of such genes may cause virus attenuation, because infected cells are more efficiently recognised and lysed by CTL, but may simultaneously enhance immunogenicity since the antigens may be better presented to the immune system.

SUMMARY

Vaccinia virus is the oldest human vaccine and the only vaccine to cause eradication of a disease. It has played a pioneering role in the development of animal virology and the study of virus-coded enzymes, and continues to provide new information about virus interactions with the host at the cellular and multicellular levels. Genetically modified vaccinia viruses are valuable research tools and have considerable potential as new live vaccines. Their development has ensured that interest in this virus will continue.

ACKNOWLEDGEMENTS

I thank Valerie Boasten for typing the manuscript and the Medical Research Council and the Wellcome Trust for support. GLS is a Lister Institute Jenner Research Fellow.

REFERENCES

Ada, G. L., Brown, F. & Schild, G. C. (1985). Recombinant vaccinia viruses as live virus vectors for vaccine antigens: Memorandum for a WHO/USPHS/NIBSC meeting. *Bulletin of the World Health Organisation,* **63**, 471.

Ahn, B. Y. & Moss, B. (1989), Capped poly (A) leaders of variable lengths at the 5′ ends of vaccinia virus late mRNAs. *Journal of Virology,* **63**, 226–32.

Andrew, M. E., Couper, B. E. H., Ada, G. L. & Boyle, D. B. (1986). Cell-mediated immune responses to influenza virus antigens expressed by vaccinia virus recombinants. *Microbiological Pathogenesis,* **1**, 443–52.

U.C.W. ABERYSTWYTH LIBRARY

Andrew, M. E. Coupar, B. E. H., Boyle, D. B. & Ada, G. L. (1987*a*). The roles of influenza virus haemagglutinin and nucleoprotein in protection: analysis using vaccinia virus recombinants. *Scandinavian Journal of Immunology,* **25**, 21–8.

Andrew, M., Boyle D., Coupar, B. E. H., Whitfield, P. L., Both, G. W. & Bellamy, A. R. (1987*b*). Vaccinia virus recombinants expressing the SA11 rotavirus VP7 glycoprotein gene induce serospecific neutralising antibodies. *Journal of Virology* **61**, 1054–60.

Archard, L. C., Mackett, M., Barnes, D. E. & Dumbell, K. R. (1984). The genome structure of cowpox virus white pock variants. *Journal of General Virology,* **65**, 875–86.

Arzoglou, P., Drillien, R. & Kirn, A. (1978). Evidence for an alkaline protease in vaccinia virus. *Virology*, **95**, 211–14.

Bajszar, G., Wittek, R., Weir, J. P. & Moss, B. (1983). Vaccinia virus thymidine kinase and neighboring genes: mRNAs and polypeptides of wild-type virus and putative nonsense mutants. *Journal of Virology*, **45**, 62–72.

Baldick, C. J. & Moss, B. (1987). Resistance of vaccinia virus to rifampicin conferred by a single nucleotide substitution near the predicted NH_2 terminus of a gene encoding an M_r, 62,000 polypeptide. *Virology*, **156**, 138–45.

Ball, L. A. (1987). High frequency homologous recombination in vaccinia virus DNA. *Journal of Virology,* **61**, 1788–95.

Bangham, C. R. M., Openshaw, P. J. M., Ball, L. A., King. A. M. Q., Wertz, G. W. & Askonas, B. A. (1986), Human and murine cytotoxic T cells specific to respiratory syncyntial virus recognise the viral nucleoprotein (N) but not the major glycoprotein (G) expressed by vaccinia virus recombinants. *Journal of Immunology,* **137**, 3973–7.

Barbosa, E. & Moss, B. (1978*a*). mRNA (nucleoside-2′-)-methyltransferase from vaccinia virus. Purification and physical properties. *Journal of Biological Chemistry,* **253**, 7692–7.

Barbosa, E. and Moss, B. (1978*b*). mRNA (nucleoside-2′-)-methyltransferase from vaccinia virus. Characteristics and substrate specificity. *Journal of Biological Chemistry,* **253**, 7698–702.

Baroudy, B.M. and Moss, B. (1980). Purification and characterisation of a DNA-dependent RNA polymerase from vaccinia virions. *Journal of Biological Chemistry,* **255**, 4372–80.

Baroudy, B. M., Venkatesan, S. & Moss, B. (1982). Incompletely base-paired flip-flop terminal loops link the two DNA strands of the vaccinia virus genome into one uninterrupted polynucleotide chain. *Cell*, **28**, 315–24.

Baroudy, B. M. Venkatesan, S. & Moss, B. (1983), Structure and replication of vaccinia virus telomeres. *Cold Spring Harbor Symposium on Quantitative Biology,* **47,** 723–9.

Barrett, T., Belsham, G. J., Subbarao, S. M. & Evans, S. A. (1989). Immunization with a vaccinia virus recombinant expressing the F protein protects rabbits from challenge with a lethal dose of rinderpest virus. *Virology,* **170,** 11–18.

Bauer, W. R., Ressner, E.C., Kates, J. & Patzke, J. U. (1977), A DNA nicking-closing enzyme encapsidated in vaccinia virus: Partial purification and properties. *Proceedings of the National Academy of Sciences, USA,* **74,** 1841–5.

Baxby, D. (1981), *Jenner's Smallpox Vaccine. The Riddle of the Origin of Vaccinia Virus.* Heinemann, London.

Belsham, G. J., Anderson, E. C., Murray, P. K., Anderson, J. & Barrett, T. (1989). Immune response and protection of cattle and pigs generated by a vaccinia virus recombinant expressing the F protein of rinderpest virus. *The Veterinary Record,* **124**, 655–8.

Bennink, J. R., Yewdell, J. W., Smith, G. L., Moller, C. & Moss, B. (1984). Recombinant vaccinia virus primes and stimulates influenza virus haemagglutinin-specific cytotoxic T lymphocytes. *Nature, London,* **311,** 578–9.
Bennink, J. R., Yewdell, J. W., Smith, G. L. & Moss, B. (1986). Recognition of cloned influenza virus hemagglutinin gene products by cytotoxic T lymphocytes. *Journal of Virology,* **57**, 786–91.
Bennink, J. R., Yewdell, J. W., Smith, G. L. & Moss, B. (1987). Anti-influenza virus cytotoxic lymphocytes recognise the three viral polymerases and a nonstructural protein: Responsiveness to individual viral antigens is major histocompatibility complex controlled. *Journal of Virology*, **61**, 1098–102.
Bernards, R., Destree, A., McKenzie, S., Gordon, E., Weinberg, R.A. & Panicali, D. (1987). Effective tumour immunotherapy directed against an oncogene encoded product using a vaccinia vector. *Proceedings of the National Academy of Sciences, USA,* **84**, 6854–8.
Bertholet, C., Drillien, R. & Wittek, R. (1985). One hundred base pairs of 5′ flanking sequence of a vaccinia virus late gene are sufficient to temporally regulate late transcription. *Proceedings of the National Academy of Sciences, USA,* **82**, 2096–100.
Bertholet, C., Stocco, P., Van Meir, E. & Wittek, R. (1986). Functional analysis of the 5′ flanking sequence of a vaccinia virus late gene. The *EMBO Journal,* **5,** 1951–7.
Bertholet, C., Van Meir, E., ten Heggeler-Bordier, B. & Wittek, R. (1987). Vaccinia virus produces late mRNAs by discontinuous synthesis. *Cell,* **50,** 153–62.
Blancou, J., Kieny, M. P., Lathe, R., Lecocq, J. P., Pastoret, P. P., Soulebot, J.P. & Desmettre, P. (1986). Oral vaccination of the fox against rabies using a live recombinant vaccinia virus. *Nature, London*, **322,** 373–5.
Boone, R. F. & Moss, B. (1977). Methylated 5′ terminal sequences of vaccinia virus mRNA species made *in vivo* at early and late times after infection. *Virology,* **79**, 67–80.
Borysiewicz, L. K., Hickling, J. K., Graham, S., Sinclair, J., Cranage, M. P., Smith, G. L. & Sissons, J. G. P. (1988). Human cytomegalovirus specific cytotoxic T cells – relative frequency of stage specific CTL recognising the 72 kDa immediate early protein and glycoprotein B expressed by recombinant vaccinia viruses. *Journal of Experimental Medicine,* **168**, 919–31.
Boulter, E. A. & Appleyard, G. (1973). Differences between extracellular and intracellular forms of poxvirus and their implications. *Progress in Medical Virology,* **16**, 86–108.
Boursnell, M. E. G., Foulds, I. J., Campbell, J. I. & Binns, M. M. (1988). Non-essential genes in the vaccinia virus Hind III K fragment: a gene related to serine protease inhibitors and a gene related to the 37K vaccinia virus envelope antigen. *Journal of General Virology,* **69**, 2995–3003.
Boyle, D. B. & Coupar, B. E. H. (1986). Identification and cloning of the fowlpox virus thymidine kinase gene using vaccinia virus. *Journal of General Virology,* **67**, 1591–600.
Boyle, D. B. & Coupar, B. E. H. (1988). A dominant selectable marker for the construction of recombinant poxviruses. *Gene,* **65,** 123–8.
Boyle, D. B., Coupar, B. E. H. and Both, G. W. (1985). Multiple-cloning-site plasmids for the rapid construction of recombinant poxviruses. *Gene,* **35,** 169–77.
Bray, M., Zhao, B., Markoff, L., Eckels, K. H., Chanock, R. M & Lai. C.-J. (1989). Mice immunized with recombinant vaccinia virus expressing dengue 4 virus structural proteins with or without nonstructural protein NS1 are protected against fatal dengue virus encephalitis. *Journal of Virology,* **63**, 2853–6.

Brochier, B. M., Blancou, J., Aubert, M. F. A., Kieny, M. P., Desmettre, P. & Pastoret, P.-P. (1989). Interaction between rabies infection and oral administration of vaccinia-rabies recombinant viruses to foxes *(Vulpes vulpes)*. *Journal of General Virology*, **70**, 1601–4.

Browne, H. M., Churcher, M. J., Stanley, M. A., Smith, G. L. & Minson, A. C. (1988). Analysis of the L1 gene product of human papillomavirus type 16 by expression in a vaccinia virus recombinant. *Journal of General Virology*, **69**, 1263–73.

Broyles, S. S. & Moss, B. (1986). Homology between RNA polymerases of poxviruses, prokaryotes and eukaryotes: Nucleotide sequence and transcriptional analysis of vaccinia virus genes encoding 147-kDa and 22-kDa subunits. *Proceedings of the National Academy of Sciences, USA*, **83**, 3141–5.

Broyles, S. S. & Moss, B. (1987). Identification of the vaccinia virus gene encoding nucleoside triphosphate phosphohydrolase, a DNA-dependent ATPase. *Journal of Virology*, **61**, 1738–42.

Broyles, S. S. & Moss, B. (1988). DNA-dependent ATPase activity associated with vaccinia virus early transcription factor. *Journal of Biological Chemistry*, **263**, 10761–5.

Broyles, S. S., Yuen, L., Shuman, S. & Moss, B. (1988). Purification of a factor required for transcription of vaccinia virus early genes. *Journal of Biological Chemistry*, **263**, 10754–60.

Buller, R. M. L., Smith, G. L., Cremer, K., Notkins, A. L. & Moss, B. (1985). Decreased virulence of recombinant vaccinia virus expression vectors is associated with a thymidine kinase-negative phenotype. *Nature, London*, **317**, 813–15.

Buller, R. M. L., Chakrabarti, S., Cooper, J. A., Twardzik, D.R. & Moss, B. (1988*a*). Deletion of the vaccinia virus growth factor gene reduces virulence. *Journal of Virology*, **62**, 866–74.

Buller, R. M. L., Chakrabarti, S., Moss, B. & Fredrickson, T. (1988*b*). Cell proliferative response to vaccinia virus is mediated by VGF *in vitro* and *in vivo*. *Virology*, **164**, 182–92.

Burke, K. L., Dunn, G., Ferguson, M., Minor, P. D. & Almond, J. W. (1988). Antigen chimaeras of poliovirus as potential new vaccines. *Nature, London*, **332**, 81–2.

Cantin, E. M., Eberle, E., Baldick, C. J., Moss, B., Willey, D. E. Notkins, A. L. & Openshaw, H. (1987). Expression of herpes simplex virus 1 glycoprotein B by a recombinant vaccinia virus and protection of mice against lethal herpes simplex virus 1 infection. *Proceedings of the National Academy of Sciences, USA*, **84**, 5908–12.

Chakrabarti, S., Brechling, K. & Moss, B. (1985). Vaccinia virus expression vector: Co-expression of β-galactosidase gene provides visual screening of recombinant virus plaques. *Molecular and Cellular Biology*, **5**, 3403–9.

Chakrabarti, S., Robert-Guroff, M., Wong-Staal, F., Gallo, R.C. & Moss, B. (1986). Expression of HTLV-III envelope gene by recombinant vaccinia virus. *Nature, London*, **320**, 535–7.

Challberg, M. D. & Englund, P. T. (1979). Purification and properties of the deoxyribonucleic acid polymerase induced by vaccinia virus. *Journal of Biological Chemistry*, **254**, 7812–19.

Chambers, T. M., Kawaoka, Y. & Webster, R. G. (1988). Protection of chickens from lethal influenza infection by a vaccinia-expressed haemagglutinin. *Virology* **167**, 414–21.

Cheng, K.-C. & Moss, B. (1987). Selective synthesis and secretion of particles composed of the hepatitis B virus middle surface protein directed by a recombinant vaccinia virus: induction of antibodies to pre-S and S epitopes. *Journal* of Virology, **61**, 1286–90.

Cheng, K.-C., Smith, G. L. & Moss, B. (1986). Hepatitis B virus large surface protein is not secreted but is immunogenic when selectively expressed by recombinant vaccinia virus. *Journal of Virology,* **60**, 337–44.

Clegg, J. C. S. & Lloyd, G. (1987). Vaccinia recombinant expressing Lassa-virus internal nucleocapsid protein protects guinea pigs against Lassa fever. *Lancet,* **I**:186–7.

Cochran, M. A., Puckett, C. & Moss, B. (1985*a*). *In vitro* mutagenesis of the promoter region of a vaccinia virus gene: Evidence for tandem early and late regulatory signals. *Journal of Virology,* **54**, 30–7.

Cochran, M. A., Mackett, M. & Moss, B. (1985 *b*). Eukaryotic transient expression system dependent on transcription factors and regulatory DNA sequences of vaccinia virus. *Proceedings of the National Academy of Sciences, USA,* **82**, 19–23.

Collier, L.H. (1955). The development of a stable smallpox vaccine. *Journal of Hygiene,* **53**, 76–101.

Condit, R.C., Motyczka, A. & Spizz, G. (1983). Isolation, characterization and physical mapping of temperature-sensitive mutants of vaccinia virus. *Virology,* **128**, 429–43.

Coupar, B. E. H., Andrew, M. E., Both, G. W. & Boyle, D. B. (1986). Temporal regulation of influenza hemagglutinin expression in vaccinia. *European Journal of Immunology,* **16**, 1479–87.

Coupar, B. E. H., Boyle, D. B. & Both, G. W. (1987). Effect of *in vitro* mutations in a vaccinia virus early promoter region monitored by herpes simplex virus thymidine kinase expression in recombinant vaccinia virus. *Journal of General Virology,* **68,** 2299–309.

Cranage, M. P., Kouzarides, T., Bankier, A. T., Satchwell, S., Weston, K., Tomlinson, P., Barrell, B., Hart, H., Bell, S. E. Minson, A. C. & Smith, G. L. (1986). Identification of the human cytomegalovirus glycoprotein B gene and induction of neutralizing antibodies via its expression in recombinant vaccinia virus. *The EMBO Journal,* **5**, 3057–63.

Cranage, M. P., Smith, G. L., Bell, S. E., Hart, H., Brown, C., Bankier, A. T., Tomlinson, P., Barrell, B. G. & Minson, A. C. (1988). Identification and expression of a human cytomegalovirus glycoprotein with homology to the Epstein–Barr virus BXLF2 product, varicella-zoster virus gpIII, and herpes simplex virus glycoprotein H. *Journal of Virology,* **62**, 1416–22.

Cremer, K., Mackett, M. Wohlenberg, C., Notkins, A. L. & Moss B. (1985). Vaccinia virus recombinants expressing herpes simplex type 1 glycoprotein D prevents latent herpes in mice. *Science,* **228**, 737–40.

Dales, S. and Pogo, B. T. (1981) The poxviruses. *Virology Monographs,* **18**, 1–109.

DeLange, A. M. (1989). Identification of temperature sensitive mutants of vaccinia virus that are defective in conversion of concatemeric replicative intermediates to the mature linear genome. *Journal of Virology,* **63**, 2437–44.

DeLange, A. M., Reddy, M., Scraba, D., Upton, C. and McFadden, G. (1986). Replication and resolution of cloned poxvirus telomeres *in vivo* generates linear microchromosomes with intact viral hairpin termini. *Journal of Virology,* **59,** 249–59.

de la Salle, H., Altenburger, W., Elkiam, R., Dott, K., Dieterk, A., Drillien, R., Cuzenave, J. P., Tolstoshev, P. & Lecocq, J.P. (1985). Active gamma-carboxylated human factor is expressed using recombinant DNA techniques. *Nature, London,* **316**, 268–70.

DeMagistris, L. & Stunnenberg, H.G. (1988). Cis-acting sequences affecting the length of the poly (A) head of vaccinia virus late transcripts. *Nucleic Acids*

Research, **16**, 3141–56.

Deubel, V., Kinney, R. M., Esposito, J. J., Cropp, C. B., Vorndam, A. V., Monath, T. P. & Trent, D. W. (1988). Dengue 2 virus envelope protein expressed by a recombinant vaccinia virus fails to protect monkeys against dengue. *Journal of General Virology*, **69**, 1921–9.

Downie, A. W. (1939). Study of the lesions produced experimentally by cowpox virus. *Journal of Pathological Bacteriology*, **48**, 361–79.

Drillien, R. & Spehner, D. (1983), Physical mapping of vaccinia virus temperature-sensitive mutations. *Virology*, **131**, 385–93.

Drillien, R., Koehren, F. & Kirn, A. (1981). Host range deletion mutant of vaccinia virus defective in human cells. *Virology*, **111**, 488–99.

Drillien, R., Spehner, D., Kirn, A., Giraudon, P., Buckland, R., Wild, F. & Lecocq, J.-P. (1988). Protection of mice from fatal measles encephalitis by vaccination with vaccinia virus recombinants encoding either the haemagglutinin or the fusion protein. *Proceedings of the National Academy of Sciences, USA*, **85**, 1252–6.

Dubbs, D. R. & Kit, S. (1964). Isolation and properties of vaccinia mutants deficient in thymidine kinase-inducing activity. *Virology*, **22**, 214–25.

Earl, P. L., Jones, E. V. & Moss, B. (1986*a*). Homology between DNA polymerases of poxviruses, herpesviruses and adenoviruses: Nucleotide sequence of the vaccinia virus DNA polymerase gene. *Proceedings of the National Academy of Sciences, USA*, **83**, 3659–63.

Earl, P. L., Moss, B., Morris, R. P., Wehrl, Y. K., Nihsio, J. & Chesebro, B. (1986*b*). T-lymphocyte priming and protection against Friend leukaemia by vaccinia-retrovirus env gene recombinant. *Science*, **234**, 728–31.

Edwards, R. H., Selby, M. J., Mobley, W. C., Weinrich, S. L., Hruby, D. E. & Rutter, W. J. (1988). Processing and secretion of nerve growth factor: expression in mammalian cells with a vaccinia virus vector. *Molecular and Cellular Biology*, **8**, 2456–64.

Elango, N., Prince, G. A., Murphy, B. R., Venkatesan, S., Chanock, R. M. & Moss, B. (1986). Resistance to human respiratory syncytial virus (RSV) infection induced by immunization of cotton rats with a recombinant vaccinia virus expressing RSV G glycoprotein. *Proceedings of the National Academy of Sciences, USA* **83**, 1906–10.

Esposito, J. J. & Knight, J.C. (1985). Orthopoxvirus DNA: A comparison of restriction profiles and maps. *Virology*, **143**, 230–51.

Esposito, J., Brechling, K., Baer, G. & Moss, B. (1987). Vaccinia virus recombinants expressing rabies virus glycoprotein protects against rabies. *Virus Genes*, **1**, 7–21.

Esposito, J. J., Knight, J. C., Shaddock, J. H., Novembre, F. J. & Baer, G. M. (1988). Successful oral rabies vaccination of racoons with raccoon poxvirus recombinants expressing rabies virus glycoprotein. *Virology*, **165**, 313–16.

Esteban, M. and Holowczak, J. A. (1977*a*). Replication of vaccinia DNA in mouse L cells. III. Intracellular forms of viral DNA. *Virology*, **82**, 308.

Esteban, M. & Holowczak, J. A. (1977*b*). Replication of vaccinia DNA in mouse L cells. I. *In vivo* DNA synthesis. *Virology*, **78**, 57–75.

Esteban, M., Flores, L. & Holowczak, J. A. (1977). Model for vaccinia virus DNA replication. *Virology*, **83**, 467–73.

Estin, C. D., Stevenson, U. S., Plowman, G. D., Hu, S.-K., Sridhar, P., Hellstrom, I., Brown, J. P. & Hellstrom, K. E. (1988). Recombinant vaccinia virus against the human melanoma antigen p97 for use in immunotherapy. *Proceedings of the National Accademy of Sciences, USA*, **85**, 1052–6.

Evans, E. & Traktman, P. (1987). Molecular genetic analysis of a vaccinia virus

gene with an essential role in DNA replication. *Journal of Virology*, **61**, 3152–62.
Falkner, F. G. & Moss, B. (1988). *Escherichia coli* gpt gene provides dominant selection for vaccinia virus open reading frame vectors. *Journal of Virology*, **62**, 1849–54.
Falkner, F. G., Fuerst, T. R. & Moss, B. (1988). Use of vaccinia virus vectors to study the synthesis, intracellular localisation and action of human immunodeficiency virus trans-activator protein. *Virology*, **164**, 450–7.
Fathi, Z., Sridhar, P., Facha, R. F. & Condit, R. C. (1986). Efficient targeted insertion of an unselected marker into the vaccinia virus genome. *Virology*, **155**, 97–105.
Fenner, F. & Sambrook, J. F. (1966). Conditional lethal mutants of rabbitpox virus. II. Mutants (*p*) that fail to multiply in PA-2a cells. *Virology*, **28**, 600–9.
Fischetti, V. A., Hodges, W. M. & Hruby, D. E. (1989). Protection against streptococcal pharyngeal colonization with a vaccinia: M protein recombinant. *Science*, **244**, 1487–90.
Flexner, C., Hügin, A. & Moss, B. (1987). Prevention of vaccinia virus infection in immuno-deficient mice by vector-directed IL-2 expression. *Nature, London*, **330**, 259–61.
Flexner, C., Murphy, B. R., Rooney, J. F., Wohlenberg, C., Yuferov, V., Notkins, A. L. & Moss, B. (1988). Successful vaccination with a polyvalent live vector despite existing immunity to an expressed antigen. *Nature, London*, **335**, 259–62.
Foglesong, P. D. (1985). *In vitro* transcription of a cloned vaccinia virus gene by a soluble extract prepared from vaccinia virus-infected HeLa cells. *Journal of Virology*, **53**, 822–6.
Franke, C. A., Rice, C. M., Strauss, J. H. & Hruby, D. E. (1985). Neomycin resistance as a dominant selectable marker for selection and isolation of vaccinia virus recombinants. *Molecular and Cellular Biology*, **5**, 1918–24.
Fuerst, T. R. & Moss, B. (1989). Structure and stability of mRNA synthesised by vaccinia virus-encoded bacteriophage T7 RNA polymerase in mammalian cells. Importance of 5′ untranslated leader. *Journal of Molecular Biology*, **206**, 333–48.
Fuerst, T. R., Earl, P. L. & Moss, B. (1987). Use of a hybrid vaccinia virus-T7 RNA polymerase system for expression of target genes. *Molecular and Cellular Biology*, **7**, 2538–44.
Fuerst, T. R., Niles, E. G., Studier, F. W. & Moss, B. (1986). Eukaryotic transient-expression system based upon recombinant vaccinia virus that synthesizes bacteriophage T7 RNA polymerase. *Proceedings of the National Academy of Sciences, USA*, **83**, 8122–6.
Fuerst, T. R., Fernandez, M. P. & Moss, B. (1989). Transfer of the inducible *lac* repressor/operator system from *Escherichia coli* to a vaccinia virus expression vector. *Proceedings of the National Academy of Sciences, USA*, **86**, 2549–53.
Funahashi, S., Sato, T. & Shida, H. (1988). Cloning and characterization of the gene encoding the major protein of the A-type inclusion body of cowpox virus. *Journal of General Virology*, **69**, 35–47.
Geshelin, P. & Berns, K. I. (1974). Characterization and localization of the naturally occurring crosslinks in vaccinia virus DNA. *Journal of Molecular Biology*, **88**, 785–96.
Gilbert, J. H., Pedersen, N. C. & Nunberg, J. H. (1987). Feline leukemia virus envelope protein expression encoded by a recombinant vaccinia virus: apparent lack of immunogenicity in vaccinated animals. *Virus Research*, **7**, 49–67.
Gillard, S., Spehner, D. & Drillien, R. (1985). Mapping of a vaccinia host range sequence by insertion into the viral thymidine kinase gene. *Journal of Virology*, **53**, 316–8.

Gillard, S., Spehner, D., Drillien, R. & Kirn, A. (1986). Localization and sequence of a vaccinia virus gene required for multiplication in human cells. *Proceedings of the National Academy of Sciences, USA,* **83**, 5573–7.

Gold, P. and Dales, S. (1968). Localization of nucleotide phosphohydrolase within vaccinia virus. *Proceedings of the National Academy of Sciences, USA,* **60**, 845–52.

Golini, F. & Kates, J. R. (1985). A soluble transcription system derived from purified vaccinia virus. *Journal of Virology,* **53**, 205–13.

Gompels, U. A. and Minson, A. C. (1989). Antigenic properties and cellular localization of herpes simplex virus glycoprotein H synthesized in a mammalian cell expression system. *Journal of Virology,* **63** 4744–55.

Gotch, F., McMichael, A., Smith, G. & Moss, B. (1987). Identification of viral molecules recognized by influenza-specific human cytotoxic T lymphocytes. *Journal of Experimental Medicine,* **165**, 408–16.

Gould, K., Cossins, J., Bastin, J., Brownlee, G. G. & Townsend, A. R. M. (1989). A fifteen amino acid fragment of influenza nucleoprotein synthesised in the cytoplasm is presented to class 1 restricted cytotoxic T lymphocytes. *Journal of Experimental Medicine*, **170**, 1051–5.

Hänggi, M., Bannwarth, W. & Stunnenberg, H. G. (1986). Conserved TAAAT motif in vacinia virus late promoters: Overlapping TATA box and site of transcription initiation. *The EMBO Journal,* **5**, 1071–6.

Hashizume, S., Yoshizawa, H., Morita, M. & Suzuki, K. (1985). Properties of attenuated mutant of vaccinia virus LC16m8, derived from Lister strain. In *Vaccinia Viruses as Vectors for Vaccine Antigens,* G. V. Quinnan, ed. pp 87–100. Elsevier, New York.

Hiller, G. & Weber, K. (1985). Golgi-derived membranes that contain an acylated viral polypeptide are used for vaccinia virus envelopment. *Journal of Virology,* **55**, 651–9.

Hirt, P., Hiller, G. & Wittek, R. (1986). Localization and fine structure of a vaccinia virus gene encoding an envelope antigen. *Journal of Virology,* **58**, 757–64.

Howard, S. T. & Smith, G. L. (1989). Two early vaccinia virus genes encode polypeptides related to protein kinases. *Journal of General Virology,* **70**, 3187–20

Hruby, D. E. and Ball, L. A. (1982). Mapping and identification of the vaccinia thymidine kinase gene. *Journal of Virology,* **43**, 403–9.

Hruby, D. E., Maki, R. A., Miller, D. B. & Ball, L. A. (1983). Fine structure analysis and nucleotide sequence of the vaccinia virus thymidine kinase gene. *Proceedings of the National Academy of Sciences. USA,* **80**, 3411–15.

Hu, S.-K., Kosowski, S. G. & Dalrymple, J. M. (1986). Expression of AIDS virus envelope gene in recombinant vaccinia viruses. *Nature, London,* **320**, 537–40.

Hu, S.-L., Fultz, P. N., McClure, H. M., Eichberg, J. W., Thomas, E. K., Zarling, J., Singhal, M. C., Kosowski, S. G., Swenson, R. B., Anderson, D. C. & Todaro, G. (1987). Effect of immunization with a vaccinia-HIV env recombinant on HIV infection of chimpanzees. *Nature, London,* **328**, 721–3.

Hu, S.-L., Plowman, G. D., Sridhar, P., Stevenson, U. S., Brown, J. P. & Estin, C. D. (1988). Characterization of a recombinant vaccinia virus expressing human melanoma-associated antigen p 97. *Journal of Virology,* **62**, 176–80.

Jakeman, K. J., Smith, H. & Sweet, C. (1989). Mechanism of immunity to influenza: material and passive neonatal protection following immunization of adult ferrets with a live vaccinia-influenza virus haemagglutinin recombinant but not with recombinants containing other influenza virus proteins. *Journal of General Virology,* **70**, 1523–31.

Jenner, E. (1798). *An Inquiry into the Cause and Effects of the Variolae Vaccinae, a Disease Discovered in Some of the Western Counties of England, particularly*

Gloucestershire, and Known by the Name of Cowpox. London. (Reprinted in C.N.B. Camac, ed., *Classics of Medicine and Surgery*. p. 213, Dover, New York.).

Joklik, W. K. (1962). The purification of four strains of poxvirus. *Virology,* **18**, 9–18.

Jones, E. V. & Moss, B. (1984). Mapping of the vaccinia virus DNA polymerase gene by marker rescue and cell-free translation of selected RNA. *Journal of Virology,* **49**, 72–77.

Jones, E. V. & Moss, B. (1985). Transcriptional mapping of the vaccinia virus DNA polymerase gene. *Journal of Virology,* **53,** 312–15.

Jones, L., Ristow, S., Yilma, T. & Moss, B. (1986). Accidental human vaccination with vaccinia virus expressing nucleoprotein gene. *Nature, London* **319**, 545.

Jones, E. V., Puckett, C. and Moss, B. (1987), DNA-dependent RNA polymerase subunits encoded within the vaccinia virus genome. *Journal of Virology,* **61**, 1765–71.

Jonjiic, S., del Val, M., Keil, G. M., Reddehase, M. J. & Koszinowski, U. H. (1988). A nonstructural viral protein expressed by a recombinant vaccinia virus protects against lethal cytomegalovirus infection. *Journal of Virology* **62**, 1653–8.

Kassavetis, G. A., Zentner, P. G. & Geiduschek, E. P. (1986). Transcription at bacteriophage T4 variant late promoters. An application of a newly devised promoter-mapping method involving RNA chain retraction. *Journal of Biological Chemistry,* **261**, 14256–65.

Katach, L. T., Wasmoen, T. L. & Collett, M. S. (1988). Rift valley fever virus M segment: use of recombinant vaccinia viruses to study phlebovirus gene expression. *Journal of Virology,* **62**, 826–33.

Kates, J. R. & Beeson, J. (1970). Ribonucleic acid synthesis in vaccinia virus. II. Synthesis of polyriboadenylic acid. *Journal of Molecular Biology,* **50**, 19–23.

Kates, J. R. & McAuslan, B.R. (1967). Poxvirus DNA-dependent RNA polymerase. *Proceedings of the National Academy of Sciences, USA* **58,** 134–41.

Kempe, C. H., Fulginiti, V., Minimitani, M. & Shienfield, H. (1968). Smallpox vaccination of eczema patients with a strain of attenuated live vaccinia (CV1–78). *Pediatrics,* **42**, 980–5.

Kent, R. K. (1988). Isolation and analysis of the vaccinia virus p4B gene promoter. Ph.D. Thesis, University of Cambridge.

Kerr, S. M. & Smith, G. L. (1989). Vaccinia virus encodes a polypeptide with DNA ligase activity. *Nucleic Acids Research*,**17**, 9039–50.

Kieny, M. P., Lathe, R., Drillien, R., Spehner, D., Skory, S., Schmitt, D., Wiktor, T., Koprowski, H. & Lecocq, J.P. (1984). Expression of rabies virus glycoprotein from a recombinant vaccinia virus. *Nature, London,* **312**, 163–6.

King, A. M. Q., Stott, E. J., Langer, S. J., Young, K. K.-Y., Ball, L. A. & Wertz, G. W. (1987). Recombinant vaccinia viruses carrying the N gene of human respiratory syncytial virus: studies of gene expression in cell culture and immune response in mice. *Journal of Virology,* **61**, 2885–90.

Kinney, R. M., Esposito, J. J. Johnson, B. J. B., Roehrig, J. T., Mathews, J. H., Barret, A. D. T. & Trent, D. W. (1988). Recombinant vaccinia/venezuelan equine encephalitis (VEE) virus expresses VEE structural proteins. *Journal of General Virology,* **69**, 3005–13.

Kleiman, J. H. & Moss, B. (1975*a*). Purification of a protein kinase and two phosphate acceptor proteins from vaccinia virions. *Journal of Biological Chemistry,* **250**, 2420–9.

Kleiman, J. H. & Moss, B. (1975*b*). Characterization of a protein kinase and two phosphate acceptor proteins from vaccinia virions. *Journal of Biological Chemistry,* **250**, 2430–7.

Kost, T. A., Jones, E. V., Smith, K. M., Reed, A. P., Brown, A. L. & Miller, T. J. (1989). Biological evaluation of two distinct mRNAs within the BamHI fragment 7 of pseudorabies virus: expression of the coding regions by vaccinia virus. *Virology,* **171**, 365–76.

Kotwal, G. J. & Moss, B.(1988*a*). Vaccinia encodes a secretory polypeptide structurally related to complement control proteins. *Nature, London,* **335**, 176–8.

Kotwal, G. J. & Moss, B. (1988*b*). Analysis of a large cluster of non-essential genes deleted from a vaccinia virus terminal transposition mutant. *Virology,* **167**, 524–37.

Kotwal, G. J. & Moss, B. (1989). Vaccinia virus encodes two proteins that are structurally related to members of the plasma serine protease inhibitor superfamily. *Journal of Virology,* **63**, 600–6.

Kotwal, G. J., Hügin, W. H. & Moss, B. (1989). Mapping and insertional mutagenesis of a vaccinia virus gene encoding a 13.8 kD secreted protein. *Virology,* **171**, 579–87.

Kozak, M. (1983). Comparison of initiation of protein synthesis in prokaryotes, eukaryotes and organelles. *Microbiological Reviews,* **47**, 1–45.

Lake, J. R. & Cooper, P. D. (1980). Deletions of the terminal sequences in the genomes of the white pock (*u*) and host-restricted (*p*) mutants of rabbitpox virus. *Journal of General Virology,* **48,** 135–47.

Lakritz, N., Foglesong, P. D., Reddy, M., Baum, S., Hurwitz, J. & Bauer, W. R. (1985). A vaccinia virus DNase preparation which cross-links superhelical DNA. *Journal of Virology,* **53**, 935–43.

Lane, J. M., Ruben, F. L., Neff, J. M. & Millar, J. D. (1969). Complications of smallpox vaccination, 1968. National surveillance in the United States. *New England Journal of Medicine,* **281**, 1201–8.

Langford, C. J., Edwards, S. J., Smith, G. L., Mitchell, G. F., Moss, B., Kemp, D. J. & Anders, R. F. (1986). Anchoring a secreted *Plasmodium* antigen on the surface of recombinant vaccinia virus-infected cells increases its immunogenicity. *Molecular and Cellular Biology,* **6**, 3191–9.

Lathe, R., Kieny, M. P., Gerlinger, P., Clertant, P., Guizani, I., Cuzin, R. & Chambon, P. (1987). Tumour prevention and rejection with recombinant vaccinia virus. *Nature, London,* **326**, 878–80.

Li, Y., Luo, L., Snyder, R. M. & Wagner, R.W. (1988). Expression of the M gene of vesicular stomatitis virus cloned in various vaccinia virus vectors. *Journal of Virology,* **62**, 776–82.

McDermott, M. R., Graham, F. L., Hanke, T. & Johnson, D. C. (1989). Protection of mice against lethal challenge with herpes simplex virus by vaccination with an adenovirus vector expressing HSV glycoprotein B. *Virology,* **169**, 244–7.

Mackett, M. & Archard, L. C. (1979). Conservation and variation in orthopoxvirus genome structure. *Journal of General Virology,* **45**, 683–702.

Mackett, M. & Arrand, J. R. (1985). Recombinant vaccinia virus induces neutralizing antibodies in rabbits against Epstein–Barr virus membrane antigen gp340. *The EMBO Journal,* **4**, 3229–34.

Mackett, M. & Smith, G. L. (1986). Vaccinia virus expression vectors. *Journal of General Virology,* **67**, 2067–82.

Mackett, M., Smith, G. L. & Moss, B. (1982). Vaccinia virus: A selectable eukaryotic cloning and expression vector. *Proceedings of the National Academy of Sciences, USA,* **79**, 7415–19.

Mackett, M., Smith, G. L. & Moss, B. (1984). General method for the production and selection of infectious vaccinia virus recombinants expressing foreign genes. *Journal of Virology,* **49**, 857–64.

Mackett, M., Yilma, T., Rose, J. & Moss, B. (1985*a*). Vaccinia virus recombinants: Expression of VSV genes and protective immunization of mice and cattle. *Science,* **227**, 433–5.

Mackett, M., Smith, G. L. & Moss, B. (1985*b*). The construction and characterization of vaccinia virus recombinants expressing foreign genes. In *DNA Cloning: A Practical Approach* (D. M. Glover, ed.) pp. 191–211. I.R.L. Press, Oxford,

McLaughlin, E., Willey, D. E., Cantin, E. M. Eberle, R., Moss, B. & Openshaw, H. (1988). A recombinant vaccinia virus expressing herpes simplex virus type 1 glycoprotein B induces cytotoxic T lymphocytes in mice. *Journal of General Virology,* **69**, 1731–4.

McMichael, A. J., Michie, C. A., Gotch, F. M., Smith, G. L. & Moss, B. (1986). Recognition of influenza A virus nucleoprotein by human cytotoxic T lymphocytes. *Journal of General Virology,* **67**, 719–26.

Marchioli, C. C., Yancey, R. J., Petrovskis, E. A., Timmins, J. G. & Post, L. E. (1987). Evaluation of pseudorabies virus glycoprotein gp50 as a vaccine for Aujesky's disease in mice and swine: expression by vaccinia virus and chinese hamster ovary cells. *Journal of Virology,* **61**, 3977–82.

Martin, S. A. & Moss, B. (1975). Modification of RNA by mRNA guanylyltransferase and mRNA (guanine-7-)methyltransferase from vaccinia virions. *Journal of Biological Chemistry,* **250**, 9330–5.

Martin, S. A. & Moss, B. (1976). mRNA guanylyltransferase and mRNA (guanine-7-)methyltransferase from vaccinia virions. Donor and acceptor substrate activities. *Journal of Biological Chemistry,* **251**, 7313–21.

Martin, S. A., Paoletti, E. & Moss, B. (1975). Purification of mRNA guanylyltransferase and mRNA (guanine-7-)methyltransferase from vaccinia virus. *Journal of Biological Chemistry,* **250**, 9322–29.

Martin, S., Moss, B., Berman, P. W., Laskey, L. A. & Rouse, B. T. (1987). Mechanisms of antiviral immunity induced by a vaccinia virus recombinant expressing herpes simplex type 1 glycoprotein D: cytotoxic T cells. *Journal of Virology,* **61**, 726–34.

Martin, S., Cantin, E. & Rouse, B. T. (1989). Evaluation of antiviral immunity using vaccinia virus recombinants expressing cloned genes for herpes simplex virus type 1 glycoproteins. *Journal of General Virology,* **70**, 1359–70.

Merchlinsky, M. (1989). Intramolecular homologous recombination in cells infected with temperature-sensitive mutants of vaccinia virus. *Journal of Virology,* **63**, 2030–5.

Merchlinsky, M. & Moss, B. (1986). Resolution of linear minichromosomes with hairpin ends from circular plasmids containing vaccinia virus concatemer junctions. *Cell,* **45**, 879–84.

Merchlinsky, M. & Moss, B. (1989). Resolution of vaccinia virus DNA concatemer junctions requires late-gene expression. *Journal of Virology,* **63**, 1595–603.

Mims, C. A. & White, D. O. (1984). *Viral Pathogenesis and Immunology*, Blackwell, Oxford.

Morgan, A. J., Mackett, M., Finnerty, S., Arrand, J. R., Scullion, F. T. & Epstein, M.A. (1988). Recombinant vaccinia virus expressing Epstein–Barr virus glycoprotein gp340 protects cotton top tamarins against EB-virus induced lymphomas. *Journal of Medical Virology,* **25**, 189–195.

Morita, M., Suzuki, K. Yasuda, A., Kojima, A., Sugimoto, M. Kobayashi, H., Watanabe, K., Kajiyama, K. & Hashizume S. (1987). Recombinant vaccinia virus LC16m0 or LC16m8 that expresses hepatitis B surface antigen while preserving the attenuation of the parental virus strain. *Vaccine,* **5**, 65–70.

Morrison, D. K. & Moyer, R. W. (1986), Detection of a subunit of cellular pol.II within highly purified preparations of RNA polymerase isolated from rabbit poxvirus virions. *Cell,* **44**, 587–96.

Morrison, D. K., Carter, J. K. & Moyer, R. W. (1985). Isolation and characterization of monoclonal antibodies directed against two subunits of rabbit poxvirus-associated DNA-directed RNA polymerase. *Journal of Virology,* **55**, 670–80.

Morrison, H. G., Bauer, S. P., Lange, J. V., Esposito, J. J., McCormick, J. B. & Auperin, D. D. (1989). Protection of guinea pigs from Lassa fever by vaccinia virus recombinants expressing the nucleoprotein or the envelope glycoprotein of lassa virus. *Virology,* **171**, 179–88.

Moss, B. & Flexner, C. (1987) Vaccinia virus expression vectors. *Annual Reviews of Immunology,* **5,** 305–24.

Moss, B. & Rosenblum, E. M. (1973). Protein cleavage and poxvirus morphogenesis: Tryptic peptide analysis of core precursors accumulated by blocking assembly with rifampicin. *Journal of Molecular Biology,* **19**, 65.

Moss, B., Rosenblum, E. N., Katz, E. & Grimley, P. M. (1969). Rifampicin: a specific inhibitor of vaccinia virus assembly. *Nature, London,* **224**, 1280–4.

Moss, B., Rosenblum, E. N. & Gershowitz, A. (1975). Characterization of a polyriboadenylate polymerase from vaccinia virions. *Journal of Biological Chemistry,* **250**, 4722–9.

Moss, B., Winters, E. & Cooper, N. (1981*a*). Instability and reiteration of DNA sequences within the vaccinia virus genome. *Proceedings of the National Academy of Sciences, USA,* **78**, 1611–15.

Moss, B., Winters, E. & Cooper, J.A. (1981*b*). Deletion of a 9,000 base pair segment of the vaccinia virus genome that encodes non-essential polypeptides. *Journal of Virology,* **40**, 387–395.

Moss, B., Smith, G. L., Gerin, J. L. & Purcell, R. H. (1984). Live recombinant vaccinia protects chimpanzees against hepatitis B. *Nature, London,* **311**, 67–9.

Moyer, R. W. & Graves, R.L. (1981). The mechanism of cytoplasmic orthopoxvirus DNA replication. *Cell,* **27,** 391–401.

Moyer, R. W. & Rothe, C.T. (1980). The white pock mutants of rabbit poxvirus. I. Spontaneous host range mutants contain deletions. *Virology,* **102,** 119–32.

Moyer, R. W., Graves, R. L. & Rothe, C. T. (1980). The white pock *(u)* mutants of rabbit poxvirus. III. Terminal DNA sequence duplication and transposition in rabbit poxvirus. *Cell,* **22**, 545–53.

Munyon, W., Paoletti, E. & Grace, J. T. Jr. (1967). RNA polymerase activity in purified infectious vaccinia virus. *Proceedings of the National Academy of Sciences. USA,* **58**, 2280–8.

Muthukrishnan, S., Moss, B., Cooper, J. A. & Maxwell, E. S. (1978). Influence of 5′ terminal cap structure on the initiation of translation of vaccinia virus mRNA. *Journal of Biological Chemistry,* **253**, 1710–15.

Nagayama, A., Pogo, B. G. T. & Dales, S. (1970). Biogenesis of vaccinia: separation of early stages from maturation by means of rifampicin. *Virology,* **40**, 1039–51.

Nakano, E., Panicali, D. & Paoletti, E. (1982). Molecular genetics of vaccinia virus: demonstration of marker rescue. *Proceedings of the National Academy of Sciences, USA,* **79**, 1593–6.

Niles, E. G., Condit, R. C., Caro, P., Davidson, K., Matusick, L. & Seto J. (1986). Nucleotide sequence and genetic map of the 16kb vaccinia virus *Hind* III D frangment. *Virology,* **153**, 96–112.

Oda, K. & Joklik, W. K. (1967). Hybridization and sedimentation studies on 'early' and 'late' vaccinia messenger RNA. *Journal of Molecular Biology,* **27**, 395–419.

Olmsted, R. A., Elango, N., Prince, G. A., Murphy, B. R., Johnson, P. R., Moss B., Chanock, R. M. & Collins, P. L. (1986). Expression of F glycoprotein of respiratory syncytial virus by a recombinant vaccinia virus: Comparisons of the individual contributions of the F and G glycoproteins to host immunity. *Proceedings of the National Academy of Sciences, USA,* **83**, 7462–6.

Paez, E. & Estaban, M. (1985). Interferon prevents the generation of spontaneous deletions at the left terminus of vaccinia virus DNA. *Journal of Virology,* **56**, 75–84.

Paez, E., Pallo, S. & Esteban, M. (1987). Virus attenuation and identification of structural proteins that are selectively modified during virus persistence. *Journal of Virology,* **61**, 2642–7.

Panicali, D. & Paoletti, E. (1982). Construction of poxviruses as cloning vectors: Insertion of the thymidine kinase gene from herpes simplex into the DNA of infectious vaccinia virus. *Proceedings of the National Academy of Sciences, USA,* **79**, 4927–31.

Panicali, D., Davis, S. W., Mercer, S. R. & Paoletti, E. (1981). Two major DNA variants present in serially propagated stocks of the WR strain of vaccinia virus. *Journal of Virology,* **37**, 1000–10.

Panicali, D., Davis, S. W., Weinberg, R. L. & Paoletti, E. (1983). Construction of live vaccines by using genetically engineered poxviruses: Biological activity of recombinant vaccinia virus expressing influenza virus hemaggluntin. *Proceedings of the National Academy of Sciences, USA,* **80**, 5364–9.

Panicali, D., Grzelecki, A. & Huang, C. (1986). Vaccinia virus vectors utilizing the β-galactosidase assay for rapid selection of recombinant viruses and measurement of gene expression. *Gene,* **47**, 193–9.

Paoletti, E. and Lipinskas, B. R. (1978). Soluble endoribonuclease activity from vaccinia virus: Specific cleavage of virion-associated high molecular weight RNA. *Journal of Virology,* **26**, 822–4.

Paoletti, E. & Moss, B. (1974). Two nucleic acid-dependent nucleoside triphosphate phosphohydrolases from vaccinia virus. Nucleotide substrate and polynucleotide cofactor specificities. *Journal of Biological Chemistry,* **249**, 3281–6.

Paoletti, E., Rosemond-Hornbeak, H. & Moss, B. (1974). Two nucleic acid-dependent nucleoside triphosphate phosphohydrolases from vaccinia virus. Purification and characterization. *Journal of Biological Chemistry,* **249**, 3273–80.

Paoletti, E., Lipinskas, B. R., Samsanoff, C., Mercer, S. & Panicali, D. (1984). Construction of live vaccines using genetically engineered poxviruses: Biological activity of vaccinia virus recombinants expressing the hepatitis B surface antigen and the herpes simplex virus glycoprotein D. *Proceedings of the National Academy of Sciences, USA,* **81**, 193–7.

Patel, D. D. & Pickup, D. J. (1987). Messenger RNAs of a strongly expressed late gene of cowpox virus contains 5′-terminal poly(A) sequences. *The EMBO Journal,* **6**, 3787–94.

Patel, D. D. & Pickup, D. J. (1989). The second largest subunit of poxvirus RNA polymerase is similar to the corresponding subunits of prokaryotic and eukaryotic RNA polymerases. *Journal of Virology,* **63**, 1076–86.

Patel, D. D., Pickup, D. J. & Joklik, W. K. (1986). Isolation of cowpox virus A-type inclusions and characterization of their major protein component. *Virology,* **149**, 174–89.

Paterson, R. G., Lamb, R. A. Moss, B. & Murphy, B. R. (1987). Comparison of the relative roles of the F and HN surface glycoproteins of the paramyxovirus simian virus 5 in inducing immunity. *Journal of Virology,* **61**, 1972–7.

Payne, L. G. (1978). Polypeptide composition of extracellular enveloped vaccinia virus. *Journal of Virology,* **27**, 28–37.

Payne, L. G. (1979). Identification of the vaccinia hemagglutinin polypeptide from a cell system yielding large amounts of extracellular enveloped virus. *Journal of Virology,* **31**, 147–55.

Payne, L. G. (1980). Significance of extracellular enveloped virus in the *in vitro* and *in vivo* dissemination of vaccinia. *Journal of General Virology,* **50**, 89–100.

Pedley, C. B. & Cooper, R. J. (1987). The assay, purification and properties of vaccinia virus-induced uncoating protein. *Journal of General Virology,* **68**, 1021–8.

Pemberton, R. M., Cannon, J. J., Openshaw, P. J. M., Ball, L. A., Wertz, G. W. & Askonas, B. A. (1987). Cytotoxic T cell specificity for respiratory syncytial virus proteins: fusion protein is an important target antigen. *Journal of General Virology,* **68**, 2177–82.

Pennington, T. H. & Follett, E. A. C. (1974). Vaccinia virus replication in enucleate BSC-1 cells: Particle production and synthesis of viral DNA and proteins. *Journal of Virology,* **13**, 488–93.

Pensiero, N. M., Jennings, G. B., Schmaljohn, C. S. & Hay, J. (1988). Expression of Hantaan virus M genome segment by using a vaccinia virus recombinant. *Journal of Virology,* **62**, 696–702.

Perkus, M. E., Piccini, A., Lipinskas, B.R. & Paoletti, E. (1985). Recombinant vaccinia virus: Immunization against multiple pathogens. *Science,* **229**, 981–4.

Perkus, M. E., Panicali, D., Mercer, S. & Paoletti, E. (1986). Insertion and deletion mutants of vaccinia virus. *Virology,* **152**, 285–97.

Pickup, D. J., Ink, B. S., Parsons, B. L., Hu, W. & Joklik, W. K. (1984). Spontaneous deletions and duplications of sequences in the genome of cowpox virus. *Proceedings of the National Academy of Sciences, USA,* **81**, 6817–21.

Pickup, D. J., Ink, B. S., Hu, W., Ray, C. A. & Joklik, W. K. (1986). Hemorrhage in lesions caused by cowpox virus is induced by a viral protein that is related to plasma protein inhibitors of serine proteases. *Proceedings of the National Academy of Sciences, USA,* **83**, 7698–702.

Plucienniczak, A., Schroeder, E., Zettlmeissl, G. & Streek, R. E. (1985). Nucleotide sequence of a cluster of early and late genes in a conserved segment of the vaccinia virus genome. *Nucleic Acids Research,* **13**, 985–98.

Pogo B. G. T. (1977). Elimination of naturally occurring crosslinks in vaccinia virus DNA after viral penetration into cells. *Proceedings of the National Academy of Sciences, USA,* **74**, 1739–43.

Pogo, B. G. T. (1980). Changes in parental vaccinia virus DNA after viral penetration into cells. *Virology,* **101**, 520–4.

Pogo, B. G. T. & Dales, S. (1969). Two deoxyribonuclease activities within purified vaccinia virus. *Proceedings of the National Academy of Sciences, USA,* **63,** 820–7.

Pogo, B. G. T. & O'Shea, M. T. (1977). Further characterization of deoxyribonucleases from vaccinia virus. *Virology,* **77**, 55–66.

Pogo, B. G. T. & O'Shea, M. (1978). The mode of replication of vaccinia virus DNA. *Virology,* **84**, 1–8.

Pogo, B. G. T., O'Shea, M. & Freimuth, P. (1981). Initiation and termination of vaccinia virus DNA replication. *Virology,* **108**, 241–8.

Puckett, C. & Moss, B. (1983). Selective transcription of vaccinia virus genes in template-dependent soluble extracts of infected cells. *Cell,* **35**, 441–8.

Quinnan, G. V., Jr, ed. (1985). *Vaccinia Viruses as Vectors for Vaccine Antigens.* Elsevier, New York.

Ramshaw, I. A., Andrew, M. E., Philips, S. M., Boyle, D. B. & Coupar, B. E. H. (1987). Recovery of immunodeficient mice from a vaccinia/IL-2 recombinant infection. *Nature, London,* **329**, 545–7.

Reddy, M. K. & Bauer, W. R. (1989). Activation of the vaccinia virus nicking-joining enzyme by trypsinization. *Journal of Biological Chemistry,* **264**, 443–9.

Redfield, R. R., Wright, D. C., James, W. D., Jones, T. S., Brown, C. & Burke, M.D. (1987). Disseminated vaccinia in a military recruit with human immunodeficiency virus (HIV) disease. *New England Journal of Medicine,* **316**, 673–6.

Rice, C. M., Franke, C. A., Strauss, J. H. & Hruby D. E. (1985). Expression of Sindbis virus structural proteins via recombinant vaccinia virus: Synthesis, processing and incorporation into mature Sindbis virions. *Journal of Virology,* **56**, 227–39.

Rigby, P. W. J. (1983). Cloning vectors derived from animal viruses. *Journal of General Virology,* **64**, 255–66.

Rodriguez, J. F. & Esteban, M. (1987). Mapping and nucleotide sequence of the vaccinia virus gene that encodes a 14-kilodalton fusion protein. *Journal of Virology,* **61**, 3550–54.

Rodriguez, J. F. & Esteban, M. (1989). Plaque size phenotype as a selectable marker to generate vaccinia virus recombinants. *Journal of Virology,* **63**, 997–1001.

Rodriguez, J. F., Janeczko, R. & Esteban, M. (1985). Isolation and characterization of neutralizing monoclonal antibodies to vaccinia virus. *Journal of Virology,* **56**, 482–8.

Rodriguez, J. F., Kahn, J. S. & Esteban, M. (1986). Molecular cloning, encoding sequence, and expression of vaccinia virus nucleic acid-dependent nucleoside triphosphatase gene. *Proceedings of the National Academy of Sciences. USA,* **83**, 9566–70.

Rodriguez, J. F., Rodriguez, D., Rodriguez, J.-R., McGowan, E. B. & Esteban, M. (1988). Expression of the firefly luciferase gene in vaccinia virus: a highly sensitive gene marker to follow virus dissemination in tissues of infected animals. *Proceedings of the National Academy of Sciences, USA,* **85**, 1667–71.

Rodriguez, D., Rodriguez, J.-R., Rodrigues, J. F., Trauber, D. & Esteban, M. (1989). Highly attenuated vaccinia virus mutants for the generation of safe recombinant viruses. *Proceedings of the National Academy of Sciences, USA,* **86**, 1287–91.

Rohrmann, G. & Moss, B. (1985). Transcription of vaccinia virus early genes by template-dependent extract of purified virions. *Journal of Virology,* **56,** 349–55.

Rohrmann, G., Yuen, L. & Moss, B. (1986). Transcription of vaccinia virus early genes by enzymes isolated from vaccinia virions terminates downstream of a regulatory sequence. *Cell,* **46,** 1029–35.

Roizman, B. & Arsenakis, M. (1985). Genetic engineering of herpes simplex virus genomes for attenuation and expression of foreign genes. In *Vaccinia Viruses as Vectors for Vaccine Antigens* (G.V. Quinnan, ed.) pp. 211–24. Elsevier, New York.

Rooney, J. F., Wohlenberg, C., Cremer, K. J., Moss, B. & Notkins, A. L. (1988). Immunization with a vaccinia virus recombinant expressing herpes simplex virus type 1 glycoprotein D: Long-term protection and effect of revaccination. *Journal of Virology,* **62**, 1530–4.

Rosel, J. & Moss, B. (1985). Transcriptional and translational mapping and nucleotide sequence analysis of a vaccinia virus gene encoding the precursor of the major core polypeptide 4b. *Journal of Virology,* **56**, 830–8.

Rosel, J., Earl, P. L., Weir, J. P. & Moss, B. (1986). Conserved TAAATG sequence at the transcriptional and translational initiation sites of vaccinia virus late genes deduced by structural and functional analysis of *Hind*III H genome fragment. *Journal of Virology,* **60**, 436–49.

Roseman, N. A. & Hruby, D. E. (1987). Nucleotide sequence and transcript organization of a region of the vaccinia virus genome which encodes a constitutively expressed gene required for DNA replication. *Journal of Virology,* **61**, 1398–406.

Rosemond-Hornbeak, H. & Moss, B. (1974). Single-stranded deoxyribonucleic acid-specific nuclease from vaccinia virus. Endonucleolytic and exonucleolytic activities. *Journal of Biological Chemistry,* **249**, 3292–6.

Rosemond-Hornbeak, H., Paoletti, E. & Moss, B. (1974). Single-stranded deoxyribonucleic acid specific nuclease from vaccinia virus. Purification and characterization. *Journal of Biological Chemistry,* **249,** 3287–91.

Rota, P. A., Shaw, M. W. & Kendal, A. P. (1987). Comparison of the immune response to variant influenza type B hemagglutinins expressed in vaccinia virus. *Virology,* **161**, 269–75.

Rota, P. A., Shaw, M. W. & Kendal, A. P. (1989). Cross-protection against microvariants of influenza virus type B by vaccinia viruses expressing haemagglutinins from egg- or MDCK cell-derived subpopulations of influenza virus type B/England/222/82. *Journal of General Virology,* **70**, 1533–7.

Sam, C. K. & Dumbell, K. R. (1981). Expression of poxvirus DNA in coinfected cells and marker rescue of thermosensitive mutants by subgenomic fragments of DNA. *Annales Virologiques, Institut Pasteur, Paris,* **132E**, 135–150.

Sambrook, J. & Shatkin, A.J. (1969). Polynucleotide ligase activity in cells infected with simian virus 40, polyoma virus or vaccinia virus. *Journal of Virology,* **4** 719–26.

Schmitt, J. F. C. & Stunnenberg, H. G. (1988). Sequence and transcriptional analysis of the vaccinia virus Hind III I fragment. *Journal of Virology,* **62**, 1889–97.

Schwer, B & Stunnenberg, H.G. (1988). Vaccinia virus late transcripts generated *in vitro* have a poly (A) head. *The EMBO Journal,* **7**, 1183–90.

Schwer, B., Visca, P., Vos, J. C. & Stunnenberg, H. G. (1987). Discontinuous transcription or RNA processing of vaccinia virus late messengers results in a 5′ poly (A) leader. *Cell,* **50**, 163–9.

Shaffer, R. & Traktman, P. (1987). Vaccinia virus encapsidates a novel topoisomerase with the properties of a eucaryotic type I enzyme. *Journal of Biological Chemistry,* **262**, 9309–15.

Shida, H. (1986). Nucleotide sequence of the vaccinia virus hemagglutinin gene. *Virology,* **150**, 451–62.

Shida, H., Tochikura, T., Sato, T., Konno, T., Hirayoshi, K., Seki, M., Ito, Y., Hatanaka, M., Hinuma, Y., Sugimoto, M., Takahashi-Nishimaki, F., Maruyama, T., Miki, K., Suzuki, K., Morita, M., Sashiyama, H. & Hayami, M. (1987). Effect of recombinant vaccinia viruses that express HTLV1 envelope gene on HTLV1 infection. *The EMBO Journal,* **6**, 3379–84.

Shida, H., Hinuma, Y., Hatanaka, M., Morita, M., Kidokoro, M., Suzuki, K., Maruyama, T., Takahashi-Nishimaki, F., Sugimoto, M., Kitamura, R., Miyazawa, T. & Hayami, M. (1988). Effects and virulences of recombinant vaccinia viruses derived from attenuated strains that express the human T-cell leukemia virus type 1 envelope gene. *Journal of Virology,* **62**, 4474–80.

Shuman, S. (1989). Vaccinia DNA topoisomerase I promotes illegitimate recombination in *E. coli. Proceedings of the National Academy of Sciences, USA,* **86**, 3489–93.

Shuman, S. & Hurwitz, J. (1981). Mechanism of mRNA capping by vaccinia virus guanylyltransferase: Characterization of an enzyme-guanylate intermediate. *Proceedings of the National Academy of Sciences, USA,* **78,** 187–91.

Shuman, S. & Moss, B. (1987). Identification of a vaccinia virus gene encoding a type 1 DNA topoisomerase. *Proceedings of the National Academy of Sciences. USA,* **84**, 7478–82.

Shuman, S., Surks, M., Furneaux, H. & Hurwitz, J. (1980). Purification and characterization of a GTP-pyrophosphate exchange activity from vaccinia virions.

Association of the GTP-pyrophosphate exchange activity with vaccinia mRNA guanylyltransferase RNA (guanine-7-)methyltransferase complex capping enzyme. *Journal of Biological Chemistry,* **255**, 11588–98.

Shuman, S., Broyles, S. S. & Moss, B. (1987). Purification and characterization of a transcription termination factor from vaccinia virus. *Journal of Biological Chemistry,* **262**, 12372–80.

Shuman, S., Golder, M. & Moss, B. (1989). Insertional mutagenesis of the vaccinia virus gene encoding a type 1 DNA topoisomerase: evidence that the gene is essential for growth. *Virology,* **170**, 302–6.

Siomi, H., Shida, H., Nam, S. H., Nosaka, T., Maki, M. & Hatanaka, M. (1988). Sequence requirements for nucleolar localization of human T cell leukemia virus type 1 pX protein, which regulates viral RNA processing. *Cell,* **55**, 197–209.

Slabaugh, M. B. & Mathews, C.K. (1986). Hydroxyurea-resistant vaccinia virus: Overproduction of ribonucleotide reductase. *Journal of Virology,* **60**, 506–14.

Slabaugh, M. B. & Roseman, N.A. (1989). Retroviral protease-like gene in the vaccinia virus genome. *Proceedings of the National Academy of Sciences, USA,* **86**, 4152–5.

Slabaugh, M. B., Johnson, T. L. & Mathews, C. K. (1984). Vaccinia virus induces ribonucleotide reductase in primate cells. *Journal of Virology,* **52**, 507–14.

Slabaugh, M., Roseman, N., Davis, R. & Mathews, C. (1988). Vaccinia virus-encoded ribonucleotide reductase: Sequence conservation of the gene for the small subunit and its amplification in hydroxyurea-resistant mutants. *Journal of Virology,* **62**, 519–27.

Small, P. A., Smith, G. L. & Moss, B. (1985). Intranasal vaccination with recombinant vaccinia containing influenza haemagglutinin prevents both influenza virus pneumonia and nasal infection: intradermal vaccination prevents only viral pneumonia. In *Vaccinia Viruses as Vectors for Vaccine Antigens* (G.V. Quinnan, ed.) pp.175–178. Elsevier, New York.

Smith, G. L. & Moss, B. (1983). Infectious poxvirus vectors have capacity for at least 25,000 base pairs of foreign DNA. *Gene,* **25**, 21–8.

Smith G. L., Mackett, M. & Moss, B. (1983*a*). Infectious vaccinia virus recombinants that express hepatitis B virus surface antigen. *Nature, London,* **302**, 490–5.

Smith, G. L., Murphy, B. R. & Moss, B. (1983 *b*). Construction and characterization of an infectious vaccinia virus recombinant that expresses the influenza haemagglutinin gene and induces resistance to influenza virus infection in hamsters. *Proceedings of the National Academy of Sciences, USA,* **80**, 7155–9.

Smith, G. L., Levin, J. Z., Palese, P. & Moss, B. (1987). Synthesis and cellular location of the ten influenza virus polypeptides individually expressed by recombinant vaccinia viruses. *Virology,* **160**, 336–45.

Smith, G. L., Godson, G. N., Nussenzweig, V., Nussenzweig, R. S., Barnwell, J. & Moss, B. (1984). *Plasmodium knowlesi* sporozoite antigen: expression by infectious vaccinia virus. *Science,* **224**, 397–9.

Smith, G. L., Cheng, K.-C. & Moss, B. (1986). Vaccinia virus: an expression vector for genes from parasites. *Parasitology,* **92S**, 109–18.

Smith, G. L., de Carlos, A. & Chan, Y. S. (1989*a*). Vaccinia virus encodes a thymidylate kinase: gene sequence and transcriptional mapping. *Nucleic Acids Research*, **17**, 7581–90.

Smith, G. L., Chan, Y. S. & Kerr, S. M. (1989*b*). Transcriptional mapping and nucleotide sequence of a vaccinia virus gene encoding a polypeptide with extensive homology to yeast DNA ligases. *Nucleic Acids Research,* **17**, 9051–62.

Smith, G. L., Howard, S. T. & Chan, Y. S. (1989*c*). Vaccinia virus encodes a family of genes with homology to serine protease inhibitors. *Journal of General*

Virology, **70**, 2333–43.

Spadari, S. (1976) Properties of DNA ligase from uninfected and virus-infected HeLa cells. *Nucleic Acids Research,* **3**, 2155–67.

Spehner, D., Gillard, S., Drillien, R. & Kirn, A. (1988). A cowpox virus gene required for multiplication in Chinese hamster ovary cells. *Journal of Virology,* **62,** 1297–304.

Spencer, E., Loring, D., Hurwitz, J. & Monroy, G. (1978). Enzymatic conversion of 5′-phosphate terminated RNA to 5′-di and triphosphate-terminated RNA. *Proceedings of the National Academy of Science, USA,* **75**, 4793–7.

Spencer, E., Shuman, S. & Hurwitz, J. (1980). Purification and properties of vaccinia virus DNA-dependent RNA polymerase. *Journal of Biological Chemistry,* **255**, 5388–95.

Spriggs, M. K., Murphy, B. R., Prince, G. A., Olmsted, R. A. & Collins, P. L. (1987). Expression of the F and HN glycoproteins of human parainfluenza virus type 3 by recombinant vaccinia viruses: contributions of the individual proteins to host immunity. *Journal of Virology,* **61**, 3416–23.

Spriggs, M. K., Collins, P. C., Tierney, E., London, W. T. & Murphy, B. R. (1988). Immunization with vaccinia virus recombinants that express the surface glycoproteins of human parainfluenza virus type 3 (PIV3) protects monkeys against PIV3 infection. *Journal of Virology,* **62**, 1293–6.

Spyropolous, D. D., Roberts, B. E., Panicali, D. & Cohen, L. K. (1988). Delineation of the viral products of recombination in vaccinia virus-infected cells. *Journal of Virology,* **62**, 1046–54.

Stamatos, N. M., Chakrabarti, S., Moss, B. & Hare, D. J. (1987). Expression of polyoma virion proteins by a vaccinia virus vector: association of VP1 and VP2 with the nuclear framework. *Journal of Virology,* **61**, 516–25.

Stephens, E. B., Compans, R. W., Earl, P. & Moss, B. (1986). Surface expression of viral glycoproteins is polarized in epithelial cells infected with recombinant vaccinia viral vectors. *The EMBO Journal,* **5**, 237–45.

Stern, W. & Dales, S. (1976). Biogenesis of vaccinia: isolation and characterization of a surface component that elicits antibody suppressing infectivity and cell-cell fusion. *Virology,* **75**, 232–41.

Stott, E. J., Taylor, G., Ball, L. A., Anderson, K., Young, K. K.-Y., King, A. M. Q., & Wertz, G. W. (1987). Immune and histopathological response in animals vaccinated with recombinant vaccinia viruses that express individual genes of human respiratory syncytial virus. *Journal of Virology,* **61**, 3855–61.

Stroobant, P., Rice, A. P., Gullick, W. J., Cheng, D. J., Kerr, I. M. & Waterfield, M.D. (1985). Purification and characterization of vaccinia virus growth factor. *Cell,* **42**, 383–93.

Sugimoto, M., Yasuda, A., Miki, K., Morita, M., Suzuki, K., Uchida, N. & Hashizume, S. (1985). Gene structures of low-neurovirulent vaccinia virus LC16mO, LC16m8 and their Lister original (LO) strains. *Microbiology and Immunology,* **29**, 421–8.

Sullivan, V. (1988). Analysis of herpes simplex type 1 glycoproteins using recombinant vaccinia virus vectors. Ph.D. Thesis, University of Cambridge.

Sullivan, V. & Smith, G. L. (1987). Expression and characterization of herpes simplex virus type 1 (HSV-1) glycoprotein G (gG) by recombinant vaccinia virus: neutralisation of HSV-1 infectivity with anti-gG antibody. *Journal of General Virology,* **68**, 2587–98.

Sullivan, V. & Smith, G. L. (1988). The herpes simplex virus type 1 US7 gene product is a 66K glycoprotein and is a target for complement-dependent virus neutralisation. *Journal of General Virology,* **69**, 859–67.

Takahashi-Nishimaki, F., Susuki, K, Morita, M., Maruyama, T., Miki, K., Hashizume, S. & Sugimoto, M. (1987). Genetic analysis of vaccinia virus Lister strain and its attenuated mutant LC16m8: production of intermediate variants by homologous recombination. *Journal of General Virology*, **68**, 2705–10.

Tamin, A., Villarreal, E. C., Weinrich, S. L. & Hruby, D. E. (1988). Nucleotide sequence and molecular genetic analysis of the vaccinia virus Hind III N/M region encoding the genes responsible for resistance to α-amanitin. *Virology*, **165**, 141–50.

Tartaglia, J. & Paoletti, E. (1985). Physical mapping and DNA sequence analysis of the rifampicin resistance locus in vaccinia virus. *Virology*, **147**, 394–404.

Tartaglia, J., Piccini, A. & Paoletti, E. (1986). Vaccinia virus rifampicin-resistance locus specifies a late 63,000 Da gene product. *Virology*, **150**, 45–54.

Tengelsen, L. A., Slabaugh, M. B., Bibler, J. K. & Hruby, D. E. (1988). Nucleotide sequence and molecular genetic analysis of the large subunit of ribonucleotide reductase encoded by vaccinia virus. *Virology*, **164**, 121–31.

Thomas, G., Herbert, E. & Hruby, D. E. (1986). Expression and cell type specific processing of human preproenkephalin with a vaccinia recombinant. *Science*, **232**, 1641–3.

Tomley, F. M., Mockett, A. P. A., Boursnell, M. E. G., Binns, M. M., Cook, J. K. A., Brown, T. D. K. & Smith, G. L. (1987). Expression of the infectious bronchitis virus spike protein by recombinant vaccinia virus and induction of neutralising antibodies in vaccinated mice. *Journal of General Virology*, **68**, 2291–8.

Townsend, A., Bastin, J., Gould, K., Brownlee, G. G., Andrew, M., Boyle, D. B., Chan, Y. S. & Smith, G. L. (1988). Defective presentation to class 1 restricted CTL in vaccinia infected cells is overcome by enhanced degradation of antigen. *Journal of Experimental Medicine*, **168**, 1211–24.

Traktman, P., Sridhar, R., Condit, R. C. and Roberts, B. (1984). Transcriptional mapping of the DNA polymerase gene of vaccinia virus. *Journal of Virology*, **49**, 125–31.

Tutas, D. J. & Paoletti, E. (1977). Purification and characterization of core-associated polynucleotide 5′-triphosphatase from vaccinia virus. *Journal of Biological Chemistry*, **252**, 3092–8.

Twardzik, D. R., Brown, J. P., Ranchalis, J. E., Todaro, G. J. & Moss, B. (1985). Vaccinia virus-infected cells release a novel polypeptide functionally related to transforming and epidermal growth factors. *Proceedings of the National Academy of Sciences, USA*, **82**, 5300–4.

van Drunen Littel-van den Hurk, S., Zamb, T. & Babuik, L. A. (1989). Synthesis, cellular location and immunogenicity of bovine herpes virus 1 glycoproteins gI and gIII expressed by recombinant vaccinia virus. *Journal of Virology*, **63**, 2159–68.

Vassef, A., Mars, M., Dru, A., Plucienniczak, A., Streeck, R.E. & Beaud, G. (1985). Isolation of *cis* -acting vaccinia virus DNA fragments promoting expression of herpes simplex virus thymidine kinase by recombinant viruses. *Journal of Virology*, **55**, 163–72.

Venkatesan, S., Gershowitz, A. & Moss, B. (1980). Modification of the 5′-end of mRNA: association of RNA triphosphatase with the RNA guanylyltransferase RNA (guanine-7-)methyltransferase complex from vaccinia virus. *Journal of Biological Chemistry*, **255**, 903–8.

Venkatesan, S., Baroudy, B. M. & Moss, B. (1981). Distinctive nucleotide sequences adjacent to multiple initiation and termination sites of an early vaccinia virus gene. *Cell*, **125**, 805–13.

Venkatesan, S., Gershowitz, A. & Moss, B. (1982). Complete nucleotide sequences of two adjacent early vaccinia virus genes located within the inverted terminal repetition. *Journal of Virology,* **44**, 637–46.

Walker, B. D., Flexner, C., Paradis, T. J., Fuller, T. C., Hirsch, M. S., Schooley, R. T. & Moss, B. (1988). HIV-1 reverse transcriptase is a target for cytotoxic T lymphocytes in infected individuals. *Science,* **240**, 64–6.

Wei, C. M. & Moss, B. (1975). Methylated nucleotides block 5′-terminus of vaccinia virus mRNA. *Proceedings of the National Academy of Sciences, USA,* **72**, 318–22.

Weinrich, S. L. & Hruby, D.E. (1986). A tandemly-orientated late gene cluster within the vaccinia virus genome. *Nucleic Acids Research,* **14**, 3003–16.

Weir, J. P. & Moss, B. (1983). Nucleotide sequence of the vaccinia virus thymidine kinase gene and the nature of spontaneous frameshift mutations. *Journal of Virology,* **46**, 530–7.

Weir, J. P. & Moss, B. (1984). Regulation of expression and nucleotide sequence of a late vaccinia virus gene. *Journal of Virology,* **51**, 662–9.

Weir, J. P. & Moss, B. (1987*a*). Determination of the promoter region of an early vaccinia virus gene encoding thymidine kinase. *Virology,* **158**, 206–10.

Weir, J. P. & Moss, B. (1987*b*). Determination of the transcriptional regulatory region of a vaccinia virus late gene. *Journal of Virology,* **61**, 75–80.

Weir, J. P., Bajszar, G. and Moss, B. (1982). Mapping of the vaccinia virus thymidine kinase gene by marker rescue and by cell-free translation of selected mRNA. *Proceedings of the National Academy of Sciences, USA,* **79**, 1210–14.

Wertz, G. W., Stott, E. J., Young, K. K. Y., Anderson, K. & Ball, L.A. (1987). Expression of the fusion protein of human respiratory syncytial virus from recombinant vaccinia virus vectors and protection of vaccinated mice. *Journal of Virology,* **61**, 293–301.

Whitton, J. L., Southern, P. J. & Oldstone, M. B. A. (1988*a*). Analyses of the cytotoxic T lymphocyte responses to glycoprotein and nucleoprotein components of lymphocytic choriomeningitis virus. *Virology,* **162**, 321–7.

Whitton, J. L., Gebhard, J. R., Lewicki, H., Trishon, A. & Oldstone, M. (1988*b*). Molecular definition of a major cytotoxic T-lymphocyte epitope in the glycoprotein of lymphocytic choriomeningitis virus. *Journal of Virology,* **62**, 687–95.

Wiktor, T. J., Macfarlan, R. I., Reagan, K. J., Dietzschold, B., Curtis, P. J., Wunner, W. H., Kieny, M.-P., Lathe, R., Lecocq, J.-P., Mackett, M., Moss, B. & Koprowski, H. (1984). Protection from rabies by a vaccinia recombinant containing the rabies virus glycoprotein gene. *Proceedings of the National Academy of Sciences, USA,* **81**, 7194–8.

Wilson, E. M. & Hruby, D. E. (1987). A single stranded insertion vector (SSIV) for introducing foreign genes into the vaccinia genome. *Nucleic Acids Research,* **15**, 4690.

Wilson, E. M., Hodges, W. M. & Hruby, D. E. (1986). Construction of recombinant vaccinia virus strains using single-stranded insertion vectors. *Gene,* **49**, 207–13.

Wittek, R. & Moss, B. (1980). Tandem repeats within the inverted terminal repetition of vaccinia virus. *Cell,* **21**, 277–84.

Wittek, R., Müller, H. K., Menna, A. & Wyler, R. (1978). Length heterogeneity in the DNA of vaccinia virus is eliminated on cloning the virus. *FEBS Letters,* **90**, 41.

Wittek, R., Hänggi, M. & Hiller, G. (1984*a*). Mapping of a gene coding for major late structural polypeptide on the vaccinia virus genome. *Journal of Virology,* **49**, 371–8.

Wittek, R., Richner, B. & Hiller, G. (1984*b*). Mapping of the genes coding for the two major vaccinia virus core polypeptides. *Nucleic Acids Research,* **12**, 4835–48.

World Health Organization (1985). Guidelines for the Development of Recombinant Vaccinia Viruses for Use as Vaccines. Report of a WHO Informal Meeting, Geneva, September 6 and 7 1985. WHO, Geneva.

Wright, C. F. & Moss, B. (1987). *In vitro* synthesis of vaccinia virus late mRNA containing a 5′poly (A) leader sequence. *Proceedings of the National Academy of Sciences, USA,* **84**, 8883–7.

Yang, W.-P., Kao, S.-Y. & Bauer, W. R. (1988). Biosynthesis and post-translation cleavage of vaccinia virus structural protein VP8. *Virology,* **167**, 585–90.

Yewdell, J. W., Bennink, J. R., Smith, G. L. & Moss, B. (1985). Influenza A virus nucleoprotein is a major target antigen for cross-reactive anti-influenza A cytotoxic T lymphocytes. *Proceedings of the National Academy of Sciences, USA,* **82**, 1785–9.

Yewdell, J. W., Bennink, J. R., Mackett, M., Lefrancois, L., Lyles, D. S. & Moss, B. (1986). Recognition of cloned vesicular stomatitis virus internal and external gene products by cytotoxic virus internal and external gene products by cytotoxic T lymphocytes. *Journal of Experimental Medicine,* **163,** 1529–38.

Yilma, T., Anderson, K., Ristow, S., Brechling, K., Chakrabarti, S. & Moss, B. (1987). An infectious vaccinia virus recombinant expresses the bovine interferon gamma gene. In *Vaccines 86.* (Lerner, R. A., Chanock, R. M. and Brown, F. eds) Cold Spring Harbor Press, New York.

Yilma, T., Hsu, D., Jones, L., Owens, S., Grubman, M., Mebus, C., Yamanaka, M. & Dale, B. (1988). Protection of cattle against rinderpest with vaccinia recombinants expressing the HA or F gene. *Science,* **242,** 1058–61.

Yuen, L. & Moss, B. (1986). Multiple 3′ ends of mRNA encoding vaccinia virus growth factor occur within a series of repeated sequences downstream of T clusters. *Journal of Virology,* **60,** 320–3.

Yuen, L. & Moss, B. (1987). Oligonucleotide sequence signalling transcriptional termination of vaccinia virus early genes. *Proceedings of the National Academy of Sciences, USA.* **84,** 6418–21.

Yuen, L., Davison, A. J. & Moss, B. (1987). Early promoter-binding factor from vaccinia virus virions. *Proceedings of the National Academy of Sciences, USA,* **84,** 6069–73.

Zagury, D., Leonard, R., Fouchard, M., Reveil, B., Bernard, J., Ittele, D., Cattan, A., Zirimwabagabo, L., Kalumbu, M., Justin, W., Salaun, J.-J. & Goussard, B. (1987). Immunization against AIDS in humans. *Nature, London,* **326**, 249–50.

Zagury, D., Bernard, J., Cheynier, R., Desportes, I., Leonard, R., Fouchard, M., Reveil, B., Ittele, D., Lurhuma, Z., Mbayo, K., Wane, J., Salaun, J.-J., Goussard, B., Dechazal, L., Burny, A., Nara, P. & Gallo. R.C. (1988). A group specific anamnestic immune reaction against HIV1 induced by a candidate vaccine against AIDS. *Nature, London,* **332**, 728–31.

Zarling, J. M., Morton, W., Moran, P. A., McClure, J., Kosowski, S. G. & Hu, S.-L. (1986 *a*). T-cell responses to human AIDS virus in macaques immunized with recombinant vaccinia viruses. *Nature, London,* **323**, 344–6.

Zarling, J. M., Moran, P. A., Lasky, L. A. and Moss, B. (1986*b*) Herpes simplex virus (HSV)-specific human T-cell clones recognise HSV glycoprotein D expressed by a recombinant vaccinia virus. *Journal of Virology,* **59**, 506–9.

PROBLEMS AND PROSPECTS FOR SYNTHETIC PEPTIDE VACCINES

D. J. ROWLANDS

Department of Virology Research and Development, Wellcome Biotech, Langley Court, Beckenham, Kent BR3 3BS, UK

INTRODUCTION

A major contribution to the protection provided by vaccination against infectious diseases is the induction of antibodies capable of neutralising the pathogens involved and thus preventing the establishment of infection. The structures recognised by such antibodies are usually (particularly in the case of viruses) proteins present at the surface of the organism. Often only a small number of proteins are involved in the interaction with neutralising antibodies and in many cases a single viral protein can be sufficient to induce protective immunity. It is therefore conceivable that even simpler sub-unit vaccines could be developed and allow vaccine production to switch from a biological process to one involving synthetic chemistry.

Anderer set the scene for the development of peptide vaccines in 1963 when he showed that a short peptide sequence cleaved from the carboxy-terminal region of the coat protein of tobacco mosaic virus with trypsin could elicit virus-neutralising antibodies when coupled to a carrier protein and injected into rabbits. Subsequently it was demonstrated that the chemically synthesised peptide was similarly active (Anderer & Schlumberger, 1965). The potential for producing synthetic immunogens was confirmed and extended by Sela and his colleagues who showed that a fragment of the coat protein of the bacteriophage MS2 and its synthetic equivalent could induce antibodies reactive with the intact virus particle (Sela, 1969). Further exploration of the possibility of producing synthetic vaccines depended on two basic requirements: a convenient and reliable means of chemically synthesising peptides and the availability of amino acid sequence data for the protein under investigation. The first requirement was provided by the work of Merrifield (1963) who developed a simple solid phase process for assembling peptides. The availability of sequence information was a major barrier until the advent of DNA sequencing methods and gene cloning techniques in the mid-1970s provided the means of rapidly determining the derived amino acid sequence of any gene product.

It soon became apparent that the ability of peptides to induce antibodies capable of recognising the protein from which the sequence was derived was the rule rather than the exception (Lerner, 1982). It was thus possible to produce serological reagents that could detect protein products which, because of low expression levels or rapid processing, could not be readily detected by other means. An important caveat to this encouraging scenario from the standpoint of developing synthetic vaccines is that the majority of peptides induce antibodies which only recognise proteins in a more or less denatured state and were not, therefore, of value in providing protective immunity against a pathogen.

It is therefore important to devise means for determining or predicting the parts of a protein sequence which comprise the important antigenic sites of the native structure. A variety of methods have been used to address this problem and will be discussed in some detail. A central issue relating to this problem is that of the topographical nature of antigenic sites on proteins. It is apparent from the complex folding structure of virtually all proteins, together with the dimensions of the antigen combining sites of antibody molecules, that the great majority of epitopic sites must be conformational, i.e. composed of a combination of amino acid residues brought into juxtaposition by the tertiary folding of the protein chain but distal on the linear sequence (Barlow *et al.*, 1986). Given this situation it would be predicted that the induction of antibodies capable of binding to native proteins by simple linear peptides should be a very rare event. This is not the case and possible reasons for it will be discussed.

While synthetic peptide vaccine technology has been developing, there have been tremendous advances in our understanding of the processes involved in the induction of immune responses. In particular, the key role of T lymphocytes in protective immunity has been recognised and the basic features of their antigen recognition mechanisms understood. Broadly speaking, T cells fall into two categories. Helper T cells are required for the maximal stimulation of B cells to proliferate and mature into antibody-producing cells. The recognition signals responsible for these cellular interactions consist of short peptide sequences (Th cell epitopes) produced from the inducing antigen by poorly understood intracellular processing events and presented at the cell surface in combination with class II molecules of the major histocompatibility complex. These complexes of foreign peptide and host MHC protein are recognised by T cell receptors, proteins of the immunoglobulin family present on the surface of T cells. The practical consequences of these important interactions for synthetic vaccine production is that the constructs must contain both B cell epitopes (i.e. sequences capable of eliciting antibodies of appropriate specificities) and helper T cell epitopes. Approaches to overcome this problem are discussed.

Finally, to become a practical reality it will be essential to deliver synthetic vaccines in an acceptable formulation. Peptides are usually poorly immunogenic and in most experimental work they have been administered in potent stimulants such as Freund's complete adjuvant. This is unacceptable for practical purposes and it is therefore of great importance to develop other methods for enhancing the immunogenicity of synthetic vaccines.

At this point it is worth considering the reasons why the synthetic approach to vaccine development should be worth pursuing, given the array of problems to be overcome. Some of the potential advantages of such vaccines are as follows:

1. The product would be chemically defined and thus of readily established purity and quantity. These are rarely features of currently available vaccines.
2. Since they are chemically synthesised, there would be no possibility of contamination with the pathogen against which they are designed to protect or with unknown pathogens present in cell cultures or serum-containing media. Both have been the source of problems with conventional vaccines.
3. The product would be far more stable than conventional biological products. This would be advantageous both for prolonged storage life and for reproducibility of efficacy.
4. Downstream processing would be simple or unnecessary.
5. The stability of such products would make them ideal candidates for incorporation into slow or pulsed release capsular formulations to obviate the necessity for booster injections.
6. As we learn more about the molecular mechanisms involved in immune recognition and responses, and the mechanisms by which pathogenic organisms evade these responses, it should be possible to specifically design vaccines that will help to overcome such problems as host restriction of responsiveness and antigenic variation.

These exciting possibilities alone make the study of synthetic peptide antigens worthwhile.

IDENTIFICATION OF ANTIGENIC SITES ON PROTEINS

A preliminary requirement in the development of peptide vaccines is the identification of those regions of a protein sequence which contribute to the structure of its antigenic sites. Obviously the availability of algorithms which allowed the prediction of such regions from the primary amino acid sequence would be extremely valuable and many attempts have been made to develop such programmes. In practice the complexities of the intra- and inter-molecular interactions involved in protein folding and assembly make the sequence dependent prediction of antigenic sites extremely difficult. In addition, because of the conformation-dependent nature of the majority of antigenic sites, predictive methods could only indicate their linear components.

A variety of methods have been used to directly determine the location and composition of antigenic sites of proteins. However, all are complex and time consuming although the results obtained are more reliable than those derived by prediction methods.

Predictions from sequence data

Antigenic regions of native proteins are necessarily located at their surfaces and simple considerations would predict that the parts of proteins maximally exposed to solvent are hydrophilic in nature. This argument encouraged Hopp & Woods (1981) to develop a predictive algorithm based on hydrophilicity values assigned to each amino acid (Levitt, 1976). By averaging these values for short stretches of sequences (e.g. six residues) and repeating the process with overlapping tracts of amino acids through the sequence, a hydrophilicity profile for the entire protein is obtained. The hydrophilic peaks are predicted to coincide with antigenic determinants.

A variety of analogous methods have been developed which rely on factors such as flexibility or the immunogenicity score for individual amino acids derived from known structures. Unfortunately, in exhaustive experiments designed to test the reliability of predictive methods, Getzoff *et al.*, (1989) found that they provided no significant improvement over random selection of sequences.

Methods for the prediction of the probability of specific secondary structure motifs adopted by stretches of protein sequence have been derived empirically (Chou & Fasman, 1978). The rationale for using such predictions is that antigenic sites frequently include loops and turns in the backbone chain of the protein. Antigenic site prediction programmes which include a combination of predictive methods have been claimed to have a success rate of 65% (Biou *et al.*, 1988). However, until we know a lot more about the factors governing the folding of proteins and the physical nature of antigenic sites, all predictive methods are likely to remain empirical and unreliable.

Activity of protein fragments

A more direct approach to the identification of antigenic regions is to analyse the activity of fragments of proteins as antigens and, or, immunogens. Fragments can be generated by proteolytic digestion or chemical cleavage, and subsequently purified for analysis. This method was successfully used to predict two antigenic regions on the VP1 protein of foot and mouth disease virus by Strohmaier *et al.* (1982). Its inherent disadvantages are: (i) large quantities of protein are required; (ii) conveniently located cleavage sites may not be present; (iii) only linear epitopes can be detected; and (iv) the epitope may be cleared.

Expression of protein fragments

Protein fragments can be readily expressed as fusion proteins by linking portions of the coding DNA into expression vectors to produce fusion protein products. These can then be screened for reaction with neutralising antibodies directed against the native protein. Antigenic sequences of the glycoprotein of feline leukaemia virus were identified in this way by Nunberg *et al.* (1984) and the method is gaining wide acceptance. The advantage of this method is that, given the availability of the cloned DNA coding for the protein in question, large numbers of randomly generated fragments can be expressed and screened quickly and easily. Nucleotide sequencing can then be used to identify sequences of interest.

Sequence comparisons of antigenic variants

Many viruses are antigenically highly variable, and sequence comparisons between antigenic variants can be used to identify regions where amino acid substitutions occur frequently. It is reasonable to assume that regions of proteins of high sequence variation correspond to antigenically important sites. This method successfully identified a major antigenic site on the VP1 protein of foot-and-mouth disease virus (Bittle *et al.*, 1982) and has been used to map the antigenic structure of the haemagglutinin protein of influenza virus (Wiley *et al.*, 1981).

Monoclonal antibody resistant mutants

The high mutation rates characteristic of most viruses facilitate the isolation of variants resistant to neutralisation by the selecting monoclonal antibodies. The great majority of these mutations occur within the epitope recognised by the antibody, and nucleotide sequencing of variant viruses to identify the positions of substitutions provides a means for mapping antigenic sites. Among the many examples of the successful application of this approach there is a growing number in which the location of antigenic sites identified by mutant selection has been related to the molecular structure of the protein or virus as determined by X-ray crystallography (Wiley *et al.*, 1981; Hogle *et al.*, 1985; Rossmann *et al.*, 1985; Acharya *et al.*, 1989).

The 'Pepscan' method

This ingenious approach relies on methodology for rapid synthesis of large numbers of small peptides. The peptides are assembled on spacer molecules linked to small plastic pins. After synthesis the peptides are retained on the plastic supports and their reactivity with antibodies tested by ELISA. By producing overlapping peptides staggered at one amino acid intervals across the entire protein sequence, regions reactive with the antibodies can be readily identified (Geysen *et al.*, 1984).

The mimotope approach

The Pepscan method is, in general, restricted to the identification of linear epitopes. Most antibodies, however, recognise conformational epitopes and the mimotope method was devised to derive peptides capable of reacting with such antibodies (Geysen *et al.*, 1986). The method consists of progressively building peptides in a random fashion, starting with dimers, and using relative reactivities of the different peptides with monoclonal antibody to direct the 'evolution' of peptides with progressively higher affinities for the antibody. Since no preconceptions of the structure of the epitope recognised by the antibody are involved, both natural and unnatural amino acids can be used to develop the best fitting peptide. Remarkably active antigenic peptides have been constructed in this way but it remains to be seen whether they are equally efficient immunogens.

WHAT IS REQUIRED OF A PEPTIDE TO ELICIT PROTEIN REACTIVE ANTIBODIES?

As mentioned in the introduction, peptides frequently elicit antibodies that can recognise denatured proteins but the induction of antibodies that can recognise native protein, and thus be of biological significance, is a rarer event. In fact, the solving of the crystal structure of antibody Fab fragments complexed with proteins has shown that the area involved in antigen–antibody contact is 600–700 Å^2. Synthetic mimicry of such a large area would be exceedingly difficult and so the fact that biologically active antibodies can, in many cases, be elicited by simple linear peptides presents us with an apparent paradox. However, the relative importance of each of the residues in contact with the antibody paratope is unclear at present. Experiments with peptides suggested that, with most antibodies, six to seven amino acids are the minimum number required for antibody binding (Geysen *et al.*, 1984). Furthermore, experiments using replacement sets of peptides in which each amino acid in turn is exchanged with all possible alternatives have shown that many residues can be exchanged without affecting antibody binding and only a few are irreplaceable (Getzoff *et al.*, 1989). However, the spatial orientation of the critical binding residues must be of paramount importance in determining the specificity of antibody binding. Since the structural constraints on small peptides in an aqueous environment are minimal, it is likely that they can readily adopt the most appropriate topographical arrangement to fit the binding pocket of an antibody molecule, provided that they contain the features essential for binding. It follows that, although a particular peptide may react well with an anti-protein antibody by induced fit, the peptide will not necessarily induce a significant population of antibodies which are reactive with the intact native protein.

Regions of protein surfaces vary in their degree of flexibility or thermal lability and it is interesting that there is a correlation between the most immunogenic sites and those demonstrated to have greatest flexibility (Tainer *et al.*, 1984; Westhof *et al.*, 1984). This implies that a propensity for induced fit between antibody and antigen may be an important feature in determining the immunogenic profile of proteins. It is interesting that one of the few structures of antibody–antigen complexes which has been solved showed that some rearrangement had occurred on binding (Colman *et al.*, 1987).

Although some capacity for induced fit between antibodies and antigens may be a common feature of antigenic sites on native proteins, the potential range of antibodies inducible by a randomly structured peptide would extend far beyond the potential for induced fit with a protein epitope. Consequently, the antibodies induced by a peptide immunogen which are capable of reacting with the same sequence when structurally constrained within the context of the folding of the intact native protein will statistically represent only a small fraction of the range of anti-peptide species. The obvious approach to resolving this problem is to design peptide immunogens in which the range of conformational possibilities is restricted and includes structures mimicking the shape of the sequence in the native protein. Some attempts have been made to address this question but before discussing these it is worth considering some specific examples to illustrate the problem.

Foot-and-mouth disease virus is one of the best examples of the effectiveness of synthetic peptide immunogens. Linear peptides representing an immunodominant sequence of VP1 (amino acids 141–160) can induce high levels of virus neutralising antibodies in laboratory animals (guinea pigs) which are protected from virus challenge after a single injection of the experimental vaccine (Bittle *et al.*, 1982). Analysis of anti-peptide antisera showed that as much as 30% of the total anti-peptide antibodies were reactive with virus particles (Parry *et al.*, 1988). The probable reasons for this extraordinarily high proportion of virus-reactive antibodies in anti-peptide sera became apparent from structural studies both with the virus and with the synthetic peptides. Resolution of the structure of the virus by X-ray crystallographic methods showed that the immunodominant region of VP1 (residues 134–158) occurs as a highly disordered loop at the surface of the virus particle (Acharya *et al.*, 1989). There was insufficient information in the electron density maps to ascribe a structural motif to this part of the virus. The nature of the disorder is, of course, unknown. The possibilities include: (i) the sequence adopts several discrete conformational forms which appear as disordered statistically; (ii) the loop sequence adopts a defined conformation but is linked to the virus surface via a flexible stem structure and therefore appears disordered in space; or (iii) the whole loop region has no preferred conformational structure and is continuously flexible. Whichever of these possibilities is correct, it is likely that the disordered and highly exposed nature of this sequence is linked to the success of simple linear peptides in inducing virus reactive antibodies.

Structural studies of synthetic peptides representing FMDV loop sequences using circular dichroism, 2D nuclear magnetic resonance and molecular modelling techniques suggest that they do adopt preferred conformational forms in solution (Mascagni *et al.*, unpublished data). Furthermore, amino acid substitutions within the sequence which have a profound effect on the specificity of the anti-peptide antibodies also dramatically affect the conformation of the peptides. It would appear then that both the unusual location of the FMDV loop on the virus and the inherent ability of the peptides to adopt a restricted range of conformational forms account for their success as immunogens.

A contrasting example is provided by another picornavirus, human rhinovirus. With this system sequences have been identified which can elicit virus-neutralising responses when presented as linear synthetic peptides but the titres are much lower than those seen with the FMDV system (Francis *et al.*, 1987*b*). However, the proportion of anti-peptide antibodies which is cross reactive with virus particles in the rhinovirus system is only 1% or less (M. Francis, personal communication). This difference is probably due both to the greater conformational restriction of the antigenic sequences on the rhinovirus particle compared to the FMDV site and the propensity of the peptides to adopt either no preferred conformational forms or ones that are inappropriate to the virus structure. In this and many other systems the design of peptides which structurally mimic the *in situ* conformation of the sequence in the native protein is likely to greatly improve the cross-reactive population of antibodies.

Cyclisation is the simplest method of restricting the conformational options of peptides and there are several reports of improvement in the antigenicity of peptides by this procedure (Kennedy *et al.*, 1983; Schulze-Gahmen *et al.*, 1986). More sophisticated approaches to producing peptides of defined structure have been explored by workers at the Scripps Research Institute (Satterthwait *et al.*, 1988). The principle involved in these studies was to synthesise peptides using modified amino acids which allowed the formation of covalent bonds at defined positions in the peptide, thus producing stable equivalents of naturally occurring hydrogen bonds in the native conformation. Using these methods with the asparagine–alanine–asparagine–proline repeat sequence from the circumsporozoite protein of malaria, antibodies specific for conformational alternatives of the peptide could be elicited. Although in its infancy at present, the development of methodologies for defining and fixing the three-dimensional structure of peptide immunogens will be crucial to their broader application in vaccine technology.

Another problem which has to be considered in the development of synthetic vaccines is that of antigenic variation. This is an acute problem with many viruses, for example, influenza virus, human rhinovirus and foot-and-mouth disease virus (FMDV), and is relevant to conventional as well as to novel vaccine development. It might be anticipated that the difficulties in inducing effective immunity in the face of natural antigenic variation would be exacerbated by the use of a peptide immunogen which can only represent a small part of the virus protein surface. By the same token it might be expected that there would be rapid selection of variant virus able to escape neutralisation by antibodies which can only react to a small antigenic site on the virus, much as is seen when viruses are placed under the selection pressure of monoclonal antibodies. However, the situation may be more hopeful than these discouraging considerations would predict. With FMDV the selection of neutralisation-resistant mutants with antipeptide antiserum was accomplished only with great difficulty by repeated passage of the virus in the presence of antiserum (Parry *et al.*, 1989). Furthermore, the variant viruses obtained were fully susceptible to neutralisation by antisera from other individual animals which had received the same peptide. This is quite different from the situation in which monoclonal antibodies are used for selection, in which case fully resistant variants arise at a frequency of *ca* 1×10^4. The probable explanation for this difference is that polyclonal antipeptide antisera contain a range of antibodies recognising different aspects of the antigenic sequence and that residues that are crucial for the binding of some antibodies may be of little importance for others. Consequently, to escape neutralisation, there would need to be more than one substitution within the sequence or substitutions which result in profound conformational changes of the antigenic site. However, antiserum to a shorter peptide representing an antigenic site of poliovirus did not neutralise variant viruses selected with monoclonal antibodies although it did neutralise a range of wild type isolates (Ferguson *et al.*, 1985).

Another surprising observation with the FMDV system was that antipeptide antisera were more broadly cross reactive with naturally occurring variant viruses than antisera raised against the whole virus (Ouldridge *et al.*, 1986). This was particularly true with viruses in which it has been shown that the sequence used to make peptides is part of a topographically assembled epitope on the virus. Presumably the explanation for this phenomenon is that, by representing only a portion of a complex antigenic site, the antipeptide antibodies are not as susceptible to minor rearrangements of the components of the natural site as are antibodies elicited by the intact virus.

In addition to the broadly cross-reactive antibody responses elicited by peptides representing naturally occurring FMDV sequences, the spectrum of cross reactivities can be further altered by changes of the sequence.

Although the results of such sequence changes are unpredictable and the structural basis of their mode of action is not understood, artificial sequences have been produced which further expand the range of cross neutralising activities induced by the peptides (Brown, 1985; Parry *et al.*, 1989). One modified peptide induced protective neutralising responses to two serotypes of the virus.

THE PROBLEM OF MHC RESTRICTION

In order to elicit an effective immune response, an immunogen must include, in addition to antigenic features necessary to induce antibodies with appropriate specificities, sequences which can function as helper T cell epitopes. Helper T cells stimulate antigen-reactive B cells to proliferate and mature into antibody-secreting plasma cells and to produce a population of memory B cells. The recognition signal, between helper T cells and B cells, which facilitates the exchange of lymphokine and receptor-mediated stimuli to precipitate these maturation events in B cells, involves linear peptide sequences (T cell epitopes) derived by processing of the immunogenic protein. T cells recognise these peptide epitopes via immunoglobulin-like molecules, namely the T cell receptors, at their surfaces. The enormous array of binding specificities displayed by T cell receptors is derived by genomic rearrangements of the variable domains of the protein in much the same way as antibody diversity is generated. Although T cell receptors recognise specific peptides derived from an immunogenic protein, they only do so when the peptide is presented in combination with cell surface proteins of the major histocompatibility complex (MHC). Consequently, a limitation of the potential functioning of peptide sequences as T cell epitopes is that they must be able to associate with an MHC molecule. MHC class I molecules occur at the surface of nearly all cells in the body and are recognised by cytotoxic T cells of the CD8(T8) phenotype. Helper T cells (CD4 or T4 cells), which are important for the maturation of humoral immune responses, recognise epitopes in combination with MHC class II molecules. These are only expressed at the surface of cells involved in immune responses. Although the MHC molecules are extremely polymorphic throughout populations, each individual haplotype only codes for two or three variants, depending on the animal species. This situation results in genetic restriction of response to antigens since a sequence within a protein which can bind to the MHC molecules of one individual, and function as a T cell epitope, may not be able to bind to a different array of MHC molecules in a different individual of the same species.

The processing events that result in the presentation of peptide fragments in combination with MHC molecules are poorly understood. It has recently become apparent that two pathways resulting in MHC class I or class II presentation exist (Long & Jacobson, 1984). The pathway leading to class

I presentation appears to be restricted to proteins present in the cytosol, usually to proteins that have been translated within the cell. Antigens which are internalised by endocytosis are degraded by a pathway which results in class II presentation. In both cases the T cell epitope peptides appear to reside in a distinct groove at the tip of the MHC molecule (Bjorkman *et al.*, 1987) and it is the affinity of peptides for binding into this groove that dictates their potential functioning as T cell epitopes.

Naïve helper T cells are primed to follow a proliferative and maturation pathway when they first encounter epitopes displayed at the surface of antigen-presenting cells with a complementary specificity to the receptor molecules they bear. Many cells, such as macrophages, can function as antigen-presenting cells but the most significant *in vivo* appear to be dendritic cells (Austyn, 1987). Primed helper T cells function to stimulate B cells which display the same T epitope by dint of having acquired the antigen via the specificity of their surface immunoglobulins.

In summary, B cell epitopes and T cell epitopes usually represent different components of an immunogenic protein and both are required for the generation of a good antibody response. The obvious relevance of this situation for the development of sub-unit immunogens such as peptide vaccines is that synthetic constructs must include helper T cell epitopes in addition to B cell epitopes capable of eliciting the appropriate antibody specificities. In the situation in which a peptide is chemically coupled to, or is colinearly expressed with, a carrier protein, then helper T cell epitopes may be provided by the latter. However, to pursue the goal of a totally synthetic vaccine, much consideration must be given to the inclusion of helper T cell epitopes which can function in a sufficiently large proportion of the population and are situated appropriately in relation to the B cell epitope components to provide strong functional help to the antibody response.

Studies of synthetic peptide vaccines for FMDV illustrate the practical consequences of these considerations. A 20-amino acid sequence from the VP1 protein of the virus can induce a good and reasonably uniform virus-neutralising antibody response in outbred guinea pigs (Francis *et al.*, 1988). However, comparison of responses in inbred mice of defined MHC haplotype showed that some strains responded very well and others produced no measurable quantities of antibody, i.e. there was classical MHC restriction of responsiveness. Colinear synthesis of the FMDV sequence with additional 17 amino acid tracts, representing sequences which had previously been shown to function as T cell epitopes in those strains of mice which could not elicit a response to the FMDV peptide alone, induced a good anti-FMDV peptide response (Francis *et al.*, 1987*c*). Similar observations have been made with synthetic peptide antigens from a number of other systems (Good *et al.*, 1987; Borras-Cuesta *et al.*, 1987; Leclerc *et al.*, 1987).

These experiments formally demonstrated that the problem of genetic restriction of antibody response can be overcome by appropriate design of peptide immunogens. An interesting complication with the FMDV experiments was that, of the three 'foreign' T cell epitope constructs used, two resulted in a spectrum of antibodies recognising in total the whole of the FMDV peptide sequence but the third construct only elicited antibodies to a restricted portion of the sequence. Subsequently it has been shown that the introduction of a short stretch of glycine residues between B and T cell epitopes can abrogate antibody responses (M. J. Francis, personal communication). These observations presumably result from alterations in antigen processing which, as stated above, is still poorly understood.

An encouraging feature of the MHC restriction problem is that a number of laboratories have found T cell epitopes that are functionally active in a wide range of MHC backgrounds (Francis *et al.*, 1989; Sinigaglia *et al.*, 1988). This would reduce the number of T cell epitopes required to generate responses in an acceptable proportion of individuals.

The role of cytotoxic cells in protection against infection is unclear and probably varies with different pathogens. Their major role appears to be in the clearance of established infection but in some diseases their actions may exacerbate clinical manifestations (Lukacher *et al.*, 1984; Cannon *et al.*, 1989). Because of the nature of the processing pathway of cytotoxic T cell epitopes, inactivated vaccines, including peptide immunogens do not generally elicit cytotoxic cell responses. It remains to be seen whether peptide vaccines can be designed which will induce CTL responses and, if so, whether these will play a positive role in protection against disease.

ENHANCEMENT OF IMMUNOGENICITY

As a general rule, peptides of 20–30 amino acids are poorly immunogenic when injected in the free form even when the requirements for T cell epitopes outlined above are satisfied. For example, uncoupled FMDV peptide can induce protective responses in guinea pigs following a single inoculation but this requires *ca* 100 μg of peptide. In terms of molarity this dose is several orders of magnitude greater than the amount of non-infectious virus required to elicit a similar response. Furthermore, in order to achieve useful responses with this amount of peptide it is important that the 20 amino acid sequence is dimerised, e.g. by the oxidation of C terminal cysteine residues to form disulphide links (Francis *et al.*, 1987*a*). One of the reasons for the poor immunogenicity of peptides is probably their rapid clearance from the body. Peptides of about 20 amino acids in length can be cleared from the bloodstream by glomerular filtration in a matter of seconds. Consequently, improvements in immunogenicity of synthetic peptides can be achieved by increasing their effective size either by chemical coupling to large carrier molecules or by cross linking them using bifunctional reagents such as glutaraldehyde. A more sophisticated method for polymeric presentation of peptides has been developed which involves the synthesis of a defined multimeric core structure to which the peptide antigens are attached (Tam, 1988). This core moiety is produced by conventional peptide synthesis procedures using lysine with the modification that both the α- and ε-amino groups are blocked with the same protecting group. The consequence of this is that both amino groups can form peptide bonds during the next round of amino acid addition, so building a multi-branched carrier core in which the number of amino groups available for coupling is doubled at each step. Having constructed a poly-lysine core of the desired size, peptide sequences can be synthesised simultaneously from each amino group. Octameric peptides produced in this way are far more immunogenic than peptide monomers. Peptide epitopes have also been presented as uniform polymeric arrays by expressing the relevant sequences fused to proteins which have the innate property of self-assembly into higher order structures. A number of fusion partners have been used in this approach including surface or core antigen proteins of hepatitis B virus, the Ty particle protein of yeast and the capsid proteins of poliovirus (Delpeyroux *et al.*, 1986; Clarke *et al.*, 1987; Adams *et al.*, 1987; Burke *et al.*, 1989). Peptide sequences have been added at the N or C termini of the proteins (e.g. hepatitis B core protein and Ty protein) or introduced within the protein at positions which form surface loops (e.g. hepatitis B surface antigen and poliovirus capsid proteins).

Comparative immunogenicity studies with chimaeric constructs based on hepatitis B virus core (HBc) antigen have shown this to be a very efficient carrier system for eliciting high-level responses to peptide epitopes (Francis *et al.*, 1990). The HBc protein has a molecular weight of *ca* 21,000 and self assembles into virus-like particles of *ca* 27 nm diameter. Amino acid sequences added to either the N or C termini of the protein are located at the surface and so form high density arrays on the assembled particles. Chimaeric HBc particles including an N terminal extension representing the FMDV loop sequence are almost as immunogenic on a particle weight basis as FMD virions (Clarke *et al.*, 1987). The high density, multimeric presentation of epitopes on chimaeric HBc particles is important for immunogenicity since monomeric fusion protein elicits much lower levels of antibody (Francis *et al.*, 1990). In addition, HBc protein contains a number of powerful T cell epitopes within its sequence (Milich & McLachlan, 1987).

The antibody response to HBc-linked epitopes is 10-fold better than that obtained by coupling the peptide to keyhole limpet haemocyanin and at least 100-fold better than the response to free peptide. Furthermore, good responses can be obtained in the presence of aluminium hydroxide gel, the only adjuvant licensed for human use. Lower but significant responses are elicited following injection in the absence of adjuvant or by the intranasal or oral routes (Francis *et al.*, 1990). A thorough understanding of the reasons for the potent immunogenicity of these particles may enable the construction of chemically synthesised immunogens of comparable activity.

The insertion of peptide sequences at appropriate positions within a protein has the potential advantage that the conformational constraints imposed by such a location are likely to result in the peptide adapting a more 'native' structure. In fact, the replacement of antigenic loops in the poliovirus VP1 protein with foreign sequences has produced viable chimaeric viruses (Burke *et al.*, 1989). These can induce antibodies capable of recognising the native antigen from which the inserted sequence was derived better than those elicited by linear synthetic peptides.

Appropriate B cell epitopes, adequate T cell epitopes and multimeric assembly are all features required for inducing an antibody response. In addition, there is the question of adjuvanticity. Adjuvants are almost universally required to elicit high levels of response to inert antigen and, moreover, the choice of adjuvant can influence the isotype composition of the antibodies produced (see Bomford, this volume). The most widely used and powerful experimental adjuvant is complete Freund's adjuvant (CFA), an emulsion of mycobacterium cell walls in oil. However, the severe side-effects associated with this preparation precludes its use in commercial vaccines. Muramyl dipeptide (MDP) has been identified as one of the active ingredients of CFA and chemical linkage of this molecule to synthetic peptides has been reported to improve immunogenicity (Audibert *et al.*, 1982). MDP retains some of the pyrogenic activity of CFA but derivatives have been prepared which lack these toxic effects but still enhance immunogenicity (Allison & Byars, 1986).

The proposal to construct synthetic vaccines with built-in adjuvanting properties is highly attractive and the work with MDP is an encouraging start. However, we still need to learn a lot more about the mechanism of action of adjuvants before this approach can be developed in a rational way. An attractive possibility for the future is the incorporation of targeting features into immunogens so that they will be specifically recognised by appropriate cells of the immune response mechanism.

A feature of most vaccines, current or experimental, is that several doses are required to induce a high-level, long-lasting response. This requirement often produces great logistical problems of identifying and assembling vaccinees for the booster applications. The inherent stability of synthetic vaccines offers a solution to this problem. A portion of the antigen could be entrapped within capsules designed to release their contents weeks or months after injection and so provide the boosting doses of antigen without the requirement for further injections. Preliminary experiments in which FMDV peptide was encased within soluble glass capsules provided encouragement for the feasibility of this approach (M. Francis & C. Drake, personal communication).

MUCOSAL IMMUNITY

The portal of entry of most infectious agents is via mucosal epithelia of the gut or the respiratory tract. It would therefore seem logical that the induction of secretory immunity would be desirable in providing the first line of defence against infection. Unfortunately, parenteral inoculation of non-replicating antigens induces very little secretory antibody. Delivery of antigen directly to mucosal epithelia by the oral or intranasal routes results in better secretory responses and is being explored as an alternative method of vaccination. This approach has the added attraction that vaccinations would not involve needle injection. The additional problems in vaccine design associated with delivery of antigens to mucosal surfaces are the extra hazards of proteolytic degradation, particularly via the oral route, and the efficiency of uptake and transfer to cells of the immune system.

Possible methods for preventing degradation are: (i) the micro-encapsulation of antigen in, for example, liposomes; (ii) the design of synthetic antigens using analogues of natural amino acids; and (iii) genetic expression of peptide epitopes as fusion proteins (e.g. with flagellin) by enteric bacteria (Newton *et al.*, 1989). Enhancement of antigen uptake may be improved by incorporating specific epithelial cell binding features such as those present on bacterial pili or toxins. The β sub-units of cholera toxin, for example, induces good levels of secretory and humoral antibody by the intranasal route (Tamura *et al.*, 1988).

INDUCTION OF SPECIFIC AUTOIMMUNITY

An interesting application of synthetic peptide antigens is their potential therapeutic use by the induction of immunity to specific host molecules. An example of this approach is the development of a birth control vaccine (reviewed in Griffin, 1986). Peptides representing sequences from chimaeric gonadatrophin have been shown to elicit antibodies which react with, and neutralise, the hormone *in vivo*, resulting in sterility. Vaccines based on this principle are currently in human trials.

In other studies it has been shown that the induction of antibodies against renin can reduce blood pressure in naturally hypertensive rats (Carelli *et al.*, 1990). This is because the antibodies prevent renin from cleaving angiotensinogen to produce angiotensin. However, when the whole protein (mouse renin) was used as the immunogen there was evidence of immune complex disease. These side-effects were not observed when renin peptides were used to vaccinate the rats and reduction in blood pressure was still observed. It will be interesting to see how far the deliberate manipulation of autoimmune responses for therapeutic purposes will proceed.

INDUCTION OF 'CATALYTIC' IMMUNITY

The principal feature of antibodies induced against an antigen is that they bind to specific features of the antigen (epitopes), often with high efficiency. The beneficial effects of this union with respect to protection against a pathogen are derived in two ways. First, the binding of antibodies may interfere with vital early events in infection such as attachment to cells or uncoating in the case of viruses. Second, destruction of the pathogenic organism may proceed via the effector domains on the antibody, resulting in opsonisation and engulfment by macrophages or complement-mediated lysis. The antibody–antigen interaction itself, however, is an inert event. Enzymes, on the other hand, bind substrate in such a way as to facilitate a chemical transition. This is achieved because the binding site of an enzyme recognises not the ground energy state of the substrate but a rarer high energy form. By trapping the substrate in such an unstable form the energy barrier for completion of a chemical transformation is thereby effectively reduced. By designing antigens which mimic the transition states involved in such chemical conversions it has been possible to induce antibodies which function as enzymes. Recently the design of a high energy intermediate analogue incorporating a metal ion has resulted in the induction of antibodies capable of cleaving a peptide bond (Iverson *et al.*, 1989). The potential extension of this work is the development of synthetic antigens which induce antibodies capable not only of recognising a specific feature of a protein but of cleaving the protein at that position. It is therefore possible to envisage vaccines that will produce 'hit and run' antibodies that can recognise, inactivate and move on, an appealing prospect that relies entirely on inventive chemistry.

CONCLUSIONS

Synthetic peptides have been shown to work, in terms of inducing protective immunity, in a number of systems. However, the identification of appropriate sequences to use is difficult and their immunogenicity is generally low. Both of these problems are the subject of intensive research, principally because of the considerable perceived benefits to be derived from producing chemically defined vaccines.

Another potential advantage of synthetic immunogens is that they may be designed to induce antibodies with specificities not seen following immunisation with conventional immunogens. Examples are defined anti-immune responses, catalytic antibodies and antibodies against features of immunogens that are normally immunorecessive.

Whatever the practical outcome, the exploration of these possibilities makes for exciting science and, at the least, will lead to an improved understanding of the functioning of immune responses.

REFERENCES

Acharya, R., Fry, E., Stuart, D. I., Fox, G., Rowlands, D. J. & Brown F. (1989). The three dimensional structure of foot-and-mouth disease virus at 2.9 Å. *Nature (London)*, **337**, 709–16.

Adams, S. E., Dawson, K. M., Gull, K., Kingsman, S. & Kingsman, A. J. (1987). The expression of hybrid HIV: Ty virus like particles in yeast. *Nature (London)*, **329**, 68–70.

Allison, A. C., Byars, N. E. & Walters, R. V. (1986). Immunologic adjuvants: efficacy and safety considerations. In *Advances in Carriers and Adjuvants for Veterinary Biologics*, Nervig, R. M., Gough, P. M., Kaeberle, M. L. & Whetstone, C. A., eds, pp. 91–103, Ames Iowa: State University Press.

Anderer, F. A. (1963). Versuche zur Bestimmung der Serologisch Terminaten Gruppen des Tobakmosaikvirus. *Naturforsch. Teil B.*, **188**, 1010–14.

Anderer, F. A. & Schlumberger, H. D. (1965). Properties of different artificial antigens immunologically related to tobacco mosaic virus. *Biochimica Biophysica Acta*, **97**, 503–9.

Audibert, F., Jolivet, M., Chedid, L., Arnon, R. & Sela, M. (1982). Successful immunisation with a totally synthetic diphtheria vaccine. *Proceedings of the National Academy of Sciences, USA*, **79**, 5042–6.

Austyn, J. M. (1987). Lymphoid dendritic cells. *Immunology*, **62**, 161–70.

Barlow, D. J., Edwards, M. G. & Thornton, J. M. (1986). Continuous and discontinuous antigenic determinants. *Nature (London)*, **322**, 747–8.

Biou, V., Gibrat, J. F., Levin, J. M., Robson, B. & Garnier, J. 1988). The secondary structure prediction: combination of three different methods. *Protein Engineering*, **2**, 185–91.

Bittle, J. L., Houghten, R. A., Alexander, H., Schinnick, T. M., Sutcliffe, J. G., Lerner, R. A., Rowlands, D. J. & Brown, F. (1982). Protection against foot and mouth disease by immunisation with a chemically synthesised peptide predicted from the viral nucleotide sequence. *Nature (London)*, **298**, 30–3.

Bjorkman, P. J., Saper, M. A., Samraout, B., Bennett, W. S., Strominger, J. L. & Wiley, D. C. (1987). Structure of the human class I histocompatibility antigen, HLA-A2. *Nature (London)*, **329**, 506–11.

Borras-Cuesta, F., Petit-Camurdan, A. & Fedan, Y. (1987). Engineering of immunogenic peptides by co-linear synthesis of determinants recognised by B and T cells. *European Journal of Immunology*, **17**, 1213–16.

Brown, F. (1985). Synthetic peptides as foot and mouth disease vaccines. In *High Technology Route to Virus Vaccines*. American Society for Microbiology, p. 49–54.

Burke, K. L., Evans, D. J., Jenkins, O., Meredith, J., D'Souza, E. D. A. & Almond, J. W. (1989). A cassette vector for the construction of antigenic chimaeras of poliovirus. *Journal of General Virology*, **70**, 2475–9.

Cannon, M. J., Openshaw, P. J. M. & Askonas, B. A. (1989). Lethal pulmonary disease in respiratory syncytial virus infected mice following transfer of cytotoxic T cells. In *Vaccines 89: Modern Approaches to New Vaccines*. Lerner, R.A., Ginsberg, H. S., Chanock, R. M. & Brown, F., eds New York: Cold Spring Harbor Laboratory, p. 497–9.

Carelli, C., Fehrentz, J. A., Galen, F. X., Liu, C. F., Michel, J. B., Castro, B. & Corval, P. (1990). Anti-renin immunisation–Simply a new experimental model of a synthetic vaccine? In *Vaccine 90: Modern Approaches to New Vaccines*, Brown, F., Chanock, R., Ginsberg, H. S. & Lerner, R. A. eds, New York: Cold Spring Harbor Laboratory, in press.

Chou, P. Y. & Fasman, G. D. (1978). Prediction of secondary structure of proteins from their amino acid sequence. *Advances in Enzymology*, **47**, 145–8.

Clarke, B. E., Newton, S. E., Carroll, A. R., Francis, M. J., Appleyard, G., Syred, A. D., Highfield, P. E., Rowlands, D. J. & Brown, F. (1987). Improved immunogenicity of a peptide epitope after fusion to hepatitis B core protein. *Nature (London)*, **330**, 381–4.

Colman, P. M., Laver, W. G., Varghese, J. N., Baker, A. T., Tulloch, P. A., Air, G.M. & Webster, R.G. (1987). Three dimensional structure of a complex of antibody with influenza virus neuraminidase. *Nature (London)*, **326**, 358–63.

Delpeyroux, F., Chenciner, N., Lim, A., Malpiece, Y., Blondel, B., Crainic, R., van der Werf, S. & Streeck, R. E. (1986). A polio neutralization epitope expressed on hybrid hepatitis B surface antigen particles. *Science*, **233**, 472–5.

Ferguson, M., Evans, D. M. A., Magrath, D. I., Minor, P. D., Almond, J. W. & Schild, G. C. (1985). Induction by synthetic peptides of broadly reactive, type specific neutralizing antibody to poliovirus type 3. *Virology*, **143**, 505–15.

Francis, M. J., Fry, C. M., Rowlands, D. J., Bittle, J. L., Houghten, R. A., Lerner, R. A. & Brown, F. (1987*a*). Immune response to uncoupled peptides of foot and mouth disease virus. *Immunology*, **61**, 1–6.

Francis, M. J., Hastings, G. Z., Sangar, D. V., Clark, R. P., Syred, A., Clarke, B. E., Rowlands, D. J. & Brown, F. (1987*b*). A synthetic peptide which elicits neutralising antibody against human rhinovirus type 2. *Journal of General Virology*, **68**, 2687–91.

Francis, M. J., Hastings, G. Z., Syred, A. D., McGinn, B., Brown, F. & Rowlands, D. J. (1987*c*). Non-responsiveness to a foot and mouth disease virus peptide overcome by addition of foreign helper T cell determinants. *Nature (London)*, **330**, 168–70.

Francis, M. J., Fry, C. M., Rowlands, D. J. & Brown, F. (1988). Qualitative and quantitative differences in the immune response to foot and mouth disease virus antigens and synthetic peptides. *Journal of General Virology*, **69**, 2483–91.

Francis, M. J., Hastings, G. Z., Campbell, R. O., Rowlands, D. J., Brown, F. & Peat, N. (1989). T cell help for B cell antibody production to rhinovirus peptides. In *Vaccines 89: Modern Approaches to New Vaccines including prevention of AIDS*. Lerner, R. A., Ginsberg, H. S., Chanock, R. M. & Brown, F. eds, New York: Cold Spring Harbor Laboratory, p. 437–44.

Francis, M. J., Clarke, B. E., Hastings, G. Z., Brown, A. L., Rowlands, D. J. & Brown, F. (1990). Immune response to peptide/hepatitis B core antigen fusion proteins. In *Vaccines 90: Modern Approaches to New Vaccines*, Brown, F., Chanock, R., Ginsberg, H. S. & Lerner, R. A. eds, New York: Cold Spring Harbor Laboratory, in press.

Getzoff, E. D., Geysen, H. M., Rodda, S. J., Alexander, H., Tainer, J. A. & Lerner, R. A. (1987). Mechanisms of antibody binding to a protein. *Science*, **235**, 1191–6.

Getzoff, E. D., Tainer, J. A., Lerner, R. A. & Geysen, H. M. (1989). The chemistry and mechanism of antibody binding to protein antigens. *Advances in Immunology*, **43**, 1–97.

Geysen, H. M., Meloen, R. H. & Barteling, S. J. (1984). Use of peptide synthesis to probe viral antigens for epitopes to a resolution of a single amino acid. *Proceedings of the National Academy of Sciences, USA*, **81**, 3998–4002.

Geysen, H. M., Rodda, S. J. & Mason, T. J. (1986). *A priori* delineation of a peptide which mimics a discontinuous antigenic determinant. *Molecular Immunology*, **23**, 709–16.

Good, M. F., Maloy, W. L., Lunde, M. N., Margalit, H., Cornette, J. L., Smith,

G. L., Moss, B., Miller, L. H. & Berzofsky, J. A. (1987). Construction of synthetic immunogen: use of new T-helper epitope on malaria circumsporozoite protein. *Science*, **235**, 1059–62.

Griffin, P. D. (1986). A fertility regulating vaccine based on the carboxyl terminal peptide of the beta subunit of human chorionic gonadotrophin. In *Immunological Approaches to Contraception and Promotion of Fertility*. Talwar, G. P. ed., Plenum, 43–59.

Hogle, J. M., Chow, M., & Filman, D. J. (1985). Three dimensional structure of poliovirus at 2.9 Å resolution. *Science*, **229**, 1358–65.

Hopp, T. P. & Woods, K. R. (1981). Prediction of protein antigenic determinants from amino acid sequences. *Proceedings of the National Academy of Sciences, USA*, **78**, 3824–8.

Iverson, B. L. & Lerner, R. A. (1989). Sequence-specific peptide cleavage catalysed by an antibody. *Science*, **243**, 1184–8.

Kennedy, R. C., Dreesman, G. R., Sparrow, J. T., Culwell, A. R., Sanchez, Y., Ionescu-Matiu, I., Blaine Hollinger, F. & Melnick, J. L. (1983). Inhibition of a common human anti-hepatitis B surface antigen idiotype by a cyclic synthetic peptide. *Journal of Virology*, **46**, 653–5.

Leclerc, C., Przewlocki, G., Schutze, M. P. & Chedid, L. (1987). A synthetic vaccine constructed by copolymerization of B and T cell determinants. *European Journal of Immunology*, **17**, 269–73.

Le Francier, P., Derrien, M., Jamet, X., Choay, J., Lederer, F., Audibert, M., Parant, M., Parant, F. & Chedid, L. (1982). A pyrogenic, adjuvant-active N-acetylmuramyl-dipeptide. *Journal of Medicinal Chemistry*, **25**, 87–90.

Lerner, R. A. (1982). Tapping the immunological repertoire to produce antibodies of predetermined specificity. *Nature (London)*, **299**, 592–6.

Levitt, N. (1976). A simplified representation of protein conformation for rapid simulation of protein folding. *Journal of Molecular Biology*, **104**, 59–107.

Long, E. O. & Jacobson, S. (1989). Pathways of viral antigen processing and presentation to CTL. *Immunology Today*, **10**, 45–8.

Lukacher, A. E., Braciale, V. L. & Braciale, T. J. (1984). *In vivo* effector functions of influenza virus specific cytotoxic T lymphocyte clones is highly specific. *Journal of Experimental Medicine*, **160**, 814–26.

Merrifield, R. B. (1963). Solid phase peptide synthesis 1. The synthesis of a tetrapeptide. *Journal of the American Chemical Society*, **85**, 2149–54.

Milich, D. R., McLachlan, A., Moriarty, A. & Thornton, G. B. (1987). Immune response to hepatitis B virus core antigen (HBcAg): localisation of T cell recognition sites within HBcAg/HBeAg. *Journal of Immunology*, **139**, 1–9.

Newton, S. M. C., Jacob, C. O. & Stocker, B. A. D. (1989). Immune response to cholera toxin epitope inserted in *Salmonella* flagellin. *Science*, **244**, 70–2.

Nunberg, J. H., Rodgers, G., Gilbert, J. H. & Sneard, R. M. (1984). Method to map antigenic determinants recognised by monoclonal antibodies: localisation of a determinant of virus neutralisation of the feline leukaemia virus envelope protein GP70. *Proceedings of the National Academy of Sciences, USA*, **81**, 3675–9.

Ouldridge, E. J., Parry, N. R., Barnett, P. V., Bolwell, C., Rowlands, D. J., Brown, F., Bittle, J. L., Houghton, R. A. & Lerner, R. A. (1986). Comparison of the structures of the major antigenic sites of foot and mouth disease virus of two different serotypes. In *Vaccines 86: New Approaches to Immunization*, F. Brown, R. M. Chanock & R. A. Lerner, eds, New York: Cold Spring Harbor Laboratory, p. 45–9.

Parry, N. R., Syred, A., Rowlands, D. J. & Brown, F. (1988). A high proportion of antipeptide antibodies recognise foot and mouth disease virus particles. *Immunology*, **64**, 567–72.

Parry, N. R., Ouldridge, E. J., Barnett, P V., Clarke, B. E., Francis, M. J. Fox, J. D., Rowlands, D. J. & Brown, F. (1989). Serological prospects for peptide vaccines against foot-and-mouth disease virus. *Journal of General Virology*, **70**, 2919–30.

Rossmann, M. G., Arnold, E., Erickson, J. W., Frankenberger, E. A., Griffith, J. P., Johnson, J. E., Kamer, G., Luo, M., Mosser, A. C., Rueckert, R. R., Sherry, B. & Vriend, G. (1985). Structure of a human common cold virus and functional relationship to other picornaviruses. *Nature (London)*, **317**, 145–53.

Satterthwait, A. C., Arrhenius, T., Hagopian, R. A., Zavala, F., Nussenzweig, V. & Lerner, R. A. (1988). Conformational restriction of peptidyl immunogens with covalent replacements for the hydrogen bond. *Vaccine*, **6**, 99–103.

Schulze-Gahmen, U., Klenk, H-D. & Beyreuther, K. (1986). Immunogenicity of loop-structured short synthetic peptides mimicking the antigen site A of influenza virus haemagglutinin. *European Journal of Biochemistry*, **159**, 283–9.

Sela, M. (1969). Antigenicity: some molecular aspects. *Science*, **166**, 1365–74.

Sinigaglia, F., Guttinger, M., Kilgus, J., Doran, D. M., Matile, H., Etlinger, H., Trzeciak, A., Gillessen, D. & Pink, J. R. L. (1988). A malaria T cell epitope recognised in association with most mouse and human MHC class II molecules. *Nature (London)*, **336**, 778–80.

Strohmaier, K., Franze, R. & Adam K. H. (1982). Localisation and characterisation of the antigenic portion of foot and mouth disease protein. *Journal of General Virology*, **59**, 295–306.

Tainer, J. A., Getzoff, E. D., Alexander, H., Houghten, R. A., Olsen, A. J., Lerner, R. A. & Hendrickson, W. A. (1984). The reactivity of antipeptide antibodies is a function of the atomic mobility of sites in a protein. *Nature (London)*, **312**, 127–33.

Tam, J. P. (1988). Synthetic peptide vaccine design: synthesis and properties of a high density multiple antigenic peptide system. *Proceedings of the National Academy of Sciences, USA*, **85**, 5409–13.

Tamura, S., Samegai, Y., Kurata, H., Nagamine, T., Aizawa, C. Kurata, T. (1988). Protection against influenza virus infection by vaccine inoculated intranasally with cholera toxin B subunit. *Vaccine*, **6**, 409–13.

Westhof, E., Altschuh, D., Moras, D., Bloomer, A. C., Mondragon, A., Klug, A. & van Regenmortel, M. H. V. (1984). Correlation between segmental mobility and the location of antigenic determinants in proteins. *Nature (London)*, **311**, 123–6.

Wiley, D. C., Wilson, I. A. & Skehel, J. J. (1981). Structural identification of the antibody binding sites of Hong Kong influenza haemagglutinin and their involvement in antigenic variation. *Nature (London)*, **289**, 373–8.

IMMUNOMODULATION BY ADJUVANTS

R. BOMFORD

Wellcome Biotechnology, Langley Court, Beckenham, Kent BR3 3BS, UK.

INTRODUCTION

Vaccination has had an enormous impact on the control of virus diseases, the live attenuated vaccines such as those against smallpox or polio being particularly successful. Vaccines containing whole inactivated viruses or purified viral components are relatively less immunogenic and are usually administered together with an adjuvant. The only adjuvants at present licensed for human use are the aluminium gels such as $Al(OH)_3$, the adjuvant properties of which were discovered more than 60 years ago (Glenny *et al.*, 1926). They are in widespread use for traditionally – prepared inactivated vaccines (Aprile & Wardlaw, 1966) and have also perforce been used for modern vaccines containing recombinant proteins such as those against Hepatitis B (Jilg *et al.*, 1984; Scolnick *et al.*, 1984).

It is now possible to control the specificity of the immune response to subunit viral vaccines by the use of recombinant proteins containing protective epitopes, or even more finely with synthetic peptides (see Chapters 3 and 5 in this volume). It would be highly desirable to have a comparable degree of control over the nature of the immune response elicited by these antigens, in terms of antibody isotype and cell-mediated immunity (CMI). It is the purpose of this chapter to examine the potential of adjuvants for selective stimulation of the various components of immunity, and the relevance of this for viral vaccines. The more general aspects of adjuvants such as their chemical structure and mode of action have been reviewed elsewhere (Adam, 1985; Allison & Byars, 1986; Morein, 1988; Warren & Chedid, 1988) and the very important question of antigen presentation is covered in Chapter 2 of this volume.

THE CONTROL OF CELL-MEDIATED IMMUNITY BY ADJUVANTS

One of the observations which led Freund to develop his famous adjuvant was that the injection of a protein antigen into a tuberculous lesion in a guinea pig produces hypersensitivity to the protein itself (Dienes, 1929). Freund's complete adjuvant (FCA) consists of a water-in-oil emulsion of mineral oil, with mycobacteria suspended in the oil phase. If the mycobacteria are omitted, the resulting Freund's incomplete adjuvant (FIA) no longer induces delayed-type hypersensitivity (DTH) to protein antigens

(White, Coons & Connolly, 1955; Bomford, 1980*b*). The adjuvant activity of mycobacteria has been localised to the peptidoglycan components of the cell wall and can be reproduced with the synthetic analogue muramyl dipeptide (MDP) (Ellouz *et al.*, 1974). MDP displays adjuvant activity in the aqueous phase, but only for the antibody response. It needs to be formulated with an oily vehicle in order to elicit DTH, although the water-in-oil emulsion of FIA can be replaced by an oil-in-water emulsion (Carelli, Audibert & Chedid, 1981; Byars & Allison, 1987).

The combination of mycobacteria or their components and an oily vehicle are not the only adjuvants capable of inducing DTH. Surface-active molecules such as the quaternary amine dimethyl dioctadecyl ammonium bromide (Snippe, Belder & Willers, 1977) or the triterpene glycoside saponin (Scott *et al.*, 1984) are also effective. The latter example is particularly interesting in that the generation of DTH could be correlated to protective immunity against the protozoan parasite *Trypanosoma cruzi,* one of the many pieces of evidence that CMI may be important in protection against multi-cellular parasites (reviewed in Bomford, 1989). Saponin forms particulate structures with the proteins from enveloped viruses which have been termed immunostimulating complexes (ISCOMS) and which have displayed strong immunogenicity with a variety of viral antigens (Morein, 1988).

$Al(OH)_3$ is not an effective adjuvant for CMI (Bomford 1980*b*) and might therefore be predicted to be a relatively poor adjuvant for vaccines against viruses whose replication is controlled by CMI. In a guinea pig model of genital herpes infection, it was found that FCA induced much stronger protection than $Al(OH)_3$ (Berman *et al.*, 1985). Both adjuvants stimulated neutralising antibodies, and the titres with FCA were higher, but the levels of CMI were not measured.

A second component of CMI which may play a role in the immunological control of many virus diseases is cellular cytotoxicity mediated by T lymphocytes of the CD8 phenotype (CTL) restricted by class I antigens of the Major Histocompatibility Complex (MHC). CTL usually appear only after infection with a living virus, or immunisation with living cells but not with non-living antigens, even when they are combined with powerful adjuvants such as FCA. This difference in immunisation requirements has been explained by the hypothesis that the processing of antigens for presentation to CD4 class II-restricted or CD8 class-I restricted T lymphocytes proceeds by different pathways (Bevan, 1987). Both types of T cell recognise small peptide products of processing. The products of exogenous antigens which enter the antigen-presenting cell (APC) by pinocytosis become associated with class II antigens, whereas class I antigens only pick up the products of endogenous antigens, synthesised in the cytoplasm. A CTL response to an exogenous antigen (ovalbumin) could be elicited if the APC were allowed to pinocytose the antigen and were then subjected to hypotonic shock, causing the pinocytotic vesicles to rupture and liberate the antigen into the cytosol (Moore, Carbone & Bevan, 1988). This observation offers an important clue for the design of adjuvants for CTL, which should promote the transport of exogenous antigen to the cytosol. Liposomes might fulfil this role and a limited degree of success has been reported for CTL against influenza nucleoprotein, where it was found that liposomally associated nucleoprotein could be used for *in vitro* amplification of CTL, although not for primary *in vivo* immunisation (Hackett, Taylor & Askonas, 1983). Also, ISCOMS containing influenza virus glycoproteins stimulated CTL precursors in the lungs of mice after intranasal inoculation (Jones *et al.*, 1988). However, at present it appears that the most effective way of introducing antigens into the cytosol is to use a viral vector, such as vaccinia carrying foreign viral genes (see Chapter 4).

THE CONTROL OF ANTIBODY ISOTYPES BY ADJUVANTS

The first clearcut difference in antibody isotype dependent on the adjuvant was the appearance of IgG2 antibodies in the guinea pig after immunisation with FCA but not with FIA (White, Jenkins & Wilkinson, 1963; Benacerraf *et al.*, 1963) and FCA also preferentially induces IgG2 in the mouse (Coe, 1966). The response after immunisation with $Al(OH)_3$ in the mouse is predominantly IgG1 but $Al(OH)_3$ also induces IgE in this species much more effectively than FCA (Hamaoka *et al.*, 1974).

IgG2 is the major complement-fixing antibody isotype in the mouse (Klaus *et al.*, 1979), a role that is fulfilled by IgG1 and IgG3 in man (Bindon *et al.*, 1988). There is, therefore, a question as to whether $Al(OH)_3$, which is inefficient at inducing IgG2 in the mouse and guinea pig, might fail to stimulate IgG1 and IgG3 in man. This does not seem to be the case, at least for IgG1, since the antibody response to a Hepatitis B vaccine with $Al(OH)_3$ adjuvant was predominantly IgG1 (Mattila, Pelkonen & Makela, 1986). Interestingly, in studies of the isotype response of man to infection by enveloped viruses, it was found in the case of mumps (Linde, Granstrom & Orvell, 1987), HSV (Ljungman *et al.*, 1988) and HIV (Ljunggren *et al.*, 1988) that the antibodies to envelope proteins were mostly IgG1, whereas other isotypes were represented in the response to core proteins. This suggests that the antibody isotype is controlled by the manner of antigen presentation which could be manipulated by adjuvants. For instance, envelope presentation could be mimicked by liposomal incorporation.

THE BASIS OF DIFFERENTIAL ADJUVANT ACTION

The control of CMI and antibody isotype by adjuvants has been investigated by classical techniques of cellular immunology and two important observations have been made for both CMI and IgE formation. First, immunisation with the adjuvant which is incapable of inducing the relevant component of immunity (FIA for CMI, and FCA for IgE) renders an animal refractory to the development of that component, even when subsequently immunised with the appropriate adjuvant. Secondly, this state of refractoriness can be passively transferred by lymphocytes (Bullock, Katz & Beracerraf, 1975; Kishimoto *et al.*, 1976) and has, therefore, been supposed to be mediated by 'suppressor' cells.

The idea that differential adjuvant action is a consequence of selective induction of T cell subsets can be refined in the light of recent knowledge of the control of CMI and antibody isotype by lymphokine release from T cell subsets. In the mouse, two subsets of T helper (Th) cells have been distinguished: Th1 cells which secrete interleukin 2 (IL-2) and interferon gamma (IFN-gamma) and Th2 cells producing interleukins 4 and 5 (IL-4 and IL-5) (Coffman *et al.*, 1988). IFN-gamma is involved in the control of IgG2 production by B cells and Th1 cells mediate CMI (Cher & Mosmann, 1987). Thus, in the mouse, the two components of immunity that are selectively stimulated by FCA are under the control of the same T cell subset, Th1. Conversely, the IL-4 secreted by Th2 cells promoted IgG1 and IgE production by B cells, and these are the classes of antibody enhanced by $Al(OH)_3$. It may thus be hypothesised that $Al(OH)_3$ preferentially stimulates the triggering of Th2 cells, whereas FCA engages Th1 cells as well. The situation in man is less clearcut, in that it has not been possible to divide T cells into two classes on the basis of lymphokine production (Paliard *et al.*, 1988) and it is not yet known if the type of differential adjuvant action observed in rodents will also appear in man.

Even if differential adjuvant action can be explained at the level of T cell subsets, this still leaves open the question of how adjuvants exert their control. There is considerable evidence that the general immunostimulatory action of adjuvants is mediated via antigen-presenting cells (APC) (reviewed in Allison & Byars, 1986) and in one study direct evidence that the control of CMI is exerted at this level has been presented. Murine peritoneal exudate cells induced with FCA and haptenated with trinitrophenol (TNP) became able to stimulate contact sensitivity after i.v. injection, in contrast to TNP-haptenated cells not exposed to FCA which stimulated suppressor cells (Britz *et al.*, 1982).

The question then arises as to whether the adjuvant is targeting the antigen to different classes of APC, each responsible for antigen presentation to a different T cell subset, or whether the same population of APC can be modulated to favour a particular T cell subset, perhaps by the secretion of specific monokines. Work on the effect of adjuvants on monokine secretion has up to now been confined to interleukin 1 (IL-1), with the finding that adjuvants of many different sorts, including $Al(OH)_3$ (Mannhalter *et al.*, 1985) and MDP (Oppenheim *et al.*, 1980; Bahr, Chedid & Behbenani, 1987) stimulate its release. This is believed to be useful for general immunostimulation by adjuvants, but does not help to explain their differential action which would require selective induction of as yet possibly unidentified monokines. The principle of selective monokine induction has been established for MDP and its derivatives. MDP itself is pyrogenic, which will make it unsuitable for human vaccines, but two non-pyrogenic, derivatives have been developed, murabutide (Le Francier *et al.*, 1982) and threonyl MDP (Allison, Byars & Waters, 1986). Both molecules induce macrophages to secrete IL-1 (Damais *et al.*, 1982; Allison *et al.*, 1986) which is curious in view of the fact that IL-1 is generally considered to be an endogenous pyrogen; and in the case of murabutide, it was confirmed that the macrophage supernatants contain IL-1 activity (lymphocyte activation) but were non-pyrogenic (Damais *et al.*, 1982).

In summary, there can be no doubt that the immunomodulatory effect of adjuvants is expressed by intervention in the normal immunoregulatory network of cytokines. It might, therefore, be asked if the cytokines themselves can be exploited as selective immunological adjuvants.

CYTOKINES AS ADJUVANTS

The cytokines whose adjuvant properties have so far been investigated are IL-1, IL-2 and IFN-gamma.

IL-1 augments the antibody response to model antigens such as bovine serum albumin (BSA) or sheep red blood cells (SRBC) (Staruch & Wood, 1983; Reed *et al.*, 1989). The potentially very exciting discovery has been made that the adjuvant effect of IL-1 *in vivo* can be replaced by a synthetic nonapeptide corresponding to amino acids 163–171 (Nencioni *et al.*, 1987). On a molar basis, the peptide is equipotent to recombinant IL-1, and monoclonal antibodies to the peptide block the adjuvant activity of IL-1 (Boraschi *et al.*, 1988; 1989). However, the adjuvant effect is modest, being a 2–4-fold increase in splenic plaque-forming cells after intravenous immunisation with SRBC.

The adjuvant effect of IL-2 was originally described as a restoration of responsiveness to myoglobin in genetically low-responder mice (Kawamura, Rosenberg & Berzofsky, 1985) an effect that was subsequently reproduced with malaria sporozoite peptides (Good *et al.*, 1988). In both these studies, it was necessary to incorporate the IL-2 into FIA. Il-2 has also shown adjuvant properties in another situation of immunological hyporesponsiveness, in immunodeficient kidney dialysis patients. They had failed to respond to a commercial Hepatitis B vaccine with $Al(OH)_3$ adjuvant but anti-viral antibodies appeared when IL-2 was injected four hours after the vaccine at the same site (Meuer *et al.*, 1989). In a more conventional vaccine situation, IL-2 promoted the protection against HSV-2 in the guinea pig model conferred by immunisation with glycoprotein D

(Weinberg & Merigan, 1988). Interestingly, IL-2 had no effect on antibody titres or the *in vitro* response of lymphocytes to viral antigen, but it did promote specific cellular cytotoxicity and this was correlated with protection. This therefore provides an example of the selective stimulation by an adjuvant of a component of immunity relevant for protection against a virus. A similar situation has been described for the potentiation of protection with an inactivated rabies vaccine by repeated treatments with IL-2 after immunisation (Nunberg *et al.*, 1989). This did not increase the titre of virus neutralising antibody, and it was proposed that CMI may have been enhanced.

IFN-gamma was active as an adjuvant for a murine malaria vaccine in mice, where it promoted both humoral and CMI (Playfair & De Souza, 1987) even in immunocompromised mice (Heath *et al.*, 1989). It also enhanced the antibody response to vesicular stomatitis virus G protein (Anderson, Fennie & Yilma, 1988).

These initial successes of cytokines as adjuvants do not yet establish them as alternatives to conventional adjuvants for vaccines designed to elicit neutralising antibodies, but they could well find a place for specialist applications in overcoming hyporesponsiveness, or inducing CMI. There will be a need for further work on the optimal formulation of vaccines containing cytokines, since many of the studies have required multiple injections or emulsification in oil, presumably on account of the short biological half-life of the mediators.

THE RELATIVE EFFICACY OF ADJUVANTS CAN DEPEND ON THE ANTIGEN

Adjuvants do not stimulate the immune response to all antigens in a uniform manner (Bomford, 1980*a*). For instance, there is a clearcut difference between $Al(OH)_3$ and saponin in the mouse. $Al(OH)_3$ is superior to saponin as an adjuvant for the weak antigen TNP-BSA, but the opposite is the case for the strong antigen TNP-KLH (Bomford, 1984). The same difference was also apparent using human growth hormone as the antigen. In a genetically low-responder strain of mice, $Al(OH)_3$ stimulated an antibody response and saponin was ineffective, whereas in a high-responder strain, saponin was the better adjuvant (Bomford, Aston & Ivanyi, 1985). The basis for this difference between the adjuvants is not understood but one possible hypothesis is that it reflects the extent of antigen degradation during processing, with $Al(OH)_3$ causing greater fragmentation of the antigen and revealing peptides that would not otherwise be available for the T cell repertoire. The processing requirements of antigens are likely to differ, the extreme example being small peptides which are expected to associate directly with MHC antigens and whose immunogenicity would be destroyed by processing. There are already indications that the adjuvant requirements of small peptides may differ from those of proteins. The humoral response to a synthetic peptide from foot and mouth disease virus was stimulated more strongly by small unilamellar liposomes than by $Al(OH)_3$ (Francis *et al.*, 1985; 1987). Although the reason for this is unknown, it could once again depend on processing, in this case, the liposomes delivering the peptide to MHC molecules or B cell immunoglobulin receptors without excessive processing.

In summary, the differential action of adjuvants, whether it be expressed as selective induction of particular components of immunity or relative efficacy for different antigens, can be explained by hypotheses involving APC and T cells. A comprehensive view of adjuvant action would have to include the B cell. At the very least, a successful adjuvant formulation must preserve some antigen with its full conformational structure intact for presentation to B cells, and preferably enhance this process as well. This aspect of antigen presentation is considered in Chapter 2.

CLINICAL QUESTIONS

The majority of current vaccines against viruses operate by stimulating circulating neutralising antibodies which prevent the spread of virus from the portal of entry, and it is not yet certain that a failure to induce CMI is a limiting factor in the development of any future vaccine. Indeed, the absence of CMI, as expressed in hypersensitivity reactions, is perceived as an advantage for most vaccines. However, it is probable that CMI will be required for the control of persistent infection with enveloped viruses such as herpesviruses or human immunodeficiency virus.

The adjuvants which are known to stimulate CMI in animals are now beginning to undergo clinical evaluation in man. Phase I clinical trials have been conducted for murabutide in aqueous formulation with tetanus toxoid (Telzak *et al.*, 1986) and an MDP analogue in a water-in-oil (squalene) emulsion was used as an adjuvant for an anti-fertility vaccine (Jones *et al.*, 1988) although in this case, the objective was to raise antibodies rather than CMI against the hormone human chorionic gonadotropin. A lipophilic derivative of MDP, MTP-PE, is being assessed in a phase I clinical trial of an AIDS vaccine containing gp120 (Braun *et al.*, 1989) and the SAF-1 adjuvant, comprising threonyl MDP together with an oil-in-water emulsion of squalane incorporating the spreading agent pluronic 121 (Byars & Allison, 1987) is expected to enter human trials shortly (Allison, personal communication). It will be of the greatest interest to discover if these new adjuvant formulations can modulate the immune response in man to the same extent as in the mouse, and whether this yields dividends in terms of protective potency of the vaccine.

REFERENCES

Adam, A. (1985). *Synthetic Adjuvants. Modern Concepts in Immunology,* Vol. 1, New York: John Wiley & Sons.

Allison, A. C. & Byars, N. E. (1986). An adjuvant formulation that selectively elicitis the formation of antibodies of protective isotypes and of cell-mediated immunity. *Journal of Immunological Methods,* **95**, 157–68.

Allison, A. C., Byars, N. E. & Waters, R. V. (1986). Immunologic adjuvants: efficacy and safety considerations. In *Advances in Carriers and Adjuvants for Veterinary Biologies* (R. M. Nervig, P. M. Gough, M. L. Kaeberle & C. A. Whetstone eds), pp. 91–103, Ames, Iowa: Iowa State University Press.

Anderson, K. P., Fennie, E. H. & Yilma, T. (1988). Enhancement of a secondary antibody response to vesicular stomatitis virus G protein by IFN-gamma treatment at primary immunisation. *Journal of Immunology,* **140**, 3599–604.

Aprile, M. A. & Wardlaw, A. C. (1966). Aluminium compounds as adjuvants for vaccines and toxoids in man: A review. *Canadian Journal of Public Health,* **57**, 343–60.

Bahr, G. M., Chedid, L. A. & Behbenani, K. (1987). Induction, *in vivo* and *in vitro,* of macrophage membrane interleukin-1 by adjuvant- active synthetic muramyl peptides. *Cellular Immunology,* **107**, 443–54.

Benacerraf, B., Ovary, Z., Block, K. J. & Franklin, E. C. (1963). Properties of guinea pig 7S antibodies. I. Electrophoretic separation of two types of guinea pig 7S antibodies. *Journal of Experimental Medicine,* **117**, 937–49.

Berman, P. W., Gregory, T., Crase, D. & Lasky, L. A. (1985). Protection from genital Herpes Simplex Virus type 2 infection by vaccination with cloned type 1 glycoprotein D. *Science,* **227**, 1490–2.

Bevan, M.J. (1987). Class discrimination in the world of immunology. *Nature (London),* **325**, 192–4.

Bindon, C. I., Hale, G., Bruggemann, M. & Waldmann, H. (1988). Human monoclonal IgG isotypes differ in complement activating function at the level of C4 as well as C1q. *Journal of Experimental Medicine,* **168**, 127–42.

Bomford, R. (1980*a*). The comparative selectivity of adjuvants for humoral and cell-mediated immunity. I. Effects on the antibody response to bovine serum albumin and sheep red blood cells of Freund's incomplete and complete adjuvants, alhydrogel, *Corynebacterium parvum, Bordetella pertussis,* muramyl dipeptide and saponin. *Clinical and Experimental Immunology,* **39,** 426–34.

Bomford, R. (1980b). The comparative selectivity of adjuvants for humoral and cell-mediated immunity. II. Effect on delayed-type hypersensitivity in the mouse and guinea pig, and cell-mediated immunity to tumour antigens in the mouse of Freund's incomplete and complete adjuvants, alhydrogel, *Corynebacterium parvum, Bordetella pertussis,* muramyl dipeptide and saponin. *Clinical and Experimental Immunology,* **39**, 435–41.

Bomford, R. (1984). Relative adjuvant efficacy of Al(OH) and saponin is related to the immunogenicity of the antigen. *International Archives of Allergy and Applied Immunology,* **75**, 280–1.

Bomford, R. (1989). Adjuvants for anti-parasite vaccines. *Parasitology Today,* **5**, 41–6.

Bomford, R., Aston, R. & Ivanyi, J. (1985). Reversal of H-2-restricted hyporesponsiveness to human growth hormone by the use of aluminium hydroxide as adjuvant. *Immunogenetics,* **21**, 505–9.

Boraschi, D., Nencioni, L., Villa, L., Censini, S., Bossu, P., Ghiara, P., Presentini, R., Perin, F., Frasca, D., Doria, G., Forni, G., Musso, T., Giovarelli, M., Ghezzi, P., Bertini, R., Besedovsky, H. O., Delrey, A., Sipe, J. P., Antoni, G., Silvestri, S. & Tagliabue, A. (1988). *In vivo* stimulation and restoration of the immune response by the non-inflammatory fragment 163–171 of human interleukin-1β *Journal of Experimental Medicine,* **168,** 675–86.

Boraschi, D., Volpini, G., Villa, Nencioni, L., Scapigliati, G., Nucci, D., Antoni, G., Matteucci, G., Cioli, F. & Tagliabue, A. (1989). A monoclonalal antibody to the IL-1β peptide 163–171 blocks adjuvanticity but not pyrogenicity of IL-1 β *in vivo. Journal of Immunology,* **143**, 131–4.

Braun, D. G., Gelzer, J., Stalder, H., Van Hoogevest, P., Peil, A., Tillson, E. M., Wintsch, J. & Cruchaud, A. (1989). Klinische Phase-1-Prufung eines potentiellen AIDS-Impfstoffes der Biocine Company. *Therapiewoche Schweiz,* **5**, 197–204.

Britz, J. S., Askenase, P. W., Ptak, W., Steinman, R. M. & Gershon, R. K. (1982). Specialized antigen-presenting cells. Splenic dendritic cells and peritoneal-exudate cells induced by mycobacteria activate effector T cells that are resistant to suppression. *Journal of Experimental Medicine,* **155**, 1344–56.

Bullock, W. W., Katz, D. H. & Benacerraf, B. (1975). Induction of T-lymphocyte responses to a small molecular weight antigen. III. T-T cell interactions to determinants linked together: Suppression vs. enhancement. *Journal of Experimental Medicine,* **142**, 275–87.

Byars, N. E. & Allison, A. C. (1987). Adjuvant formulation for use in vaccines to elicity both cell-mediated and humoral immunity. *Vaccine,* **5**, 223–8.

Carelli, C., Audibert, F. & Chedid, L. (1981). Persistent enhancement of cell-mediated and antibody immune responses after administration of muramyl dipeptide derivatives with antigen in metabolizable oil. *Infection and Immunity,* **33**, 312–14.

Cher, D. J. & Mosmann, T. R. (1987). Two types of murine helper T cell clones. II. Delayed type hypersensitivity is mediated by Th1 clones. *Journal of Immunology,* **138**, 3688–94.

Coe, J. E. (1966). 7S-gamma-1 and 7S-gamma-2 antibody response in the mouse. I. Influence of strain, antigen and adjuvant. *Journal of Immunology,* **96**, 744–53.

Coffman, R. L. Seymour, B. W. P., Lebman, D. A., Hiraki, D. D., Christiansen, J. E., Shrader, B., Cherwinski, H. M., Savelkoul, H. F. J., Finkelman, F. D., Bond, M. W. & Mosmann, T. R. (1988). The role of helper T cell products in mouse B cell differentiation and isotype regulation. *Immunological Reviews,* **102**, 5–28.

Damais, C., Riveau, G., Parant, M., Gerota, J. & Chedid, L. (1982). Production of lymphocyte activating factor in the absence of endogenous pyrogen by rabbit or human leukocytes stimulated by a muramyl dipeptide derivative. *International Journal of Immunopharmacology,* **4**, 451–62.

Dienes, L. (1929). Technic of producing tuberculin type of sensitization with egg white in tuberculous guinea pigs. *Journal of Immunology,* **17,** 531–8.

Ellouz, F., Adam, A., Ciorbaru, R. & Lederer, E. (1974). Minimal structural requirements for adjuvant activity of bacterial peptidoglycan derivatives. *Biochemical and Biophysical Research Communications,* **59,** 1317–25.

Francis, M. J., Fry, C. M., Rowlands, D. J., Brown, F., Bittle, J. L., Houghten, R. A. & Lerner, R. A. (1985). Immunological priming with synthetic peptides of foot-and-mouth disease virus. *Journal of General Virology,* **33**, 2347–54.

Francis, M. J., Fry, C. M., Rowlands, D. J., Bittle, J. L., Houghten, R. A., Lerner, R. A. & Brown, F. (1987). Immune response to uncoupled peptides of foot-and-mouth disease virus. *Immunology*, **61**, 1–6.

Glenny, A. T., Pope, C. G., Waddington, H. & Wallace, V. (1926). The antigenic value of the toxin-antitoxin precipitate of Ramon. *Journal of Pathology and Bacteriology*, **29**, 31–40.

Good, M. F., Pombo, D., Lunde, M. N., Maloy, W. L., Halenbeck, R., Koths, K., Millen, L. & Berzofsky, J. (1988). Recombinant human IL-2 overcomes genetic non-responsiveness to malaria sporozoite peptides. Correlation of effect with biologic activity of IL-2. *Journal of Immunology*, **141**, 972–7.

Hackett, C. J., Taylor, P. M. & Askonas, B. A. (1983). Stimulation of cytotoxic T cells by liposomes containing influenza virus or its components. *Immunology*, **49**, 255–63.

Hamaoka, T., Newburger, P. E., Katz, D. H. & Benacerraf, B. (1974). Hapten-specific IgE antibody response in mice. III. Establishment of parameters for generation of helper T cell function regulating the primary and secondary responses of IgE and IgG B lymphocytes. *Journal of Immunology*, **113**, 958–73.

Heath, A. W., Devey, M. E., Brown, I. N., Richards, C. E. & Playfair, J. H. L. (1989). Interferon-gamma as an adjuvant in immunocompromised mice. *Immunology*, **67**, 520–4.

Jilg, W. M., Schmidt, G., Zoulek, B., Larbeer, B., Wilske, B. & Deinhardt, F. (1984). Clinical evaluation of a recombinant hepatitis B vaccine. *Lancet*, **ii**, 1174–5.

Jones, P. D., Tha-Hla, R., Morein, B., Lovgren, K. & Ada, G. L. (1988). Cellular immune responses in the murine lung to local immunization with influenza A virus glycoproteins in micelles and immunostimulatory complexes (ISCOMS). *Scandinavian Journal of Immunology*, **27**, 645–52.

Jones, W. R., Bradley, J., Judd, S. J., Denholm, E. H., Ing, R. M. Y., Mueller, U. W., Powell, J., Griffin, P. D. & Stevens, V. C. (1988). Phase I clinical trial of a World Health Organisation birth control vaccine. *Lancet*, **i**, 1295–8.

Kawamura, H., Rosenberg, S. A. & Berzofsky, J. A. (1985). Immunization with antigen and interleukin 2 *in vivo* overcomes *Ir* gene low responsiveness. *Journal of Experimental Medicine*, **162**, 381–6.

Kishimoto, T., Hirai, Y., Suemara, M. & Yamamura, Y. (1976). Regulation of antibody response in different immunoglobulin classes. I. Selective suppression of anti-DNP IgE antibody response by preadministration of DNP-coupled mycobacterium. *Journal of Immunology*, **117**, 396–404.

Klaus, G. G. B., Pepys, M. B., Kitajima, K. & Askonas, B. A. (1979). Activation of mouse complement by different classes of mouse antibody. *Immunology*, **38**, 687–95.

Le Francier, P., Derrien, M., Jamet, X., Choay, J., Lederer, F., Audibert, M., Parant, M., Parant, F. & Chedid, L. (1982). Apyrogenic, adjuvant-active N-acetylmuramyl-dipeptides. *Journal of Medicinal Chemistry*, **25**, 87–90.

Linde, A., Granstrom, M. & Orvell, C. (1987). Immunoglobulin class and immunoglobulin G subclass enzyme-linked immunosorbent assay compared with microneutralization assay for serodiagnosis of mumps infection and determination of immunity. *Journal of Clinical Microbiology*, **25**, 1653–8.

Ljunggren, K., Broliden, P-A., Morfeldt-Manson, L. & Wahren, B. (1988). IgG subclass response to HIV in relation to cellular cytotoxicity at different clinical stages. *Clinical and Experimental Immunology*, **73**, 343–7.

Ljungman, P., Zetterqvist, L., Sundqvist, V-A., Jeansson, S., Heimdahl, A. & Wahren, B. (1988). Lymphocyte and IgG responses to different herpes simplex-virus antigens in patients with frequent HSV-1 reactivations. *Clinical and Experimental Immunology,* **72**, 207–10.

Mannhalter, J. W., Neychev, H. O., Zlabinger, G. J., Ahmad, R & Eibl, M. M. (1985). Modulation of the human immune response by the non-toxic and non-pyrogenic adjuvant aluminium hydroxide: effect on antigen uptake and antigen presentation. *Clinical and Experimental Immunology,* **61**, 143–51.

Mattila, P. S., Pelkonen, J. & Makela, O. (1986). Proportions of IgG subclasses in antibodies formed during vaccination with Hepatitis B surface antigen. *Scandinavian Journal of Immunology,* **24,** 699–714.

Meuer, S. C., Dumann H., Mayer Zum Buschenfelde, K. H., & Kohler, H. (1989). Low dose interleukin 2 induces systemic immune responses against HBsAg in immunodeficient non-responders to Hepatitis B vaccination. *Lancet,* **i**, 15–18.

Moore, M. W., Carbone, F. R., & Bevan, M. J. (1988). Introduction of soluble protein into the class I pathway of antigen processing and presentation. *Cell,* **54**, 777–85.

Morein, B. (1988). The ISCOM antigen-presenting system. *Nature, London,* **332**, 287–8.

Nencioni, L., Villa, L., Tagliabue, A., Antoni, G., Presentini, R., Perin, F., Silvestri, S. & Boraschi, D. (1987). *In vivo* immunostimulating activity of the 163–171 peptide of human IL-1β. *Journal of Immunology,* **139**, 800–4.

Nunberg, J. H., Doyle, M. V., York, S. M. & York, C. J. (1989). Interleukin 2 acts as an adjuvant to increase the potency of inactivated rabies virus vaccine. *Proceedings of the National Academy of Sciences of the USA,* **86**, 4240–3.

Oppenheim, J. J., Togawa, A., Chedid, L. & Mizel, S. (1980). Components of mycobacteria and muramyl dipeptide with adjuvant activity induce lymphocyte activating factor. *Cellular Immunology,* **50,** 71–81.

Paliard, X., de Waal Malefijt, R., Yssel, H., Blanchard, D., Chretien, I., Abrams, J., de Vries, J. & Spits, H. (1988). Simultaneous production of IL-2, IL-4 and IFN-gamma by activated human $CD4^+$ and $CD8^+$ T cell clones. *Journal of Immunology,* **141,** 849–55.

Playfair, J. H. L., & de Souza, J. B. (1987). Recombinant gamma interferon is a potent adjuvant for a murine malaria vaccine in mice. *Clinical and Experimental Immunology,* **67**, 5–10.

Reed, S. G., Pihl, D. L., Conlon, P. J., & Grabstein, K. H. (1989). IL-1 as adjuvant. Role of T cells in the augmentation of specific antibody production by recombinant human IL-1α. *Journal of Immunology,* **142**, 3129–33.

Scolnick, E. M., McLean, A. A., West, D. J., McAleer, W. J., Miller, W. J. & Buynak, E. B. (1984). Clinical evaluation in healthy adults of a hepatitis B vaccine made by recombinant DNA. *Journal of the American Medical Association,* **251**, 2812–15.

Scott, M. T., Bahr, G., Moddaber, F., Afchain, D. & Chedid, L. (1984). Adjuvant requirements for protective immunization of mice using a *Trypanosoma cruzi* 90K cell surface glycoprotein. *International Archives of Allergy and Applied Immunology,* **74**, 373–7.

Snippe, H., Belder, M. & Willers, J. M. N. (1977). Dimethyl dioctadeayl ammonium bromide as adjuvant for delayed hypersensitivity in mice. *Immunology,* **33**, 931–6.

Staruch, M.J. & Wood, D.D. (1983). The adjuvanticity of interleukin-1 *in vivo*. *Journal of Immunology,* **130**, 2191–4.

Telzak, E., Wolff, S. M. Dinarello, C. A., Conlon, T., El Kholy, A., Bahr, G. M., Choay, J. P., Morin, A. & Chedid, L. (1986). Clinical evaluation of the immunoadjuvant murabutide, a derivative of MDP, administered with a tetanus toxoid vaccine. *Journal of Infectious Diseases,* **153**, 628–33.

Warren, H. S. & Chedid, L. A. (1988). Future prospects for vaccine adjuvants. *CRC Critical Reviews in Immunology*, **8**, 83–101.

Weinberg, A. & Merigan, T. C. (1988). Recombinant interleukin 2 as an adjuvant for vaccine induced protection. Immunization of guinea pigs with Herpes Simplex virus subunit vaccines. *Journal of Immunology*, **140**, 294–9.

White, R. G., Coons, A. H. & Connolly, J. M. (1955). Studies on antibody production. IV. The role of wax fractions of *Mycobacterium tuberculosis* in adjuvant emulsions on the production of antibody to egg albumin. *Journal of Experimental Medicine*, **102**, 83–95.

White, R. G., Jenkins, G. C. & Wilkinson, P. C. (1963). The production of skin sensitizing antibody in the guinea pig. *International Archives of Allergy*, **22**, 156–65.

FACTORS INFLUENCING UPTAKE OF PRE-SCHOOL IMMUNISATION

CATHERINE S. PECKHAM AND HELEN E. BEDFORD

Department of Paediatric Epidemiology, Institute of Child Health, 30 Guilford Street, London WC1N 1EH, UK

INTRODUCTION

Immunisation has had a dramatic impact in reducing the prevalence of infectious diseases; this has led to a worldwide decline in mortality and morbidity in childhood. The effectiveness of immunisation in preventing disease has been clearly demonstrated in populations with high rates of immunisation. However, despite this success, immunisation rates, even within the United Kingdom, fall short of the levels required to eliminate the major vaccine-preventable diseases of childhood (diphtheria, tetanus, polio, measles, pertussis, rubella and tuberculosis).

In order to achieve this goal by the year 2000, the European region of The World Health Organization has set a target of 90% vaccine coverage by 1990 for all two year-old children (Begg & Noah, 1985). Immunisation rates in England are increasing, with 87% of children immunised against diphtheria, tetanus and polio. However, rates for measles and pertussis, although improving, remain well below the target and in 1987–88 were 76% and 73% respectively (Fig. 1). Of particular concern is the wide variation in vaccine uptake between districts, with rates for measles vaccine ranging from as low as 24% in some to as high as 96% in others. With a quarter of all two year-olds remaining unimmunised, and even larger proportions in some areas, outbreaks of these infectious diseases will continue to occur and the goal of elimination will not be achieved. This has been well demonstrated in the United States where, despite high levels of immunisation at school entry and a marked decline in the number of notifications, measles outbreaks continue to occur, particularly in the inner city areas among unvaccinated pre-school children (Hutchins *et al.*, 1989).

The recent introduction in the United Kingdom of the combined triple vaccine against measles, mumps and rubella (MMR) for children in the second year of life, has highlighted the necessity of achieving high vaccine uptake. The aim of this new approach is to eliminate measles, mumps and rubella from the community, and, for this to be realistic, an uptake of 95% has to be achieved. Less satisfactory levels would allow the continued circulation of the diseases in the community, albeit at reduced levels, with sporadic outbreaks and the possibility of delaying the acquisition of

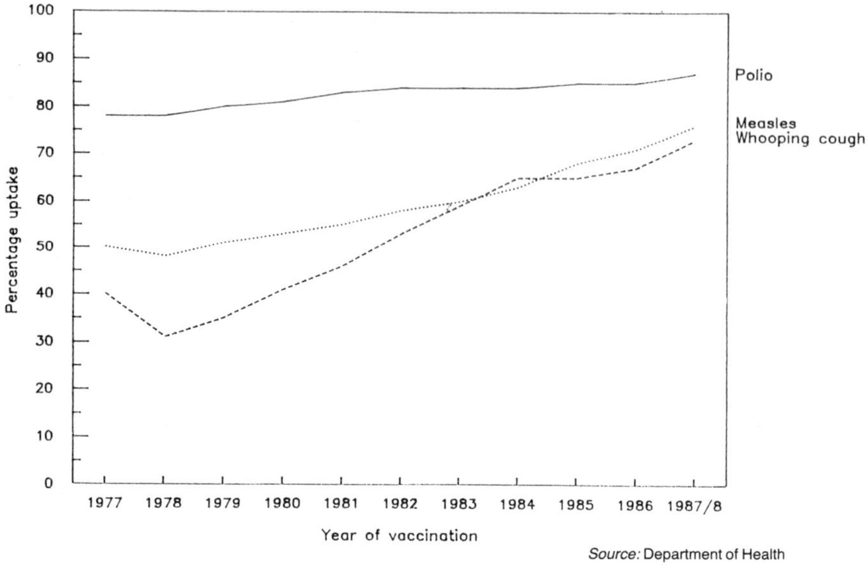

Fig. 1. Percentage of children immunised by the end of their second year, England 1977–87/88.

infection to an age when the complications could be more serious (Nokes & Anderson, 1988). Should this situation occur with rubella, for example, it could lead to an increase in the number of children born with the congenital rubella syndrome.

DISEASE MORBIDITY

As an increasing number of children are immunised and the childhood preventable diseases become less prevalent in the community, parents and health professionals often fail to appreciate the morbidity associated with these diseases (Dalphinis, 1986; Berkeley, 1983). A general practitioner is unlikely ever to see a child with measles encephalitis and possibly even a child with severe whooping cough. As a result the motivation to ensure that children are protected against these diseases becomes less urgent, and increasing emphasis and reinforcement has to be given to the importance of maintaining high levels of immunisation.

MEASLES AND PERTUSSIS

Measles is highly communicable and remains a significant cause of morbidity in children. There are between 60000 and 100000 notifications each year but the actual figure is likely to be double (Clarkson & Fine, 1985). For every 1000 notifications there are an estimated 10 hospital admissions, 10 neurological complications and 40 respiratory complications. Studies conducted in 1963 and 1976 showed no change in the severity of complications over this period of time (Miller, 1964, 1978). About 20 children die each year from measles or from its later complication, sub-acute sclerosing panencephalitis, and over half these deaths are in previously healthy children older than the recommended age for measles vaccine (Miller, 1983). This is an unacceptable level of morbidity for a vaccine-preventable disease. Yet many parents and some health professionals continue to regard measles as a trivial disease and as an inevitable part of growing up.

In 1986 there were 36500 notifications of pertussis. This figure, which probably represents less than 40% of actual cases, is likely to double during epidemics which occur in four-yearly cycles. At the present time (late in 1989) there has been an increase in notifications as predicted (Preston, 1984). Pertussis is highly communicable and children are likely to suffer significant respiratory complications, sometimes requiring hospitalisation (Pollock, Miller & Lobb, 1984). Each year there are between one and five reported deaths from pertussis in England and Wales. It is important not to consider the effects of pertussis solely in terms of mortality since the severe physical and psychological consequences of a protracted and debilitating illness may be considerable for the child and family (Johnston *et al.*, 1985).

MEASLES AND PERTUSSIS VACCINES

Measles vaccine, which was first introduced in 1968, is safe and provides 95% protection against the disease. Although mild side-effects including rash or low grade fever are common, serious complications such as febrile convulsions are rare (Miller, 1983). Convulsions are more common, by a factor of at least 10, after natural measles than after measles vaccine. In the United States Axnick, Shavel & Witte (1969) estimated that, between 1963 and 1968, measles vaccine averted 9730000 cases of measles, saved 973 lives, prevented 3224 cases of mental retardation and saved 1624000 days of hospitalisation.

Pertussis vaccine has an efficacy of 80–90% following a full primary course and, in those not fully protected, the disease is usually less severe. In the mid-1970s, public anxiety about the possibility of the vaccine being associated with brain damage, resulted in a dramatic decline in uptake from 77% in 1974 to 30% in 1978. As a result two major epidemics of whooping cough followed. Immunisation rates have been climbing slowly since this vivid reminder of the seriousness of whooping cough and reassurance about the safety of the vaccine. The risk of complications following natural pertussis infection far outweigh those associated with the vaccine (Hinman, 1984).

NATIONAL IMMUNISATION STUDY

A review of research on factors influencing vaccine uptake in childhood revealed that the success of any immunisation programme depends on a number of important and interrelated factors, involving health professionals, parents and the organisation of immunisation provision within districts and general practices (Y. D. Senturia, unpublished observations, 1987). As most of the studies reviewed were based on small, non-random samples the results cannot be generalised. In order to determine the influence of these and other factors on the uptake of immunisation at a national level, an investigation on a larger representative population was commissioned by Action Research for the Crippled Child at the end of 1986 (Peckham *et al.*, 1989). The specific purpose was to investigate the attitudes of health professionals and parents to immunisation and to determine the reasons for poor uptake nationally, as a basis for recommending strategies for improving vaccine coverage for the preventable diseases of childhood.

The study was conducted in 16 health districts in England and Wales; eight with the lowest and eight with the highest uptake of measles and pertussis vaccine. Data were collected during 1987–88 prior to the introduction of the combined measles, mumps and rubella vaccine, and before the publication of the latest memorandum on Immunisation against Infectious Disease (Joint Committee on Vaccination and Immunisation, 1988).

A random sample of 88 general practices was included. With the general practitioners' agreement, parents of two year-old children were asked to complete a questionnaire seeking information on their attitudes to measles and pertussis and the vaccines, on their child's immunisation status, and on family and social circumstances. In addition, information was sought from health visitors and from general practitioners concerning their attitudes to immunisation and knowledge of contraindications. Characteristics of the practices and districts were also gathered. As response rates were high (87% of 3871 parents, 96% of 542 health visitors and 82% of 1397 general practitioners returned questionnaires) the findings can be generalised.

FACTORS INFLUENCING UPTAKE OF IMMUNISATION

Health professionals

Health professionals' attitudes to the prevention of disease have been shown to influence vaccine uptake. In 1974, negative attitudes to pertussis immunisation were considered to be an important reason for the poor vaccine uptake (43%) in one district (Wilkinson, Tylden-Patterson & Gould, 1979). In the present study there was no relationship between professionals' attitudes and vaccine uptake. This is not surprising, since, although attitudes to the degree of severity and infectivity of measles and pertussis varied, most considered the vaccines to be effective and to have few risks. The overwhelming majority (97%) said that they encouraged immunisation.

Many studies have shown that health professionals are poorly informed about contraindications to immunisation (Hull, 1981; Wilkinson, 1986), and have assumed that this contributes to low vaccine uptake. The National Study confirmed these findings. For example, nearly half the general practitioners wrongly considered measles vaccine to be contraindicated for a child who develops a rash after eating eggs. A small number (9%) would even have advised measles vaccine for a child being treated for leukaemia. This confusion in the interpretation of possible contraindications was supported by parents' reports of why their child had not been immunised. Of the 526 children who had not received pertussis vaccine, 43% had been advised against it by a doctor or health visitor. In at least 40% of these cases the apparent reason for this advice did not constitute a valid contraindication. Similar findings have been documented in other studies (Klein, Morgan & Wanborough-Jones, 1989). Yet, it has been estimated that only 3.5% of children have a genuine contraindication to pertussis vaccine (Nicholl & Jenkinson, 1988). In this study a general practitioner's knowledge of contraindications was directly associated with vaccination rates in their practice; those who were well informed had high vaccine uptake rates. However, there was no such relationship with health visitors' knowledge of contraindications.

Parents

The factors that influence an individual parent in the decision to have their child immunised have been described in Becker's Health Belief Model (1974). This suggests that, if a parent considers their child to be at risk from a serious disease such as measles and they know that a safe and effective vaccine is readily available, they are likely to have the child immunised. Factors including age, social class and external influences such as health education and advice from health professionals, will modify their perceptions. Health education should enable a person to make health choices based on accurate information. Those providing health education must be well informed about contraindications to immunisation. Parents must be given clear, and consistent information about the benefits as well as about the risks of immunisation. The dangers of a child not being immunised must be fully appreciated.

In this study parental attitude emerged as an extremely important factor in determining whether a child was immunised. Those parents who thought that diseases were serious and infectious and the vaccines were safe and effective were much more likely to be immunised than those who thought otherwise. It was also evident that only a small change in attitude could have a marked effect on immunisation uptake. However, many studies have shown that parents do not recall having received any advice about immunisation from health professionals (Blair, Shave & McKay, 1985; Conway & Phillips, 1989). Similarly, in this study, over a quarter of parents said that no one had influenced their decision.

Parent-held child health records

Parent-held child health records have been used successfully in some districts and there is good evidence to suggest that parents do not lose them (Saffin & MacFarlane, 1988). The introduction of a standard record for all children would have a significant impact on vaccine uptake. Records should include details of the recommended immunisation schedule and be used to record the date and type of all vaccinations given and reasons for those withheld. Changes in attitude and behaviour are more likely to be achieved when parents are given increased responsibility for their child's health. A record of the child's immunisation should also be kept by the health district to enable recall and monitoring of uptake.

Organisation of immunisation provision

It was evident from this study that immunisation may not be sufficiently accessible for certain groups. Immunisation uptake was lower in children from large families and those where there was a chronically ill child. The organisation of immunisation provision needs to be flexible enough to cater for the needs of all groups. In certain circumstances, such as for children living in homeless families' accommodation, it may be necessary to offer domiciliary immunisation (Conway, 1988). Such initiatives are being facilitated by the increasing involvement of health visitors and nurses in the administration of vaccines (Jefferson, Steight & MacFarlane, 1987).

Ross has described how the motivated general practice, committed to immunising all babies born in the practice can achieve excellent immunisation rates (Ross, 1986). Commitment alone, however, is not the whole solution. In this study the organisation of immunisation provision within a practice emerged as a significant factor. General practices that adopted a team approach to the administration of immunisation, where responsibility for obtaining consent, giving injections and recalling children was shared with nurses and health visitors, had a higher vaccine uptake than those where the general practitioner retained sole responsibility for these aspects of immunisation provision. This kind of approach will enable flexibility of provision with increased opportunities for immunisation.

Opportunistic immunisation

An opportunistic approach to immunisation increases further the flexibility of service provision. This can be extended outside the general practice to include all health professionals involved in the care of children. Every contact with a doctor or health visitor, or hospital or clinic visit, is an opportunity for a child's immunisation status to be reviewed and for those children whose immunisations are not up to date to be immunised. Inner city casualty departments have been recognised as providing primary care (Jackson, 1980) and, for many children, attendance at hospital may be the only contact they have with health professionals. This should be used to its fullest advantage. An opportunistic approach to immunisation also reinforces the importance of immunisation among parents and health professionals. However, without the widespread use of parent-held child health records this approach is not feasible.

The role of computers

Many districts now use computers for recording and scheduling immunisation. It has been shown that districts with computerised systems have better uptake rates than those without (Begg & White, 1987). One author claimed that in his district the computer system was instrumental in maintaining high levels of immunisation (Bussey & Harris, 1977). Whilst it is clear that a computer system used appropriately can be a valuable asset in administering a vaccine programme, it is difficult to prove cause and effect and there remains a large range in vaccine uptake among computerised districts (Begg, Gill & White, 1989). There were too few districts included in this study to allow valid comparisons, but it was interesting to note that the uncomputerised districts had difficulty in providing information on the immunisation status of children included in the study. Indeed, in one district no data could be provided for 19% of the sample. This has obvious implications for the accurate evaluation of vaccine uptake.

Review of immunisation records

Review of the immunisation records of all children entering nurseries and schools is important and would ensure that all unimmunised or partially immunised children are identified. Those children who have not been immunised or whose immunisations are not up to date should be immunised before entry provided there are no valid contraindications or parental objections. This would not only protect the individual child but also prevent transmission of infection to other children, who may either be too young to be immunised or have valid contraindications to immunisation, such as leukaemia.

In the United States notifications of measles have declined dramatically, largely as a result of laws requiring proof of immunity before school entry (Anon, 1989). There are now fewer notifications per year in the whole of the United States than in the English West Country (Fig. 2).

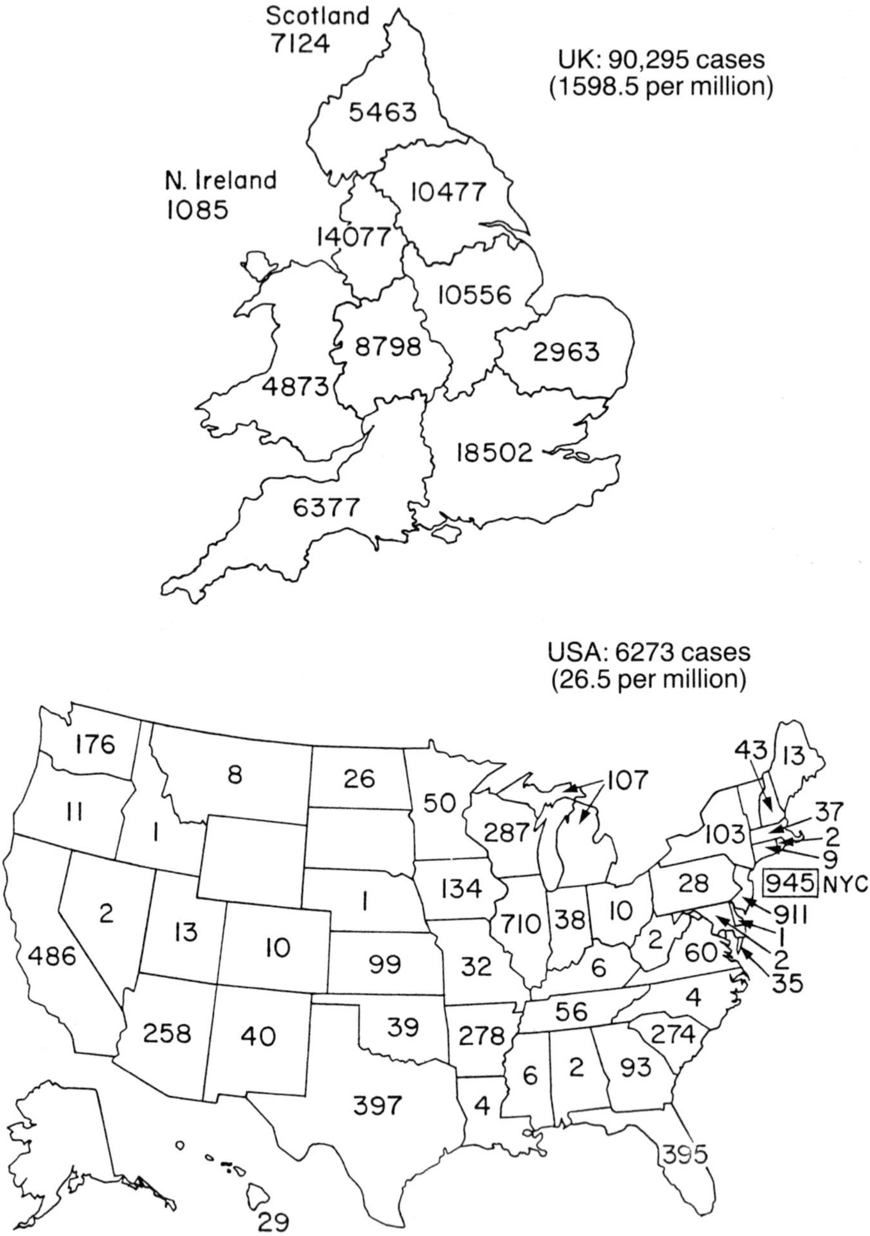

Source: Office of Population Censuses and Surveys
morbidity and mortality weekly
prepared by Communicable Disease Surveillance Centre

Fig. 2. Provisional notifications of measles, United Kingdom and United States 1986.

The introduction of similar legislation in this country is viewed by some as controversial, but in this study over half the parents, general practitioners and health visitors were in favour of such a measure. Indeed, 40% of the parents who approved had not had their children immunised and many stated that this would provide the motivation they required. As immunisation rates increase, it may become important to discuss the role of immunisation linked to school entry as a means of maintaining high uptake levels.

CONCLUSION

In this chapter, factors which have been shown to determine whether or not a child is immunised have been discussed. There is clearly no single solution to the problem of achieving and maintaining high vaccine uptake. However, it is important to recognise that the majority of parents, 99% in this study, give consent for their child to be included in the immunisation programme. Yet, a significant number remain unprotected. We must redouble our efforts to ensure that, at the earliest possible age, children are afforded their right to protection against the preventable diseases of childhood.

ACKNOWLEDGEMENTS

This study was funded by Action Research for the Crippled Child to whom we are indebted. We would like to thank all the districts, health professionals and parents involved for their co-operation with this study. Thanks also to Dr Yvonne Senturia and Dr Tony Ades for their invaluable contributions.

REFERENCES

Anonymous (1989). Measles prevention: supplementary statement. *Journal of the American Medical Association*, **261**, 827 and 831.

Axnick, W. W., Shavel, S. M. & Witte, J. J. (1969). Benefits due to immunization against measles. *Public Health Reports*, **84**, 673–80.

Becker, M. H. (1974). The health belief model and sick role behaviour. *Health Education Monographs*, **2**, 409–19.

Begg, N. T. & Noah, N. D. (1985). Immunisation targets in Europe and Britain. *British Medical Journal*, **291**, 1370–1.

Begg, N. & White, J. (1988). A survey of pre-school immunisation programmes in England and Wales. *Community Medicine*, **10**, 344–50.

Begg, N. T., Gill, O. N. & White, J. M. (1989). COVER (Cover of Vaccination Evaluated Rapidly): Description of the England and Wales Scheme. *Public Health*, **103**, 81–9.

Berkeley, M. I. K. (1983). Measles – The effect of attitudes on immunisation. *Health Bulletin*, **41**, 141–7.

Blair, S., Shave, N. & McKay, J. (1985). Measles matters, but do parents know? *British Medical Journal*, **290**, 623–4.

Bussey, A. L. & Harris, A. S. (1977). Immunisation levels – need they all decline? *Lancet*, **ii**, 970–1.

Clarkson, J. A. & Fine, P. E. M. (1985). The efficiency of measles and pertussis notification in England and Wales. *International Journal of Epidemiology*, **14**, 153–68.

Conway, J. ed. (1988). *Prescription for Poor Health*. London Food Commission/ Maternity Alliance/SHAC/Shelter.

Conway, S. P. & Phillips, R. R. (1989). Morbidity in whooping cough and measles. *Archives of Disease in Childhood*, **64**, 1442–45.

Dalphinis, J. (1986). Do immunisation defaulters know enough about immunisation? *Health Visitor*, **59**, 342–4.

Hinman, A. R. (1984). Pertussis and pertussis vaccines – reanalysis of benefits risks and costs. *Journal of the American Medical Association*, **251**, 3109–13.

Hull, D. (1981). Interpretation of the contraindications to whooping cough vaccine. *British Medical Journal*, **283**, 1231–3.

Hutchins, S. S., Escolan, J., Markowitz, L. E., Hawkins, C., Kimbler, A., Morgan, R. A., Preblud, S. R. & Orenstein, W. A. (1989). Measles outbreak among unvaccinated pre-school aged children. *Pediatrics*, **83**, 369–74.

Jackson, J. (1980). Paediatric Primary Care in Inner London. *Journal of the Royal College of General Practitioners*, **30**, 520–8.

Jefferson, N., Sleight, G. & MacFarlane, A. (1987). Immunisation of children without a doctor present. *British Medical Journal*, **294**, 423–4.

Johnston, I. D. A., Hill, M., Anderson, H. & Lambert, H. (1985). Impact of whooping cough on patients and their families. *British Medical Journal*, **290**, 1636–8.

Joint Committee on Vaccination and Immunisation (1988). *Immunisation Against Infectious Disease*. HMSO. London.

Klein, N., Morgan, K. & Wanborough-Jones, M. H. (1989). Parents' beliefs about vaccination: the continuing propogation of false contraindications. *British Medical Journal*, **298**, 1687.

Miller, C. L. (1978). Severity of notified measles. *British Medical Journal*, **1**, 1253–5.

Miller, C. L. (1983). Current impact of measles in the United Kingdom. *Reviews of Infectious Diseases*, **5**, 427–32.

Miller, D. L. (1964). Frequency of complications of measles. *British Medical Journal*, **2,** 75–8.

Nicoll, A. & Jenkinson, D. (1988). Decision making for routine MMR and whooping cough immunisation. *British Medical Journal,* **297**, 405–7.

Nokes, D. J. & Anderson, R. M. (1988). Measles, mumps and rubella vaccine: What coverage to block transmission? *Lancet*, **ii**, 1374.

Peckham, C., Bedford, H., Senturia, Y. & Ades, A. (1989). *National Immunisation Study: Factors Influencing Immunisation Uptake in Childhood*. Action Research for the Crippled Child Report. Horsham

Pollock, T. M., Miller, E. & Lobb, J. (1984). Severity of whooping cough in England before and after the decline in pertussis immunisation. *Archives of Disease in Childhood*, **59**, 162–5.

Preston, N. W. (1984). Pertussis vaccination – the time to act is now. *Pulse of Medicine*, **44**, 38–9.

Ross, S. K. (1986). Childhood immunoprophylaxis: Achievements in a Glasgow Practice. *Health Bulletin*, **44**, 370–3.

Saffin, K. & MacFarlane, A. (1988). Parent-held child health and development records. *Maternal and Child Health*, 288–91.

Wilkinson, J. R. (1986). Measles Immunisation-Contraindications perceived by General Practitioners in one health district. *Public Health*, **100**, 144–8.

Wilkinson, P., Tylden-Patterson, L. & Gould, J. (1979). Professional attitudes towards vaccination and immunisation within the Leeds Area Health Authority. *Public Health*, **93**, 11–15.

PART II: CHEMOTHERAPY

THE IMPORTANCE OF PATHOGENESIS IN THE TREATMENT OF DISEASES CAUSED BY VIRUSES

KENNETH McINTOSH

Division of Infectious Diseases, Children's Hospital, and Department of Pediatrics, Harvard Medical School, 300 Longwood Avenue, Boston, MA 02115, USA

INTRODUCTION

The successful development of antiviral chemotherapy for a specific disease has two fundamental requirements: a pharmacologically active and satisfactorily non-toxic antiviral compound, and adequate knowledge of the pathogenesis of the disease. Much effort is devoted to the former requirement since it is so intuitively obvious that the success of antiviral therapy depends on the quality of the drugs available. It is not quite so obvious, however, that, because viral diseases are complex mixtures of virus-determined events and host responses and because antiviral chemotherapy has its effect primarily (and sometimes solely) upon the former, a foreknowledge of the pathogenesis of viral infections can be an important guide to successful therapeutics. Since most antiviral chemotherapy is directed at inhibition of viral replication, it will be successful only if this replication has direct or indirect disease-producing effects at the time the treatment is being delivered. This is not to say that the host's contribution to illness cannot also be the target of treatment: there are many more or less successful 'treatments' for viral infections which target host responses (e.g. antihistamines for acute upper respiratory tract infections or steroids for shingles), and more precise information on pathogenesis will be of value in improving our therapeutic approach on this side as well.

This paper will focus on pathogenesis in the application of antiviral chemotherapy. It will attempt to outline some general principles concerning aspects of disease pathogenesis which are important in the treatment of viral disease, both on an organismal and a cellular level, and will illustrate those principles by considering several specific diseases in which they apply.

GENERAL PRINCIPLES

Tempo of illness

In general, the slower the tempo of a viral infection, the more likely it is that antiviral chemotherapy will have a measurable beneficial effect on the course of the illness. Common, self-limited viral infections can be considered to have four stages: incubation period, prodrome, acute illness, and recovery. Patients seek help from the physician usually at the time of acute illness, at the peak of symptoms or shortly thereafter. By this time, the infection has been active in the host for days to weeks, and the immune response has begun. In brief illnesses the virus may have all but disappeared from the host, and the interval between this moment and clinical recovery may be only a matter of one or two days. The common cold and labial herpes are two examples of rapid diseases in which it has been difficult, even with potent antiviral compounds, to demonstrate a beneficial effect (Merigan *et al.*, 1973; Spruance *et al.*, 1982; Fiddian *et al.*, 1983). Acute viral diarrhoea is a similar illness which, should we develop potent antiviral compounds, might present similar problems. The reasons for the rapid evolution of these diseases are probably various. Many are local mucosal infections which produce disease at the site of viral entry into the body and stimulate rapid local mucosal curative immune responses. Some are reinfections, where the virus meets an accelerated anamnestic immune response and shedding is brief and at low titre. Still others are due to reactivation of virus from latency, and immunity has been actively stimulated by frequent prior reactivation events. In such instances most investigators have found that it is difficult to demonstrate a beneficial effect of treatment, and prophylaxis is more obviously effective (Saral *et al.*, 1981, 1983; Hayden *et al.*, 1986).

In contrast, slower diseases with prolonged phases of viral replication can be profoundly affected by potent antivirals. The treatment of primary genital herpes simplex infection with acyclovir is a good example of a disease in which treatment has a welcome and clear-cut beneficial effect, shortening the period of virus shedding by 1–2 weeks and accelerating healing by as much as a week (Bryson *et al.*, 1983; Mindel *et al.*, 1982). Lassa fever is another example of a primary infection with a prolonged course, which responds dramatically to antiviral treatment, with reduction in mortality from 55% to 5% (McCormick *et al.*, 1986). Congenital viral infections are also likely to prove responsive to antiviral therapy. It follows from these examples that viral infections in the immunosuppressed host are frequently good targets for antiviral therapy, since they tend on the whole to have prolonged courses.

The balance of virally induced and immunologically induced pathology

Illnesses in which the predominant pathology is viral cytopathology can be expected to respond to antiviral treatment more readily than those in which the predominant pathology is immunologic. There are extreme examples of each of these, some of which will be discussed below. As in the discussion of the tempo of disease, viral infections in the immunodeficient host, where immunopathology is minimised, can be expected both to need antiviral therapy and to benefit from it. In contrast, some diseases are triggered by a viral infection, but the pathogenesis is almost entirely determined by the activity of the immune system. Acute post-infectious encephalitis is probably one example of this, and virus-induced asthma another. It can be expected that these diseases will respond poorly to even the best and most potent antivirals.

In between these extremes are the majority of viral infections, illnesses such as measles, enteroviral meningitis, viral bronchiolitis, chicken pox, viral gastroenteritis and many others. For most of these illnesses it is known, and for all it is suspected, that the pathology and the symptoms are produced by a combination of viral cytopathology and immunopathology. In addition, in most it is known or suspected that the balance of viral and immunological factors shifts as the disease progresses. Antiviral therapy can be expected to lessen or remove only the viral portion of this combination, and perhaps to shorten and reduce some immunologically determined pathologic effects as a consequence. The efficacy of the chemotherapy will depend on the timing of its onset in relation to the linked evolutions of the infection and the immune response. In general, earlier treatment can be expected to be more effective, and later less. In this respect, antiviral chemotherapy differs from antibacterial, since the latter, while better earlier, is rarely too late to achieve at least some effect.

The state of the virus at the cellular level

Just as it is important to track the course of the viral infection at the organismal level, it is essential, for appropriate application of antiviral chemotherapy, to understand the state of the virus at the cellular level. For certain viruses this does not seem, at least at the present time, to be an issue: variations in the pathogenesis of infection at the cellular level are not now known to be important in human diseases due to enteroviruses, myxoviruses, parainfluenzaviruses, rotaviruses and several others (further knowledge may change this situation, however). It is, however, for several other viruses profoundly important: herpes simplex, varicella zoster virus, Epstein–Barr virus, measles virus and human immunodeficiency virus, to name a few. It also seems obvious that a drug which has a profound suppressive effect on viral replication will have a very different effect on virus in the state of latency. And it is likely that similar differences will be seen between antiviral effects on a virus like measles which can produce a rapidly destructive infection or a very slow and profoundly modulated infection in a disease such as subacute sclerosing panencephalitis. An example of the importance of pathogenesis at the cellular level which has immediate application to antiviral chemotherapy at the present time is the dichotomy of productive and proliferative replication by Epstein Barr virus, since the former is sensitive to acyclovir and the latter is not (Datta *et al.*, 1980). It appears very likely that the several states in the life cycle of human immunodeficiency virus will, similarly, have important implications for the efficacy of antivirals in AIDS.

Alterations in pathogenesis as a consequence of treatment

Treatment of a viral infection with an antiviral compound rarely has simple consequences. In other words, we can expect that there will not simply be a reduction in the amount of virus with a consequent beneficial effect on the course of the illness. Simple toxic effects aside, antivirals can influence the pathogenesis of disease both directly and indirectly.

First, a potent antiviral effect may carry with it a reduction in the antigenic stimulus and a consequent reduction in the immune response. It is possible that this is an explanation of the shortened latent period between recurrent episodes of oral herpes in post-bone marrow transplant patients treated with acyclovir (Wade *et al.*, 1984). It may also be responsible for the occasional late recurrences of neonatal herpes simplex infection in treated infants (Gutman, Wilfert & Eppes, 1986). This may also be an explanation for the reduced immunologic response to respiratory syncytial virus in ribavirin-treated patients (Taber *et al.*, 1983; Rosner *et al.*, 1987). Alternatively, ribavirin may have a directly suppressive effect on the mucosal immune system (Powers, Peavy & Knight, 1982).

Second, many antiviral compounds (and likely more to come in the future) induce the formation of resistant phenotypes (Crumpacker *et al.*, 1982; Larder, Darby & Richman 1989; Hayden *et al.*, 1989). The consequences of the emergence of these strains depends very much on their individual pathogenicity, on the relation of the host to the virus at the time of their emergence, and on their capacity to spread to susceptible contacts. In certain infections of immunosuppressed patients continued viral replication may lead to prolonged and complicated infection (Crumpacker *et al.*, 1982). In immunologically normal patients and normally rapid diseases such as influenza, they may have no physiologic consequence in the treated host but may have important epidemiologic consequences in contacts (Hayden *et al.*, 1989).

Responses to antiviral chemotherapy may provide clues to pathogenesis

It does not seem an exaggeration to say that we know very little about the pathogenesis of any human viral illness. Since necessity dictates that research on new antiviral compounds proceed despite this ignorance, it seems likely that intelligent inference from the results of these effects will provide important clues to the pathogenesis of the diseases under treatment. Nowhere is this more likely than in the treatment of human immunodeficiency virus infections, in which the pressure to develop new agents is intense, and ignorance of pathogenetic mechanisms, despite extensive writing and many intelligent ideas on the subject (Ho, Pomerantz & Kaplan 1987), is profound. An example which may be worth pursuing further is the apparent reversal of neuropsychological deterioration in children and adults by intravenous zidovudine (Pizzo *et al.*, 1988; Yarchoan *et al.*, 1987). If this can be substantiated in further studies, it may provide important clues concerning the mechanism of the encephalopathy in AIDS.

EXAMPLES OF THE IMPORTANCE OF PATHOGENESIS IN THERAPY

Respiratory infections

Respiratory viral infections are probably one of the simpler groups of viral infections with regard to pathogenesis. The virus enters the body by infecting cells on one of the mucosal surfaces, usually nasal or conjunctival and, after a relatively brief incubation period of 2–5 days, induces local symptoms through cellular destruction. The immune response is predominantly mucosal and rapid, and takes the form of locally secreted specific IgA, IgG and IgM antibody (McIntosh *et al.*, 1978; Yanagihara & McIntosh, 1980; Murphy *et al.*, 1986). There is also, presumably, a local cellular immune response which, depending on the particular virus involved, may be essential in halting virus replication and promoting recovery (Fishaut, Tubergen & McIntosh, 1980). Lower tract disease is probably due to spread of the virus along the mucosal surface to the bronchi, bronchioles and, occasionally, alveoli through aspiration of droplets of infected secretions.

Although there may be a viremic phase of some respiratory viral infections (Domurat *et al.*, 1985; Naficy, 1963), it does not seem to be important in the pathogenesis of the respiratory illness.

There are, however, some large holes in our knowledge. Probably one of the largest, and the one most pertinent to the issues under discussion here, is how much disease is due to the direct destructive effect of the virus and how much to the disturbances produced by the host response. It seems likely that the answer to this question is not simple, since the distribution of pathogenetic events between the virus and the host probably changes as the illness progresses, with early symptoms being largely virus-induced and late ones more host-induced.

Respiratory syncytial virus infection is one of the best studied models. Even in infants undergoing their first infections, evidence of host immunity was frequently found in the first available samples obtained on the day of admission to hospital (McIntosh, McQuillin & Gardner, 1979). In infants, RSV disappears from the airway as the local immune response matures (McIntosh *et al.*, 1978, 1979), although there is some disagreement about how fast this occurs (Hall *et al.*, 1975). There is not, however, a perfect correspondence between disappearance of virus and recovery: in some infants recovery precedes virologic cure, and in some the opposite is the case (Hall, Hall & Speers, 1979).

In symptomatic reinfections the titre of virus in secretions is lower and the period of excretion is shorter (Mills *et al.*, 1971). In most children they are also milder, but in the particular case of infants and children who have a predisposition to recurrent wheezing, RSV and other respiratory viruses can 'trigger' an attack of asthma, with consequent severe respiratory symptoms (McIntosh *et al.*, 1973). For RSV, such attacks are probably largely due to reinfections, since most such children are out of infancy and have experienced one or more previous RSV infections.

The consequences for antiviral chemotherapy are important. They are reflected in the results of various trials of aerosolised ribavirin, amantidine and rimantidine in the treatment of RSV and influenza virus infections (Wingfield, Pollack & Grunert, 1969; Galbraith *et al.*, 1971; Knight *et al.*, 1981; Hall *et al.*, 1983) and of locally applied interferon for rhinovirus infections (Merigan *et al.*, 1973). First, the therapeutic effects demonstrated in all of these studies are not striking. Presumably one reason for this is that the infections are relatively rapid, and the subjects do not usually present themselves to the attention of the investigators until they are at a point where the immune response is already active and is itself both a potent antiviral force and a contributing cause of symptoms. Second, as with other viral diseases, early application of antiviral treatment is essential to the demonstration of a beneficial effect. This would seem to be particularly true for respiratory infections for the reasons given. As mentioned above, in rapidly evolving viral diseases it is frequently easier to demonstrate the value of prophylaxis than therapy. This has certainly been the case for amantidine in influenza A infections (Dolin *et al.*, 1982; Galbraith *et al.*, 1969) and interferon in rhinovirus infections (Hayden *et al.*, 1986; Douglas *et al.*, 1986).

The two extremes of the virus–host balance are also seen in respiratory viral infections. Patients with cellular immune defects have chronic, progressive pneumonias with prolonged viral shedding (Fishaut *et al.*, 1980). Such patients will presumably benefit the most from antiviral chemotherapy, if it can be delivered to the site of viral replication and has sufficient potency (McIntosh *et al.*, 1984). At the other end of the spectrum are those patients with virus-induced asthma. Hospital-based studies of the role of viruses in childhood asthma have clearly demonstrated that by the time symptoms are severe enough to merit admission to hospital, the virus is usually no longer present in the respiratory tract (Disney, Matthews & Williams, 1971). Such children are unlikely to benefit from even the best antivirals.

Encephalitis

Information on the pathogenesis of viral encephalitis is only now beginning to accumulate, and it has important implications for treatment. The situation is complex. At the very simplest level, certain forms of human viral encephalitis can be singled out in which the neurologic pathology is mostly due to virus-induced cytolysis. Herpes simplex encephalitis is such a disease, and probably also (although less well studied) eastern equine encephalitis and certain other arbovirus-induced encephalitides. Likewise, encephalitis in the immunosuppressed host (including progressive multifocal leukoencephalopathy) belongs in this category. At the other end of the spectrum are virus-induced Reye's encephalopathy and the group of post-infectious encephalitides such as those following measles, vaccinia infection and influenza. In between are the common encephalitides caused by varicella virus (Johnson & Milbourn, 1970). It is not clear where rabies or HIV-induced encephalopathy fit into this scheme. Certain of these diseases deserve further discussion.

The central nervous system syndromes following measles virus infection are good examples since they have been fairly well studied and illustrate certain aspects of this spectrum. Although antiviral therapy has not been systematically examined in any of these diseases, they serve as useful models from which to discuss the encephalitides caused by the herpes family viruses.

Recent studies of post-measles encephalitis have concluded that this disease is, in fact, probably not accompanied by active viral replication in the central nervous system (Johnson *et al.*, 1984). This conclusion followed several pieces of indirect evidence: that most (although not all) attempts to culture measles virus directly from brain tissue on biopsy or autopsy have failed; that there is no evidence of a local immune response, either in the cerebrospinal fluid (CSF) or in lymphocytes recovered from the CSF (Griffin *et al.*, 1989); and that antibodies to myelin basic protein are found in the CSF, and their presence recalls the similar presence of such antibodies in experimental allergic encephalitis and the encephalitis accompanying repeated doses of brain-grown inactivated rabies vaccine (Koprowski & LeBell, 1950). In further studies, the same authors found soluble CD8 in the CSF of patients with post-measles encephalitis and a distinct absence of soluble IL2 receptor (Griffin *et al.*, 1989). The meaning of these findings is not clear, but it is of interest that others have found abundant soluble IL2 receptor in the CSF of patients with herpes simplex encephalitis (Boutin *et al.*, 1987). If the conclusion of these authors is correct, and post-measles encephalitis does not involve active viral replication and virus-induced cytopathology, then antiviral chemotherapy has no role to play in the treatment of postmeasles encephalitis.

In contrast, there is evidence for active measles virus replication in the central nervous system of immunodeficient patients who develop subacute measles inclusion body encephalitis (MIBE) and in patients with subacute sclerosing panencephalitis (SSPE). The former is a peculiar form of encephalitis which occurs exclusively in immunosuppressed patients, usually during treatment for leukaemia, lymphoma, or other malignant disorders. The encephalitis follows clinical measles by five weeks to six months and takes either a subacute or acute course, frequently ending in death. In this disease abundant evidence of virus can be found by routine histology, in both glial cells and neurons (Aicardi *et al.*, 1977; Murphy & Yunis, 1976), accompanied by widespread grey matter necrosis. Inflammation is variably present, perhaps as a reflection of variable immunodeficiency. The genetic nature of the virus has been studied in one case (Cattaneo *et al.*, 1987), and it was found to have multiple mutations and an altered transcriptional pattern, with downstream messages very poorly represented in comparison to control lytic measles virus strains. Although cases treated with high-dose ribavirin have not been described, it seems quite possible that chemotherapy would have a favourable effect on the course of this disease: the illness is slow but relentless, immune pathology would appear to be minimal (the disease occurring only in the immunosuppressed), and the outcome is clear-cut and severe. On the other hand, as discussed below, the genetic alterations in virus recovered from the brain of one case are cause for concern about the possible efficacy of ribavirin.

Subacute sclerosing panencephalitis is another interesting special case. In this instance the incubation period averages six years, and the immunologic system is apparently intact. Although the virus has been conclusively demonstrated in the glial and neuronal cells of the central nervous system, there is abundant evidence that it is defective. Like the virus strain from MIBE it shows evidence of extensive mutation and exhibits diminished numbers of downstream transcripts (Cattaneo *et al.*, 1988). It is also inefficient, requiring two weeks, rather than two days to produce sufficient protein to detect by immunoblot. The progression of the illness, unlike that in MIBE, is slow, perhaps because of the marked alteration of the viral genome, perhaps because of the persistence of a strong immune response. In this instance there have been several attempts to treat with oral ribavirin, but with no evident success (Ogle *et al.*, 1989). It may be that, because ribavirin's principal site of action on negative-stranded RNA viruses appears to be at the transcriptional level (Toltzis, O'Connell & Patterson, 1988), the drug can exert little effect on a virus whose transcriptional and protein synthetic machinery is already crippled.

In paediatrics, one of the most commonly identified types of meningoencephalitis is that accompanying varicella (Johnson & Milbourn, 1970). Even in the normal child there is a spectrum of this illness which invites speculation concerning the pathogenesis of the central nervous system injury and the susceptibility of the disease to antiviral chemotherapy. At the viral end of the spectrum, there is pure varicella meningitis, with headache, vomiting but no other signs of encephalitis or encephalopathy, and a CSF formula which includes numerous mononuclear leukocytes, a mildly elevated protein and a normal glucose. At the immune end of the spectrum is an illness which is very like post-measles encephalitis, with onset days to weeks after the resolution of the rash, cortical involvement, with lethargy, confusion and seizures, and a CSF formula which may be normal or may contain only minimal elevation of protein and a few mononuclear cells. This disease is considered to respond to treatment with corticosteroids, and analogies are frequently drawn with both post-measles encephalitis and experimental allergic encephalomyelitis (Bauman & Bergman, 1984).

Between these two extremes is the most common form of varicella-associated encephalitis, with onset at any time between several days before the rash (which is unusual) to the most common time, about 2–7 days after the onset of the rash and before its full resolution. The signs are usually cerebellar, rather than cortical, a feature which is unusual in other virus-associated encephalitides, although occasionally seen. The disease, particularly if severe, prompts inquiries about the use of acyclovir. In the absence of good information about pathogenesis and any information from clinical trials, many consultants look at the state of the rash, judge from its morphology whether there is still active viral replication, project from there to the central nervous system, and base their judgement on such clinical reasoning (Grose, 1986). This disease has not been systematically studied, with regard either to its pathogenesis or to its susceptibility to antiviral treatment, and it is clear that more information in both these areas would have practical consequences.

It is interesting that zoster encephalitis (as opposed to varicella encephalitis) has many features which indicate that replicating virus in the central nervous system is important in the pathogenesis. This disease has two opposing relationships to the immune system: on the one hand it tends to occur in the immunosuppressed or the very elderly; and on the other hand it is usually accompanied by a brisk rise in serum antibody to varicella-zoster virus. What is known about its pathogenesis would dictate that the virus starts from a position of latency within the central nervous system, with activation taking the form of productive infection and cell destruction. Virus can be isolated from the cerebrospinal fluid, and there is evidence for an antiviral immune response within the nervous system itself (Andiman, White-Greenwald & Tinghitella, 1982). Like zoster in the absence of overt meningoencephalitis, clinicians treat it with both corticosteriods and acyclovir, although controlled efficacy trials have not been published. It would seem not unreasonable to recommend the use of both drugs together in instances where the illness was diagnosed early.

Herpes simplex encephalitis is a disease in which, in spite of large gaps in our understanding of its pathogenesis, there is evidence from histopathology that much of the destruction of brain tissue is the consequence of viral cytopathology, and that the brisk inflammatory reaction is a response to necrosis brought about by viral replication. It is not clear why this should be so, and why the virus so clearly predominates while the immune response is so unimportant pathologically. The mystery is compounded by the fact that encephalitis following reactivation is not clinically or pathologically different from encephalitis during primary infection (Leider *et al.*, 1965). Nevertheless, the widespread cellular destruction caused by virus replication allows the prediction that viral chemotherapy will be valuable and efficacious, and several elegantly designed and executed studies have proven this to be so (Whitley *et al.*, 1986; Skoldenberg *et al.*, 1984).

Human immunodeficiency virus infections

In both treatment and exploration of pathogenesis, we are at an early stage in our investigations of infections by human immunodeficiency virus (HIV). Much of what could be said about the interaction of antivirals with disease mechanisms is strictly speculative, largely because there is so little clear information about pathogenesis. It is nevertheless perhaps worth listing those aspects of HIV infection which appear to respond to antiviral chemotherapy and those which do not, in the hope that critical examination of these disease parameters in the future may shed light on pathogenesis and guide further efforts at treatment.

Zidovudine, the only antiviral compound which has been examined carefully in controlled trials, has a clear effect on the level of p24 antigen in the serum or plasma (Fischl *et al.*, 1987). At the same time, it has no discernible effect on the positivity of blood cultures. This may be because the methodology for quantitating HIV in the blood is still primitive, but the consistency of the finding is striking, and it seems likely that it has some other explanation.

Zidovudine also has a temporary effect on the number of circulating CD4+ cells, but this soon wears off and CD4+ cell numbers decrease at a rate not significantly different from controls (Fischl *et al.*, 1987). This temporary finding may be due to the emergence of zidovudine-resistant mutants (Larder *et al.*, 1989), but this theory has not been explored. The correlation between the effect of zidovudine on T4 count and clinical freedom from opportunistic infections is also not clear, and this also bears further study.

Patients receiving zidovudine report an increase in appetite and sense of well-being. As mentioned above, children receiving the drug administered intravenously by continuous-infusion and adults receiving it orally show improvement in higher brain function (Pizzo *et al.*, 1988; Yarchoan *et al.*, 1987), although this conclusion must remain tentative until it has been confirmed with larger numbers. The implication of both these findings, however, is that replicating virus in cells other than CD4+ lymphocytes, possibly members of the macrophage family, possibly other cells, is pathogenetically important both in the encephalopathy and in the gastrointestinal manifestations of the infection. On the other hand, zidovudine might have nonspecific effects unrelated to its antiviral activity which account for these findings.

Research on the effects of antiviral drugs will go on hand in hand with research on the pathogenesis of HIV infections. It seems very likely that knowledge about each will contribute to knowledge about the other, but it will be important to design experiments and construct hypotheses with this interaction in mind so as to accelerate progress toward truly effective treatment.

TREATMENT OF HERPESVIRUS INFECTIONS: AN EXAMPLE OF THE RELEVANCE OF PATHOGENESIS AT THE SUBCELLULAR LEVEL

Since herpesviruses can hold several widely differing relationships with the infected cell, and since the virus, in these different states, varies in susceptibility to antiviral chemotherapy, it is important, in planning treatment protocols, to consider these relationships in various clinical herpesvirus syndromes.

One of the most relevant subcellular variations in herpesvirus biology is the difference in strategy between productive, or cytolytic, and proliferative Epstein–Barr virus replication (Pagano & Datta, 1982). Unlike herpes simplex virus-infected cells, EBV-infected cells do not selectively phosphorylate acyclovir (Colby *et al.*, 1981), and the virus does not appear to code for its own thymidine kinase. The selective action of acyclovir triphosphate on EBV DNA replication appears to lie in the exquisite sensitivity of the EBV polymerase to its action, in relation to the sensitivity of cellular polymerases (Datta *et al.*, 1980). The viral DNA polymerase, however, appears to function only in productive infections, whereas in proliferative infections, where the viral DNA exists in a circular form and replicates only at the same rate as the cellular DNA, production of new DNA appears to be dependent entirely upon cellular polymerases and is very resistant to inhibition by acyclovir (Sixbey & Pagano, 1985). Thus, it could be predicted (and was by Pagano, 1987) that the antiviral action of acyclovir would be apparent only in productively infected cells and would be minimal, and probably subclinical, in proliferatively infected cells.

Since the time of that prediction it has been shown that in infectious mononucleosis acyclovir can inhibit, and even halt completely, shedding of EBV from infected pharyngeal cells during the interval of time it is administered (Andersson *et al.*, 1986). During the same interval the level of infection of circulating B cells in the peripheral blood remains unchanged (Yao *et al.*, 1988), and at the end of treatment shedding of EBV from the pharynx reappears at the same level as it was before the drug was started. Interestingly, it has been difficult to show a measurable effect on individual signs and symptoms in spite of the dramatic effect on virus shedding. Certain signs, such as liver and spleen size, were completely unaffected. When all symptoms (fever, weight loss, sore throat, tonsillar swelling and subjective assessment of sickness) were considered together, a small but significant effect of the treatment could be shown (Andersson *et al.*, 1986). It is of interest that, in contrast, several studies have shown a dramatic effect of corticosteroids on symptoms (Prout & Dalrymple, 1965; Bolden, 1972). There are no published studies of combined therapy with steroids and acyclovir, and these should probably be done.

In contrast to infectious mononucleosis, and again in agreement of the prediction of Pagano (1987), the EBV-induced lymphoproliferation after organ transplantation cannot be either prevented or treated with acyclovir (Zutter *et al.*, 1988). Presumably this frequently fatal form of infection depends primarily on the proliferative mode of viral replication, which, in turn, depends on the action of cellular DNA polymerases poorly sensitive to the action of acyclovir.

A number of interesting questions arise in the wake of these dramatic examples of the specificity of acyclovir and the differing subcellular pathogeneses of EBV infections. One of them involves the interesting lymphocytic interstitial pneumonia (LIP) which appears commonly during the course of HIV infection in children (Rubinstein *et al.*, 1986). There is some evidence, albeit not absolutely convincing, that this disease is in some way related to pulmonary infection with EBV virus, and that a peculiar relationship pertains between EBV and the variable and progressive immunodeficiency induced by HIV after vertical transmission (Andiman *et al.*, 1985). Like infectious mononucleosis, this disease also responds to corticosteroids (Rubinstein *et al.*, 1988), and we might guess therefore that antiviral treatment would have a lesser clinical effect. Nevertheless, there are important questions. Would a combination of steroids and acyclovir be effective in treatment? Would acyclovir, administered prophylactically in infants and children with both HIV and EBV infection, prevent the development of LIP? The answers to these questions depend on the pathogenesis of this disease: its relationship to EBV, the mode of replication of EBV if it is indeed involved, and the mechanism of the lymphoproliferation. If lymphoproliferation is due to infection of B- (and perhaps T-) cells with EBV, with consequent cell division as is seen in some other lymphoproliferative syndromes in immunodeficient hosts (Zutter *et al.*, 1988), acyclovir would appear to have little role in treatment or prevention. If, on the other hand, the proliferating lymphocytes are reacting to productive EBV infection of other cells, acyclovir (along with corticosteroids) might be helpful. These questions deserve investigation in the clinical setting.

Since the major activity of acyclovir is on the action of the viral polymerase, it could be predicted that it would have no effect on the state of latency of the herpes viruses. This was dramatically proven in the studies of prophylactic oral acyclovir to prevent recurrences of genital herpes in subjects with histories of frequent clinical symptoms. As long as the drug was administered, the frequency of these recurrences dropped dramatically, usually to zero or almost zero. As soon as the drug was stopped, recurrences resumed, with a latency and frequency identical to that seen during the trial in the placebo recipients (Douglas *et al.*, 1984; Straus *et al.*, 1984). This result was expected and predictable. Less predictable has been the universal experience that treatment of primary genital herpes infections with acyclovir has had no effect on the establishment of latency and the frequency of recurrence (Bryson *et al.*, 1983; Mindel *et al.*, 1984). This finding may shed light on the pathogenetic events in the establishment of latency during or after genital herpes simplex infection. Because acyclovir had a dramatic effect on the level of virus replication which began 48 hours after the drug was started, it appears likely that latency is established early in the course of infection, perhaps during the pre-eruptive phase or the first 5–6 days of the eruptive phase.

CONCLUSION

This brief review of a large topic is aimed to convince the reader that pathogenesis is an important consideration in the treatment of viral diseases. It has taken only a few of many possible examples to illustrate these points. As antiviral chemotherapy develops, it is likely that the importance of these interrelationships will grow. We can expect to have better, safer and more potent drugs to inhibit viral replication. On the exciting frontier of immunotherapy, our increasingly sophisticated knowledge of the beneficial and harmful portions of the immune response, with specific information on the cells involved (whether, for example, CD8+ or CD4+ cytolytic cells) and the mediator paths triggered to complete the inflammatory response, will allow hitherto unimagined interventions at the level of host response. All of this will depend on our increasing knowledge of disease mechanisms.

REFERENCES

Aicardi, J., Goutieres, F., Arsenio-Nunes, M. L. & Lebon, P. (1977). Acute measles encephalitis in children with immunosuppression. *Journal of Paediatrics*, **59**, 232–9.

Andersson, J., Britton, S., Ernberg, I., Andersson, U., Henle, W., Skoldenberg, B. & Tisell, A. (1986). Effect of acyclovir on infectious mononucleosis: A double-blind, placebo-controlled study. *Journal of Infectious Diseases*, **153**, 283–90.

Andiman, W. A., Eastman, R., Martin, K., Katz, B. Z., Rubinstein, A., Pitt, J., Pahwa, S. & Miller, G. (1985). Opportunistic lymphoproliferations associated with Epstein–Barr viral DNA in infants and children with AIDS. *Lancet*, **ii**, 1390–3.

Andiman, W. A., White-Greenwald, M. & Tinghitella, T. (1982). Zoster encephalitis: Isolation of virus and measurement of varicella-zoster-specific antibodies in cerebrospinal fluid. *American Journal of Medicine*, **73**, 769–72.

Bauman, M. L. & Bergman, I. (1984). Postvaricella Encephalitis. *Archives of Neurology*, **41**, 556–8.

Bolden, K. J. (1972). Cortiscosteroids in the treatment of infectious mononucleosis. *Journal of the Royal College of General Practitioners*, **22**, 87–95.

Boutin, B., Matsuguchi, L., Lebon, P., Ponsot, G., Arthuis, M. & Nelson, D. L. (1987). Soluble IL-2 receptors in acute and subacute encephalitis. *Annals of Neurology*, **22**, 658–61.

Bryson, Y. J., Maryanne, D., Lovett, M., Aguna, G., Taylor, S., Cherry, J. D., Johnson, B. L., Wiesmeier, E., Growdon, W., Creagh-Kirk, T. & Keeney, R. (1983). Treatment of first episodes of genital herpes simplex virus infection with oral acyclovir: a randomized double-blind controlled trial in normal subjects. *New England Journal of Medicine*, **308**, 916–21.

Cattaneo, R., Rebmann, G., Baczko, K., ter Meulen, V. & Billeter, M. A. (1987). Altered ratios of measles virus transcripts in diseased human brains. *Virology*, **160,** 523–6.

Cattaneo, R., Schmid, A., Eschle, D., Baczko, K. & ter Meulen, V. (1988). Biased hypermutation and other genetic changes in defective measles viruses in human brain infections. *Cell*, **55**, 255–65.

Colby, B. M., Furman, P. A., Shaw, J. E., Elion, G. B. & Pagano, J. S. (1981). Phosphorylation of acyclovir [9-(2-hydroxyethoxymethyl)guanine] in Epstein–Barr virus infected lymphoblastoid cell lines. *Journal of Virology*, **38**, 606–11.

Crumpacker, C. S., Schnipper, L. E., Marlowe, S. I., Kowalsky, P. N., Hershey, B. J. & Levin, M. J. (1982). Resistance to antiviral drugs of herpes simplex virus isolated from a patient treated with acyclovir. *New England Journal of Medicine*, **306**, 344–6.

Datta, A. K., Colby, B. M., Shaw, J. E. & Pagano, J. S. (1980). Acyclovir inhibition of Epstein–Barr virus replication. *Proceedings of the National Academy of Sciences USA*, **77**, 5163–6.

Disney M. E., Matthews, R. & Williams, J. D. (1971). The role of infection in the morbidity of asthmatic children admitted to hospital. *Clinical Allergy*, **1**, 399–406.

Dolin, R., Reichman, R. C., Madore, H. P., Maynard, R., Linton, P. N. & Jones, J. W. (1982). A controlled trial of amantadine and rimantadine in the prophylaxis of influenza A infection. *New England Journal of Medicine*, **307**, 580–4.

Domurat, F., Roberts, N. J. Jr., Walsh, E. E. & Dagan, R. (1985). Respiratory syncytial virus infection of human mononuclear leukocytes *in vitro* and *in vivo*. *Journal of Infectious Diseases*, **152**, 895–902.

Douglas, J. M., Critchlow, C., Benedetti, J. H., Mertz, G. J., Connor, J. D., Hintz, M. A., Fahnlander, A., Remington, M., Winter, C. & Corey, L. (1984). A double-blind study of oral acyclovir for suppression of recurrences of genital herpes simplex virus infection. *New England Journal of Medicine*, **310**, 1551–6.

Douglas, R. M., Moore, B. W., Miles, H. D., Davies, L. M., Graham N. M. H., Ryan, P., Worswick, D. A. & Albrecht, J. K. (1986). Prophylactic efficacy of intranasal alpha$_2$-interferon against rhinovirus infections in the family setting. *New England Journal of Medicine*, **314**, 66–70.

Fiddian, A. P., Yeo, J. M., Stubbings, R. & Dean, D. (1983). Successful treatment of herpes labialis with topical acyclovir. *British Medical Journal*, **286**, 1699–701.

Fischl, M. A., Richman, D. D., Grieco, M. H., Gottlieb, M. S., Volberding, P. A., Laskin, O. L., Leedom, J. M., Groopman, J. E., Mildvan, D., Schooley, R. T., Jackson, G. E., Durack, D. T., King, D. and the AZT Collaborative Working Group. (1987). The efficacy of azidothymidine (AZT) in the treatment of patients with AIDS and AIDS-related complex. *New England Journal of Medicine*, **317**, 185–91.

Fishaut, M., Tubergen, D. & McIntosh, K. (1980). Cellular response to respiratory viruses with particular reference to children with disorders of cell-mediated immunity. *Journal of Pediatrics*, **96**, 179–86.

Galbraith, A. W., Oxford, J. S., Schild, G. C., Potter, C. W. & Watson, G.

I. (1971). Therapeutic effect of adamantanamine hydrochloride in naturally occurring influenza A2/Hong Kong infection. *Lancet,* **ii**, 113–15.

Galbraith, A. W., Oxford, J. S., Schild, G. C. & Watson, G. I. (1969). Protective effect of 1-adamantanamine hydrochloride on influenza A_2 infections in the family environment. *Lancet,* **ii**, 1026–8.

Griffin, D. E., Ward, B. J., Jacregui, E., Johnson, R. T. & Vaisberg, A. (1989). Immune activation in measles. *New England Journal of Medicine*, **320**, 1667–72.

Grose, C. (1986). Pathogenesis of central nervous system disease caused by varicella virus. *Pediatric Infectious Disease Journal,* **5,** 607–8.

Gutman, L. T., Wilfert, C. M. & Eppes, S. (1986). Herpes simplex virus encephalitis in children: analysis of cerebrospinal fluid and progressive neurodevelopmental deterioration. *Journal of Infectious Diseases*, **154**, 415–21.

Hall, C. B., Douglas, R. G. & Geiman, J. M. (1975). Quantitative shedding patterns of respiratory syncytial virus in infants. *Journal of Infectious Diseases*, **132**, 151–6.

Hall, C. B., Hall, W. J. & Speers, D. M. (1979). Clinical and physiologic manifestations of bronchiolitis and pneumonia: outcome of respiratory syncytial virus. *American Journal of Diseases in Children*, **133**, 798–802.

Hall, C. B., McBride, J. T., Walsh, E. E., Bell, D. M., Gala, C. L., Hildreth, S., Ten Eyck, L. G. & Hall, W. J. (1983). Aerosolized ribavirin treatment of infants with respiratory syncytial viral infection: a randomized double-blind study. *New England Journal of Medicine*, **308**, 1443–7.

Hayden, F. G., Albrecht, J. K., Kaiser, D. L. & Gwaltney, J. M. (1986). Prevention of natural colds by contact prophylaxis with intranasal alpha$_2$-interferon. *New England Journal of Medicine*, **314**, 71–5.

Hayden, F. G., Belshe, R. B., Clover, R. D., Hay, A. J., Oakes M. R. & Soo, W. (1989). Emergence and apparent transmission of rimantadine-resistant influenza A virus in families. *New England Journal of Medicine*, **231**, 1696–702.

Ho, D. D., Pomerantz, R. J. & Kaplan, J. C. (1987). Pathogenesis of infection with human imunodeficiency virus. *New England Journal of Medicine*, **317** 278–86.

Johnson, R. T., Griffin, E. E., Hirsch, R. L., Wolinsky, J. S., Roedenbeck, S., DeSoriano, I. L. & Vaisberg, A. (1984). Measles encephalomyelitis – clinical and immunologic studies. *New England Journal of Medicine*, **310**, 137–41.

Johnson, R. & Milbourn, P. E. (1970). Central nervous system manifestations of chickenpox. *Canadian Medical Association Journal*, **102**, 831–4.

Knight, V., McClung, H. W., Wilson, S. Z., Waters, B. K., Quarles, J. M., Cameron, R. W., Greggs, S. E., Zerwas, J. M. & Couch, R. B. (1981). Ribavirin small-particle aerosol treatment of influenza. *Lancet,* **ii**, 945–9.

Koprowski, H. & LeBell, I. (1950). The presence of complement-fixing antibodies against brain tissue in sera of persons who had received antirabies vaccine treatment. *American Journal of Hygiene*, **51**, 292–9.

Larder, B. A., Darby, G. & Richman, D. D. (1989). HIV with reduced sensitivity to zidovudine (AZT) isolated during prolonged therapy. *Science*, **243**, 1731–4.

Leider, W., Magoffin, R. L., Lennette, E. H. & Leonards, L. N. R. (1965). Herpes-simplex-virus encephalitis: its possible association with reactivated latent infection. *New England Journal of Medicine*, **273**, 341–7.

McCormick, J. B., King, I. J., Webb, P. A., Scribner, C. L., Graven, R. B., Johnson, K. M., Elliott, L. H. & Williams, R. B. (1986). Lassa Fever: effective therapy with ribavirin. *New England Journal of Medicine*, **314**, 20–6.

McIntosh, K., Ellis, E. F., Hoffman, L. S., Lybass, T. G., Eller, J. J. & Fulginiti, V. A. (1973). The association of viral and bacterial respiratory infections with exacerbations of wheezing in young asthmatic children. *Journal of Pediatrics*, **82**, 578–90.

McIntosh, K., Kurachek, S. C., Cairns, L. M., Burns, J. C. & Goodspeed, B. (1984). Treatment of respiratory viral infection in an immunodeficient infant with ribavirin aerosol. *American Journal of Diseases of Children*, **138**, 305–8.

McIntosh, K., Masters, H. B., Orr, I., Chao, R. K. & Barkin, R. M. (1978). The immunologic response to infection with respiratory syncytial virus in infants. *Journal of Infectious Diseases*, **138**, 24–32.

McIntosh, K., McQuillin, J. & Gardner, P. S. (1979). Cell-free and cell-bound antibody in nasal secretions from infants with respiratory syncytial virus infection. *Infection and Immunity*, **23**, 276–81.

Merigan, T. C., Reed, S. E., Hall, T. S. & Tyrrell D. A. J. (1973). Inhibition of respiratory virus infection by locally applied interferon. *Lancet*, **i**, 563–67.

Mills, J., Van Kirk, J. E., Wright, P. F. & Chanock, R. M. (1971). Experimental respiratory syncytial virus infection of adults. *Journal of Immunology*, **107**, 123–30.

Mindel, A., Adler, M. W., Sutherland, S. & Fiddian, A. P. (1982). Intravenous acyclovir treatment for primary genital herpes. *Lancet*. **i**, 697–700.

Mindel, A., Weller, I. V. D., Faherty, A., Sutherland, S., Hindley, D., Fiddian, A. P. & Adler, M. W. (1984). Prophylactic oral acyclovir in recurrent genital herpes. *Lancet*, **ii**, 57–9.

Murphy, B. R., Graham, B. S., Prince, G. A., Walsh, E. E., Chanock, R. M., Karzon, D. T. & Wright, P. F. (1986). Serum and nasal-wash immunoglobulin G and A antibody response of infants and children to respiratory syncytial virus F and G glycoproteins following primary infection. *Journal of Clinical Microbiology*, **23**, 1009–14.

Murphy, J. V. & Yunis, E. J. (1976). Encephalopathy following measles infection in children with chronic illness. *Journal of Pediatrics*, **88**, 937–42.

Naficy, K. (1963). Human influenza infection with proved viremia. *New England Journal of Medicine*, **269**, 964–6.

Ogle, J. W., Toltzis, P., Parker, W. D., Alvarez, N., McIntosh, K., Levin, M. J. & Lauer, B. A. (1989). Oral ribavirin therapy for subacute sclerosing panencephalitis. *Journal of Infectious Diseases*, **159**, 748–50.

Pagano, J. S. (1987). Pathogenesis and treatment of Epstein–Barr virus infection. *Journal of Experimental Pathology*, **3**, 441–8.

Pagano, J. S. & Datta, A. K. (1982). Perspectives on interactions of acyclovir with Epstein–Barr and other herpes viruses. *American Journal of Medicine*, *Acyclovir Symposium*, pp. 18–26.

Pizzo, P. A., Eddy, J., Falloon, J., Balis, F. M., Murphy, R. F., Moss, H., Wolters, P., Brouwers, P., Jarosinski, P., Rubin, M., Broder, S., Yarchoan, R., Brunetti, A., Maha. M., Nusinoff-Lehrman, S. & Poplack, D. G. (1988). Effect of continuous intravenous infusion of zidovudine (AZT) in children with symptomatic HIV infection. *New England Journal of Medicine*. **219**, 880–96.

Powers, C. N., Peavy, D. L. & Knight, V. (1982). Selective inhibition of functional lymphocyte subpopulations. *Antimicrobial Agents and Chemotherapy*, **22**, 108–14.

Prout, C. & Dalrymple, W. (1965). A double-blind study of eighty-two cases of infectious mononucleosis treated with corticosteroids. *Journal of the American College Health Association*, **15**, 62–6.

Rosner, I. K., Welliver, R. C., Edelson, P. J. & Geraci-Cardullo, K. (1987). Effect of ribavirin therapy on respiratory syncytial virus-specific IgE and IgA responses after infection. *Journal of Infectious Diseases*, **155**, 1043–7.

Rubinstein, A., Bernstein, L. J., Charytan, M., Krieger, B. Z. & Ziprkowski, M. (1988). Corticosteroid treatment for pulmonary lymphoid hyperplasia in children with the acquired immune deficiency. *Pediatric Pulmonology*, **4**, 13–17.

Rubinstein, A., Morecki, R., Silverman, B., Charytan, M., Kreiger, B. Z., Andiman, W., Ziprkowski, M. N. & Goldman, H. (1986). Pulmonary disease in children with acquired immune deficiency syndrome and AIDS-related complex. *Journal of Pediatrics*, **108**, 498–503.

Saral, R., Ambinder, R. F., Burns, W. H., Angelopulos, C. M., Griffin, G. E., Burke, P. J. & Lietman, P. S. (1983). Acyclovir prophylaxis against herpes simplex virus infection in patients with leukemia. *Annals of Internal Medicine*, **99**, 773–6.

Saral, R., Burns, W. H., Laskin, O. L., Santos, G. W. & Lietman, P. S. (1981). Acyclovir prophylaxis of herpes-simplex-virus infections. *New England Journal of Medicine*, **305**, 63–7.

Sixbey, J. W. & Pagano, J. S. (1985). Epstein–Barr virus transformation of human B lymphocytes despite inhibition of viral polymerase. *Journal of Virology*, **53**, 299–301.

Skoldenberg, B., Forsgren, M., Alestig, K., Bergstrom, T., Burman, L., Dahlqvist, E., Forkman, A., Fryden, A., Lovgren, K., Norlin, K., Norrby, R., Stenkvist, E. O., Stiernstedt, G., Uhnoo, I. & DeVahl, K. (1984). Acyclovir versus vidarabine in herpes simplex encephalitis. *Lancet*, **ii**, 707–11.

Spruance, S. L., Schnipper, L. E., Overall, J. C., Jr., Kern, E. R., Webster, B., Modlin, J., Wenerstrom, G., Burton C., Arndt, K. A., Chiu, G. L. & Crumpacker, C. S. (1982). Treatment of herpes simplex labialis with topical acyclovir in polyethylene glycol. *Journal of Infectious Diseases*, **146**, 85–90.

Straus, S. E., Takiff, H. E., Seidlin, M., Bachrach, S., Linnger, L., DiGiovanna, J. J., Western, K. A., Smith, H. A., Nusinoff-Lehrman, S., Creagh-Kirk, T. & Alling, D. W. (1984). Suppression of frequently recurring genital herpes. *New England Journal of Medicine*, **312**, 1545–50.

Taber, L. H., Knight, V., Gilbert, B. E., McClung, H. W., Wilson, S. Z., Norton, J., Thurson, J. M., Gordon, W. H., Atmar, R. L. & Schlaudt, W. R. (1983). Ribavirin aerosol treatment of bronchiolitis associated with respiratory syncytial virus infection in infants. *Journal of Pediatrics*, **72**, 613–18.

Toltzis, P., O'Connell, K. & Patterson, J. L. (1988). Effect of phosphorylated ribavirin on vesicular stomatitis virus transcription. *Antimicrobial Agents and Chemotherapy*, **32**, 492–7.

Wade, J. C., Day, L. M., Crowley, J. J. & Meyers, J. D. (1984). Recurrent infection with herpes simplex virus after marrow transplantation: role of the specific immune response and acyclovir treatment. *Journal of Infectious Diseases*, **149**, 750–6.

Whitley, R. J., Alford, C. A., Hirsch, M. D., Schooley, R. T., Luby, J. P., Aoki, F. Y., Hanley, D., Nahmias, A. J., Soong, S. J. & the NIAID Collaborative Antiviral Study Group. (1986). Vidarabine versus acyclovir therapy in herpes simplex encephalitis. *New England Journal of Medicine*, **314**, 144–9.

Wingfield, W. L., Pollack, D. & Grunert, R. R. (1969). Therapeutic efficacy of amantadine HCl and rimantadine HCl in naturally occurring influenza A2 respiratory illness in man. *New England Journal of Medicine*, **281**, 579–83.

Yanagihara, R. & McIntosh, K. (1980). Secretory immunological response in infants and children to parainfluenza virus types 1 and 2. *Infection and Immunity*, **30**, 23–8.

Yao, Q. Y., Ogan, P., Rowe, M., Wood, M. & Rickinson, A. B. (1988). The Epstein–Barr virus–host balance in acute infectious mononucleosis patients receiving acyclovir anti-viral therapy. *International Journal of Cancer*, **43**, 61–6.

Yarchoan, R., Berg, G., Brouwers, P., Fischl, M. A., Spitzer, A. R., Wichman, A., Grafman, J., Thomas, A. V., Safai, B., Brunetti, A., Perno, C. F., Schmidt, P. J., Larson, S. M., Myers, C. E. & Broder, S. (1987). Response of human-immunodeficiency-virus-associated neurological disease to 3′-azido-3′-deoxythymidine. *Lancet*, **i**, 132–5.

Zutter, M. M., Martin, P. J. Sale, G. E., Shulman, H. M., Fisher, L., Thomas E. D. & Durnam, D. (1988). Epstein–Barr virus lymphoproliferation after bone marrow transplantation. *Blood*, **72**, 520–9.

VIRUS REPLICATION AND STRATEGIES FOR SPECIFIC INHIBITION

GRAHAM DARBY

Molecular Sciences Department, Wellcome Research Laboratories, Beckenham, Kent BR3 3BS, UK

INTRODUCTION

Virus particles are essentially inert, and with few exceptions are rather fragile objects whose existence is short-lived and unproductive, unless they can penetrate the defences of a living cell. Once access is gained to the complex machinery of the cell, they can achieve spectacular results. A single virus particle can initiate a sequence of events which leads to production of vast numbers of new particles and usually to the death of the cell. Furthermore, spread of the infection in the multicellular host can have disastrous consequences for the whole organism. It is this ability of viruses to impair functions of complex organisms, including plants, animals and especially man himself, which provides us with a powerful incentive to devise means to thwart their activities.

The virus particle

Viruses are a heterogeneous group of agents which differ widely in composition, structure and replication strategies (Matthews, 1982). However, they share many common features. Each virus possesses a genome with a limited set of genes. The genome may be composed of either DNA or RNA, it can be single or double stranded and there may be several individual segments (McGeoch, 1984). Associated with the genome may be one or more virus-coded enzymes, which are required for the transcription (or reverse transcription) of the genetic material (Munyon, Paoletti & Grace, 1967; Baltimore, 1970; Temin & Mitzutani, 1970).

The genome is contained within a shell (the capsid) which is composed of a regular array of protein subunits, and this may in turn be surrounded by a membrane which is studded with virus-coded glycoproteins appearing as 'spikes' on the surface. The glycoproteins are normally anchored into the membrane by a hydrophobic transmembrane domain (Von Heijne, 1981).

While it is usually a simple matter to destroy the infectivity of free virus using mild chemical treatment, detergents, heat or radiation, once an infection is established in a cell or in a multicellular host the virus is inaccessible. We therefore need to devise more subtle strategies to block the multiplication of viruses, and to do this we need to understand in detail their replication mechanisms.

Replication

Although the details of virus replication processes differ considerably between virus groups, there are a number of aspects which are common to them all. At its simplest the replication of all viruses can be broken down into three stages. The first stage is the *invasion* of the host cell and this involves the attachment of the virus to the cell surface, its transport to the interior of the cell and its partial or complete breakdown.

The next stage is the *eclipse*, so-called because early workers with bacterial viruses were able to show that immediately after invasion of the bacterial cell no infectious virus could be detected (Doermann, 1952). Although their early view was that this period represented a quiescent phase in the growth cycle, we now know that it is in fact a period of intense activity. Virus genes are expressed and the genome of the virus is replicated. In addition, virus structural components accumulate. An important feature of this phase is that it is tightly regulated either by virus-specific proteins or in some cases by the physiological state of the cell.

The final phase is the *maturation* and *release* of new infectious virus particles into the extracellular environment, or occasionally, if cells are in intimate contact, directly into adjacent cells.

Strategies for intervention

In the remainder of this chapter I shall consider in more detail each stage in the infectious process and highlight those events which might provide us with suitable targets for antiviral drugs. In considering these events and potential strategies for intervention it is worthwhile remembering that our major objective is to block virus replication without adversely affecting uninfected cells or the whole organism. To achieve this the target must be carefully selected and should satisfy a number of criteria.

Specificity

Viruses rely heavily on cellular functions for their replication. However, the ideal target will be virus-specific, either a reaction catalysed by a virus-coded protein, a regulatory event involving a virus-specific nucleic acid sequence, or some other event requiring participation of a virus-specific product.

Dependence

Whatever the target chosen, inhibition of that event should abort the infection process. The target event must therefore be one which is essential for virus multiplication.

Selectivity

Finally, the target event should be distinct from any which occurs in uninfected cells since we wish to avoid interfering with normal cell function. This is the most difficult criterion to satisfy and we often have to be satisfied with a compromise, accepting that we may have to exploit subtle differences between virus and cellular processes.

Having identified suitable target events, the various strategies for intervention will be considered and, where possible, examples of inhibitors exploiting these strategies will be cited. I do not intend to provide a comprehensive catalogue of antiviral agents, but will use specific examples to illustrate the approaches. The examples will be restricted to viruses of clinical importance, and it is not coincidence that many will relate to HIV. The global impact of AIDS has ensured major investment in new research programmes designed to search for anti-HIV drugs and the immense sense of urgency engendered by the threat of this disease has ensured that every feasible strategy is being explored.

INVASION

The approach

The first stage in the invasion of the host cell is contact between the virus and the cell. Although a crucial stage in the infection process, this meeting relies on chance events. A swimmer gulping down a mouthful of water contaminated with poliovirus from another person's faeces, a grandmother with a weeping cold sore kissing her newly acquired grandchild or an i.v. drug user sharing a dirty needle with an HIV-infected friend are all events which can bring a virus into contact with a new susceptible host (Melnick, Gerba & Wallis, 1978; Wildy, Field & Nash, 1982; Marmor *et al.*, 1987). They are not events which require the direct intervention of any virus functions and so are not targets for antiviral drugs. However, it must be obvious that other forms of intervention are possible at this stage – but that is not the topic for this chapter.

Attachment

In contrast to the random processes which bring virus and cell into close proximity, attachment to, and penetration of, the cell require several very specific processes (for review, see Marsh & Helenius, 1988). The first is an interaction between a surface component of the virus (glycoprotein spike or capsid protein) and a specific component on the cell surface. This event is one which satisfies all the criteria for an antiviral target. However, although the specificity of this process has been recognised for many years, the details of the interactions have been worked out in relatively few cases. Perhaps the best understood is that between the influenza virus haemagglutinin (HA), a trimeric glycoprotein spike embedded in the influenza membrane, and the terminal N-acetyl neuraminic acid (sialic acid) residue on a cell surface oligosaccharide (Wilson, Skehel & Wiley, 1981; Weis *et al.*, 1988).

Influenza HA/sialic acid interaction

Clearly, if the interaction between HA and sialic acid can be prevented, then the infection will be blocked at a very early stage. Two possible strategies are immediately obvious: either coat the receptors on the cell surface so that they are inaccessible to the HA, or coat HA so that it cannot bind to the cell surface receptor. Although either strategy would be expected to block the infection, working on the principle that it is better to tackle the virus-specific product rather than tamper with the cell, the preferable target is HA. It is also likely to be an easier target to tackle because we start with the knowledge that a relatively simple molecule (sialic acid) will interact with the appropriate site on the protein, and that provides us with a good chemical starting point. In contrast, the site on the protein which interacts with sialic acid is complex and involves several non-contiguous regions of the HA1 polypeptide (Weis *et al.*, 1988).

It has been shown that sialic acid itself will block the interaction of influenza with its receptor on the surface of red cells, and, although high concentrations of the sugar are required, there has been some success in developing synthetic sialic acid glycosides as more effective inhibitors (Pritchett *et al.*, 1987).

This approach illustrates one of the most common tactics employed in antiviral drug discovery. If a protein is known to have an essential function in virus growth, and if that function involves an interaction with a small molecule, then analogues of that molecule may inhibit the protein's action. Once an inhibitor is identified, the usual approach is to produce analogues in a systematic fashion and identify features of the molecule important for its inhibitory properties. Such data can then be used to refine inhibitors in a more rational way and to direct chemical syntheses. This approach using 'Quantitative Structure Activity Relationships' (QSAR) is extremely powerful. However, it is hoped that, in the future it will be supplemented by structural information derived from X-ray crystallography. This would put the whole task on a more rational basis, but, more importantly, additional potential binding sites on the protein not identified by the analogues synthesised might be recognised and exploited. This is an approach which is feasible now with HA since both the structure of the protein and its complex with sialic acid have been recently solved (Wilson *et al.*, 1981; Weis *et al.*, 1988).

HIV gp120/CD4 interaction

Some of the most exciting work under-way in the attachment area at the moment is with HIV where it is known that the initial interaction between the virus and the host cell involves an association between the T cell marker, CD4, and the surface glycoprotein of the virus gp120 (Dalgleish *et al.*, 1984; Klatzman *et al.*, 1984). Here a very direct approach has been used. It was reasoned that if all virus-associated gp120 could be complexed with free CD4 molecules, the gp120 would be unable to attach to CD4 receptors on the cell surface. The first stage was to produce a soluble form of CD4, lacking the membrane anchoring region of the natural protein (Fischer *et al.*, 1988; Hussey *et al.*, 1988; Deen *et al.*, 1988; Traunecker, Lüke & Karjalainenk, 1988). This was achieved by genetic manipulation and expression of the truncated CD4 gene in heterologous systems. It was shown that CD4 produced in this way could indeed block virus infectivity in tissue culture systems. Unfortunately this material has the disadvantage that it is rapidly degraded *in vivo* by natural proteases (Capon *et al.*, 1989). However, the second generation of these molecules has already been produced in an attempt to solve this problem (Capon *et al.*, 1989). The soluble portion of CD4 has been linked to the Fc regions of an IgG molecule and this creates a chimera with properties similar to antibody, including longer *in vivo* half-life. The chimera also has the advantage that immunological events initiated by the Fc region of antibodies such as complement-mediated lysis and antibody-dependent cytotoxicity can be triggered by this molecule. A secondary effect would therefore be the destruction by the immune system of infected cells expressing gp120 on their surface.

There are some real concerns over this approach. Apart from the difficulties of delivering rather fragile protein molecules to appropriate sites – oral administration is ruled out because they would be unlikely to survive the hostile environment of the GI tract – there is the more serious concern that long-term administration of such molecules could elicit an autoimmune response.

It may be that a third generation of CD4 derived molecules will be required. The regions of the CD4 peptide which interact with gp120 have been localised to two disulphide-bridged loops lying between residues 25 and 93, and 139 and 168 (Jameson *et al.*, 1988) and a small region in one of the loops (residues 48 to 51) Clayton *et al.* (1988). A peptide spanning this critical region (residues 25 to 58) has been shown to inhibit virus attachment (Jameson *et al.*, 1988). However, a modified peptide derived from a different region (residues 83 to 94) of the same loop has also been shown to block infection (Lifson *et al.*, 1988). It is possible that these small molecules might pose far fewer problems as therapeutics than the use of large unstable, immunogenic proteins. However, it is likely that peptide mimics will be needed if we are to have orally available drugs which can survive the environment of the human GI tract.

Recently a different approach using CD4 has been developed which is designed not to block the infectious process directly but rather to target the infected cell. This is properly viewed as an extension of the work on CD4/Fc chimeras. Toxins may be linked to CD4 molecules which bind to gp120 on the surface of infected cells. The toxins are then taken up by the cells which are subsequently killed. This approach has been used successfully in culture with both ricin (Till *et al.*, 1988) and *Pseudomonas* exotoxin (Chaudhary *et al.*, 1988), but it remains to be seen whether these 'golden bullets' carrying powerful toxins will be sufficiently specific and safe enough for *in vivo* use.

Apart from preventing cell surface interactions directly, there is theoretically another strategy which could be employed with HIV. This relies on the observation that gp120 is only loosely attached to the virus surface through a protein/protein interaction with a second virus glycoprotein, gp41. It is gp41 which is anchored into the virus envelope. If gp120 could be displaced, thus stripping the virus of the molecule which attaches it to the host cell surface, this could render the virus non-infectious. It has already been shown that mutations in the N-terminus of gp120 or gp41 can disrupt the interaction, and viruses carrying such mutations are non-infectious for T-lymphocytes (Kowalski *et al.*, 1987). The strategy of disrupting protein/protein interactions is not one which has yet been applied with any success to antiviral drug discovery. However, there is some evidence that it may be viable, and, if so, could be applicable to many different systems where such interactions are important, such as virus assembly, gene regulation or inhibition of multi-subunit enzymes. We have seen above that peptides can be used to disrupt the gp120/gp41 interaction. Also it has been shown that the herpes simplex virus-specific enzyme, ribonucleotide reductase, can be inactivated by disrupting the interaction between its two subunits. This is brought about by treatment with a short peptide derived from the C-terminus of the larger subunit (Frame, Marsden & Dutia, 1985). The disruption of protein/protein interactions is a feasible strategy for drug development but it remains to be fully exploited. The ability to disrupt protein/protein interactions with small molecules may ultimately be possible only where there are a relatively small number of points of contact between two protein subunits.

The various strategies discussed above are illustrated in Figure 1.

Penetration and uncoating

Penetration of the cell requires transport of the virus across the plasma membrane and delivery into the appropriate cellular compartment (i.e. cytoplasm or nucleus). Uncoating of the virus involves removal of the envelope (if the virus has one) and either partial breakdown of the capsid so that the virion-associated enzymes can acquire their essential substrates and their products (e.g. mRNA) can be released into the cell, or release of free nucleic acid molecules from the capsid. In practice, penetration

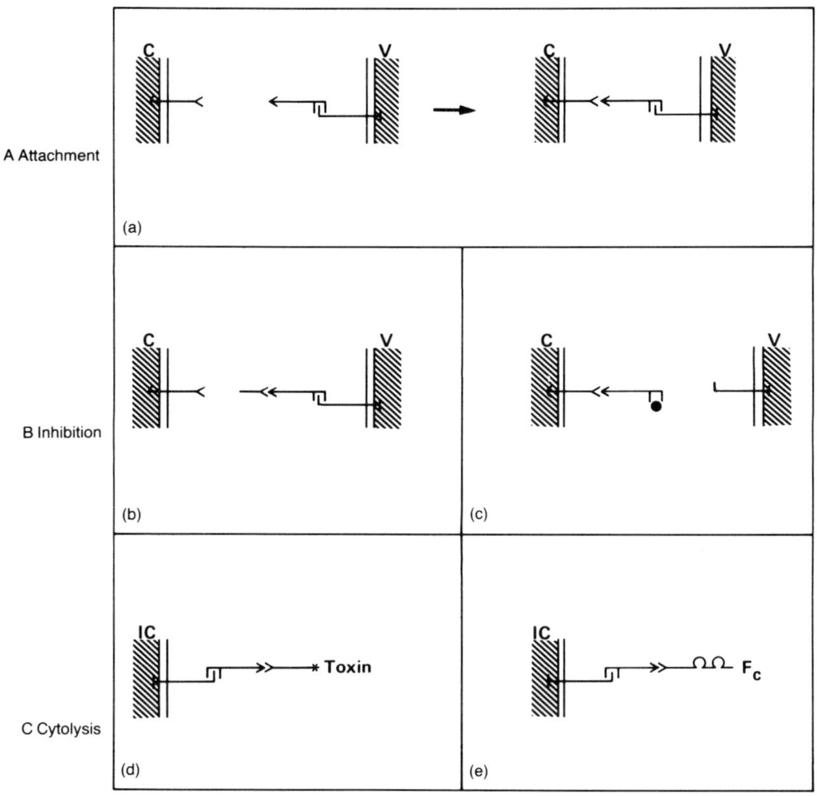

Fig. 1. Antiviral strategies based on gp41/gp120/CD4 interactions: (a) the normal attachment process involving a specific interaction between gp120 and CD4; (b) blockade of the gp120 receptor binding region with soluble CD4; (c) disruption of gp120/gp41 interaction; (d) attack on an infected cell with virus glycoproteins on its surface using soluble CD4 linked to toxin; (e) attack on an infected cell with virus glycoproteins on its surface using soluble CD4 linked to F_c. The CD4 toxin and F_c chimeras also inhibit infectivity through direct interactions with the virus particle. Key: CD4 ; gp120 ; gp41; soluble CD4 ; C uninfected cell; V virus particle; IC infected cell.

and uncoating go hand in hand and it is rarely possible to separate them into discrete categories.

The majority of these processes appear to be carried out by normal cellular machinery and are therefore not amenable to selective inhibition. However, one virus-specific event which has been observed in widely different virus groups is a dramatic conformational change in a surface glycoprotein or capsid protein.

In the case of influenza virus such a conformation shift is seen in the haemagglutinin glycoprotein. During maturation of influenza, the haemagglutinin, which is synthesised as a single polypeptide precursor, HA_0, is cleaved to form two chains, HA_1 and HA_2, and this cleavage is essential for virus infectivity (Klenk *et al.*, 1975; Lazarowitz & Choppin, 1975). Cleavage allows a pH-dependent conformational change which occurs during entry following envelopment of the virus in an endosome. It exposes the hydrophobic N-terminal region of HA_2 which then burrows into the vesicle membrane and mediates fusion of the virus and vesicle membranes (Huang *et al.*, 1980; Skehel *et al.*, 1982). This results in the release of the virus core into the cytoplasm from where it travels to the nucleus in which the virion-associated RNA polymerase transcribes the genomic negative-stranded RNA segments.

Another example of a conformational change in a capsid protein occurs in the picornaviruses. There is clear evidence for such a change during the penetration process and it appears to be necessary for uncoating and release of the virion RNA.

It is a considerable challenge to devise strategies which might block such conformational changes. However, an important aspect of the search for antiviral drugs is the role played by serendipity. While seeking compounds which would inhibit rhinovirus replication, scientists from several different pharmaceutical companies stumbled on a number of compounds which bound to virus and inhibited an event extremely early in virus replication (Bauer *et al.*, 1981; Ishitsuka *et al.*, 1982; Zerial *et al.*, 1985; Diana *et al.*, 1985; Otto *et al.*, 1985). However, these compounds did not prevent attachment to the cell surface and appeared to work by blocking the uncoating process. Subsequently, it was shown by the excellent work of Rossman and his colleagues (Smith *et al.*, 1986*b*) that these molecules interact with the capsid protein VP1 and occupy a cavity beneath the surface of the virion which is accessible via a surface pore, and it is believed that these molecules may act by preventing the conformation change required to release virus nucleic acid. Despite the elegant work on these compounds, none has yet progressed to the clinic. Nevertheless, it follows that it is feasible to block essential conformational changes in proteins by binding small molecules which act as 'space fillers'.

Other penetration and uncoating processes certainly involve conformation changes in surface glycoproteins or capsid proteins and this will be an area for intense study in the future.

THE ECLIPSE

Once the virus particle has released its genome or has been partially uncoated (destroying its infectivity) we enter the eclipse phase of the infection.

As we have seen (Section on Invasion) this phase is a period of intense activity in the cell and it will be convenient to consider it in three stages: events dependent upon the parental genome and virion associated enzymes, genome replication and, finally, regulation.

Virion-associated enzymes

The earliest events in the eclipse phase are directed towards the supply of virus-specific mRNA molecules which specify the synthesis of virus proteins. Many of the enzymes involved in these processes are unique to virus-infected cells and therefore should be ideal targets for selective drugs. The only exceptions are in the nuclear DNA viruses where the cell's transcription machinery (modified by virus factors) transcribes the genomic DNA, and in the case of the positive strand RNA viruses where the genome is immediately translated by host ribosomes. However, none of the unique transcriptional enzymes have yet been successfully exploited as targets for antivirals.

One virion-associated polymerase which has been attacked successfully is the reverse transcriptase of HIV. The only drug licensed to date for the treatment of AIDS (zidovudine) is a potent inhibitor of reverse transcription (Furman *et al.*, 1986; Ono *et al.*, 1986). Once it was apparent that a retrovirus was implicated in AIDS, the search was on for virus inhibitors. Previous experience with the antiherpes drug acyclovir (see Genome Replication, below) had shown that nucleoside analogues could be used to block DNA replication, and in particular that lack of a 3′-OH group on the sugar led to premature chain termination. The obvious place to start looking for potential inhibitors was therefore amongst nucleoside analogues with modified 3′-carbon atoms in the deoxyribose moiety.

The first success was achieved very rapidly when it was shown that the analogue 3′-azido-2′-3′-dideoxythymidine (AZT) was a powerful inhibitor of animal retroviruses and shortly thereafter of HIV replication (Mitsuya *et al.*, 1985). Subsequent studies have shown that only small quantities of AZT are converted in the cell to the triphosphate of AZT, which is the active form of the drug, but this is an extremely potent and selective inhibitor of the reverse transcriptase *in vitro* (Furman *et al.*, 1986).

Success with this compound has now been translated to the clinic where the drug has been shown to benefit AIDS and ARC patients (Fischl *et al.*, 1987) and, more recently, certain categories of asymptomatic individuals. This success fuelled interest in compounds of this type and several others are now, or soon will be, in clinical trials. These include the 2′-3′-dideoxycytidine (Mitsuya & Broder, 1986), 2′-3′-dideoxyinosine (Mitsuya & Broder, 1986), and D4T (2′-3′-dideoxy-2′-3′dehydrothymidine) (Baba *et al.*, 1987) all of which share a common feature in that they lack the 3′-OH group essential for elongation of the DNA chain (Fig. 2).

One difficulty with targeting nucleosides at reverse transcriptase is the need to satisfy the requirement for an analogue which will be phosphorylated by the cellular kinases. One way of partially resolving this problem is to deliver nucleotide analogues rather than nucleosides. Recently De Clercq and his co-workers have described a series of phosphonomethoxyethyl

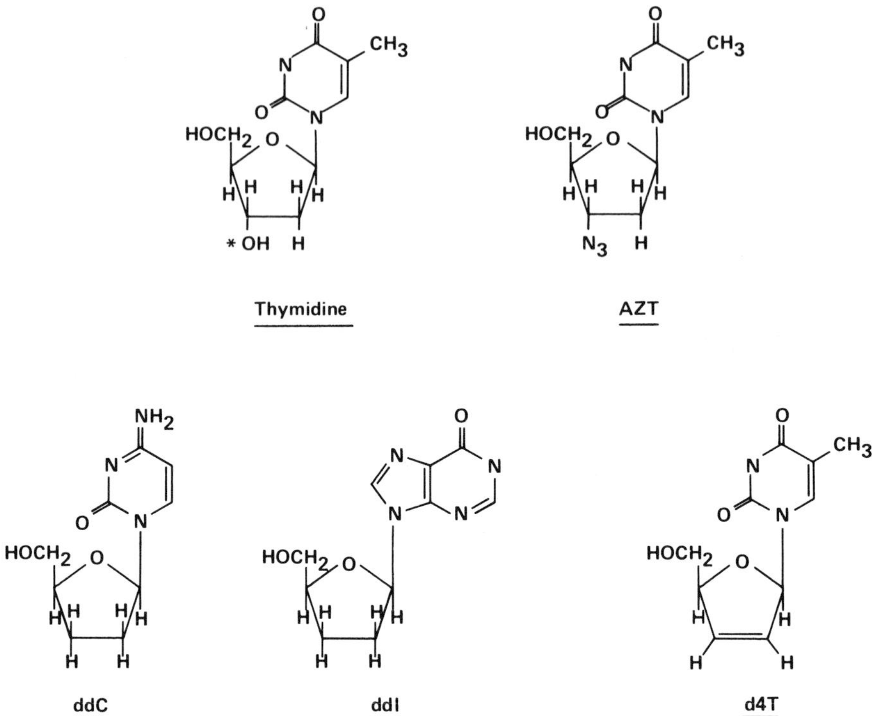

Fig. 2. Comparison of thymidine and 3′-deoxynucleosides. * This indicates the 3′-OH group in the normal 2′-deoxynucleoside. The nucleoside analogues are all inhibitors of HIV and all are presumed to have a similar mechanism of action, i.e. phosphorylation by cell enzymes to the triphosphate followed by interaction with reverse transcriptase and inhibition of DNA synthesis by chain termination.

purine derivatives with good anti-HIV activities and a broad spectrum of activities against other viruses. Apparently these compounds have selectivity for virus-infected cells (De Clercq *et al.*, 1986; De Clercq, 1988). One of these compounds, HPMPA, has been shown to be further phosphorylated by cell enzymes although the role of the phosphorylated derivatives in blocking DNA synthesis has not yet been elucidated (Votruba *et al.*, 1987).

The above approaches would again be greatly facilitated if the structural architecture of the reverse transcriptase were known and its binding and active site residues identified. Some progress has been made towards identifying functional regions of the protein using site-directed mutagenesis (Larder *et al.*, 1987; Lowe *et al.*, 1988) but, so far, attempts to obtain crystallographic data of high resolution on the recombinant protein have been progressing slowly. Clearly, an understanding of the detailed structure/function relationships with this protein would enable a more rational approach to the design of nucleoside analogue inhibitors. However, knowledge of this type would also open the door to the design of non-nucleoside inhibitors which could potentially circumvent some of the problems seen with the nucleosides.

Genome replication

A central feature of the eclipse phase is the replication of the virus genome and for almost all viruses this requires the participation of a virus-coded nucleic acid polymerase. The only exceptions to this rule are some of the small DNA viruses such as the papovaviruses, where DNA replication appears to be carried out by cellular polymerases – although the process is certainly controlled to a degree by virus-coded proteins.

To date, the most successful antiviral drugs are those targeted to the virus polymerases. The reverse transcriptase of HIV, the target for zidovudine, has already been described. However, the first truly selective antiviral to be discovered was the anti-herpes nucleoside analogue acyclovir (Elion *et al.*, 1977). This compound acts by inhibiting DNA replication by a mechanism involving chain termination (Furman *et al.*, 1979). However, as we have seen, for any nucleoside analogue to act at the level of nucleic acid synthesis there is a prior requirement for activation of the analogue by phosphorylation. (See Figure 3.)

The activation of acyclovir occurs with efficiency only in virus infected cells and the initial stage in the activation process appears to be the main focus of this selectivity. Phosphorylation of acyclovir to its monophosphate is dependent upon the virus enzyme, thymidine kinase (Fyfe *et al.*, 1978). The broad substrate specificity of this herpes simplex specific enzyme allows phosphorylation of the compound which otherwise would remain inert in the cell (Elion *et al.*, 1977). Once monophosphate is produced, cell enzymes convert the nucleotide to its triphosphate.

Not only is the activation of acyclovir a selective process requiring the virus-specific kinase, but, also, the action of the triphosphate has a degree of selectivity (Derse *et al.*, 1981), the virus DNA polymerase appearing more sensitive to inhibition by this analogue than cellular polymerases.

Thus, in the case of acyclovir, we have a compound which is selectively activated in infected cells and which, when activated, acts selectively to block genome replication. This 'double dose' of selectivity contributes to the safety of the drug which has now been used extensively in the clinic and it certainly represents a strategy which is extremely attractive. The activity of acyclovir against other human herpesviruses such as varicella-zoster (chicken pox:VZV) and cytomegalovirus is not as impressive as its anti-herpes simplex activity but, at least in the case of varicella there is the potential for adopting a similar strategy with a different analogue since that virus also encodes a thymidine kinase. Indeed, the analogue

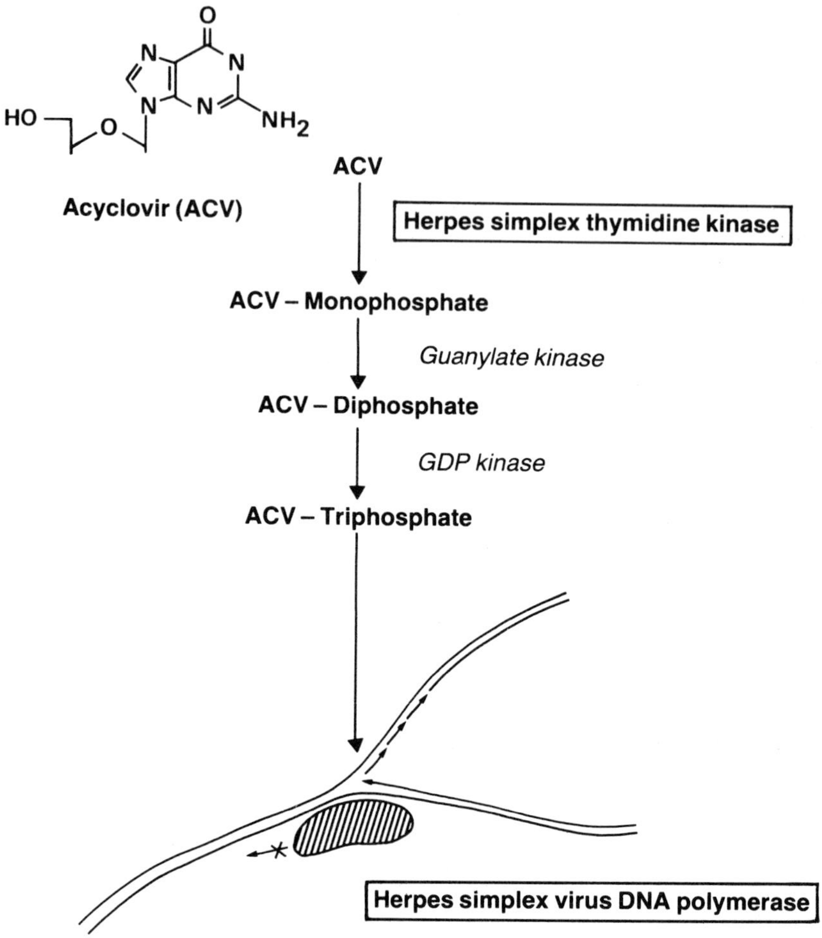

Fig. 3. Mode of action of acyclovir. Acyclovir (structure shown inset) is phosphorylated initially by herpes simplex thymidine kinase to the monophosphate. Further phosphorylation to triphosphate is by cell enzymes. Acyclovir triphosphate is incorporated into the growing DNA strand and this results in chain termination.

1-β-D-arabinofuranosyl-E-5(2-bromovinyl)uracil (BVaraU) has already been shown to be an extremely potent inhibitor of VZV replication (Shigeta *et al.*, 1983) and its mode of action is similar to that of ACV. One distinction is that BVaraUTP, while being a potent inhibitor of VZV DNA synthesis, appears not to be incorporated into virus DNA (Yokota *et al.*, 1989).

In the period preceding genome replication there are other enzymes involved in nucleotide metabolism which could potentially be exploited for the selective modification and activation of nucleoside analogues. Such enzymes include the ribonucleotide reductase of HSV (Cohen, 1972) and the thymidylate synthase peculiar to varicella-zoster (Thompson *et al.*, 1987). The former is required for replication (Preston, Palfreyman & Dutia, 1984) and some progress has already been made in discovering inhibitors for it (Frame *et al.*, 1985).

In addition to the enzymes of nucleoside metabolism and the nucleic acid polymerases themselves there are other proteins involved in replication of the HSV genome. Seven separate virus-coded proteins are necessary for the replication of virus DNA (Wu *et al.*, 1988; Challberg, 1986) and are presumably involved in a replication complex. Any of these proteins are potential targets for antivirals and there is scope for developing agents which will either disrupt the complex or inhibit any of the various enzyme activities.

A similar situation is likely to apply to other replication systems. It is already known, for example, that, in the case of influenza virus, the replication complex involves at least three different virus-coded proteins (Braam, Ulamanen & Krug, 1983). Genome replication is therefore likely to continue as a favourite hunting ground for antiviral drug designers for many years to come.

Regulation and gene expression

Gene expression is tightly regulated in most virus systems so that proteins are produced during the infection cycle at the time they are required and in appropriate amounts. Protein synthesis itself is carried out by the cellular translation machinery but it is directed by virus-specific mRNA. The pattern of protein synthesis is regulated both at the level of transcription and also at the level of translation, and regulation may involve both virus and cellular factors. This has not been a productive area in the search for antivirals but the complex interplay of proteins and nucleic acids should provide many opportunities for intervention. To illustrate the potential, I shall discuss briefly the state of our knowledge relating to the expression of HIV genes.

The regulation of HIV

Like all retroviruses, the RNA genome of HIV is converted to a DNA intermediate which is subsequently integrated into the cell chromosome. HIV genes are transcribed from virus DNA. The state of the cell itself has an important role in HIV regulation since we see a fully productive infection only in lymphocytes which have been stimulated to divide. In the stationary cell the infection remains latent. However, this is only the tip of the iceberg with HIV since, in addition to the dependence on the

U.C.W. ABERYSTWYTH LIBRARY

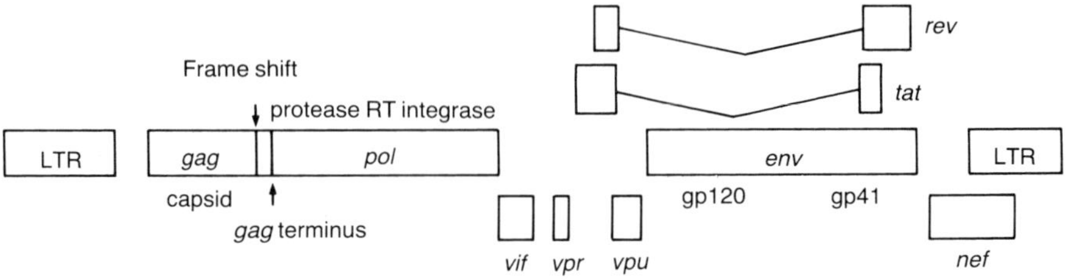

Fig. 4. Genome organisation of HIV. The organisation of HIV genes in the proviral DNA is shown.

cell cycle, there is also a complex network of virus-coded regulatory proteins, both activators and repressors, which control gene expression.

The organisation of the HIV genome is illustrated in Figure 4. The most studied of the regulatory proteins is the transactivator (*tat*) which is a small (M_r14 000) protein. Mutants defective in *tat* function are non-viable (Dayton *et al.*, 1986; Fischer *et al.*, 1986) and so it is a potential drug target. The protein is required for the stable expression of mRNA and it is believed to function through an interaction with a region of strong secondary structure at the 5′-end of the virus genome (Rosen, Sodroski & Hazeltine, 1985; Cullen, 1986; Wright *et al.*, 1986; Peterlin *et al.*, 1986). Although the precise mechanism of action is not known, it appears to affect translation as well.

A second well-studied protein is the *rev* gene product. Again, this appears to be required for productive infection, since in its absence an abnormal pattern of RNA splicing is seen (Feinberg *et al.*, 1986; Knight, Flowerfelt & Ghrayeb, 1987) and there is a failure to synthesise the capsid proteins and surface glycoproteins of the virion. The mRNA molecules contain *cis*-acting repression sequences (CRS) which prevent their accumulation and translation in the absence of the *rev* product (Rosen *et al.*, 1988). There is a further sequence required for *rev* function and that is the CAR sequence in the glycoprotein gene.

There are at least four other genes in the same region of the genome as *tat* and *rev* but their functions are less well understood. The *vif* gene product may be a suitable target for antivirals since it appears to be required for the production of infectious virions (Strebel *et al.*, 1987). *Vpu* and *nef* would seem to be inappropriate targets since not only are they are not essential for replication but, in fact, they appear respectively to enhance the cytotoxicity of the virus and increase its rate of replication.

In the area of regulation and gene expression there are clearly many proteins which provide suitable drug targets, but it is also clear that many of the processes involve interactions with specific sequences in the virus genome. This latter area has received considerable attention in the last few years with the advent of rapid, reliable methods for synthesising oligodeoxyribonucleotides since the most straightforward approach to building a molecule to block a nucleic acid sequence is to make its complement.

'Anti-sense' oligonucleotides

The term 'anti-sense' is used to include any sequence which is complementary to a functional sequence in the genome, either coding or regulatory. In the past three years there has been a spate of papers dealing with use of oligonucleotides to inhibit virus replication, and these papers illustrate the range of approaches which may be feasible.

Zamecnik *et al.* (1986) used normal oligodeoxynucleotides complementary either to splice sites in HIV RNA, or to the site which binds lysyl tRNA, the primer for reverse transcription. T'so and his colleagues used modified oligodeoxynucleotides (methylphosphonates) targeted to VSV coding sequences (Agris *et al.*, 1986) or HSV splice sites (Smith, C. C. *et al.*, 1986a). The use of modified oligonucleotides was to try to reduce the nuclease sensitivity and increase the availability of these compounds. These themes were further explored by several groups (Matsukura *et al.*, 1987; Agrawal *et al.*, 1988; Sarin *et al.*, 1988) who used phosphorothioate and methyl phosphonate analogues to block HIV replication. However, it was clear in the case of phosphorothioates (Matsukura *et al.*, 1987) that inhibition was due not only to the 'anti-sense' character of the oligonucleotides but also to an inherent property of phosphorothioate analogues. These studies have set the stage and provide convincing evidence that oligonucleotides can have interesting biological effects.

A new dimension was added by Zerial, Thuong & Hélène (1987) who were able to increase the binding of oligonucleotides to their complementary sequences by adding a covalently linked intercalating agent to the oligonucleotide terminus. These oligonucleotides which were targeted to common sequences at the 5′-ends of influenza genome RNA fragments were effective inhibitors of influenza replication. Clearly this is a theme which can be expanded considerably since other types of functional molecules could be added to the oligonucleotides which would then target them to the appropriate nucleic acid sequence.

The use of oligonucleotides is an approach with considerable potential but there are enormous hurdles ahead if they are ever to be developed into clinically useful antivirals. Perhaps the most serious would be setting up oligonucleotide synthesis on an industrial scale and the inevitable sky-high costs of such production.

MATURATION AND RELEASE

The final stages of the infection process are those of assembly and maturation and the release of new virus particles into the extracellular environment (Stephens & Compans, 1988). These processes are grouped together because, like attachment and penetration, some aspects of assembly and release tend to be indistinguishable from each other. This is especially true of those enveloped viruses which assemble their capsids in close proximity to the cell membrane and where the final stage of the assembly process, the acquisition of the outer membrane, is also the process which releases the virus from the cell. This budding process, appears to be directed by an association between virus capsid proteins and the intracellular domains of the virus glycoproteins which aggregate in the plasma membrane.

There are many stages during these processes where in theory intervention is possible although so far there are few examples which can be cited.

Protein/protein interactions

The major interactions directing the assembly of capsids are direct protein/protein interactions. These may be interactions between structural subunits of the particle or more transient interactions between structural subunits and maturation proteins which are not incorporated into the mature virus. The interactions between these are analogous to the interactions between the protein subunit of enzyme or the interactions between gp120 and gp41 in the HIV envelope, and the potential strategies for disrupting these interactions are also analogous.

Nucleic acid/protein interactions

The other important interactions in capsid assembly are those between nucleic acids and proteins. Packaging of nucleic acids into newly formed capsids generally involves the specific recognition of sequences in the genome by capsid proteins. Such 'packaging signals' have been identified in herpesviruses, retroviruses and many others and they may be suitable targets for antiviral compounds. As with protein/protein interactions, the approaches to disrupting nucleic acid/protein interactions in capsid assembly will parallel the strategies used to disrupt similar interactions between nucleic acid and protein which are involved in genome expression and regulation.

Post-translational modifications

There are two important modifications of virus proteins which occur during the maturation of infectious particles. One is the modification of surface proteins by glycosylation and the other is the cleavage of large precursor proteins by proteolytic enzymes. The former process is carried out exclusively by cellular enzymes whereas, in contrast, the latter is frequently performed by virus-coded enzymes.

Glycosylation

Although glycosylation is carried out by cellular enzymes, there is a long tradition of attempting to block production of fully infectious virus by inhibiting this process. The inhibitors of glycosylation include, for example, natural sugar analogues such as 2-deoxyglucose which is known to inhibit herpes simplex replication (Courtney, Steiner & Benyesh-Melnick, 1973; Spivack, Prusoff & Tritton, 1982). The compounds block the addition of sugar residues to the glycoproteins. As would be expected, their toxicities preclude their use as antiviral drugs.

There has been considerable excitement recently in the HIV area over a series of compounds which function not by blocking addition of new residues but rather by blocking cleavage of sugar residues from glycoproteins, the final stage in the maturation of infectious HIV. Castanospermine is an inhibitor of d-glucosidase 1 which removes the terminal glucose residues of surface oligosaccharides. In cell culture this appeared to be a promising inhibitor of HIV with little cytotoxicity (Walker *et al.*, 1987). However, it has been shown that castanospermine has appreciable toxicity in a mouse system (Ruprecht *et al.*, 1989). This is more than is seen with retrovir which is a compound with far more potent antiviral activity, and illustrates the inherent danger in targeting antivirals at cellular rather than at viral functions.

Proteolytic cleavage

A number of viruses undergo proteolytic cleavage during late stages of their maturation (for review see Kräusslich & Wimmer, 1988). Influenza virus haemagglutinin precursor must be cleaved to generate HA1 and HA2. If this does not happen, the particles are non-infectious since it is required to produce the hydrophobic N-terminal region of HA2 which is necessary for membrane fusion. Unfortunately this cleavage is carried out by cell rather than by virus-coded enzymes. However, in several other systems virus-coded proteases are required for the cleavage of proteins in immature virions. These were first discovered in the picornaviruses but, more recently, attention has focused on the protease encoded by HIV.

One of the components of the HIV *pol* gene encodes an aspartic protease (Seelmeier *et al.*, 1988). The *gag* gene which encodes the capsid proteins, and the *pol* gene which encodes in addition to the protease, a reverse transcriptase and integrase (see Fig. 4), are expressed as a polyprotein following an infrequent frame shift close to the C-terminus of the *gag* gene. The protease is able to self-cleave from this polyprotein precursor and the free protease is also able to cleave the *gag* precursor to its constituent proteins. However, it is believed that these reactions occur in immature particles and are part and parcel of the maturation process.

Recently it has been shown that if the protease is inactive – this can be achieved by site-specific mutagenesis of the active site of the enzyme – particles may still be produced but they appear non-infectious (Kohl *et al.*, 1988). Thus protease cleavage is essential for proper maturation of the virus and the protease is a suitable target for development of antiviral drugs.

Already progress has been made in this direction. The structure of the HIV protease was derived by Navia *et al.* (1989) (although subsequent work by Wlodawer *et al.* (1989) has pointed out several important inaccuracies in that structure) and should aid the rational design of inhibitors. Furthermore, it has already been shown that pepstatin, a peptide analogue, is a potent inhibitor of the enzyme (Seelmeier *et al.*, 1988). Most known protease inhibitors are peptide analogues with non-cleavable bonds in place of the normal cleavage site, and, although it is unlikely that such molecules will be sufficiently robust to be developed as anti-HIV drugs, the study of their interactions with the protease will feed back useful information into the drug design process.

There are sufficient differences between the specificity of the virus protease and cellular aspartic proteases to believe that development of selective inhibitors of the virus enzyme will be possible. One difficulty yet unanswered is whether the protease in the immature virus particles will be accessible to such inhibitors. That is a question which cannot be answered satisfactorily until we have such inhibitors.

SUMMARY AND CONCLUSIONS

During the virus growth cycle there are many key events which are mediated by virus-coded enzymes, many interactions between virus proteins or between virus and cellular proteins which are essential for virus growth, and many interactions between virus-specific nucleic acid sequences and virus or host cell proteins. All of these processes are potential targets for intervention and it is in the area of HIV research where we see their exploitation most clearly.

Antiviral research is a new area. It was as recent as 1974 that the first truly selective antiviral was discovered (acyclovir) and it is only in this decade that we have seen real benefit derived clinically from the use of such drugs. There is now a sense of optimism amongst those involved in this exciting area of science, and the explosion of HIV onto the scene in the early 1980s has provided immense impetus to the whole area.

In the years since the Second World War the major approach to the discovery of antivirals was that of screening. However, in parallel, an enormous investment was made in fundamental virology and this has paid off in terms of our understanding of viruses and their life cycles. We are now poised to convert that understanding into compounds of therapeutic value for the treatment of virus disease. If progress is proportional to investment, then we should watch the HIV area for the next major developments in antivirals.

REFERENCES

Agrawal, S., Goodchild, J., Civeira, M. P., Thornton, A. H., Sarin, P. S. & Zamecnik, P. S. (1988). Oligodeoxynucleoside phosphorothioates as inhibitors of human immunodeficiency virus. *Proceedings of the National Academy of Sciences, USA*, **85**, 7079–83.

Agris, C. H., Blake, K. R., Miller, P. S., Reddy, M. P. & T'so, P. O. P. (1986). Inhibitors of VSV protein synthesis and infection by sequence specific oligonucleoside methylphosphonates. *Biochemistry*, **25**, 6268–75.

Baba, M., Pauwels, R., Herdewijn, P., De Clercq, E., Desmyter, J., & Vandeputte, M. (1987). Both 2′,3′-dideoxythymidine and its 2′,3′-unsaturated derivative (2′,3′-dideoxythymidinene) are potent and selective inhibitors of human immunodeficiency virus replication *in vitro*. *Biochemical and Biphysical Research Communications*, **142**, 128–34.

Baltimore, D. (1970. RNA-dependent DNA polymerase in virions of RNA tumour viruses. *Nature, London*, **266**, 1209–11.

Bauer, D. J., Selway, J. W. T., Batchelor, J. F., Tisdale, M., Caldwell, J. C. & Young, D. A. B. (1981). 4′,-6-Dichloroflavan (BW683C) a new antirhino-virus compound. *Nature, London*, **292**, 369–73.

Braam, J., Ulamanen, I. & Krug, R. M. (1983). Molecular model of a eukaryotic transcription complex: function and movements of influenza P proteins during capped RNA primed transcription. *Cell*, **34**, 609–18.

Capon, D. J., Chamow, S. M., Mordenti, J., Masters, S. A., Gregory, T., Mitsuya, H., Byron, R. A., Lucas, C., Wurmy, F. M., Groopman, J. E., Broder, S. & Smith, D. H. (1989). Designing CD4 immunoadhesions for AIDS therapy. *Nature, London*, **337**, 531.

Challberg, M. D. (1986). A method for identifying the viral genes required for herpesvirus DNA replication. *Proceedings of the National Academy of Sciences, USA*, **83**, 9094–8.

Chaudhary, V. K., Mizukami, T., Fuerst, T. R., FitzGerald, D. J., Moss, B., Paston. I & Berger, E. A. (1988). Selective killing of HIV infected cells by recombinant human CD4 – pseudomonas exotoxin hybrid protein. *Nature, London*, **335**, 369–72.

Clayton, L. K., Hussey, R. E., Skinbrich, R., Ramachandran, H., Husain, Y. & Reinherz, E. L. (1988). Substitution of murine for human CD4 residues identifies amino acids critical for HIV-gp120 binding. *Nature, London*, **335**, 363–6.

Cohen, G. H. (1972). Ribonucleotide reductase activity of synchronised KB cells infected with herpes simplex virus. *Journal of Virology*, **9**, 408–18.

Courtney, R. J. Steiner, S. M. & Benyesh-Melnick, M. (1973). Effects of 2-deoxy-D-glucose on herpes simplex virus replication. *Virology*, **52**, 447–55.

Cullen, B., (1986). Trans-activation of human immunodeficiency virus occurs via a bimodal mechanism. *Cell*, **46**, 973–82.

Dalgleish, A. G., Beverley, P. C. L., Clapham, P. R., Crawford, D. H., Greaves, M. F. & Weiss, R. A. (1984). The CD4 antigen is an essential component of the receptor for the AIDS retrovirus. *Nature, London*, **312**, 763–7.

Dayton, A. I., Sodroski, J. G., Rosen, C. A., Goh, W. C. & Hazeltine, W. A. (1986). The trans-activator gene of human T-cell lymphotropic virus type III is required for replication. *Cell*, **44**, 941–7.

De Clercq, E. (1988). Recent advances in the search for selective antiviral agents. *Advances in Drug Research*, **17**, 1–59.

De Clercq, E., Holy, A., Rosenberg, I., Sakuma, T., Balzarini, J. & Mandgel, P. C. (1986). A novel selective broad spectrum anti-DNA virus agent. *Nature, London*, **323**, 464–7.

Deen, K. C., McDougal, J. S., Inacker, R., Folena-Wasserman, G., Arthos, J., Rosenberg, J., Maddon, P. J., Axel, R. & Sweet, R. W. (1988). A soluble form of CD4 (T4) protein inhibits AIDS virus infection. *Nature, London*, **331**, 82–4.

Derse, D., Cheng, Y-C., Furman, P. A., St. Clair, M. H. & Elion, G. B. (1981). Inhibition of purified human and herpes simplex virus induced DNA Polymerases by 9-(2-hydroxyethoxymethyl) guanine triphosphate. Effects on primer template function. *Journal of Biological Chemistry*, **256**, 11447–51.

Diana, G. D., McKinlay, M. A., Brisson, C. J., Zalay, E. S., Miralles, J. V. & Salvador, U. J. (1985). Isoxazoles with antipicornavirus activity. *Journal of Medicinal Chemistry*, **28**, 748–52.

Doermann, A. H. (1952). The intracellular growth of bacteriophages. I. Liberation of intracellular bacteriophage T_4 by premature lysis with another phage or with cyanide. *Journal of General Physiology*, **35**, 645–53.

Elion, G. B. Furman, P. A., Fyfe, J. A., DeMiranda, P., Beauchamp, L. & Schaeffer, H. J. (1977). Selectivity of an antiherpetic agent 9-(2-hydroxyethoxymethyl)guanine. *Proceedings of the National Academy of Sciences, USA*, **74**, 5716–20.

Feinberg, M. B., Jarrett, F., Aldovini, A., Gallo, R. C. & Wong-Staal, F. (1986). HTLV-III expression and production involve complex regulation at the levels of splicing and translation of viral RNA. *Cell*, **46**, 807–17.

Fischer, R. A., Bertonis, J. M., Meier, W., Johnson, V. A., Costopoulos, D. S., Liu, T., Tizard, R., Walker, B. D., Hirsch, M. S., Schooley, R. T. & Flavell, R. A. (1988). HIV is blocked *in vitro* by recombinant soluble CD4. *Nature, London*, **331**, 76–8.

Fisher, A. G., Feinberg, M. B., Josephs, S. F., Harper, M. E., Marselle, L. M., Reyes, G., Gonda, M. A., Aldovini, A., Debouk, C., Gallo, R. C. & Wong-Staal, F. (1986). The *trans*-activator gene of HTLV-III is essential for virus replication. *Nature, London*, **320**, 367–71.

Fishl, M. A., Richman, D. D., Grieco, M. H., Gottlieb, M. S., Volberding, P. A., Laskin, O. L., Leedon, J. M., Groopman, J. E., Mildvan, D., Schooley, R. T., Jackson, G. G., Dirack, D. T., King, D. and the AZT collaborative working group (1987). The efficiency of azidothymidine (AZT) in the treatment of patients with AIDS and AIDS-related complex. *New England Journal of Medicine*, **317**, 183–91.

Frame, M. C., Marsden, H. S. & Dutia, B. M. (1985). The ribonucleotide reductase induced by herpes simplex virus type 1 involves minimally a complex of two polypeptides (136K and 38K). *Journal of General Virology*, **66**, 1581–7.

Furman, P. A., St. Clair, M. H., Fyfe, J. A., Rideout, J. L., Keller, P. M. & Elion, G. B. (1979). Inhibition of herpes simplex virus-induced DNA polymerase activity and viral DNA replication by 9-(2-hydroxyethoxymethyl)guanine and its triphosphate. *Journal of Virology*, **32**, 72–7.

Furman, P. A., Fyfe, J. A., St. Clair, M. H., Weinhold, K., Rideout, J. L., Freeman, G. A., Nusinoff-Lehrman, S., Bolognesi, D. P., Broder, S., Misuya, H. & Barry, D. W. (1986). Phosphorylation of 3′-azido-3′deoxythymidine and selective interaction of the 5′-triphosphate with human immunodeficiency virus reverse transcriptase. *Proceedings of the National Academy of Sciences, USA*, **83**, 8333–7.

Fyfe, J. A., Keller, P. M., Furman, P. A., Miller, R. L. & Elion, G. B. (1978). Thymidine kinase from herpes simplex virus phosphorylates the new antiviral compound 9-(2-hydroxyethoxymethyl)guanine. *Journal of Biological Chemistry*, **253**, 8721–7.

Huang, R. T. C., Wahn, K., Klenk, H.-D., & Rott, R. (1980). Fusion between cell membrane and liposomes containing the glycoproteins of influenza virus. *Virology*, **104**, 294–302.

Hussey, R. E., Richardson, N. E., Kowalski, M., Brown, N. R., Chang, H-C., Siliciano, R. F., Dorfman, T., Walker, B., Sodroski, J. & Reinherz, E. L. (1988). A soluble CD4 protein selectively inhibits HIV replication and syncytium formation. *Nature, London*, **331**, 78–81.

Ishitsuka, H., Ninomiya, Y. T., Oshawa, C., Fujiu, M. & Suhara, Y. (1982). Direct and specific inactivation of rhinovirus by chalcone R_0 09–0410. *Antimicrobial Agents & Chemotherapy*, **22**, 617–21.

Jameson, B. A., Pao, P. E., Kong, L. I., Hahn, B. H., Shaw, G. M., Hood, L. E. & Kent, S. B. H. (1988). Location and chemical synthesis of a binding site for HIV-1 and the CD4 protein. *Science*, **240**, 1335–9.

Klatzman, D., Champagne, E., Chameret, S., Gruest, J., Guetard, D., Hercend, T., Gluckman, J-C. & Montagnier, L. (1984). T-Lymphocyte T4 molecule behaves as a receptor for human retrovirus LAV. *Nature, London*, **312**, 767–8.

Klenk, H-D., Rott, R., Orlich, M. & Blödorn, J. (1975). Activation of influenza A by trypsin treatment. *Virology*, **68**, 426–39.

Knight, D. M., Flowerfelt, F. A., Ghrayeb, J. (1987). Expression of the art/trs protein of HIV and study of its role in viral envelope synthesis. *Science*, **236**, 837–40.

Kohl, N. E., Emini, E. A., Schleif, W. A., Davis, L. J., Heinbach, J. C., Dixon, R. A. F. Scolnick, E. M. & Sigal, I. S. (1988). Active human immuno-deficiency virus protease is required for viral infectivity. *Proceedings of the National Academy of Sciences, USA*, **85**, 4686–90.

Kowalski, M., Potz, J., Basiripour, L., Dorfman, T., Goh, W. C., Terwilliger, A. D., Dayton, A., Rosen, C., Hazeltine, W. & Sodroski, J. (1987). Functional regions of the envelope glycoprotein of human immunodeficiency virus type 1. *Science*, **237**, 1351–5.

Kräusslich, H.-G. & Wimmer, E. (1988). Viral proteinases. *Annual Review of Biochemistry*, **57**, 701–54.

Larder, B. A., Purifoy, D. J. M., Powell, K. L. & Darby, G. (1987). Site-specific mutagenesis of AIDS virus reverse transcriptase. *Nature, London*, **327**, 716–17.

Lazarowitz, S. G. & Choppin, P. W. (1975). Enhancement of influenza A and B viruses by proteolytic cleavage of the hemagglutinin polypeptide. *Virology*, **68**, 440–54.

Lifson, D. J., Hwang, K. M., Nara, P. L., Blair, F., Padgett, M., Dunlop. N. M. & Eiden, L. E. (1988). Synthetic CD4 peptide derivatives that inhibit HIV infection and cytopathicity. *Science*, **241**, 712–16.

Lowe, D. M., Aitken, A., Bradley, C., Darby, G. K., Larder, B. A., Powell, K. L., Purifoy, D. J. M., Tisdale, M. & Stammers, D. K. (1988). HIV-1 reverse transcriptase: Crystallisation and analysis of domain structure by limited proteolysis. *Biochemistry*, **27**. 8884–9.

McGeoch, D. J. (1984). The nature of animal virus genetic material. In *The Microbe 1984*, Mahy, B. W. J. & Pattison, J. R. eds, pp. 75–107, Cambridge: Cambridge University Press.

Marmor, M., Des Jarlais, D. C., Cohen, H., Friedman, S. R., Beatrice, S. C., Dubin, N., el-Sadr, W., Mildvan, D., Yancovitz, S., Mathur, U., *et al.*, (1987). Risk factors for infection with human immunodeficiency virus among intravenous drug abusers in New York City. *Aids*, **1**, 39–44.

Marsh, M. & Helenius, A. (1988). Virus entry into animal cells. *Advances in Virus Research*, **35**, 107–52.

Matsukura, M., Shinozuka, K., Zon, G., Misuya, H., Reitz, M., Cohen, J. S.

& Broder, S. (1987). Phosphorothioate analogues of deoxyribonucleotides. Inhibition of replication and cytopathic effects of human immunodeficiency virus. *Proceedings of the National Academy of Sciences, USA*, **84**, 7706–10.

Matthews, R. E. F. (1982). Classification and nomenclature of viruses: fourth report of the international committee on taxonomy of viruses. *Intervirology*, **17**, 1–199.

Melnick, J. L., Gerba, C. P. & Wallis, C. (1978). Viruses in water. *Bulletin of the World Health Organization*, **56**, 499–508.

Mitzuya, H. & Broder, S. (1986). Strategies for antiviral therapy in AIDS. *Nature, London*, **325**, 773–8.

Mitsuya, H., Weinhold, K. J., Furman, P. A., St.Clair, M. H., Lehrman, S. N., Gallo, R. L., Bolognesi, D., Barry, D. W. & Broder, S. (1985). 3′-azido-3′-deoxythymidine (BWA509U): an antiviral agent that inhibits infectivity and cytopathic effects of human T-lymphotropic virus type III/lymphadenopathy-associated virus *in vitro*. *Proceedings of National Academy of Sciences, USA*, **82**, 7096–100.

Munyon, W. E., Paoletti, E. & Grace, J. T. (1967). RNA polymerase activity in purified infectious vaccinia virus. *Proceedings of the National Academy of Sciences, USA*, **58**, 2280–8.

Navia, M. A., Fitzgerald, P. M. D., McKever, B. M., Leu, C.-T., Heimbach, J. C., Herber, W. K., Sigel, I. S., Parke, P. L. & Springer, J. P. (1989). Three-dimensional structure of aspartyl protease from human immunodeficiency virus HIV-1. *Nature, London*, **337**, 615–20.

Ono, K., Ogasawara, M., Iwata, Y., Nakane, H., Fuji, T., Sawai, K. & Saneyshi, M. (1986). Inhibition of reverse transcriptase activity by 2′,3′-dideoxy thymidine 5′-triphosphate. *Biochemical and Biophysical Research Communications*, **140**, 498–507.

Otto, M. J., Fox, M. P., Fancher, M. J., Kuhot, M. F., Diana, G. D. & McKinlay, M. A. (1985). *In vitro* activity of WIN 51711, a new broad spectrum antipicornavirus drug. *Antimicrobial Agents & Chemotherapy*, **27**, 883–6.

Peterlin, B. K., Luciw, P. A., Barr, P. J. & Walker, M. D. (1986). Elevated levels of mRNA can account for the transactivation of human immunodeficiency virus. *Proceedings of the National Academy of Sciences, USA*, **83**, 9734–8.

Preston, V. G., Palfreyman, J. W. & Dutia, B. M. (1984). Identification of a herpes simplex virus type 1 polypeptide which is a component of the virus induced ribonucleotide reductase. *Journal of General Virology*, **65**, 1457–66.

Pritchett, T. J., Brossmer, R., Rose, U. & Paulson, J. C. (1987). Recognition of monovalent sialosides by influenza virus H_3 haemagglutinin. *Virology*, **160**, 502–6.

Rosen, C. A., Sodroski, J. G. & Hazeltine, W. A. (1985). The location of cis-acting regulatory sequences in the human T-cell lymphotropic virus type III (HTLV III/LAV) long terminal repeat. *Cell*, **41**, 813–23.

Rosen, C. A., Terwilliger, E. F., Dayton, A. E., Sodroski, J. G. & Hazeltine, W. A. (1988). Intragenic *cis* -acting *art* gene responsive sequences of the human immunodeficiency virus. *Proceedings of the National Academy of Sciences, USA*, **85**, 2071–5.

Ruprecht, R. M., Mullaney, S., Andersen, J. & Branson, R (1989). *In vivo* analysis of castanospermine, a candidate antiretroviral agent. *Journal of Acquired Immune Deficiency Syndromes*, **2**, 149–57

Sarin, P. S., Agrawal, S., Civeira, M. P., Goodchild, J., Ikeuchi, T. & Zamecnik, P. C. (1988). Inhibition of acquired immunodeficiency syndrome virus by oligodeoxynucleoside methyl phosphonates. *Proceedings of the National Academy of Sciences, USA*, **85**, 7448–51.

Seelmeier, S., Schmidt, H., Turk, V. & Van der Helm, K. L. (1988). Human immunodeficiency virus has an aspartic type protease that can be inhibited by pepstatin. *Proceedings of the National Academy of Sciences, USA*, **85**, 6612–16.

Shigeta, S., Yokota, T., Iwabuchi, T., Baba, M., Konno, K., Ogata, M. & De Clercq, E. (1983). Comparative efficiency of anti-herpes drugs against various strains of varicella zoster virus. *Journal of Infectious Diseases*, **147**, 576–84.

Skehel, J. J., Bayley, P. M., Brown, E. B., Martin, S. R., Waterfield, M. D., White, J. M., Wilson, I. A. & Wiley, D. C. (1982). Changes in the conformation of influenza virus hemagglutinin at the pH optimum of virus-mediated membrane fusion. *Proceedings of the National Academy of Sciences*, USA, **79**, 968–72.

Smith, C. C., Aurelian, L., Reddy, M. P., Miller, P. S. & T'so, P. O. P. (1986a). Antiviral effect of an oligo (nucleoside phosphonate) complementary to the splice junction of herpes simplex virus type 1 immediate early pre mRNAs 4 and 5. *Proceedings of the National Academy of Sciences, USA*, **83**, 2787–91.

Smith, T. J., Kremer, M. J., Luo, M., Vriend, G., Arnold, E., Kramer, G., Rossman, M. G., McKinlay, M. A., Diana, G. D. & Otto, M. J. (1986b). The site of attachment to human rhinovirus 14 for antiviral agents that inhibit uncoating. *Science*, **223**, 1286–93.

Spivack, J. G., Prusoff, W. H. & Tritton, T. R. (1982). Inhibition of herpes simplex virus replication by methyl daunosamine. *Antimicrobial Agents & Chemotherapy*, **22**, 176–9.

Stephens, E. B. & Compans, R. W. (1988). Assembly of animal viruses at cell membranes. *Annual Review of Microbiology*, **42**, 489–516.

Strebel, K. Daugherty, D., Clouse, K., Cohen, D., Folles, T., Martin, MA. (1987). The HIV 'A' (*sor*) gene product is essential for virus infectivity. *Nature, London*, **328**, 728–30.

Temin, H. M. & Mitzutani, S. (1970). RNA-dependent DNA polymerase in virions of Rous sarcoma virus. *Nature, London*, **266**, 1211–13.

Thompson, R., Honess, R. W., Taylor, J., Morran, J. & Davison, A. J. (1987). Varicella-zoster virus specifies a thymidylate synthase. *Journal of General Virology*, **68**, 1449–55.

Till, M. A., Ghetie, V., Gregory, T., Patzer, E. J., Porter, J. P., Uhr, J. W., Capon, D. S. & Viletta, E. S. (1988). HIV infected cells are killed by rCD4-ricin A side chain. *Science*, **242**, 1166–8.

Traunecker, A., Lüke, W. & Karjalainenk, (1988). Soluble CD4 molecules neutralize human immunodeficiency virus type 1, *Nature, London*, **331**, 84–6.

Von Heijne, G. (1981). Membrane proteins – the amino acid compositions of membrane-penetrating segments. *European Journal of Biochemistry*, **120** 275–8.

Votruba, I., Bernaerts, R., Sakuma, T., De Clercq, E., Merta, A., Rosenberg, I. & Holy, A. (1987). Intracellular phosphorylation of broad spectrum anti-DNA virus agent (S)-9-(3-hydroxy-2-phosphonylmethoxypropyl)adenine and inhibition of viral DNA synthesis. *Molecular Pharmacology*, **32**, 524–9.

Walker, B. D., Kowalski, M., Goh, W. C., Kozarsky, K., Krieger, M., Rosen, C., Rohrschneider, L., Hazeltine, W. H. & Sodroski, J. (1987). Inhibition of human immunodeficiency virus syncytium formation and virus replication by castanospermine. *Proceedings of the National Academy of Sciences, USA*, **84**, 8120–4.

Weis, W., Brown, J. H., Cusack, S., Paulson, J. C., Skehel, J. J. & Wiley, D. C. (1988). Structure of the influenza virus haemagglutinin complexed with its receptor, sialic acid. *Nature, London*, **333**, 426–31.

Wildy, P., Field, H. J. & Nash, A. A. (1982). Classical herpes latency revisited. In *Virus Persistence*, Mahy, B. W. J., Minson, A. C. & Darby, G. K. eds, *Society for General Microbiology Symposium 35* pp. 133–67. Cambridge: Cambridge University Press.

Wilson, I. A., Skehel, J. J. & Wiley, D. C. (1981). Structure of the haemagglutinin of influenza virus at 3Å resolution. *Nature, London*, **289**, 366–73.

Wlodawer, A., Miller, M., Jaskolski, M., Sathyanarayana, B. K., Baldwin, E., Weber, I. T., Selk, L. M., Clawson, L., Schneider, J. & Kent, S. B. H. (1989). Conserved folding in retroviral proteases: Crystal structure of a synthetic HIV-1 protease. *Science*, **245**, 616–21.

Wright, C. K., Felber, B. K., Paskalis, H. & Pavlakis, G. N. (1986). Expression and characterisation of the transactivator of HTLV-III/LAV virus. *Science*, **234**, 988–92.

Wu, C. A., Nelson, N. J., McGeoch, D. J. & Challberg, M. D. (1988). Identification of herpes simplex virus type 1 genes required for origin-dependent DNA synthesis. *Journal of Virology*, **62**, 435–43.

Yokota, T., Konno, K., Mori, S., Shigeta, S., Kumagai, M., Watanabe, Y. & Machida, H. (1989). Mechanism of selective inhibition of varicella zoster virus replication by 1-β-D-arabinofuranosyl-E-5-(2-bromovinyl)uracil. *Molecular Pharmacology*, **36**, 312–16.

Zamecnik, P. C., Goodchild, J., Taguchi, Y. & Sarin, P. (1986). Inhibition of replication and expression of human T-cell lymphotropic virus type III in cultured cells by exogenous synthetic oligonucleotides complementary to viral RNA. *Proceedings of the National Academy of Sciences, USA*, **83**, 4143–6.

Zerial, A., Thuong, N. T., Hélène, C. (1987). Selective inhibition of the cytopathic effect of type A influenza virus by oligodeoxynucleotides covalently linked to intercalating agent. *Nucleic Acids Research*, **15**, 9909–19.

Zerial, A., Werner, G. H., Phillpotts, R. J., Willmann, J. S., Higgins, P. G. & Tyrrell, D. A. J. (1985). Studies on 44081 RP., a new antirhinovirus compound, in cell cultures and in volunteers. *Antimicrobial Agents & Chemotherapy*, **27**, 846–50.

CHEMOTHERAPY OF INFLUENZA AND RESPIRATORY VIRUSES

J. S. OXFORD

Department of Medical Microbiology,
The London Hospital Medical College,
London E1 2AD, UK

INTRODUCTION

Many important viral diseases of children and adults including those spread by the respiratory route have been controlled, or are amenable to prevention, with inactivated or live attenuated vaccines; excellent examples are measles, rubella, poliomyelitis and smallpox (reviewed by Oxford & Oberg, 1985). The known advantages of harnessing or stimulating the body's own immune defences to combat a potential infection make this the method of choice in preventive medicine. Moreover, vaccines are relatively cheap, safe, and the immunity induced may be lifelong. But some viruses and particularly those causing infections of the upper and lower respiratory tract are, for various reasons, not amenable to this approach. They include influenza, parainfluenza viruses, rhino or common cold viruses of temperate, tropical and sub-tropical areas. The reasons for the failure of vaccine to combat certain respiratory virus infections can be summarised as firstly the multiplicity of virus serotypes (e.g. rhinoviruses) and secondly the ability of the virus to undergo antigenic changes (e.g. influenza). Table 1 summarises the properties of these 'target' respiratory viruses. Influenza viruses, both A and B, undergo considerable antigenic variation, sometimes in excess of five amino acid changes per year in a single hypervariable region of the haemagglutinin surface position and thus can circumvent both vaccine-induced and natural immunity (Wilson, Skehel & Wiley, 1981; Robertson *et al.*, 1985). Many serotypes of rhinoviruses co-exist in the community and induction of artificial or natural immunity to such a large number of antigenic variants is unlikely to be achieved. Similarly, respiratory syncytial virus exists as different serotypes, as do the related parainfluenza viruses. However, vaccines are not always problem free and an attempt to prevent respiratory syncytial virus infection immunisation with an inactivated vaccine resulted in enhanced disease (Kim *et al.*, 1969).

What, therefore, are the prospects of prophylaxis or therapy with antiviral chemicals? The synthesis of amantadine (Davies *et al.*, 1964), acyclovir (Schaeffer *et al.*, 1978) and zidovudine (Mitsuya & Broder, 1986) which inhibit respectively influenza virus, herpes viruses and human

Table 1. *The myriad of viruses causing pathology of the respiratory tract*

Viruses causing respiratory tract infection	Biological features	Vaccines	Antiviral agents
Upper respiratory tract			
Rhinoviruses	Small positive stranded RNA viruses. 150 serotypes	None	Enviroxime, chalcone, recombinant α-2a interferon*
Coronaviruses	Large enveloped RNA^+ viruses	None (human)	None
Lower respiratory tract			
Influenza A, B and C	Large enveloped RNA viruses. Negative strand. Antigenic variants	Inactivated	Amantadine,* rimantadine,* ribavirin*
Adenoviruses	Small DNA viruses	Experimental live attenuated	None
Parainfluenza types I, II, III and IV	Large enveloped negative strand RNA viruses	Experimental inactivated	None
Respiratory syncytial virus	Large enveloped negative strand RNA viruses: antigenic variants	Experimental live attenuated	Ribavirin*

* Licensed antiviral; clinically active.

immunodeficiency virus type 1, are major achievements of antiviral chemotherapy. Therefore it is now quite clear that antiviral chemotherapy is an achievable goal, even with respiratory viruses. However, in the wider context of antimicrobial chemotherapy these inhibitors do not approach penicillin in their spectra of activity or degree of inhibition and are more comparable with some of the first antibacterial agents such as salvarsan. Also, to our knowledge, no antiviral compound has yet been able to inhibit the replication of any virus completely. A proportion of virus particles always appears to be able to circumvent the drug-induced blockade. We shall discuss this problem in rather more detail below in terms of 'drug resistance'.

There is no doubt about the urgent need for effective preventative or curative therapy for respiratory viruses (WHO, 1980). Respiratory viruses pose a continuing threat to human life and endeavour throughout the world. A very diverse range of viruses occupies both the upper and lower reaches of the respiratory tract (Fig. 1). It is the combination of extreme biological diversity among the pathogenic agents and extensive physiological differences between the upper and lower respiratory tract (such as rapid clearance of locally administered drugs and relative inaccessibility to neutralising IgG) which has resulted in the present dearth of successful antiviral agents and of vaccines.

Antiviral chemotherapy is not a theoretical exercise in drug–virus interactions but rather a practical topic with a defined objective: to search for and find specific inhibitors which in clinical practice will have a use in either preventing symptoms normally induced by a virus or, more unlikely, preventing infection by a virus. In this brief review I shall emphasise current knowledge about the chemistry of antiviral drugs against respiratory viruses, and summarise the laboratory and clinical effectiveness of certain of these compounds whilst endeavouring to explore how new compounds could be discovered. This is a daunting task involving the application of chemistry and molecular biology but, already, 100,000 chemicals have been synthesised and the first lead compound may exist. Many exciting developments have been recorded with the use of crude or recombinant interferons but these will not be further discussed here.

Where it is of some merit I shall widen the discussion to include antivirals against herpes or HIV. Indeed, the worldwide importance of AIDS and the intense scientific curiosity about the causative virus will undoubtedly lead to entirely new approaches in the search for antivirals and these technical advances will also, it can be safely predicted, help in the search for inhibitors of respiratory viruses (see Darby, this volume).

THE CHEMISTRY OF ANTIVIRAL COMPOUNDS

There appear to be few chemical restrictions on the types of compound which inhibit virus replication, at least in cell cultures in the laboratory

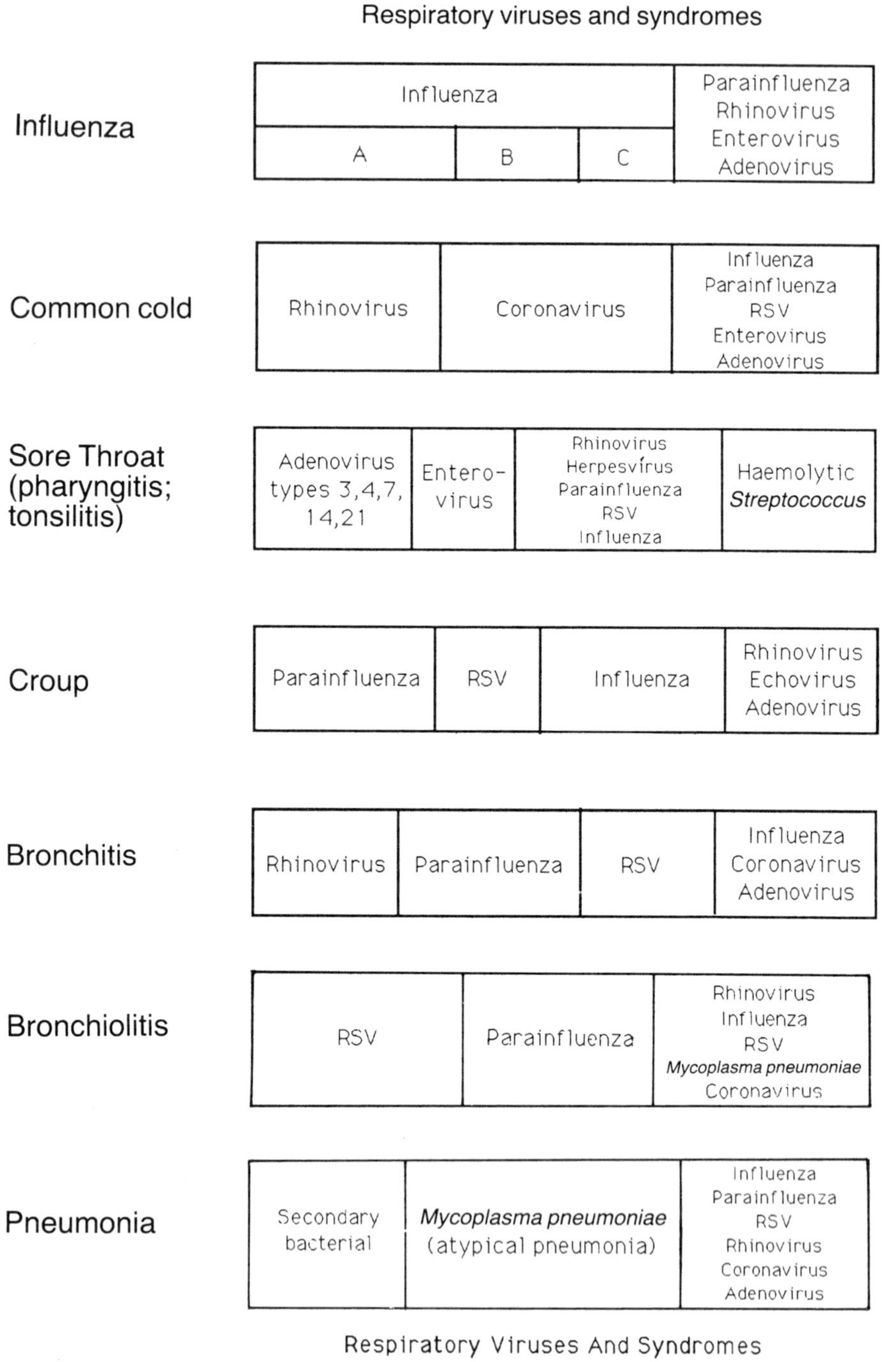

Fig. 1. The range of respiratory viruses correlated with clinical disease (from Stuart-Harris, Schild & Oxford *Influenza, the Virus and the Disease*, Edward Arnold, 1985, with permission.) Note that the relative size of the above blocks is related to the proportion of viruses causing the disease.

(Fig. 2). The molecules include natural products of plants such as alkaloids, and chalcones (Ishitsuka *et al.*, 1982b), sulphated oligosaccharides and polyanions such as heparin with relatively high molecular weights (M_r 10,000), synthetic oligonucleotides (Zamecnik *et al.*, 1986) and short peptides of intermediate values, low molecular weight benzimidazoles (Gwaltney, 1968; O'Sullivan *et al.*, 1969, flavan and flavones (Selway, 1986, Ishitsuka *et al.*, 1982b), guanidines and chemically simple amines and thiosemicarbazones (Gladych *et al.*, 1969). None of these compounds has been 'designed' as an antiviral: all have been found as a result of random screening. But, following the identification of a 'lead' compound, considerable chemical ingenuity has been and is still used to synthesise the more effective derivatives. A particularly fruitful source of antiviral compounds, respiratory viruses not excluded, are the analogues of purines and pyrimidines.

Thousands of analogues of the naturally occurring nucleosides have now been synthesised and tested in the laboratory and most pharmaceutical companies have large libraries of such compounds. These nucleoside analogues are particularly useful to chemists since they can be modified extensively at the sugar moiety to include ribose and deoxyribose (as in normal nucleosides), arabinose, xylose, substituted sugars (OH replaced by F, Cl, NH_2, etc), acyclic sugar analogues and carbocyclic analogues where the O of the furanose ring is replaced by CH_2. Furthermore, the pyrimidine or purine may be modified chemically in many ways to give fewer or additional nitrogen atoms, different hetero-atoms or ring sizes, substituents in the 5 or 6 position of pyrimidines or the 8 position of purines and linkage to the sugars via a carbon rather than a nitrogen atom. In addition to purine and pyrimidine nucleosides, ara-nucleosides and ara-nucleotides have been synthesised and also amino-nucleosides and aza-heterocyles and their nucleosides (Swallow & Kampfner, 1985).

It will be appreciated that even a single atomic substitution may change an active to an inactive molecule or vice versa. But, frustratingly, it has been a common experience that, *after* the initial discovery of an active molecule, subsequent chemical modification more often leads to compounds with less antiviral activity than the reverse. It is not often appreciated that many of these compounds are difficult to synthesise even by experienced chemists, and may require many months of work. Often, therefore, only small quantities are available for *in vitro* tests and in only a few cases is enough synthesised for animal experiments. Ribavirin is a example of the use clinically of a nucleoside analogue against influenza A and B viruses and respiratory syncytial virus (Gilbert & Knight, 1986; Hall *et al.*, 1983*a*,*b*).

Nucleoside analogues are not necessarily the most important inhibitors of influenza and, for example, considerable attention has been focused on pyrophosphate analogues such as phosphonoformate (Foscarnet) which specifically inhibits influenza RNA transcriptase activity (Oberg, 1983).

Amantadine

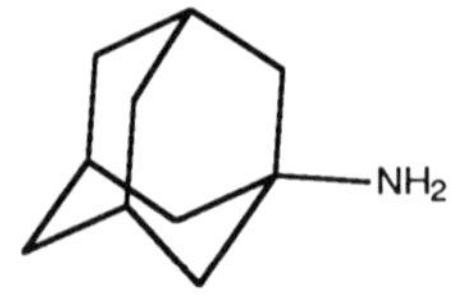

Rimantadine

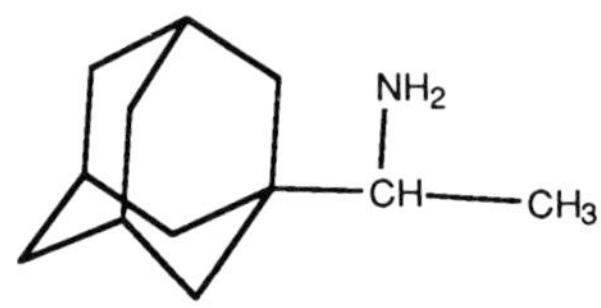

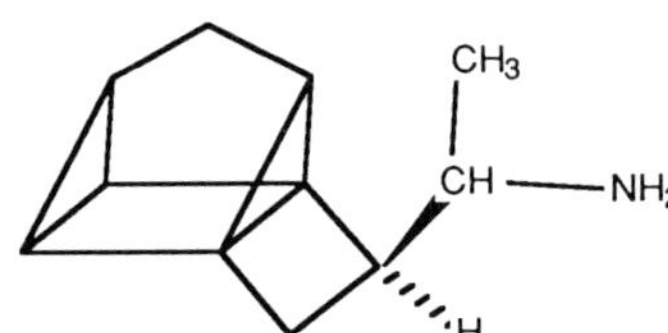

8-aminoethyl-tetracyclononane hydrochloride

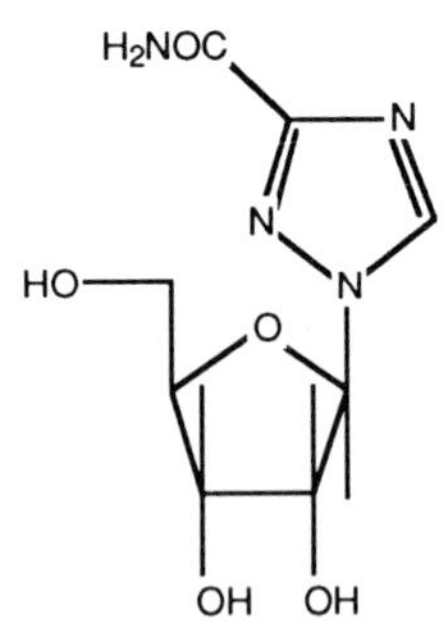

Ribavirin

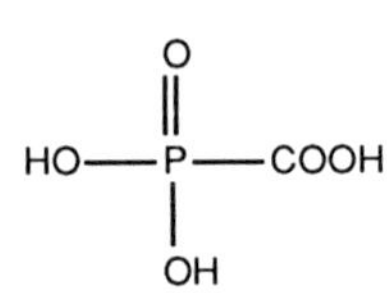

Phosphonoformic acid

Fig. 2. Chemical structures of antiviral agents against respiratory viruses.

Fig. 2 (continued)

4′,6-Dichloroflavan

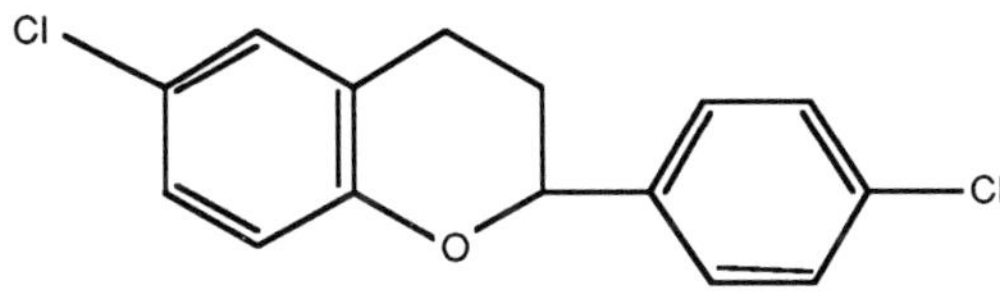

4′-Ethoxy-2′-hydroxy-4,6′-dimethyoxychalcone

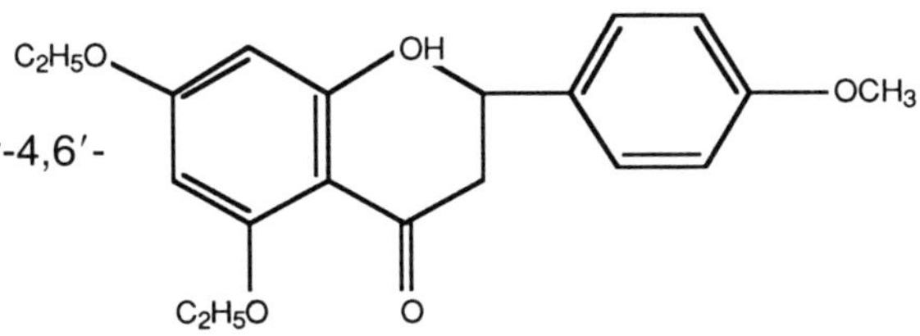

4′5-Dihydroxy-3,3′7-trimethoxyflavone

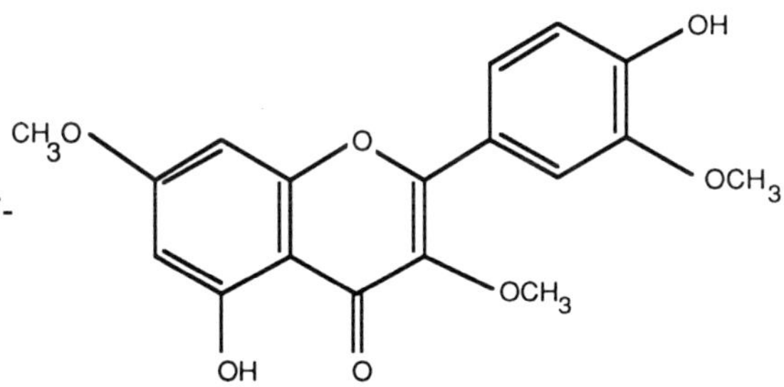

Enviroxime

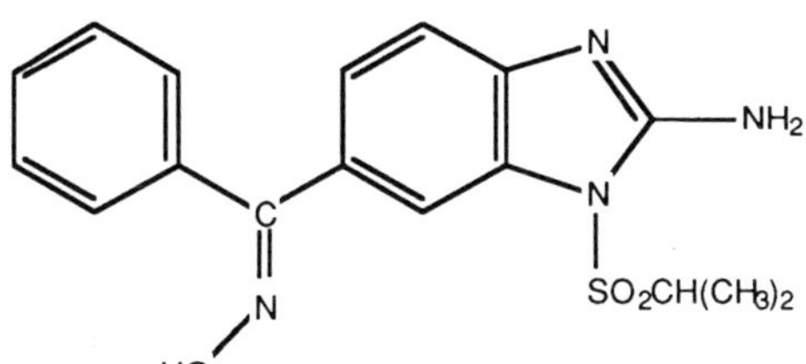

Disoxaril

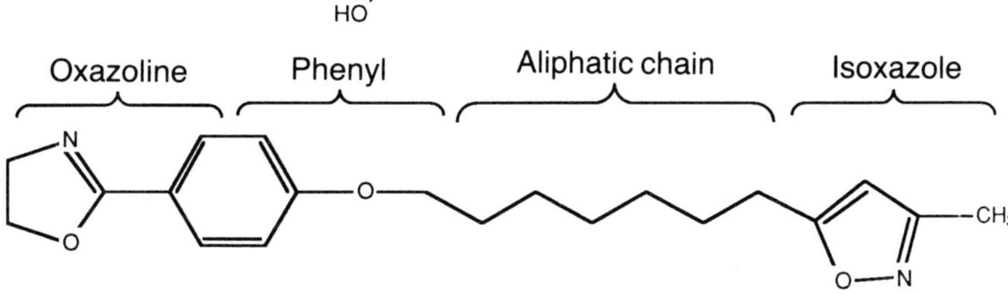

But one of the first antivirals against a respiratory virus to be extensively tested in humans was the primary acyclic amine, amantadine or Symmetrel (Oxford & Galbraith, 1983) which inhibits influenza A virus but not influenza B virus replication. Addition of a methyl grouping (rimantadine) alters the pharmacological distribution of the drug and prevents entry into the brain, thus reducing the particular side effect described as 'jitteriness' but without affecting the antiviral activity or spectrum. Many hundreds of similar amantadine molecules have been synthesised, including most recently a tetracyclononane ring analogue of rimantadine (see Fig. 2) (Swallow & Kampfner, 1985), but none, with the possible exception of the latter, greatly exceeds the original compound in its inhibitory effect against influenza A virus.

POINTS OF ACTION OF ANTIVIRALS DURING THE REPLICATION CYCLE OF A VIRUS

Table 2 summarises the various stages of a virus life cycle which are amenable to blockade with inhibitory molecules. It has been found that literally tens of thousands of molecules can inhibit virus replication in cell culture. In general, the more complex are the regulatory mechanisms of a virus, the more easy is the discovery of inhibitory molecules. Surprisingly, with certain complex viruses, one of the major practical problems is to decide precisely which of the many inhibitory molecules discovered should be investigated more thoroughly. A generously estimated ratio of the activity of antiviral compounds in cell culture, animal models and humans is 1000 : 10 : 1. This decision of which compound should proceed to the necessary protracted and expensive clinical testing stage is particularly difficult in the case of rhinoviruses, where an easily available animal model does not exist at present but where many compounds can be shown in the laboratory to inhibit the replication of the virus in cell culture.

CELL-FREE VIRUS AS A TARGET

It might be expected that viruses are most vulnerable when in the extracellular or transmission phase of their life cycle. Indeed, it is precisely at this point that many vaccines exert their effects via the induction of neutralising antibody. But, surprisingly, few antiviral compounds are known to inhibit or inactivate extracellular virus. A good example of molecules acting at this stage is the series of chemicals which bind to the external VP1 coat protein of picornaviruses, and typical examples are the arildone series of antiviral inhibitors (Tyrrell & Al Nakib, 1988) dichloroflavan (Tisdale & Selway, 1984; Swallow & Kampfner, 1985) and chalcone (Ishitsuka *et al.*, 1982*a*). These compounds are considered to stabilise the virus particle itself (Ninomiya *et al.*, 1984) and so prevent the uncoating stage whereby the virus releases its nucleic acid intracellularly. In an interesting study

Table 2. *The different steps in virus replication which are susceptible to inhibitors*

Target	Antiviral	Virus inhibited
1. Virus adsorption	Dextran sulphate	HIV-1, herpes
	Short peptides	HIV-1, paramyxoviruses, influenza, measles
2. Viral penetration and uncoating	Amantadine*	Influenza A
	Rimantadine*	
3. Virus-induced enzymes		
Reverse transcriptase	Phosphonoformate*	HIV-1
	Azidothymidine*	
	Dideoxycytidine*	
DNA polymerase	Acyclovir*	Generalised herpes and genital infections
	Trifluorothymidine*	Herpes eye infections
	Idoxuridine*	Herpes eye infections
	Gancyclovir*	Cytomegalovirus infections (e.g. pneumonia)
	Phosphonoformate*	Cytomegalovirus infections of eye
RNA transcriptase	Phosphonoformate, thiosemicarbazones	Influenza A and B
Neuraminidase	FIAC	Influenza A and B
4. Viral nucleic acid replication	Ribavirin*	Influenza A and B, respiratory syncytial
5. Viral mRNA stability	Interferon	Many viruses including influenza A and B
6. Virus release	Interferon*	Many viruses including rhinoviruses and influenza A and B
	Amantadine*	Influenza A
7. Free virus particle	Dichloroflavan, WIN compounds; disoxaril chalcones RMI 15, 731 R61837 3-methoxy-6-[4-(3-methyl phenyl)-1-piperazinyl]pyridazine	Rhinoviruses

* A licensed antiviral – the other compounds listed are undergoing experimental clinical testing.

of the interaction of a virus with an antiviral (and the first to be carried out at the atomic level using X-ray diffraction studies) Smith *et al.* (1986) described the binding of WIN 51711 (Disoxaril: Otto *et al.*, 1985) and WIN 52084 within a small canyon at the base of the receptor binding pocket of rhinovirus type 14. It is hypothesised that the antiviral molecule blocks the entry into the inner part of the virion of ions which would normally participate in the process of uncoating of the rhinovirus RNA soon after the virion has entered the cell. Similarly the cell receptor binding site on influenza HA has been identified following X-ray crystallography studies of the protein and is situated as a shallow saucer-like indentation at the tip of the molecule and in close proximity to the important antigenic site B. The precise configuration and spatial relationships of amino acids constituting the sides and floor of the site is known but to date this knowledge has not been exploited to develop antiviral inhibitors.

That chemicals binding to the receptor-binding site of a respiratory virus can act as inhibitors has been clearly demonstrated, as alluded to above, with human rhinovirus type 14 and a series of compounds consisting of a 3-methylisoxazole group connecting to a seven membered aliphatic chain and a 4-oxazolinyl phenoxy group. X-ray crystallography has enabled the interaction of these drugs with rhinovirus coat proteins to be studied in great detail. However, it should be emphasised that the compounds were first found in a random biological screen. The X-ray crystallography data allowed a scientific analysis of molecular changes in the inhibitor which could abrogate virus inhibitory activity and, further, pointed to a hypothesis of the mode of action of the drugs: 72 molecules of WIN 51711 bind per virion. The electron density of WIN 51711 was unequivocal with the oxazoline and phenoxy groups (see Fig. 2) appearing as two flattened bulges separated from the bulge of the iso-oxazole group. The oxazoline phenoxy (OP) groups were located in a hydrophilic region at an opening on the floor of the canyon of the receptor binding site. The OP end may orientate by hydrogen bonding between the nitrogen in the oxazoline ring and a hydrogen of Asn^{219} amide group in VP1. The methyl isoxazole group inserts itself into the hydrophobic interior of the VP1 barrel. In this case, as mentioned above, the compound could act by blocking entrance of ions into the channel which normally acts as a conduit to the inside of the virion. An interesting analysis was performed to determine the minimum inhibitory concentration (MIC) of WIN 51711 with two human rhinoviruses, HRV-2 and 14. Also the amino acid composition of their receptor binding pockets was compared with those of other picornaviruses including meningovirus, and foot and mouth disease virus of cattle (FMDV) which are not inhibited by WIN 51711. A possible explanation for this is the replacement of Asn^{219} in the WIN 51711-insensitive viruses by serine or glycine which does not allow correct orientation of oxazoline group of the drug. In HRV-2, Val^{188} and Val^{191} have both been changed to leucine which would sterically hinder insertion of the isoxazole group and aliphatic chain into the hydrophobic pocket. Concomitantly the MIC value of WIN 51711 for HRV-2 increased compared to the MIC value for HRV-14. This classic study is now taken as an example of a future direction of selection of antiviral drugs (Rossman, 1988) but still the virology and pharmacology aspects and correlates are very important since, for example, the X-ray study could not give an indication of whether a compound selected for very tight binding to the canyon would penetrate the cell plasma membrane to reach the virus or would be absorbed after oral administration and reach the tissues of the upper respiratory tract.

VIRUS ADSORPTION

Although practical success has not yet been achieved in the clinic, there has been considerable theoretical interest in compounds able to block virus adsorption to susceptible cells. However, innovative approaches with HIV could lead to new inhibitors of other viruses including those of the respiratory tract. Short peptides have been synthesised homologous with the sequence of the putative receptor binding site of the CD4 molecule and the HIV-1 binding protein (gp 120). These peptides are expected to block the interaction of the receptor region and the CD4 molecule, without interrupting other receptor functions of the CD4. Such short peptides, and, indeed, soluble CD4, itself inhibit virus replication in the laboratory (Traunecker, Luke & Karjalainen, 1988 (and see chapter by Darby)). The use of soluble CD4 as a decoy for HIV-1 is of particular interest. In a similar way short peptides homologous to the glycoproteins of RSV, parainfluenza, measles and influenza virus but with additional 'protecting groups' to reduce their destruction by cellular proteolytic enzymes were synthesised. These appeared to act early in the virus-cellular receptor interaction (Choppin, Richardson & Scheid, 1983) but lack of activity of influenza-specific peptides in animal models aborted their further development or testing in humans.

Other compounds act in a more non-specific manner. Polyanions such as heparin or dextran sulphate alter the electrical charge of cells and thus prevent adsorption of viruses to otherwise susceptible cells in culture (de Clercq, 1989). Presumably these compounds do not have to penetrate the cell membrane to exert their antiviral effect and so cell toxicity is avoided. On the other hand, other toxicity problems may arise, for example, in the case of heparin, the anti-blood clotting effects.

VIRUS ENTRY AND UNCOATING

An understanding of the mechanism by which a virus enters the cell and uncoats it is of pivotal importance in a logical search for certain inhibitors.

Viruses such as influenza and certain flaviviruses enters by viroplexis or engulfment. Immediately thereafter, whilst in a cytoplasmic endosome (vacuole), a virus protein catalyses a fusion event whereby the virus lipid membrane contacts and fuses with the intracellular vacuole membrane. This fusion is mediated by a particular sequence (fusion motif) of hydrophobic amino acids along one of the surface glycoproteins of the virus. It will be recognised that a compound interrupting fusion would prevent uncoating and blockade virus replication at this vital stage of the virus life cycle. In the case of influenza virus, particularly, there is a further target for antivirals because the fusion sequence located on the virus HA protein can only act following a three-dimensional re-arrangement of the HA molecule itself (Wilson, Skehel & Wiley, 1981). This major change in the HA, whereby the HA trimer is thought to open out like the petals of a flower, probably occurs only at a low pH of around 5.5 found in lysosomal vacuoles. Antiviral compounds such as amantadine appear to inhibit influenza virus replication at least in part by raising the pH of the cytoplasmic vacuole and thus preventing virus-induced fusion and hence virus uncoating but this may be an oversimplification (see below).

Other respiratory viruses such as measles, paramyxoviruses and respiratory syncytial virus enter cells by a virus-induced fusion process at the plasma membrane, so-called 'fusion from without'. These viruses are therefore particularly susceptible to inhibitors of the fusion sequence *per se*, such as short peptides, which may act extracellularly as discussed above.

It is quite possible that perturbation of lipids in the cell membrane could lead to inhibition of virus entry. Amantadine-like molecules can alter the phase transition temperature of cellular lipids and certain analogues could exert antiviral effects in this matter.

It is worth mentioning at this stage that the precise mode of action of many antivirals at the molecular level has been the subject of intense study and that often the initial indication of the point of inhibition has been shown to be an oversimplification. With amantadine, for example, the mode of action is likely to be complex, with high concentrations of drug acting via the HA whilst at low drug concentrations interaction with M2 protein becomes more important (Hay *et al.*, 1986). A current hypothesis is that M2 protein forms a transmembrane channel involved in reducing the pH gradient across the membrane of normally acidic vesicles. Amantadine may interfere with this ion flux function either directly by interfering with M2 protein or possibly indirectly via an interaction with HA (Hay 1989).

VIRAL NUCLEIC ACID TRANSCRIPTION AND TRANSLATION AND VIRUS RELEASE

Most antiviral drugs known today appear to act by inhibiting one or other of the mechanisms which viruses have developed for transcribing their genetic information into mRNA or, alternatively, for replicating the input parental genome to produce the necessary multiple genomic copies for the thousands of progeny viruses to be released from the infected cell. Some compounds may affect RNA splicing enzymes or RNA capping enzymes. Ribavirin triphosphate, for example, inhibits the GTP-dependent capping of the 5′ end of viral mRNAs including vaccinia and VSV (reviewed by Gilbert & Knight, 1986). Influenza is known to require *de novo* synthesis of capped host mRNAs to prime transcription of viral mRNA. Most of these virus inhibitors are nucleoside analogues, but until X-ray crystallographic studies can be performed on viral enzymes involved in RNA or DNA replication the precise mechanism of inhibition will not be known.

In a more simple way it is known that some of the enzymes which mediate viral RNA and DNA replication and transcription are zinc metalloenzymes and have a characteristic sequence of amino acids, called a 'zinc finger', where zinc binds to the viral protein. Antiviral compounds have been found which bind to the zinc itself and hence inhibit enzyme function. However, comparable metalloenzymes exist in a normal cell and these potential antiviral compounds can also cause cell toxicity (Perrin & Stunzi, 1983).

As an example of what could be termed 'quasi-designed antivirals' we chose the then newly discovered influenza virus particle RNA transcriptase as a target for the selection of new antivirals to influenza A and B viruses. In particular, we searched for lead compounds which could bind to a putative zinc metalloenzyme. We had demonstrated the association of zinc with highly purified virions but this was not conclusive evidence that the virion RNA polymerase was a zinc metalloenzyme. Nevertheless, we identified several classes of molecules such as thiosemicarbazones and phenanthrolines which possessed viral RNA transcriptase enzyme inhibiting activity and which were concomitantly effective chelators of zinc (Table 3). Chemical alteration of the molecules to abrogate chelation of zinc also reduced the inhibitory effects on the influenza RNA transcriptase. We were unable at the time to demonstrate effective inhibition of influenza virus replication *in vitro*.

Ribavirin triphosphate as well as Foscarnet has been shown to directly inhibit influenza RNA polymerase enzyme activity. Ribavirin may inhibit initiation and elongation steps of viral mRNA.

The processing of new virus polypeptides, including the addition or removal of sugar residues, may also be vulnerable to inhibitors and, surprisingly, in view of the number of glycoproteins in normal cells, some antiviral selectivity can be achieved, as with α-glycosidase inhibitors and HIV

Table 3. *Effect of heterocyclic thiosemicarbazones and analogues on RNA-dependent RNA polymerase activity of influenza B/LEE/40 virus*

Compound	Concentration of compound (nm) inhibiting B/LEE RNA polymerase activity by 50%	Chelation activity
Isatin 3-thiosemicarbazone	0.02 ± 0.004	+
Isatin 3-selenosemicarbazone	0.03 ± 0.008	+
Methyl isatin 3-thiosemicarbazone	0.10	+
Isatin 4′-methylthiosemicarbazone	0.10	+
Isatin 4′-ethylthiosemicarbazone	0.04	+
Isatin 3-semicarbazone	>0.5	0
Isatin 3-amidinohydrazone	>0.5	0
2-acetylpyridine thiosemicarbazone	0.002 ± 0.001	+
3-acetylpyridine thiosemicarbazone	1.0	0
4-acetylpyridine thiosemicarbazone	1.0	0
2-acetylpyridine semicarbazone	1.0	0

(Walker *et al.*, 1987). This activity may depend on quantitative differences in the amount of sugar on the viral glycoproteins compared to 'normal' cell glycoproteins.

The details of how viruses are released from infected cells are rather obscure but this stage may be amenable to inhibition. Palese & Schulman (1977) were among the first to attempt to inhibit the release of influenza virus by designing inhibitors of the viral enzyme neuraminidase, but none of these compounds was specific enough in its activity to have any marked effect in culture or in animal models.

THE SEARCH FOR NEW ANTIVIRAL COMPOUNDS

Without exception, all the antivirals now known, probably extending to several tens of thousands of molecules, were discovered by random searching. Most laboratories utilise a biological test system whereby new molecules are added to tissue culture cells at a range of dilutions (100 μg/ml to 0.1 μg/ml) and these drug-treated cells (and untreated cells as a control) are infected with a known multiplicity of infectious virus (commonly the respiratory viruses influenza A and B, respiratory syncytial virus, a representative rhinovirus and clinically important viruses such as HSV-1, HSV-2, CMV and HIV-1 or 2 are included in such biological screens). It will be appreciated that some potentially effective compounds may be missed because variation occurs in virus susceptibility itself, whilst some antivirals, particularly if they have to be phosphorylated or metabolised to the active antiviral intracellularly, may not act in all tissue culture cells.

Much more attention is now being directed towards 'designer' drugs. Can the data of virus nucleic acid sequences, amino acid sequences of viral proteins, X-ray crystallographic studies of virus proteins and enzymes be exploited and used to develop new antivirals? Already there are some excellent examples published of the potential of virus polypeptide or nucleic acid data for selection of antivirals, the first related to short peptides (Choppin *et al.*, 1983) and the second to sequence specific oligonucleotides (Zamecnik *et al.*, 1986). We have already discussed the first use of X-ray crystallography with VP1 of rhinoviruses (Smith *et al.*, 1986), but in no case has any animal model or human volunteer studies been performed with these 'designer' drugs. The potential use of oligonucleotides as antivirals is particularly pertinent because they may provide an avenue for actual excision from or destruction of a viral genome in an infected cell – the ultimate goal in chemotherapy of HIV or HSV. Experiments have been carried out with virus inhibitory oligonucleotides with RNA viruses such as VSV. The principle is to synthesise an oligonucleotide of perhaps 19 nucleotides with a sequence complementary to the chosen sequences in the viral nucleic acid. This length of oligonucleotide ensures specificity, because the chance of a comparable sequence occurring in a normal cell is extremely remote. The second principle is to add groupings such as methyl or sulphur at the phosphodiester linkage to prevent the oligonucleotide from being degraded by cellular enzymes. Finally, additional molecules may be attached at the ends of the oligonucleotide to enhance either interaction or, more interestingly, inactivation or even destruction of the viral nucleic acid. The 'anti-sense' oligonucleotides may act by binding to viral messenger RNAs themselves at the translation stage. It is surprising that such large molecules are taken up by intact cells, but they do appear to exert a specific antiviral effect *in vitro*.

A truly scientific approach to the discovery of antiviral compounds against influenza virus or indeed any respiratory virus requires detail of the functioning of all virus proteins, the replication strategy for genome, and the biology of virus entry and release. To consider the practical example of influenza we now need to know the precise nature or functioning of the three polymerase proteins PA, PB1 and PB2, and also non-structural proteins, NS1 and NS2, which presumably have important roles to play in viral replication. Assay systems for the binding of polymerase proteins to host RNA have been described and could be modified for large scale antiviral screening, but again this procedure would initially involve a random screen of an antiviral library of compounds to discover a lead compound.

RESISTANCE OF VIRUSES TO INHIBITORS

A disappointing feature of antiviral chemotherapy has been the failure of any antiviral molecule to date to completely blockade virus replication. Rather, antiviral activity is quantitated as a 100- or 1000-fold reduction in virus titre, which, while being very significant, still allows some virus particles to emerge unscathed. This may have important consequences in immunocompromised persons where normal immune mechanisms may not be present to eradicate this residual virus population. Do these few virus particles differ biologically or genetically from the major portion of the virus population and are these particles drug-resistant mutants? Unfortunately, at present, compared to the knowledge accumulated over three decades of the use of the tonnes of antibiotics worldwide and analysis in laboratories of millions of microbe samples for drug resistance, very little work has been done in the area of drug resistance to viruses. *A priori* because of the high copy error inherent in RNA replication it might be theorised that viruses such as influenza could easily develop the genetic potential to evade inhibition by a particular drug. In probability, drug-resistant mutants would even pre-date the clinical use of an antiviral. There is certainly evidence that influenza viruses can be isolated from countries of the world where the drug in question, namely amantadine, has never been used clinically (Heider *et al.*, 1981), but influenza is a notorious traveller, and drug resistance could have emerged in the UK, USA or USSR and travelled to Berlin, there to be isolated. Again, in theory, this genetic diversity and evolution is 10,000-fold greater in RNA than DNA viruses and so there may exist magnitudes of difference between these virus groups as regards potential selection and emergence of drug-resistant viruses (Oxford *et al.*, 1986). A biological feature of drug-resistant RNA viruses would seem to be the occurrence of mixtures of sensitive and resistant variants (Heider *et al.*, 1981, Pemberton *et al.*, 1986, Hay *et al.*, 1985). Our studies of naturally occurring drug resistant influenza A viruses in the UK were interpreted in terms of mixtures, since from a single 'resistant' virus (A/England/641/82:H3N2) both sensitive and partially resistant clones were isolated (Tables 4 and 5). The clinical significance of such resistance is not known at present (Oxford, Field & Reeves, 1986; Belshe *et al.*, 1988). We have studied a group of influenza A (H3N2) viruses isolated from Christ's Hospital School during an outbreak of influenza A (H3N2) virus where amantadine was used chemoprophylactically and therapeutically (Davies *et al.*, 1988). Within two days of the outbreak, amantadine was administered to the remaining uninfected boys in the school. It should be mentioned that Christ's Hospital School is a community of some 800 persons where previous outbreaks of influenza had been very well documented and studied and high attack rates of around 40% encountered. The administration of amantadine resulted in a precipitous drop in the number of newly diagnosed cases of influenza (Fig. 3), but of particular interest was the continued progress of the outbreak at a very low level. At present we are quantitating the susceptibility of over 80 virus isolates to inhibition by amantadine in an attempt to establish the role, if any, of drug resistant viruses in the progress of the outbreak (Oxford, Harmon, Grilli & Smith, unpublished data). To date, however, we have isolated only a minority of drug-resistant viruses. Most

Table 4. *Sensitivity of influenza A (H1N1) and (H3N2) isolates to amantadine*

Virus isolate	Type/ subtype	Log$_{10}$ reduction in virus infectivity titre in cultures containing amantadine (μg/ml)* 2.5	15	25	Mean log$_{10}$ reduction
E/6/78	H1N1	3.67	0.83	NT	2.25
E/611/78	H1N1	3.00	2.67	2.51	2.73
E/16/78	H1N1	2.01	2.83	1.50	2.11
E/929/78	H3N2	0.63	1.94	1.50	1.36
E/500/78	H3N2	0.70	0.34	1.00	0.68
E/684/82	H3N2	1.70	1.44	0.67	1.27
E/652/82	H3N2	1.00	2.23	0.87	1.37
E/641/82	H3N2	0	0	0	0
E/151/81	H3N2	0	0.53	0.67	0.40
E/640/82	H3N2	0	NT	2.56	1.28
E/639/82	H3N2	0	NT	3.73	1.86

* Titrated in pieces of chicken allantois-on-shell (data from Pemberton *et al.*, 1986).

Table 5. *Sensitivity to amantadine of clones of amantadine-sensitive and -resistant influenza (H3N2) virus strains*

Virus isolate	Clone number	Log$_{10}$ reduction in titre in presence of 25 μg/ml amantadine	Sensitivity
E/641/82	1	1.83	pR
(H3N2)	2	3.40	S
	3	1.50	pR
	4	4.34	S
E/Chr/756/80	1	0.55	pR
(H3N2)	2	0.17	pR
	3	1.06	pR
	4	1.09	pR
	5	0.67	pR
	6	2.00	S
	7	3.80	S
	8	2.00	S
	9	2.70	S
	10	0	R

* Resistant (R) = no reduction in virus titre in presence of amantadine. Partially resistant (pR) = 0–2 log$_{10}$ reduction in presence of amantadine. Sensitive (S) = >2 log$_{10}$ reduction in presence of amantadine (data from Pemberton *et al.*, 1986).

isolates were fully or partially susceptible to *in vitro* inhibition by the drug.

Eggers & Rosenwirth (1988) have described the isolation of poliovirus resistant to arildone (4-[6-(2-chloro-4-methoxyphenoxy)hexyl]-3,5-heptanedione and, by peptide mapping, located a lesion in the VP1 polypeptide.

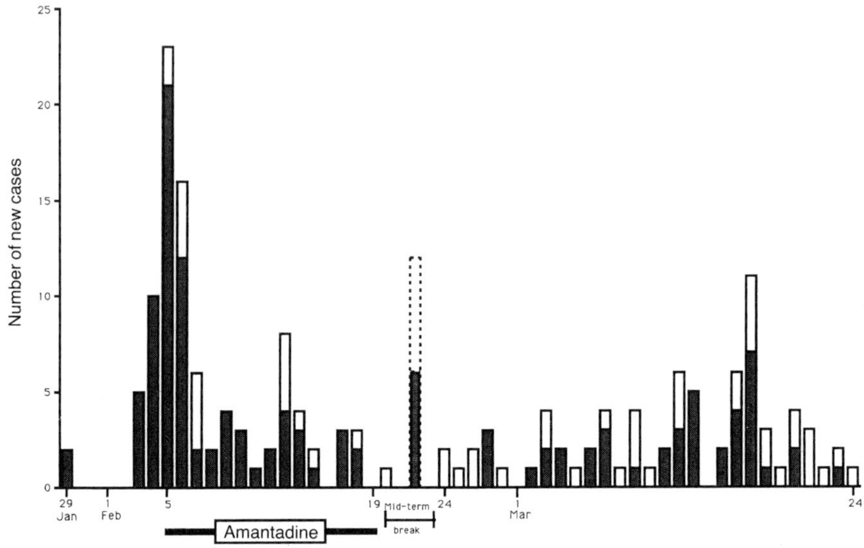

Fig. 3. Introduction of amantadine during an outbreak of influenza A (H3N2) virus at Christ's Hospital School. The number of new cases of clinical influenza each day are shown and those with laboratory confirmation of influenza A H3N2 are shaded (from Davies *et al.*, 1988).

Thus it may be assumed that arildone interacts with epitopes on the capsid in a similar way to the structurally related WIN compounds described above. An arildone-dependent poliovirus also has been selected in the laboratory (Loddo & La Colla, 1983). Cross-resistance is exhibited by compounds binding directly to rhinovirions (Ishitsuka *et al.*, 1986).

Interestingly, genetic analysis of drug-resistant viruses has given clues and confirmatory data on the mode of action of the drug Acylovir-resistant mutants of herpes simplex virus have genetic lesions in the thymidine kinase or DNA polymerase genes, indicating the participation of these two gene products in the mode of action of the drug (Field & Darby, 1980). Similarly, amantadine-resistant influenza A viruses have genetic lesions in the HA1 (Scholtissek & Faulkner, 1969) and/or M2 genes (Lubeck *et al.*, 1978), again highlighting the potential importance of these two structural proteins in the functioning of amantadine in the virus-infected cell (Hay *et al.*, 1985). Such studies with drug-resistant influenza A viruses have indicated the complex mode of action of amantadine, perhaps at two different stages of the virus life cycle. The studies of Hay and his group (Hay, 1989) particularly have shown that amantadine may act both at the early infection stage prior to the onset of primary transcription and late in the infection by blocking the release of virus particles. The latter effect may be mediated via a specific alteration of the HA sequence in drug-resistant viruses. The same group of workers have identified alterations in the primary sequence of M2 in drug-resistant viruses, perhaps indicating interference with a necessary interaction between HA or M2 proteins. Drug-resistant viruses have mutations at residues 27, 30, 31 or 34 which are all within the hydrophobic membrane spanning domain of the M2. Mutations in M2 render virus completely resistant whereas mutations in HA reduce the inhibitory effects of amantadine. Amantadine is known to have anticholinergic effects and it can be hypothesised that the drug blocks ion flux across the membrane of normally acidic vesicles (Warnick *et al.*, 1982) and that a normal function of M2 is to form a transmembrane channel across these membranes. Of course, an important aspect of this work is that quite a wide range of other molecules are known to act as channel blockers and some may also inhibit influenza virus replication (Lear *et al.* 1988).

From the practical point of view it may now be advantageous to test the ability of combinations of antivirals to inhibit different stages of viral replication and hence to reduce the emergence of drug-resistant variants as seen in clinical practice with the slowly replicating microbe *Mycobacterium tuberculosis*. Indeed, this synergistic approach has commenced with HIV chemotherapy, but as an attempt to reduce toxicities of individual antivirals rather than to prevent directly the emergence of drug-resistant virus.

THE CLINICAL NEED FOR ANTIVIRALS AGAINST RESPIRATORY VIRUSES

Influenza A and B viruses are the single major cause of severe lower respiratory tract infections throughout the world (Stuart-Harris, Schild & Oxford, 1985). For example, in the UK alone 17,000 excess deaths were caused by influenza A virus in the winter of 1975–76. The two viruses also cause severe economic disruption and in the same year there were 900,000 excess new claims for sickness benefit in the UK alone. Not all sections of the community are at equal risk of mortality. Certain ‘special risk’ groups including older persons, diabetics and persons with chronic heart, kidney or respiratory disease have been defined and are at tenfold higher risk of mortality following an infection with either influenza A or B virus. They are the major target for vaccines and chemotherapeutic agents.

A further virus causing severe lower respiratory tract infection, but particularly in infants and young children, is respiratory syncytial virus (RSV). For example, in the USA alone, RSV causes 90000 hospitalisations and 4500 deaths per year. RSV is the most frequent cause of bronchiolitis and pneumonia in infants. Like influenza virus, RSV causes outbreaks worldwide but probably with greater regularity than influenza with, for example, yearly outbreaks in temperate climates for three months each winter. A unique feature of the epidemiology of RSV is the ability of the virus to re-infect (Glezen *et al.*, 1986). Since natural immunity seems shortlived it could be surmised that vaccination would be unsuccessful in controlling this disease. Antiviral chemotherapy is expected to play an important role in prevention and cure of RSV.

A minor galaxy of viruses cause upper respiratory infection including 150 or so small RNA viruses, the rhinoviruses, whilst additional important viruses are the coronaviruses and four types of parainfluenza virus. In the USA alone in 1985, colds accounted for 17% of all episodes of acute illness requiring medical attention whilst 23 million days of lost work were recorded (National Center for Health Statistics, 1986; Dixon, 1985). The antigenic variability amongst the rhinoviruses would appear to argue against the usefulness of vaccines were it not for the presence of certain common amino acid sequences in the virus attachment site (Abraham & Colonno, 1984). Although the latter respiratory viruses are more commonly inhabitants of the upper respiratory tract, they predispose to serious secondary bacterial infections of the paranasal sinuses and middle ear and possibly contribute to increased mortality in third world countries. The first antiviral drug to suppress clinical signs has been described recently (Al-Nakib *et al.*, 1989).

With the possible exception of interferon which is active against some rhinoviruses (Merigan *et al.*, 1973; Douglas & Moore, 1986; Hayden *et al.*, 1986; Hayden & Gwaltney, 1982), an antiviral with broad spectrum activity against respiratory viruses has not been discovered. Many compounds have been described which inhibit influenza or rhinoviruses, but not both. Even compounds which inhibit influenza often have selective effects since amantadine inhibits type A influenza viruses but not types B or C. Possibly the broadest spectrum antiviral agent against respiratory viruses is the nucleoside analogue ribavirin which inhibits influenza A and B viruses and respiratory syncytial virus.

PRACTICAL EXAMPLES OF CHEMOTHERAPY OF VIRUSES WHICH CAUSE LOWER RESPIRATORY TRACT INFECTION

Influenza

From the time of the great influenza pandemic in 1918 chemotherapists have attempted to find and use selective inhibitors of the virus, but not until the discovery of the primary amine amantadine (Davies *et al.*, 1964) did effective chemoprophylaxis and even chemotherapy become a reality. It had been known for some time that simple amines and ammonium compounds inhibited the replication of influenza virus in tissue culture and in eggs but successful clinical application was only achieved with amantadine, presumably because its unusual 'birdcage' structure (Fig. 2) confers pharmacological stability. Initially Jackson *et al.* (1963) showed that the drug could prevent infection and symptoms in volunteers. It should be remembered that all these and subsequent clinical studies were pioneering in the sense that no drug had previously been shown to be active clinically against a respiratory virus. The properties of an ideal antiviral are summarised in Table 6.

Table 6. *Ideal requirements of an antiviral agent against respiratory viruses*

Absence of toxicity and side effects
Broad antiviral spectrum to inhibit influenza A and B, respiratory syncytial virus, rhinoviruses and coronaviruses*
Administered orally
Cheap to synthesise and heat stable for storage
Chemical variation possible to cope with subsequent emergence of drug-resistant viruses

* Ribavirin and interferon partially meet this requirement.

Table 7. *General practice (UK) study: effect of amantadine on incidence of clinical influenza in contacts of index cases (after Galbraith* et al., *1969)*

		All cases*			Confirmed serologically		
Treatment	Number of families	Number	%	p value	Number	%	p value
Amantadine	13	2/55	3.6		0/48	0	
Placebo	22	12/85	14.1	0.07	10/69	14.5	0.05–0.01

* Clinical diagnosis.

A study in conjunction with the Royal College of General Practitioners on amantadine (Galbraith *et al.*, 1969) was the first trial planned as a practical use of an antiviral, as well as a scientifically controlled study. Medical practitioners throughout the UK were asked to identify clinical cases of influenza and take blood and nasal wash samples from this 'index case'. The index case was left untreated but immediate family contacts were given amantadine or placebo daily for 10 days. Thus the trial monitored spread of virus from the untreated index case to the rest of the family and whether amantadine given prophylactically could prevent this spread of infection. The trial found that 14% of members of the families in the placebo group were infected compared to 3.6% in the amantadine group. However, none of these persons in the amantadine group had serologically confirmed influenza whereas 10 of 69 persons (14.5%) in the placebo group had serologically confirmed influenza. Thus amantadine used in a practical way successfully prevented spread of influenza in persons in a family setting (Table 7).

Numerous clinical trials, both in volunteers given attenuated viruses and in the community where persons were infected with naturally virulent influenza A viruses, fully confirmed or extended the initial study of prophylactic effect of amantadine. Of particular significance has been the observation over the succeeding two decades that all such types of influenza A virus so far encountered in humans, namely H2N2, H1N1 and H3N2, have been inhibited both in the laboratory and also in clinical trials. Oral rimantadine (Hall *et al.*, 1987) and low dose amantadine have both prevented infections with recent influenza A strains.

Arguably the most significant discovery, at least for the future development of antivirals against respiratory viruses, was that of the *therapeutic* effect of amantadine. In the initial studies volunteers who were infected with influenza A virus and were showing the first signs of clinical influenza were given amantadine or placebo. The influenza illness, including objective parameters such as temperature, subsided more rapidly in persons to whom amantadine was administered (reviewed by Oxford & Galbraith, 1983). Subsequently, Galbraith and his colleagues planned the first therapeutic trial on naturally infected patients in families and confirmed the therapeutic effect of oral amantadine administered not more than 24 hours after the first influenza symptoms appeared. Perhaps not surprisingly the use of the antiviral later than this time failed to abrogate symptoms of the disease. It should be mentioned that prior to these trials most virologists considered that, concomitant with respiratory symptomatology, virus replication would reach a peak and therefore the therapeutic use of an inhibitor at this stage of the disease would be fruitless. Therefore, the demonstration of the therapeutic effect of amantadine, as well as being of considerable scientific interest, also emphasises the true necessity of actual experiments and an avoidance of too much emphasis on hypothesis. Few studies have actually been performed on the quantity of influenza virus excreted in the lower respiratory tract of infected humans. Post-mortem examination of trachea specimens by immunofluorescence and cytology during the influenza A H2N2 pandemic did show that virus replication was patchy and that many cells remained uninfected. This observation would give credence to the concept of therapeutic intervention since antiviral drugs would be expected to prevent spread of virus further down the respiratory tract.

The precise degree of effectiveness of amantadine or indeed any other antiviral against respiratory virus infection has yet to be calculated. It is quite possible that early use of the compound could prevent spread of the influenza A virus into the lower respiratory tract and lungs and so abrogate mortality. Pharmacological experiments both in animals and humans have shown a preferential accumulation of amantadine in respiratory tissue including the lung. To date such clinical investigations of possible effects on mortality have not been undertaken and are urgently needed.

At present amantadine is not widely used in clinical medicine. For the future development of new antivirals it is important to understand how this situation has arisen. In my opinion three contributory reasons are: concern about the mild side reactions of jitteriness noticed by 7% of patients taking the drug, the problem of self-administration (the ideal situation for rapid use of the drug on contact with an infected person) and the presumed absence of very dramatic clinical effects such as prevention of mortality. In an event, which must be unique in the history of antimicrobials, two

Table 8. *Recommended usage in prophylaxis and therapy of amantadine and rimantadine*

1. There should be clear epidemiological and virological evidence of influenza A infection in the community.
2. Amantadine or rimantadine should be used prophylactically each day for 6–8 weeks.
3. Groups recommended for special consideration:
   (a) High-risk persons, such as diabetics, the elderly, and those with chronic heart, respiratory or circulatory disease, whether vaccinated or not.
   (b) Special community groups such as public transport employees, doctors, nurses.
   (c) Persons who may wish to be given prophylaxis or therapy.
4. Drug prophylaxis should not replace vaccination because amantadine does not inhibit influenza B virus.
5. Therapy should be initiated within 24 hours of first symptoms and given for 10 subsequent days.

Note: These recommendations are an amalgam of those first published following the NIAID meeting (1980) and the Vienna WHO meeting (Bektimirov *et al.*, 1985).

scientific meetings were convened to make suggestions about the clinical use of amantadine (Table 8). Overall the scientific panels recommended more extensive clinical use of amantadine and rimantadine (Bektimirov *et al.*, 1985; NIAID, 1980).

The discovery and progression to clinical use of rimantadine, a drug closely related to amantadine, is not without relevance and interest to the future selection of new drugs. Firstly rimantadine differs mainly from the parent molecule by the addition of a methyl group, but this is sufficient to significantly alter the pharmacological distribution of the drug without affecting antiviral action. Thus rimantadine cannot penetrate the blood–brain barrier and so does not induce any neurological side effects sometimes noted with amantadine. Tissue and plasma levels are normally twofold higher following oral administration of rimantadine (compared to amantadine) and thus dosage is reduced twofold to 100 mg daily. The rimantadine compound was first discovered in the USA at the du Pont laboratories but its apparent advantages in the clinic were first noted in the USSR, where the drug has since been used extensively. In the last five years scientific interest in rimantadine has been revived in the USA and the compound is expected to be licensed shortly for clinical use. In spite of very large searches by different groups throughout the world no amantadine-like molecule with a significantly improved efficacy has been discovered.

The nucleoside analogue ribavirin is the next most effective inhibitor of influenza A virus replication and infection. Ribavirin has the added advantage of a broad antiviral spectrum (Roberts & Laskin, 1988) including influenza A (Moore & Couch, 1984) and B viruses (McClung *et al.*, 1983) as well as respiratory syncytial virus (RSV) (Hruska *et al.*, 1980). In clinical terms the therapeutic margin is rather small (a two- to three-fold increase in drug dose compared to the dose used in antiviral therapy gives detectable toxicity) and the clinical benefit is less than amantadine for influenza A virus infections. But just as amantadine acted as a novel pioneer drug for a respiratory virus, ribavirin was also pioneering as a nucleoside analogue for possible long-term prophylaxis and therapy. When the compound was first introduced, only relatively toxic nucleoside analogues such as iododeoxyuridine and adenine arabinoside had been used in antiviral chemotherapy for the treatment of serious herpes virus infections. Now, 15 years later and with the additional clinical experience with the extensively used antiviral nucleoside analogue, acyclovir, it is generally acknowledged that not all analogues of naturally occurring purines or pyrimidines are inherently toxic for replicating mammalian cells.

Nevertheless, because of the relatively low therapeutic margin, ribavirin has been more extensively tested in aerosol form than other antivirals. Aerosol administration has the theoretical advantage of direct deposition of the drug onto the tissues of the upper and lower respiratory tract (Knight *et al.*, 1981; Wilson *et al.*, 1984) without compromising the function of other organs where the presence of the drug is, in any case, not required for antiviral action. A potential disadvantage of aerosol use is the complicated equipment required, particularly for use of ribavirin in infants suffering from RSV.

Ribavirin is the only drug to have shown effectiveness in the clinic against potentially serious RSV infections although other compounds may be effective *in vitro* (Kawana *et al.*, 1987) or in animal models (Tidwell *et al.*, 1984). Pioneering studies were carried out by Hall *et al.* (1983*a, b*) in controlled clinical trials firstly in adults and then in children. The studies involved normal infants who were hospitalised with pneumonia or bronchiolitis or infants with more serious underlying conditions, placing them at high risk from serious RSV infection. Infants receiving ribavirin showed a significantly greater amelioration of their illness than those in the placebo group and a greater increment of improvement on the first day of therapy as well as on subsequent days (Fig. 4). These important studies and those of Taber *et al.* (1983) show that antiviral therapy of RSV is possible and give encouragement and impetus in the search for other inhibitors of this otherwise uncontrolled virus infection.

CONCLUSION

Progress in the search for, and development of, antiviral inhibitors against respiratory viruses has been slow but this, in my opinion, reflects the difficulties of the problem and not a lack of interest by chemists, virologists and clinicians. An effective inhibitor of a respiratory virus (influenza A) was not discovered until 1964 and it has taken two decades to establish clinical testing parameters and correlates. The first drug clinically active against rhinovirus infection has only just been found. Perhaps expectations of

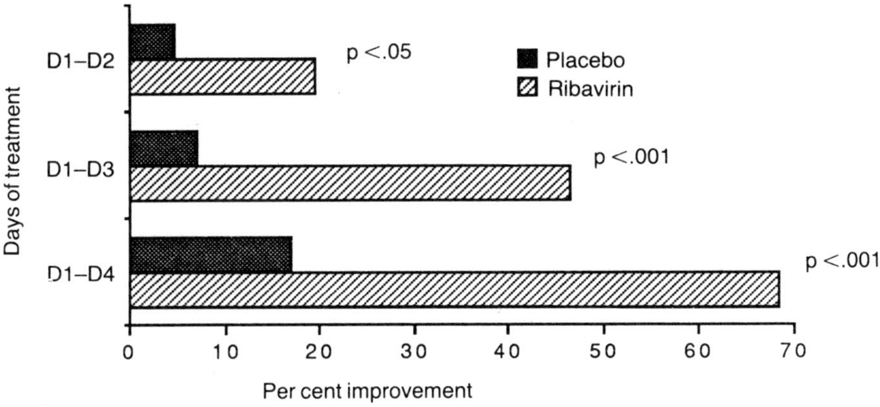

Fig. 4. Per cent improvement in illness severity scores from day 1 to day 2, 3 or 4 for infants with RSV infection treated with ribavarin or placebo (Hall C.B. pers. comm.).

clinicians and chemotherapists alike were too high and dramatic effects of antivirals against respiratory infection cannot be achieved. However, further clinical trials are urgently needed now to investigate whether amantadine or ribavirin can prevent mortality caused by influenza A virus.

The urgency of combating the AIDS epidemic and the success of the antiviral, azidothymidine, in prolonging life of infected persons in the absence of any vaccine, has resulted in a surge of interest in the potential of antiviral chemotherapy. The race to develop new inhibitors of this usually fatal disease has meant that routes are now being explored which were not followed before on the grounds of potential toxicity. A further lesson from the studies of antiviral chemotherapy of HIV has been the speed with which clinicians have resorted to the use of drug combinations, both to reduce the emergence of drug-resistant virus and also to exploit antiviral synergy and hence to reduce concentrations and possible toxicities of administered drugs.

Perhaps the most exciting work will come from the combination of protein and nucleotide chemistry where new lead compounds can be rapidly exploited and subtly different molecules rapidly identified.

The lessons of two decades of work on the development of antivirals against respiratory disease should not be forgotten. There is no replacement for excellent laboratory virology and pharmacology; tissue culture studies precede investigations in animal models and these before clinical trials in humans. The final lesson from experience with antivirals to respiratory disease is the central importance of the well-designed, double blind placebo-controlled trial together with the close collaboration of a virology laboratory. Finally, as seen with HIV-1, the search for new drugs against respiratory viruses will lead to a greater understanding of the virology and pathology caused by the agents themselves. At the moment there seems no alternative to widening the search for ever increasing numbers of chemicals. Chekov, in the closing lines of Uncle Vanya, succinctly summarises the position when he concludes 'we shall work and we shall have no rest'.

ACKNOWLEDGEMENTS

Financial assistance is received from the Oretta Bartolomei Research Foundation, Florence and Ciba-Geigy (UK) Ltd. Scientific collaboration is with E. Grilli and J. Smith (P.H.L.S. Laboratory, Guildford, Surrey, UK).

REFERENCES

Abraham, G. & Colonno, R. J. (1984). Many rhinovirus serotypes share the same cellular receptor. *Journal of Virology*, **51**, 340–5.

Al-Nakib, W., Higgins, P. G., Barrow, G. I., Tyrrell, D. A. J., Andries, K., Bussche, G. V., Taylor, N. & Janssen, P. A. J. (1989). Supression of colds in human volunteers challenged with rhinovirus by a new synthetic drug (R61837). *Antimicrobial Agents and Chemotherapy*, **33**, 522–5.

Bektimirov, T. A., Douglas, R. G., Dolin, R., Galasso, G. J., Krylov, V. F. & Oxford, J. S. (1985). Current status of amantadine and rimantadine as anti-influenza A agents: memorandum from a WHO meeting. *Bulletin of the World Health Organization*, **63**, 51–6.

Belshe, R. B., Hall Smith, M., Hall, C. B., Betts, R. & Hay, A. J. (1988). Genetic basis of resistance in rimantadine emerging during treatment of influenza virus infection. *Journal of Virology*, **62**, 1508–12.

Choppin, P. W., Richardson, C. D. & Scheid, A. (1983). Analogues of viral polypeptides which specifically inhibit viral replication. In *Problems of Antiviral Therapy*, Stuart-Harris, C. H. & Oxford, J. S., eds, pp. 13–25, London: Academic Press.

de Clercq, E. 1989. Potential drugs for the treatment of AIDS. *Journal of Antimicrobial Chemotherapy*, **23**, suppl. A 35.

Davies, J. R., Grilli, E. A., Smith, J. A. & Hoskins, T. W. (1988). Prophylactic use of amantadine in a boarding school outbreak of influenza A. *Journal of the Royal College of General Practitioners*, **38**, 346–8.

Davies, W. L., Grunert, R. R., Miles, H. B., Davies, L. M., Graham, N. M. H., Ryan, P., Warwick, D. A. & Albrecht, J. K. (1964). Antiviral activity of l-adamantanamine lamantadine. *Science*, **144**, 862–3.

Dixon, R. E. (1985). Economic costs of respiratory tract infections in the United States. *American Journal of Medicine*, **78**, 45–51.

Douglas, R. M. & Moore, B. W. (1986). Prophylactic efficacy of intranasal alpha-interferon against rhinovirus infections in the family setting. *New England Journal of Medicine*, **314**, 65–70.

Eggers, H. J. & Rosenwirth, B. (1988). Characterization of an arilodone resistant poliovirus 2 mutant with an altered capsid protein VP1. *Antiviral Research*, **9**, 23–35.

Field, H. J., & Darby, G. (1980). Pathogenicity in mice of strains of herpes simplex virus which are resistance to acyclovir *in vitro*. *Antimicrobial Agents and Chemotherapy*, **17**, 209–16.

Galbraith, A. W., Oxford, J. S., Schild, G. C. & Watson, G. I. *et al* (1969). Protective effect of 1-adamantanamine hydrochloride on influenza A2 infections in the family environment. *Lancet*, **ii**, 1026–8.

Gilbert, B. E. & Knight, V. (1986). Biochemistry and clinical applications of ribavirin. *Antimicrobial Agents and Chemotherapy*, **30**, 201–5.

Gladych, J. M. Z., Hunt, J. H., Jack, D., Haff, R. F., Boyle, J. J., Stewart, R. C. & Ferlauto, R. J. (1969). Inhibition of rhinovirus by isatin thiosemicarbazone analogues. *Nature, London*, **221**, 286–7.

Glezen, W. P., Taber, L. H., Frank, A. L. & Kasel, J. A. (1986). Risk of primary infection and reinfection with respiratory syncytial virus. *American Journal of Diseases of Childhood*, **140**, 543–6.

Gwaltney, J. M. Jr. (1968). The spectrum of rhinovirus inhibition by 2-(α-hydroxybenzyl)-benzimidazole and D-(-)-2-(α-hydroxybenzyl)-benzimidazole HCl (33393). *Proceedings of the Society of Experimental Biology and Medicine*, **129**, 665–73.

Hall, C. B., Walsh, E. E., Hruska, J. F., Betts, R. F. & Hall, W. J. (1983*a*). Ribavirin aerosol treatment of experimental respiratory syncytial viral infection in young adults: a controlled double-blind study. *Journal of the American Medical Association*, **249**, 2666–70.

Hall, C. B., McBride, J. T., Walsh, E. E., Bell, D. M., Gala, C. L., Hildreth, S., Ten Eyck, L. G. & Hall, W. J. (1983*b*). Aerosolized ribavirin treatment of infants with respiratory syncytial virus infection: a randomized double-blind study. *New England Journal of Medicine*, **308**, 1443–7.

Hall, C. B., Dolin, R., Gala, C. L., Markovitz, D. M., Zang, Y. Q., Madore, P. H., Disney, F. A., Tapley, W. B., Green, J. L., Francis, A. B. & Pichichero, M. E. (1987). Children with influenza A infection: Treatment with rimantadine. *Paediatrics*, **80**, 275–82.

Hay, A. J. (1989). The mechanism of action of amantadine and rimantadine against influenza viruses. In *Concepts in Viral Pathogenesis*, Notkins, A. L. & Oldstone, M. B. A. eds., pp. 359–67, New York: Springer-Verlag.

Hay, A. J., Wolstenholme, A. J., Skehel, J. J. & Smith, M. H. (1985). The molecular basis of the specific anti-influenza action of amantadine. *The EMBO Journal*, **4**, 3021–4.

Hay, A. J., Zambon, M. C., Wolstenholme, A. J., Skehel, J. J. & Smith, M. H. (1986). Molecular basis of resistance of influenza A viruses to amantadine. *Journal of Antimicrobial Chemotherapy*, **18**, (Suppl.B), 19–29.

Hayden, F. G. & Gwaltney, J. M. Jr. (1982). Prophylactic activity of intranasal enviroxime against experimentally induced rhinovirus type 39 infection. *Antimicrobial Agents and Chemotherapy*, **21**, 892–7.

Hayden, F. G., Albrecht, J. K., Kaiser, D. L. & Gwaltney, J. M. Jr (1986). Prevention of natural colds by contact prophylaxis with intranasal alpha$_2$-interferon. *New England Journal of Medicine*, **314**, 71–5.

Heider, H. E., Adamczyk, B., Presber, H. W., Schroeder, C., Feldblum, R. & Indulen, M. K. (1981). Occurrence of amantadine and rimantadine resistant influenza A virus strains during the 1980 epidemic. *Acta Virologica*, **25**, 395–400.

Hruska, J. F., Bernstein, J. M., Douglas, R. G. Jr. & Hall, C. B. (1980). Effects of ribavirin on respiratory syncytial virus *in vitro*. *Antimicrobial Agents and Chemotherapy*, **17**, 770–5.

Ishitsuka, H., Ninomiya, Y. T., Ohsawa, C., Fujiu, M. & Suhara, Y. (1982*a*). Direct and specific inactivation of rhinovirus by chalcone Ro 09–0410. *Antimicro-*

bial Agents and Chemotherapy, **22**, 617–21.
Ishitsuka, H., Ohsawa, C., Ohiwa, T., Umeda, I. & Suhara, Y. (1982*b*). Antipicornavirus flavone Ro 09–0179. *Antimicrobial Agents and Chemotherapy*, **22**, 611–16.
Ishitsuka, H., Ninomiya, Y. & Suhara, Y. (1986). Molecular basis of drug resistance to new antirhinovirus agents. *Journal of Antimicrobial Chemotherapy*, **18**, 11–18.
Jackson, G. G., Muldoon, R. L. & Akers, L. W. (1963). Serological evidence for prevention of influenza infection in volunteers by an influenzal drug adamantanamine hydrochloride. *Antimicrobial Agents and Chemotherapy*, 703–7.
Kawana, F., Shigeta, S., Hosoya, M., Suzuki, H. & De Clercq, E. (1987). Inhibitory effects of antiviral compounds on respiratory syncytial virus replication *in vitro*. *Antimicrobial Agents and Chemotherapy*, **31**, 1225–30.
Kim, H. W., Canchola, J. G., Brandt, C. D., Pyles, G., Chanock, R. M., Jensen, K. & Parrott, R. H. (1969). Respiratory syncytial virus disease in infants despite prior administration of antigenic inactivated vaccine. *American Journal of Epidemiology*, **89**, 422–34.
Knight, V., Wilson, S. Z., Quarles, J. M., Greggs, S. E., McClung, H. W., Waters, B. K., Cameron, R. W., Zerwas, J. M. & Couch, R. B. (1981). Ribavirin small particle-aerosol treatment of influenza. *Lancet*, **ii**, 945–9.
Lear, J. D., Wasserman, Z. R. & Degrado, W. F. (1988). Synthetic amphiphilic peptide models for protein ion channels. *Science*, **240**, 1177–81.
Loddo, B. & La Colla, P. (1983). Development of drug resistance and dependence in viruses. In *International Encyclopedia of Pharmacology and Therapeutics, Antiviral Chemotherapy*, Shugar D., Ed. vol. 1, pp. 185–222, Oxford: Pergamon.
Lubeck, M. D., Schulman, J. L. & Palese, P. (1978). Susceptibility of influenza A viruses to amantadine is influenced by the gene coding for M protein. *Journal of Virology*, **28**, 710–16.
McClung, H. W., Knight, V., Gilbert, B. E., Wilson, S. Z., Quarles, J. M. Jr, Divine, G. W., Couch, R. B., Gordon, W. H., Thurston, J. M., Almar, R. L. & Schlautt, W. R. (1983). Ribavirin aerosol treatment of influenza B virus infection. *Journal of the American Medical Association*, **249**, 2671.
Merigan, T. C., Reed, S. E., Hall, T. S. & Tyrrell, D. A. J. (1973). Inhibition of respiratory virus infection by locally applied interferon. *Lancet* **i**, 563–7.
Mitsuya, H. & Broder, S. (1986) Inhibition of the *in vitro* infectivity and cytopathic effects of human T lymphotrophic virus type III/lymphadenopathy-associated virus (HTLV-III/LAV) by 2′,3′-dideoxynucleosides. *Proceedings of the National Academy of Sciences, USA*, **83**, 1911–15.
National Center for Health Statistics. (1986). Current Estimates from the National Health Service Interview Survey. United States, 1985. *Vital and Health Statistics Series 10*, no. 160.
National Institute of Allergy and Infectious Diseases (1979). Amantadine: Does it have a role in the prevention and treatment of influenza? *Annals of Internal Medicine*, **92**, 256–8.
Ninomiya, Y., Ohsawa, C., Aoyama, M., Umeda, I., Suhara, Y. & Ishitsuka, H. (1984). Antivirus agent, Ro 09–0410, binds to rhinovirus specifically and stabilizes the virus conformation. *Virology*, **134**, 269–76.
Oberg, B. (1983). Antiviral effect of phosphonoformate. *Pharmacology and Therapeutics*, **19**, 387–415.
O'Sullivan, D. G., Dane, D. S., Pantic, D. & Briggs, M. (1969). Protective action of benzimidazole derivatives against virus infections in tissue culture and *in vivo*. *Lancet*, **i**, 446–8.
Otto, M. J., Fox, M. P., Fancher, M. J., Kuhrt, M. F., Diana, G. D. & McKinlay,

M. A. (1985). *In vitro* activity of WIN 51711, a new broad-spectrum antipicornavirus drug. *Antimicrobial Agents and Chemotherapy*, **27**, 883–6.

Oxford, J. S., Field, H. J. & Reeves, D. S. eds. (1986). Drug resistance in viruses, other microbes and eukaryotes. *Journal of Antimicrobial Chemotherapy*, **18**, Suppl. W, 232 pp.

Oxford, J. S. & Galbraith, A. (1983). In *International Encyclopedia of Pharmacology and Therapeutics, Antiviral Chemotherapy*, Shugar, D. ed., vol. 1, pp. 169–253, Oxford: Pergamon.

Oxford, J. S. & Oberg, B. (1985). *Conquest of Viral Disease: A Topical Review of Vaccines and Antivirals*. Amsterdam: Elsevier Biomedical Press.

Palese, P. & Schulman, J. L. (1977). In *Chemoprophylaxis and Virus Infections of the Respiratory Tract*, Oxford, J. S., ed., vol. 1, p. 189, Cleveland, Ohio: CRC Press.

Pemberton, R. M., Jennings, R., Potter, C. W. & Oxford, J. S. (1986). Amantadine resistance in clinical influenza A (H3N2) and (H1N1) virus isolates. *Journal of Antimicrobial Chemotherapy*, **18**, 135–40.

Perrin, D. D. & Stunzi, H. (1983). In *International Encyclopedia of Pharmacology and Therapeutics, Antiviral Chemotherapy*, Shugar, D., ed. Vol. 1, Oxford: Pergamon.

Roberts, R. B. & Laskin, O. L. (1988). In *HIV and Other Highly Pathogenic Viruses*, Smith, P. A., ed., p.95, San Diego: Academic Press.

Robertson, J. S., Naeve, C.W., Webster, R. G., Bootman, J. S., Newman, R., & Schild, G. C. (1985). Alterations in the hemagglutining associated with adaptation of influenza B virus to growth in eggs. *Virology*, **143**, 166–74.

Rossman, M. G. (1988). Antiviral agents targeted to interact with viral capsid proteins and a possible application to human immunodeficiency virus. *Proceedings of the National Academy of Sciences, USA*, **85**, 4625–7.

Schaeffer, H. J., Beauchamp, L., de Miranda, P., Elion, G. B., Bauer, D. J. & Collins, P. (1978), 9-(2-hydroxyethoxymethyl) guanine activity against viruses of the herpes group. *Nature*, London, **272**, 573–85.

Scholtissek, C. & Faulkner, GP. (1969). Amantadine-resistant and sensitive influenza A strains and recombinants. *Journal of General Virology*, **44**, 807–15.

Selway, J. W. T. (1986). Antiviral activity of flavones and flavans. *Progress of Clinical and Biological Research*, **213**, 521–36.

Smith, T. J., Kremer, M. J., Luo, M., Vriend, G., Arnold, E., Kamer, G., Rossman, M. G., McKinlay, M. A., Diana, G. D. & Otto, M. J. (1986). The site of attachment in human rhinovirus 14 for antiviral agents that inhibit uncoating. *Science*, **233,** 1286–93.

Stuart-Harris, C. H., Schild, G. C. & Oxford, J. S. (1985). In *Influenza, the Viruses and the Disease*, London: Edward Arnold.

Swallow, D. L. & Kampfner, G. L. (1985). The laboratory selection of antiviral agents. In *Antiviral Chemotherapy and Interferon*, Tyrrell, D. A. J. & Oxford, J. S., eds, p. 322, Edinburgh: Churchill Livingstone.

Taber, L. H., Knight, V., Gilbert, B. E. *et al.* (1983). Ribavirin aerosol treatment of bronchiolitis due to respiratory syncytial virus infection in infants. *Pediatrics*, **72**, 613–18.

Tidwell, R. R., Geratz, J. D., Clyde, W. A. Jr., Rosonthal, K. U. & Dubovi, E. J. (1984) Suppression of respiratory syncytial virus infection in cotton rats by bis (5-amidino-2-benzimidazolyl) methane. *Antimicrobial Agents and Chemotherapy*, **26**, 591–3.

Tisdale, M. & Selway, J. W. T. (1984). Effect of dichloroflavan (BW683C) on the stability and uncoating of rhinovirus type 1B. *Journal of Antimicrobial Chemotherapy*, **14**, 97–105.

Traunecker, A., Luke, W & Karjalainen, K. (1988). Soluble CD4 molecules neutralize human immunodeficiency virus type 1. *Nature, London*, **331**, 84–6.

Tyrrell, D. A. J. & Al-Nakib, W. (1988). Prophylaxis and treatment of rhinovirus infections. In *Clinical Use of Antiviral Drugs*, de Clercq, E., ed., pp. 241–76.

Walker, B. D., Kowalski, M., Goh, W. C., Kozarsky, K., Krieger, M., Rosen, C., Rohrschneider, L., Hazeltine, W. A. & Sodroski, J. (1987). Inhibition of human immunodeficiency virus syncytium formation and virus replication by castanospermine. *Proceedings of the National Academy of Sciences, USA*, **84**, 8120–4.

Warnick, J. E., Maleque, M. A., Bakry, N., Eldefrawi, A. T. & Albuquerque, E. X. (1982). Structure-activity relationships of amantadine. 1. Interaction of the N-alkyl analogues with the ionic channels of the nicotinic acetylcholine receptor and electrically excitable membrane. *Molecular Pharmacology*, **22**, 82–93.

Wilson, I. A., Skehel, J. J. & Wiley, D. C. (1981). Structure of the haemagglutinin membrane glycoprotein of influenza virus at 3Å resolution. *Nature, London*, **289**, 366–73.

Wilson, S. Z., Gilbert, B. E., Quarles, J. M. Jr, Knight, V., McClung, H. W., Moore, R. V. & Couch, R. B. (1984). Treatment of influenza A (H1N1) virus infection with ribavirin aerosol. *Antimicrobial Agents and Chemotherapy*, **26**, 200–3.

World Health Organization (1980). Viral respiratory diseases, report of a WHO scientific group. *WHO Technical Report Series*, **642**, 7–49.

Zamecnik, P. K., Goodchild, J., Toguchi, Y. & Sarin, P. S. (1986). Inhibition of replication and expression of human T cell lymphotropic virus type III in cultured cells by exogenous synthetic oligonucleotides complementary to viral RNA. *Proceedings of the National Academy of Sciences, USA*, **83**, 4143–6.

MANAGEMENT OF CHRONIC HEPATITIS VIRUS INFECTION

HOWARD C. THOMAS

St Mary's Hospital and Medical School, Imperial College of Science, Technology and Medicine, University of London, UK.

INTRODUCTION

With the availability of synthetic nucleoside analogues, lymphoblastoid and recombinant alpha interferons and of molecular probes to monitor viral replication, progress in the treatment of chronic hepatitis B (HBV), delta virus (HDV) and non-A, non-B hepatitis has been rapid.

HBV INFECTION

Recovery from HBV infection is dependent on the integrated activity of the patient's immune system. Chronic HBV infections arise because of defects in these defences. These must be understood before logical therapies can be developed.

Acute hepatitis B infection

HBV is not cytopathic and the liver damage is caused by immune lysis of infected hepatocytes. Analysis of the inflammatory infiltrate demonstrates the presence of NK and cytotoxic T-cells (Eggink *et al.*, 1982). Viral antigens are seen on the surface of the hepatocyte (Gudat *et al.*, 1975) and these in association with the class I MHC glycoproteins, make the cell a target for cytotoxic T-cell lysis (Doherty & Zinkernagel, 1975). Studies in patients with chronic HBV infection suggest that HBc and HBe antigens are important targets (Eddleston *et al.*, 1982; Pignatelli *et al.*, 1987). Hepatocytes usually express very little MHC class I glycoprotein (Thomas *et al.*, 1982) but, in the early stage of acute HBV infection, following the production of alpha-interferon, MHC expression on hepatocytes increases and co-incidentally transaminases rise (Fig. 1). Interferon also activates a series of enzymes, including the 2-5A oligo-adenylate synthetase system, which leads to inhibition of viral protein synthesis (Fig. 2). These changes would be expected to produce an antiviral state in uninfected regenerating liver cells, preventing re-infection. This, along with development of virus neutralising antibodies, should prevent short- and long-term relapse or re-infection.

Chronic hepatitis B infection

It is likely that there are many different reasons why patients develop the chronic carrier state.

Chronic HBV infection following exposure in neonatal life

Ninety per cent of babies born to chronically infected HBV 'e' antigen-positive mothers, become infected, and over 90% of these develop a chronic carrier state (Beasley *et al.*, 1981). These infants probably receive a large inoculum of virus from maternal blood before or during birth, and by close contact with secretions soon after birth. The reason for these infants failing to clear the virus is unknown. HBe antigen, a low molecular weight soluble protein, passes across the placenta and may suppress the development of the cellular immune response to the nucleocapsid proteins which are the target during immune clearance of infected hepatocytes (Fig. 3). Other factors such as the immaturity of the neonatal immune system (Nash, 1985), the presence of immunosuppressive factors such as alphafetoprotein, and the large inoculum of the virus, may also contribute to the high increase of chronic infection resulting from exposure at or before birth.

Chronic HBV infection following exposure after the neonatal period

In marked contrast to the situation at birth, only 5% of subjects infected after this period develop chronic infection. There are probably several different causes for the development of chronic infection at this stage of life. One defect, which has recently been documented, is the production of sub-normal quantities of alpha interferon by peripheral blood mononuclear cells (Ikeda *et al.*, 1986*a*). These individuals also have evidence of abnormal activation of their hepatocytes by interferon. Levels of hepatic 2-5A oligoadenylate synthetase, an enzyme induced by interferon, are only marginally elevated (Ikeda *et al.*, 1986*b*), and MHC Class I proteins, which are also known to be induced by interferon, are present in very low density on infected hepatocytes (Montano *et al.*, 1982). This appears to be due to the ability of HBV within the hepatocyte to selectively reduce the responsiveness of the cell to interferons. *In vitro* transfection of HBV into cells in tissue culture, makes them specifically non-responsive to interferon: they remain susceptible to lysis by Sindbis virus and MHC induction does not occur (Onji *et al.*, 1989).

Management of chronic HBV

There is no established place in treatment of chronic HBV infection for regimens involving solely anti-inflammatory or immunosuppressive drugs (Scullard *et al.*, 1981; Weller *et al.*, 1982b).

A variety of synthetic and natural antiviral compounds have been developed and are being evaluated in clinical trials.

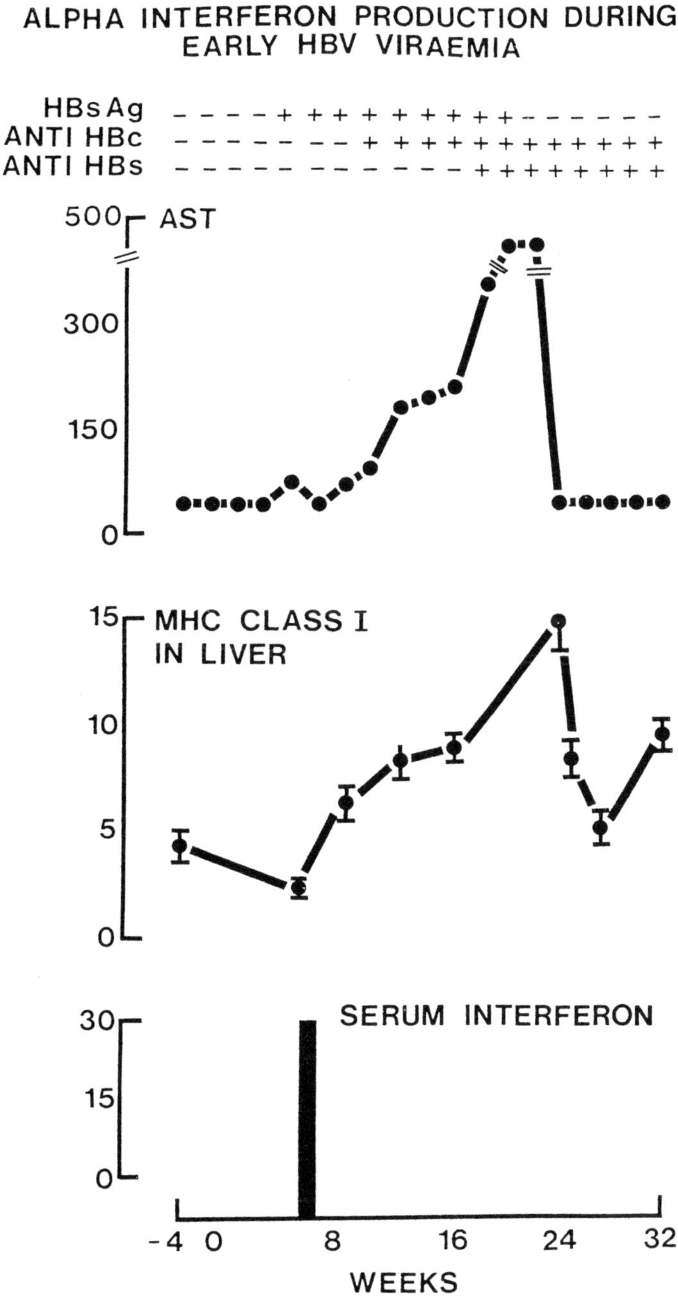

Fig. 1. Acute HBV infection in the chimpanzee. Markers of virus infection appear early, coinciding with a detectable pulse of circulating interferon. This is followed by enhanced MHC display on hepatocytes. Liver damage (rising AST) coincides with the appearance of host immunity to HBV (IgM anti-HBc). AST: aspartate aminotransferase (SGOT).

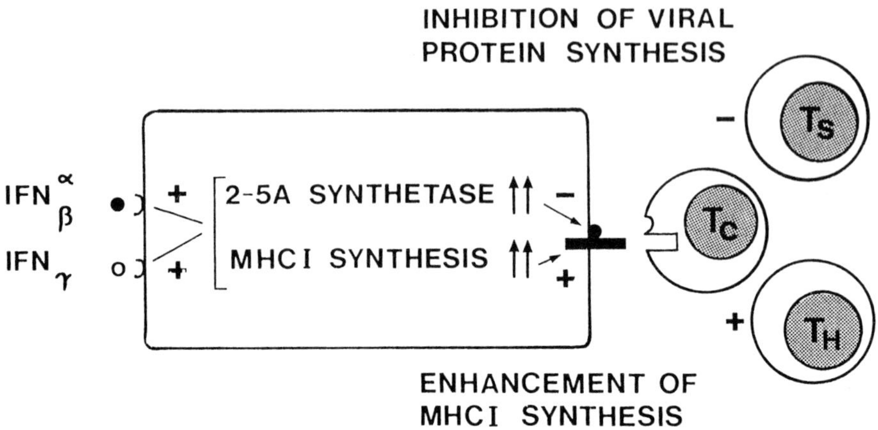

Fig. 2. Effect of interferon (IFN) on virus-infected cells. Alpha and beta IFN acting through a common receptor and gamma IFN acting through a separate receptor, activate several enzyme systems including 2-5A oligo-adenylate synthetase. This enzyme catalyses the production of oligo-adenylates which activate an endogenous ribonuclease leading to cleavage of viral and cellular RNA. Interferon also causes enhanced expression of MHC class I proteins on the hepatocyte surface, facilitating recognition of virus-infected cells by the cellular immune mechanisms of the host. HBe and HBc antigens are probably the target of this component of the host immune response. Tc, Ts, Th: cytotoxic, suppressor and helper T cells. (+) and (−) denote enhancement and suppression, respectively.

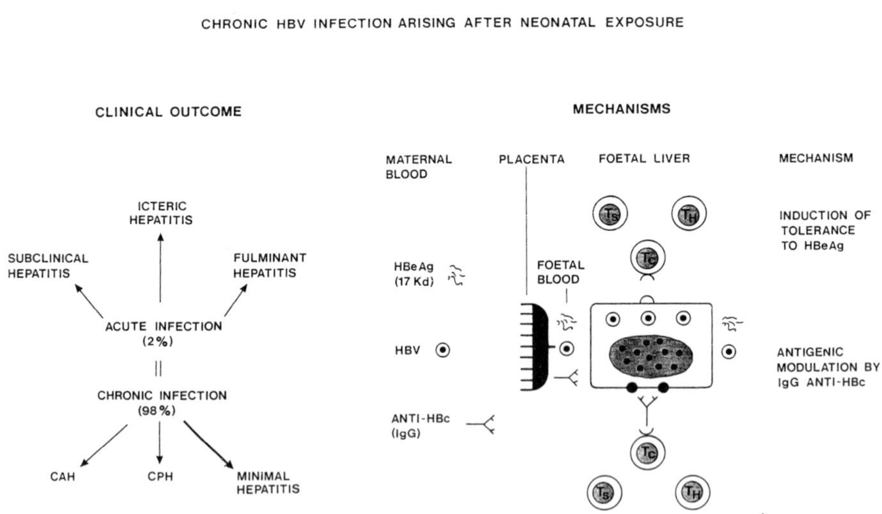

Fig. 3. Neonatal HBV infection: postulated mechanism of viral persistence. It is proposed that HBe antigen crosses the placenta from the maternal blood and induces tolerance to HBe antigen which is the target of the cellular immune response. Maternal IgG anti-HBc also crosses the placenta into the neonatal circulation. HBV infection of the neonatal liver is thus facilitated as maternal IgG blocks recognition of virus-infected cells by cytotoxic T cells. Early exposure to soluble virus protein (HBe) may induce a state of antigenic tolerance to the virus with specific suppressor cells inhibiting the host defence mechanism.

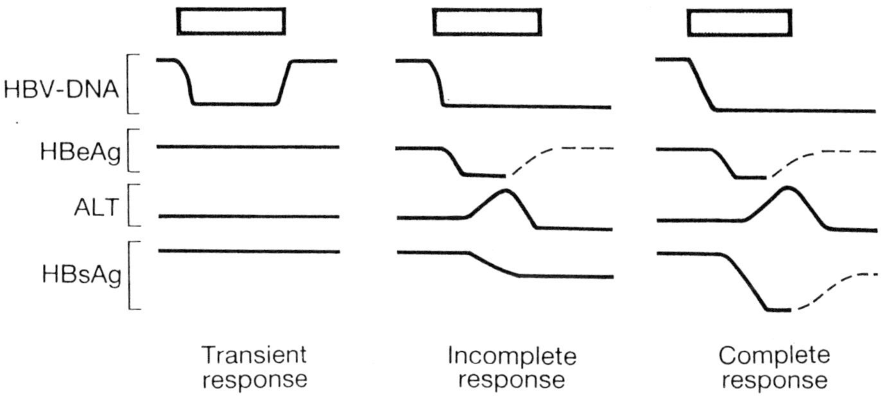

Fig. 4. Pattern of response to antiviral treatment.

INDICATIONS FOR THERAPY (NOT DEFINITIVELY IDENTIFIED)

(i) Infectivity – HBe antigen and HBV-DNA positivity.
(ii) Evidence of progressive liver disease – histological evidence of chronic active hepatitis.

CONTRAINDICATIONS FOR THERAPY

(i) Evidence of decompensated cirrhosis.

GOALS OF THERAPY

(i) Inhibition of HBV replication (by direct effect or by stimulation of the host antiviral response).
(ii) Long-term control of inflammatory necrosis of hepatocytes.
(iii) Prevention of malignant transformation of hepatocytes.

RESPONSES TO ANTIVIRAL THERAPY (FIG. 4)

Transient response

Inhibition of HBV replication – loss of HBV-DNA and HBV-DNA polymerase – during therapy *but* return of these markers on cessation of therapy.

Incomplete response

Sustained inhibition of HBV replication – loss of HBV-DNA and HBV-DNA polymerase – continuing after cessation of therapy, conversion from HBeAg to anti-HBe *but* continued HBs antigenaemia due to translation of integrated HBV-DNA.

Complete response

Sustained inhibition of HBV replication – loss of HBV-DNA and HBV-DNA polymerase – continuing after cessation of therapy, with permanent seroconversion from HBeAg and HBsAg to anti-HBe and anti-HBs.

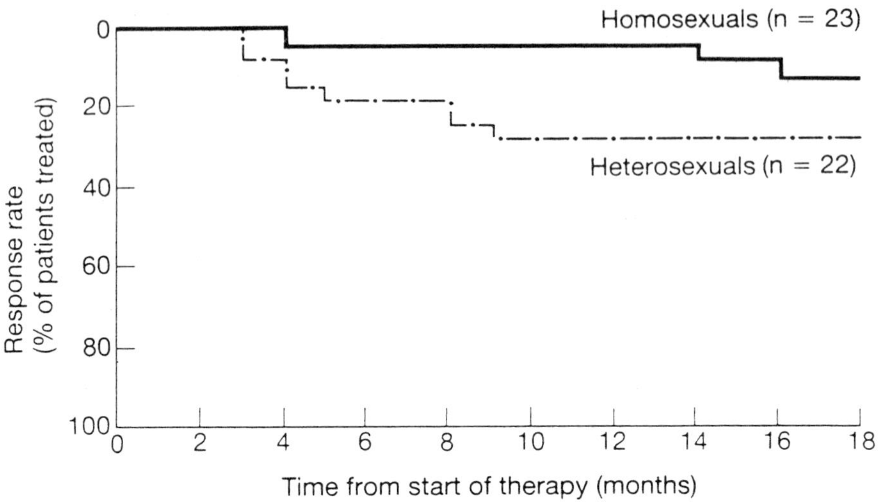

Fig. 5. Response to ARA–AMP in homosexual and heterosexual men.

ACTIVE DRUGS

Adenine arabinoside

This is a potent inhibitor of HBV replication but its clinical usefulness is limited by poor aqueous solubility. In one small, randomised, controlled study a ten-day course produced sustained HBe antigen/antibody seroconversion in 40% of cases (Bassendine *et al.*, 1981).

Adenine arabinoside monophosphate

An equally effective inhibitor of HBV replication which can be given by IM bolus injection (Scullard *et al.*, 1981; Weller *et al.*, 1982a). A course of 10 mg/kg/24 h for three days followed by 5 mg/kg/24 h for a further 23 days has been evaluated in three controlled studies (Hoofnagle *et al.*, 1984; Weller *et al.*, 1985; Trepo *et al.*, 1984). Variability of response may be related to heterogeneity of the patient population being studied. In one study, none of 24 homosexual HBV carriers responded whereas 10 out of 22 heterosexual (predominantly Southern European) carriers did (Thomas *et al.*, 1986) (Fig. 5).

Acyclovir

This is active in low concentrations in herpes simplex infected cells because of phosphorylation to active metabolities by a virus encoded thymidine kinase. This enzyme is not present in HBV infected cells so that this mode of action does not operate. Higher concentrations (near the maximum tolerated dose) do partially inhibit HBV replication (Weller *et al.*, 1983), but this has been shown to be clinically not useful.

Type I interferon (alpha and beta)

Both natural and recombinant DNA produced preparations have been examined.

Leucocyte alpha IFN

This has been shown to inhibit HBV replication when given daily in moderate doses (5–10 megaunits/m^2/day over 3 to 6 months (Greenberg *et al.*, 1976). A randomised controlled study showed that low doses for 4 to 6 weeks produced no long-term responses (Schalm *et al.*, 1982).

Recombinant alpha IFN

This given daily in high doses inhibits HBV replication but is toxic (Smith *et al.*, 1983). Moderate doses (10 megaunits/m^2 daily or thrice weekly) are better tolerated and can be given for three to six months (McDonald *et al.*, 1987). Preliminary analysis of one randomised controlled study shows 45% of patients clearing HBe antigen (McDonald *et al.*, 1987; Brook *et al.*, 1989*a*).

Lymphoblastoid alpha IFN

This given thrice weekly for 3 to 6 months in moderate doses has been shown in two randomised controlled studies to produce HBe antigen clearance in 26 to 88% of cases, significantly more often than occurred spontaneously in the control groups (Barbara *et al.*, pers. comm. 1986; Alexander *et al.*, 1987; Brook *et al.*, 1989*b*). The response rates were very variable because of heterogeneity of the patients, particularly severity of the liver disease at the start of therapy. The main response rate is similar to that seen with adenine arabinoside monophosphate (ARA-MP) in heterosexual patients (Lok *et al.*, 1985) but in homosexual patients ARA-MP does not work whereas interferon-alpha does. Additional analysis of some of these studies shows that Chinese carriers (presumably infected from birth) show no response, whereas European carriers exhibit a 55% seroconversion rate (Thomas & Scully, 1986) (Fig. 6). Further studies reveal that six months of therapy is no better than three (Fig. 7) (Scully *et al.*, 1987).

Beta interferons

Natural and recombinant types have been used and shown to have similar activity to alpha interferon. No controlled studies are reported.

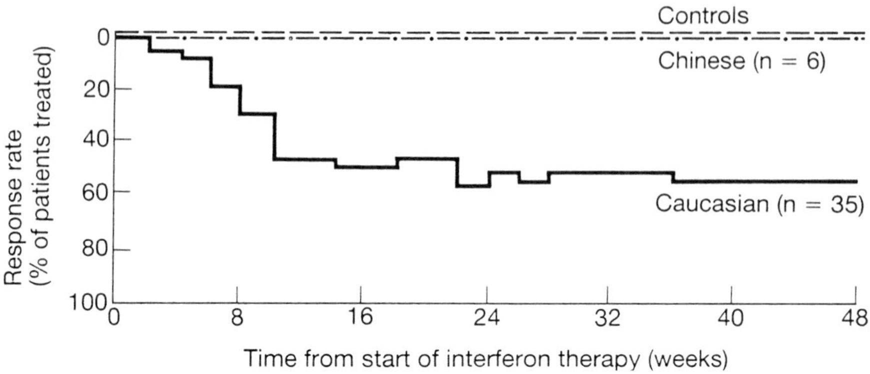

Fig. 6. Response to alpha interferon in neonatal (Chinese) and adult (Caucasian) acquired chronic HBV infection (Scully *et al.*, 1986).

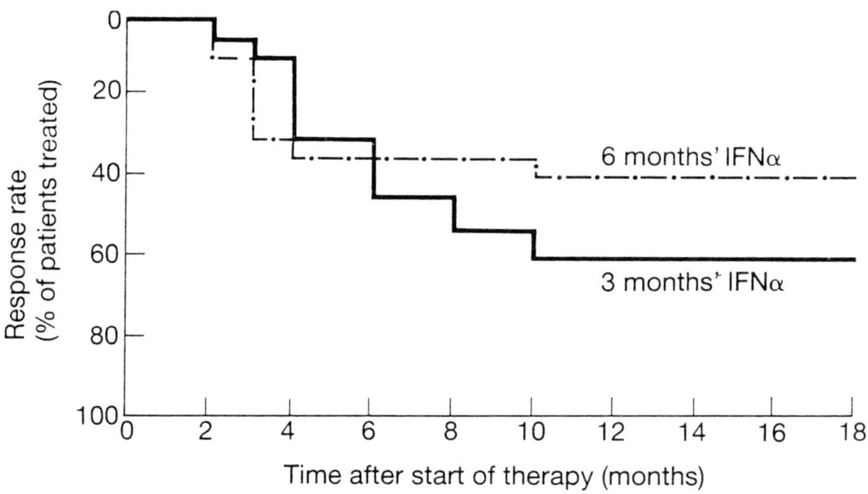

Fig. 7. Response to alpha interferon: comparison of 3 and 6 months therapy (Scully *et al.*, 1986).

Type II interferon (gamma)

Both natural and recombinant interferons are available. These are predominantly immunostimulants with less direct antiviral activity. Inhibition of HBV replication does occur after several months' therapy. Optimal dosaging is not established.

Interleukin II

This is a lymphokine which amplifies immune function. It has been tried safely in individual patients but its place has not been established.

Combination therapy

Short course prednisolone plus antiviral therapy (ARA-MP or interferon alpha)

Withdrawal of immunosuppressant therapy after several months or years, may result in HBeAg seroconversion (Scullard *et al.*, 1981; Weller *et al.*, 1982). However, after short courses of prednisolone, withdrawal does not precipitate seroconversion (Hoofnagle *et al.*, 1986). When followed by ARA-MP or alpha interferon, 60–70% seroconversion rates have been reported in small controlled studies (Perrillo *et al.*, 1985; Yokosuka *et al.*, 1985) and in one of these (Yokosuka *et al.*, 1985) the results were particularly encouraging in that the patients had probably been infected from birth, a group of patients that are resistant to treatment with interferon alone. Care must be taken when withdrawing steroids particularly in patients with severe liver disease: severe exacerbations have been recorded.

Acyclovir and alpha interferon

Preliminary studies did not include appropriate controls and therefore summative or synergistic effects could not be conclusively identified (Schalm *et al.*, 1985).

General comment

The mechanism of action of interferon in inducing seroconversion hepatitis is unknown. Several changes have been noted both in the immune system and in the infected hepatocyte (Fig. 8). An increase in MHC Class I protein display (Pignatelli *et al.*, 1986) and alteration in viral protein expression (Thomas *et al.*, 1986) in the hepatocyte occur within 24 hours of the start of therapy. An increase in the helper/cytotoxic suppressor cell ratio occurs at 6–8 weeks when the seroconversion hepatitis starts and is not seen in those that do not respond.

Changes in humoral immunity also occur. Patients who respond to therapy either have IgM anti-HBc present before treatment or develop it during the course of therapy (Chen *et al.*, 1989). Changes in cell-mediated immunity may also occur but have not been documented.

Inhibition of viral replication, associated with HBe antigen/antibody seroconversion, is usually long lasting and accompanied, after a transient exacerbation, by amelioration of the inflammatory liver disease (Brook *et al.*, 1989b) (Fig. 9). Reactivation may occur particularly in homosexual patients with or without HIV infection (Davis & Hoofnagle, 1985).

Patients treated early in the course of the disease will lose HBs antigen from their serum if cessation of HBV replication is achieved (Fig. 9(a)). If integration of HBV sequences into the host cell genome has occurred before starting treatment, HBs antigenaemia will continue after cessation

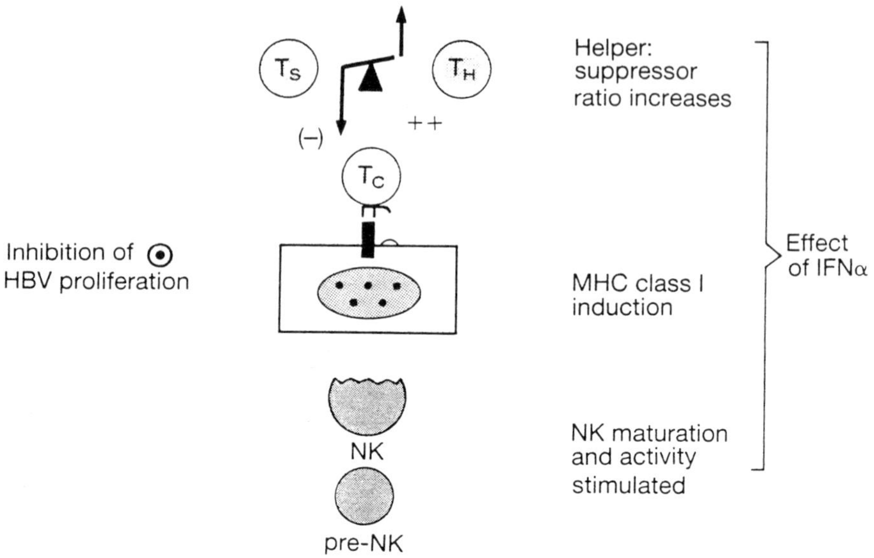

Fig. 8. Changes occurring during interferon therapy.

of detectable HBV replication (Fig. 9(b)). However, inflammatory liver disease will ameliorate in these patients if long-term inhibition of HBV replication is achieved.

Evidence that successful antiviral therapy improves hepatic histology

Several investigations have shown improvement in alanine aminotransferase levels following clearance of HBe Ag and HBV-DNA from the serum. A review of 50 treated and 25 untreated control patients has conclusively shown a reduction in inflammatory necrosis of liver cells and a diminished rate of development of cirrhosis in patients responding to therapy with loss of markers of HBV replication (Brook *et al.*, 1990).

The need to select patients for treatment

Infectivity and progressive liver disease are the major reasons for attempting therapy with either adenine arabinoside monophosphate or the alpha interferons. Currently, interferons are the most satisfactory therapy, but less than 50% of patients respond and the side-effects of interferon are substantial. For this reason, both clinical and serological features, which predict a greater chance of response, have been sought. The clinical features include marked inflammatory activity, high transaminases and more

(a)

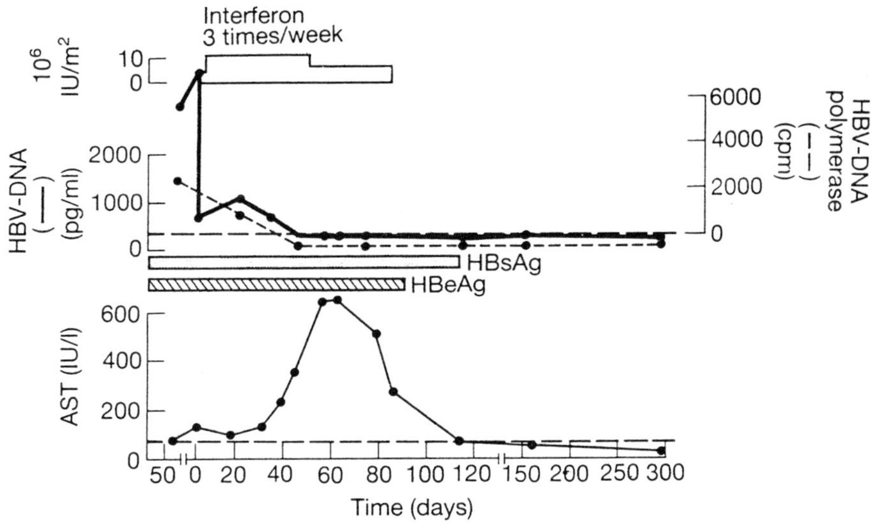

(b)

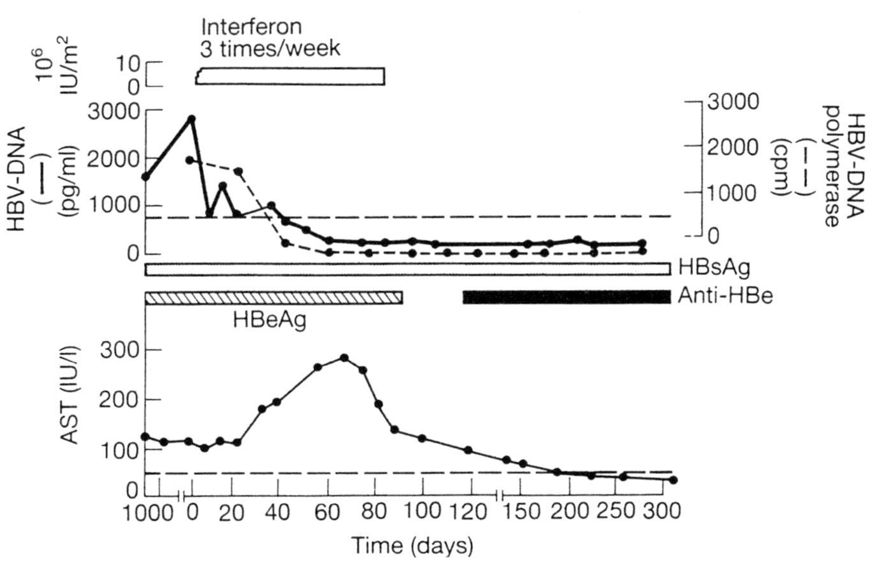

Fig. 9. (a) *Complete response:* inhibition of HBV replication and clearance of HBs antigen. (b) *Incomplete response:* inhibition of HBV replication associated with a transient exacerbation and then normalisation of the liver function tests. Note the patient remains HBsAg positive, presumably because of the presence of integrated HBV sequences. Benefits to the patient include reduced infectivity and cessation of inflammatory liver disease.

recently the presence of IgM anti HBc (Chen *et al.*, 1989) and low serum HBV-DNA (McDonald *et al*, 1987).

HDV INFECTION

HBV carriers superinfected with HDV show a marked exacerbation of their disease. Many progress from minimal liver disease to established cirrhosis within a few years. The disease is common in Central Africa and in the Mediterranean Basin. In the Western World it is only seen in haemophiliacs and drug addicts.

The cloning of the virus has provided molecular probes capable of detecting delta virus RNA both in serum and liver tissue. Preliminary studies by three groups have shown that delta virus replication can be inhibited by moderate doses of alpha interferon given thrice weekly for several weeks (Rosina *et al.*, 1987; Hoofnagle *et al.*, 1987; Thomas *et al.*, 1987).

The liver function tests and liver histology show a more variable response to interferon therapy. In patients treated early in the course of the disease, before the development of cirrhosis, when the hepatitis delta virus RNA falls in the serum, the liver transaminases fall. However, in patients treated later in the course of the disease and in those with established cirrhosis the improvement in liver function is more variable in spite of the reproducible fall in HDV-RNA in serum.

The duration of therapy required to produce long-term control of this agent is currently unknown. After 3 months' therapy, on cessation of interferon, the hepatitis delta virus replication again increases in the majority of cases.

NON-A, NON-B VIRUS INFECTION

There are at least two parenterally transmitted non-A, non-B hepatitis viruses. These can be differentiated by different incubation periods.

Short incubation non-A, non-B

This virus has not been isolated and characterised. Preliminary studies in haemophiliac and agammaglobulinaemic (Thompson *et al.*, 1987) patients and in carriers of the haemophilia gene have shown a marked improvement in transaminases after several weeks interferon therapy. Both lymphoblastoid and recombinant interferon are active against this agent and the dosage required to maintain control of the inflammatory activity is between 2 and 3 million units/m^2 given thrice weekly. When treatment is withdrawn after 3 months of therapy the transaminases again become abnormal. Long-term therapy may, therefore, be required. Side-effects have been minimal and this therapy has now been continued for at least one year in a small number of cases.

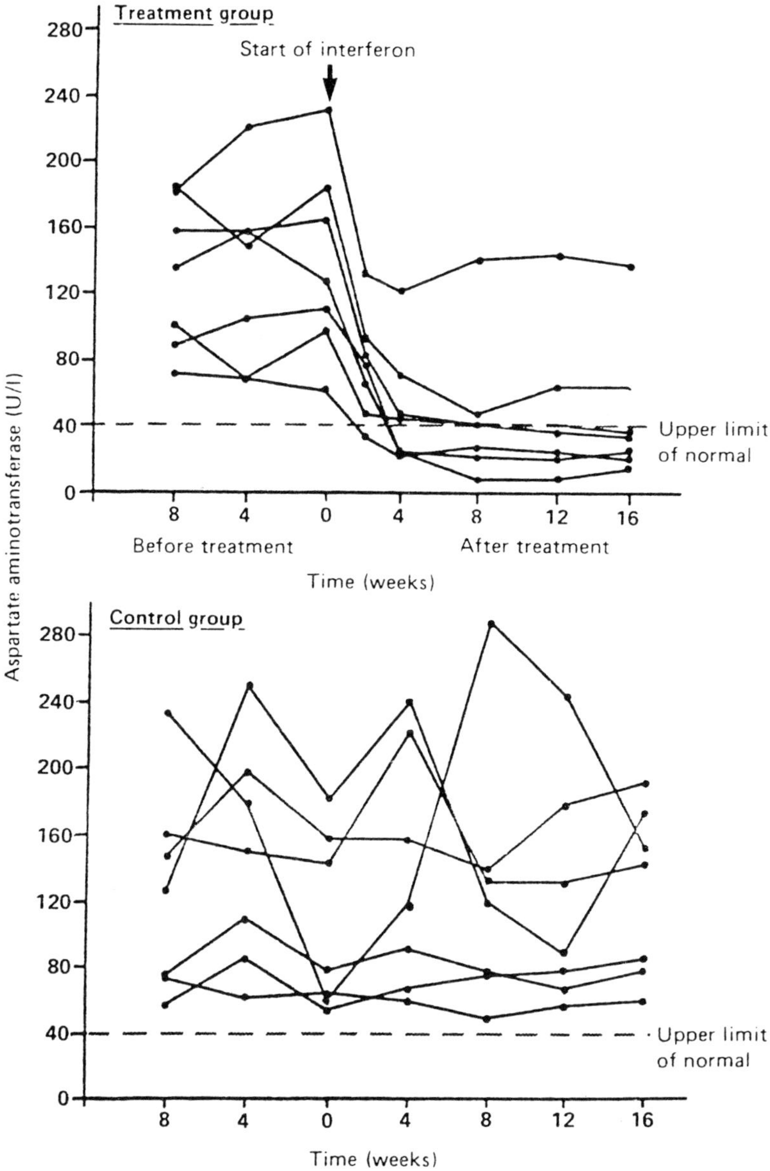

Fig. 10. Preliminary results of a randomised controlled trial of lymphoblastoid interferon in the treatment of chronic non-A–non-B viral hepatitis.

Long incubation non-A, non-B

These patients have an incubation period of 6 to 7 weeks. The virus has now been isolated and provisionally named hepatitis C. Preliminary data show that the virus is sensitive to lipid solvents, has an RNA genome of 9.5 kb and its genomic structure is similar to the flaviviruses.

Recombinant (Hoofnagle *et al.*, 1987) and lymphoblastoid (Jacyna *et al.*, 1989) interferon have been given to small numbers of patients and once again a rapid improvement in liver function tests has been observed. The minimal dose required to maintain this improvement is 1.5 to 3 million units/m^2 thrice weekly. In one randomised, controlled trial liver function tests returned to normal by 6 weeks after the commencement of interferon (Jacyna *et al.*, 1989) (Fig. 10). Once again relapse is seen with withdrawal of interferon and long-term maintenance therapy may be required. Randomised, controlled trials are continuing to determine the practicality of this long-term therapy.

CONCLUSION

It is now clear that both synthetic and natural antiviral compounds, including lymphoblastic and recombinant alpha-interferon, have a place in the management of chronic viral hepatitis. The optimal regimens and duration of treatment remain to be determined.

REFERENCES

Alexander, G. J. M., Brahm, J., Fagan, E. A., Smith, H. M., Daniels, H. M., Eddleston, A. L. W. F. & E. Williams, R. (1987). Loss of HBsAg with interferon therapy in chronic hepatitis B virus infection. *Lancet*, **ii**, 66–8.

Bassendine, M. F., Chadwick, R. E., Salmeron, J., Shipton, V., Thomas, H. C. & Sherlock, S. (1981). Adenine arabinoside therapy HBsAg-positive chronic liver disease: a controlled study. *Gastroenterology*, **80**, 1016–22.

Beasley, R. P., Hwang, L-Y & Lin, C-C. (1981). Hepatitis B immune globulin (HBIG) efficacy in the interruption of perinatal transmission of hepatitis B carrier state. *Lancet*, **ii**, 388–93.

Brook, M. G., Petrovic, L., McDonald, J. A., Scheuer, P. J. & Thomas, H. C. Evidence of histological improvement following anti-viral treatment in chronic hepatitis B virus carriers and identification of histological features that predict response. *Hepatology* (in press).

Brooks, M. G., McDonald, J. A., Karayiannis, P., Caruso, L., Forster, G., Harris, J. R. W. & Thomas, H. C. (1989a). Randomised controlled trial of interferon alpha 2A for the treatment of chronic HBV infection: factors that influence response. *Gut*, **30**, 1116–22.

Brook, M. G., Chan, E., Yap, I, Karayiannis, P., Lever, A. M. L., Jacyna, M., Main, J. & Thomas, H. C. (1989*b*). Randomised controlled trial of lymphoblastoid interferon alpha in Europid men with chronic hepatitis B virus infections, *British Medical Journal*, **299**, 652–6.

Chen, G., Karayiannis, P., McGarvey, M. J., Lever, A. M. L., McDonald, J. A., Scully, L. J. Kanatakis, S. & Thomas, H. C. (1989). Subclasses of anti-HBe in chronic HBV carriers: changes during treatment with interferon and predictors of response. *Gut*, **30**, 1123–8.

Davis, G. L. & Hoofnagle, J. H. (1985). Reactivation of chronic type B presenting as acute hepatitis. *Annals In Medicine*, **102**, 762–5.

Doherty, P. C. & Zinkernagel, R. M. (1975). A biological role for the major histocompatibility antigen. *Lancet*, **i**, 1405–9.

Eddleston, A. W. L. F., Mondelli, M., Mieli-Vergani, G. & Williams, R. (1982). Lymphocyte cytotoxicity to autologous hepatocytes in chronic hepatitis B virus infection. *Hepatology*, **2**, 122S-7S.

Eggink, H. F., Houthoff, H. J., Huitema, S., Gips, C. H. & Poppema, S. (1982). Cellular and humoral immune reactions in chronic active liver disease. Lymphocyte subsets in liver biopsies of patients with untreated idiopathic auto-immune hepatitis, chronic active hepatitis B and primary biliary cirrhosis. *Clinical and Experimental Immunology*, **50**, 17–24.

Greenberg, H. B., Pollard, R. B., Lutwick, L. I., Gregory, P. B., Robinson, W. S. & Merigan, T. C. (1976). Effect of human leukocyte interferon on hepatitis B virus infection in patients with chronic active hepatitis. *New England Journal of Medicine*, **295**, 517–22.

Gudat, F., Bianchi, L., Sonnabend, W., Thiel, G., Aenishaenslin, W. & Stadler, C. A. (1975). Pattern of core and surface expression in liver tissue reflects state of immune response in hepatitis B. *Journal of Laboratory Investigation*, **321**, 1–9.

Hoofnagle, J. H., Hanson, R. G., Minuk, G. Y., Pappas C., Schafer, D. F., Dusheiko, G. M., Straus, S. E., Popper, H. & Jones, E. A. (1984). Randomised controlled trial of adenine arabinoside monophosphate for chronic type B hepatitis. *Gastroenterology*, **86**, 150–7.

Hoofnagle, J. H., Mullen, K. D., Jones, B., Rustgi, V., Di-Bisceglie, A., Peters, M., Waggoner, J. G., Yoon Park, R. N. & Jones, E. A. (1986) Treatment of chronic non-A, non-B hepatitis with recombinant human alpha interferon. *New England Journal of Medicine*, **315**, 1575–8.

Hoofnagle, J. H, Mullen, K., Peters, M., Avigen, M., Park, Y., Waggoner, J., Gervi, L., Hoyer, B., & Smedile A. (1987). Treatment of chronic delta hepatitis with recombinant human alpha interferon. In *The Hepatitis Delta Virus*, pp. 291–8. Alan R. Liss Inc.

Ikeda, T., Lever, A. M. L. & Thomas, H. C. (1986a). Evidence for a deficiency of interferon alpha production in patients with chronic HBV infection acquired in adult life. *Hepatology*, **6**, 962–5.

Ikeda, T., Pignatelli, M., Lever, A. M. L. & Thomas, H. C. (1986b). Relationship of HLA protein display to activation of 2–5A synthetase in HBe antigen or anti-HBe positive chronic HBV infection. *Gut*, **27**, 1498–501.

Jacyna, M., Brook, G., Loke, R. H. T., Crossey, M., Murray-Lyon, I. M. & Thomas, H. C. (1989). A controlled trial of lymphoblastoid interferon in chronic Non A, Non B hepatitis. *British Medical Journal*, **298**, 80–2.

Lok, A. S. F., Novick, D. M., Karayiannis, P., Dunk, A., Sherlock, S. & Thomas, H. C. (1985). A randomised study of effect of adenine arabinoside monophosphate (4 or 8 week courses) versus lymphoblastoid interferon on hepatitis B virus (HBV) replication. *Journal of Hepatology*, **5**, 1132–58.

McDonald, J. A., Caruso, L., Karayiannis, P., Scully, L. J., Harris, J. R. W., Forster, G. & Thomas, H. C. (1987). Diminished responsiveness of male homosexual chronic HBV carriers with HIV antibodies to recombinant alpha interferon. *Hepatology*, **7**, 719–23.

Montano, L., Miescher, G. C., Goodhall, A. H., Wiedmann, K.., Janossy, G. & Thomas, H. C. (1982). Hepatitis B virus and HLA display in the liver during chronic hepatitis B virus infection. *Hepatology*, **2**, 557–61.

Nash, A. A. (1985). Tolerance and suppression in virus diseases. *British Medical Bulletin*, 41–5.

Novick, D. M., Lock, A. S. F. & Thomas, H. C. (1984). Diminished responsiveness of homosexual men to antiviral therapy for HBsAg-positive chronic liver disease. *Hepatology*, **1**, 29–35.

Onji, M., Lever, A. & Thomas, H. C. (1989). Hepatitis B virus reduces the sensitivity of cells to interferon. *Hepatology*, **9**, 92–6.

Perrillo, R. P., Regerstein, F. G., Bodicky, C. J., Campbell, C., Sanders, G. E. & Sunwoo, Y. C. (1985). Comparative efficacy of ARA-MP in prednisolone withdrawal followed by ARA-MP in treatment of HBV induced CAH. *Gastroenterology*, 780–6.

Pignatelli, M., Waters, J., Lever, A. M. L., Brown, D., Iwarson, S., Schaff, Z., Gerety, R. & Thomas, H. C. (1986). HLA Class I antigens on hepatocyte membrane: Increased expression during interferon therapy of chronic hepatitis B. *Hepatology*, **6**, 349–53.

Pignatelli, M., Waters, J., Lever, A. M. L. Iwarson, S., Gerety, R. & Thomas, H. C. (1987). Cytotoxic T-cell responses to the nucleocapsid proteins of HBV in chronic hepatitis. *Hepatology*, **4**, 15–21.

Rosina, F., Saracco, G., Lattore, V., Quartarone, V., Rizzetto, M., Verme, G., Trinchero, P., Sansalvadore, F. & Smedile, A. (1987). In *The Hepatitis Delta Virus*, pp. 299–303. Alan R Liss. Inc.

Schalm, S. W. & Heijtink, R. A. (1982). Spontaneous disappearance viral replication and liver cell inflammation in HBs-Ag positive chronic active hepatitis: results of placebo vs. interferon trial. *Hepatology*, **2**, 791–4.

Schalm, S. W., Heijtink, R. A., Van Buuren, H. R. & de Man, R. A. (1985). Acyclovir enhances the antiviral effect of interferon in chronic hepatitis B. *Lancet*, **17**, 358–60.

Scullard, G. H., Pollard, R. B., Smith J. L., Sacks, S. L., Gregory, P. B., Robinson, W. S. & Merigan, T. C. (1978). Antiviral treatment of chronic hepatitis B virus infection: changes in viral markers with interferon combined with adenine arabindoside. *Journal of Infectious Diseases*, **143**, 772–83.

Scullard, G. H., Smith, C. I., Merigan, T. C., Robinson, W.. & Gregory, P. B. (1981). Effects of immunosuppressive therapy on viral markers in chronic active hepatitis B. *Gastroenterology*, **81**, 987–91.

Scully, L. J., Lever, A. M. L., Yap, I., Pignatelli, M., Thomas, H. C. (1986). Identification of factors influencing response rate to antiviral therapy of chronic HBV infection. *Journal of Hepatology*, **3**, S291–9.

Scully, L. J., Shein, R., Karayiannis, P., McDonald, J. A. Thomas, H. C. (1987). Lymphoblastoid interferon therapy of chronic HBV infection. A comparison of 12 v 24 weeks of thrice weekly treatment. *Journal of Hepatology*, **5**, 51–8.

Smith, C. I., Weissberg, J., Bernhardt, L., Gregory, P. B., Robinson, W. S. & Merigan, T. C. (1983). Acute Dane particle suppression with leukocyte recombinant interferon in chronic HBV infection. *Journal of Infectious Diseases*, **148**, 907–13.

Thomas, H. C. & Scully, L. J. (1986). Antiviral therapy in chronic hepatitis B virus infection. *British Medical Bulletin*, **41**, 374–80.

Thomas, H. C., Farci, P., Shein, R., Karayiannis, P., Smedile, A., Caruso, L. & Gerin, L. (1987). Inhibition of hepatitis delta virus (HDV) replication by lymphoblastoid human interferon. In *The Hepatitis Delta Virus*, pp 277–90. Alan R Liss. Inc.

Thomas, H. C., Shipton, U. & Montano, L. (1982). The HLA system: its relevance to the pathogenesis of liver disease. In *Progress in Liver Disease*, **16**, pp. 517–27, New York: Grune & Stratton.

Thompson, B. J., Doran, M., Lever, A. M. L. & Webster, A. D. B. (1987).

Alpha interferon therapy for non-A, non-B hepatitis transmitted by gammaglobulin replacement therapy. *Lancet*, **i**, 539–41.

Trepo, C., Hantz, O. & Ouzan, D. (1984). Therapeutic efficacy of ARA-MP in symptomatic HBe Ag positive C A H: a randomised, placebo control study. (Abstract 193) *Journal of Hepatology*, **4**, 1055.

Weller, I. V. D., Bassendine, M. F., Craxi, A., Fowler, M.. F., Monjardino, J., Thomas, H. C. & Sherlock, S. (1982a). Successful treatment of HBs and HBeAg positive chronic liver disease: prolonged inhibition of viral replication by highly soluble adenine arabinoside 5'-monophosphate (ARA-MP). *Gut*, **23**, 717–23.

Weller, I. V. D., Bassendine, M. F., Murray, A. K., Craxi, A., Thomas, H. C. & Sherlock, S. (1982b). The effects of prednisolone/azathioprine in chronic hepatitis B viral infection. *Gut*, **23**, 650–5.

Weller, I. V. D., Carreno, V., Fowler, M. J. F., Monjardino, J., Markinen, D., Varghese, Z., Sweny, P., Thomas, H. C. & Sherlock, S. (1983). Acyclovir in hepatitis B antigen-positive chronic liver disease: inhibition of viral replication and transient renal impairment with iv bolus administration. *Journal of Antimicrobial Chemotherapy*, **11**, 223–31.

Weller, I. V. D., Lok, A. S. F., Mindel, A., Karayiannis, P., Galpin, S., Monjardino, J., Sherlock, S. & Thomas, H. C. (1985). A randomised controlled trial of adenine arabinoside 5' -monophosphate (ARA-AMP) in chronic hepatitis B virus infection. *Gut*, **26**, 745–51.

THERAPY – HIV

DOUGLAS D. RICHMAN

University of California San Diego, Infectious Diseases-111F, VA Medical Center, 3350 La Jolla Village Drive, San Diego, CA 92161, USA

An effective antiretroviral drug was identified and licensed only five years after the first published description of AIDS and only three years after the isolation of the aetiological agent, HIV. The unprecedented speed of this medical achievement is gratifying. This drug, AZT, is neither completely suppressive of disease nor free from toxicity. The impetus for more and better drugs is thus great. AZT and the majority of the first promising candidate compounds are nucleoside analogues. This situation is not surprising for a number of reasons. Because of the utility of nucleoside analogues for cancer chemotherapy, and for the therapy of herpesviruses infections, large collections of such compounds and much expertise in nucleoside synthetic chemistry were already available. Nucleosides will remain promising candidate compounds; nevertheless, the remarkably speedy advance in the understanding of the molecular biology of HIV has resulted in an extensive set of programmes to identify compounds that interfere with each of the possible virus-specific targets in the replication of HIV (Fig. 1). Compounds with antiretroviral activity *in vitro* and the targets against which they are directed (Table 1) will be reviewed first. The clinical experience with antiviral therapy for HIV infection will then be summarised.

COMPOUNDS WITH ANTIRETROVIRAL ACTIVITY *IN VITRO*

Compounds that inhibit binding and penetration

The critical first step in the life cycle of HIV is the high affinity binding of gp120, the surface glycoprotein on the virion or infected cell, to the CD4 receptor molecule on the host cell. Compounds that specifically interfere with this interaction should inhibit infection. In fact, the inhibition of infection with monoclonal antibodies to CD4 was the first approach used to identify the host cell receptor to HIV (Dalgleish *et al.*, 1984; Klatzmann *et al.*, 1984; McDougal *et al.*, 1986).

The use of neutralising antibodies has been pursued by administering plasma from seropositive donors in two small, open studies to patients with advanced AIDS resulting in clearance of p24 antigenemia and enhance-

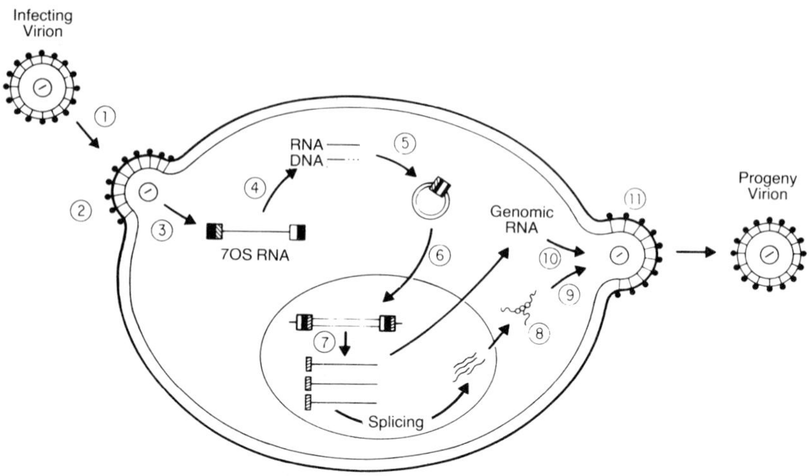

Key

1 Binding
2 Penetration
3 Uncoating
4 Reverse transcription
5 Circularization
6 Integration
7 Transcription
8 Translation
9 Post-translational processing
10 Assembly
11 Budding

Fig. 1 *Retroviral replication cycle and potential sites for specific antiviral therapy.* Eleven arbitrarily identified steps in retroviral replication are shown. Each virus-specific component of this cycle is a potential target for selective antiviral therapy. The proposed targets of potential antiretroviral drugs are indicated in Table 1.

ment of neutralising and anti-p24 antibody levels (Jackson *et al.*, 1988*b*; Karpas *et al.*, 1988). Further studies are needed to identify if this approach is antigen-specific and in fact beneficial, and if so, what is the mechanism of action. Another theoretical approach is the use of anti-idiotype antibody to the CD4 binding site; this antibody binds to the reciprocal binding site on gp120 (Dalgleish *et al.*, 1988).

An alternative approach to inhibit the interaction between gp120 and CD4 is to use portions of the CD4 molecule that bind gp120. CD4 is a glycoprotein with a hydrophobic, membrane-anchoring amino acid sequence at the carboxy terminus and a gp120 binding side at the amino terminus (Richardson *et al.*, 1988; Landau, Warton & Littman, 1988; Jameson *et al.*, 1988). The construction by recombinant DNA technology of

Table 1. *Antiretroviral drugs with documented sites of action*

Attachment–penetration	Dextran Sulphate Soluble CD4
Reverse transcription	Nucleoside analogues Foscarnet Template-binding oligonucleotides HPA-23 Rifabutin
Transcription	Ribavirin
Translation	Antisense oligonucleotides Interferons?
Post-translational processing	Glycosylation inhibitors Protease inhibitors
Assembly/budding	Interferons?

expression vectors that encode and express a CD4 polypeptide truncated between these two sites results in a soluble secreted polypeptide that inhibits HIV-1 infectivity at approximately 10 μg/ml (Bertonis *et al.*, 1988; Hussey *et al.*, 1988; Deen *et al.*, 1988; Traunecker, Luke & Karjalainen, 1988; Smith *et al.*, 1987). Shorter sequences of CD4 representing the gp120 binding site have been prepared by synthesis of oligopeptides. These also block infectivity, but hundredfold higher concentrations are required because the oligopeptides initially examined have lower binding affinities than the larger polypeptides (Lifson *et al.*, 1988). The first phase I studies of recombinant soluble CD4 were initiated in late 1988. Theoretical concerns about the ability of recombinant soluble CD4 to interfere with normal immune functions, to cross the blood–brain barrier, or to induce an antibody response have been raised but should not inhibit careful phase I studies.

This concept has been carried one step further by generating a recombinant human CD4 *Pseudomonas* exotoxin hybrid protein with the rationale of selectively targeting a lethal toxin to HIV-infected cells expressing gp120 (Chaudhary *et al.*, 1988). There are many steps between this theoretically elegant approach and therapeutic utility. Other modifications are also under investigation. For example, a chimaeric molecule containing the amino terminal binding site of CD4 and the Fc end of the IgG heavy chain appears to retain the gp120 affinity of the former and the prolonged serum half-life and several other characteristics of the latter (Capon *et al.*, 1989). Clinical studies with this molecule were scheduled to commence in late 1989. An additional attribute of CD4 preparations is the synergy with AZT observed in antiviral assays *in vitro* (Johnson *et al.*, 1989).

A synthetic octapeptide, termed peptide T, of which four amino acids of the octapeptide are threonine (abbreviated T, hence peptide T) (Pert *et al.*, 1988), has been reported to block infectivity of HIV and binding between radiolabelled gp120 and CD4 in the 100 nM range (Pert *et al.*, 1986). Several investigators have been unable to demonstrate antiviral activity with peptide T (Barnes, 1987*a*); nevertheless, further investigations including phase I studies have been initiated to assess the potential of this compound.

In addition to agents that inhibit specific, virally defined functions during the steps of binding and penetration, a class of compounds that act to block binding non-specifically have been identified. It has been appreciated for decades that polycations can enhance viral infectivity *in vitro* and polyanions can inhibit infectivity of many viruses including retroviruses (Toyoshima & Vogt, 1969; Takemoto & Spicer, 1965; De Somer *et al.*, 1968). Dextran sulphate is a polysulphated polymer of glucose. Dextran sulphate blocks the binding of HIV-1 to CD4 positive target cells (Mitsuya *et al.*, 1988; Baba *et al.*, 1988*b*). The drug inhibits the expression of viral antigen and syncytium formation of HIV-1 and HIV-2 infected cells at concentrations between 1 and 10 μg/ml, while minimal cellular toxicity is seen at 1000 μg/ml (Ueno & Kuno, 1987; Ito *et al.*, 1987*a*; Mitsuya *et al.*, 1988; Baba *et al.*, 1988*b*). Thus dextran sulphate has an extremely high selectivity index (therapeutic ratio *in vitro*). Active preparations have contained approximately 18% sulphur by molecular weight. Removal of sulphate groups eliminates antiviral activity (Mitsuya *et al.*, 1988; Bertonis *et al.*, 1988). A wide range of molecular weight species can be synthesised and preparations represent a heterogeneous mixture of molecular weights. Preparations with a peak at M_r 5000 and 8000 appeared equally effective (Ito *et al.*, 1987*a*).

Other polyanionic compounds inhibit the replication of HIV *in vitro*. These include sulphated polysaccharides extracted from animal sources (heparin) (Balzarini *et al.*, 1986*a*, *b*; Ito *et al.*, 1987*a*; Baba *et al.*, 1988*b*), sea algae (Nakashima *et al.*, 1987*a*; Yoshida *et al.*, 1988), and liquorice (Tochikura *et al.*, 1988). These compounds and dextran sulphate inhibit membrane fusion following contact between infected and uninfected cells in contrast to nucleoside analogues like AZT (Tochikura *et al.*, 1988). It is also likely that antiviral activity can be obtained below that concentration which anticoagulates blood (Baba *et al.*, 1988*b*). It is possible that other non-polymeric but large polyanionic compounds with antiretroviral activity may also belong in this class of compounds that act at an early step. Such compounds include aurintricarboxylic acid (Balzarini *et al.*, 1986*a*, *b*), Evan's Blue (Balzarini *et al.*, 1986*a*, *b*) and suramin, which is reviewed later. These compounds and the sulphated polymers inhibit the activity of cell-free reverse transcriptase (Balzarini *et al.*, 1986*a*, *b*; Mitsuya *et al.*, 1988; Nakashima *et al.*, 1987*a*; Chu *et al.*, 1988). It is unlikely that such compounds traverse the cell membrane well. In addition, these compounds inhibit host cell DNA polymerase α equally well (Mitsuya *et* al., 1988). Consequently, they are likely to act at the surface interactions of virus, and cell enzyme inhibition is probably not relevant beyond the test tube. This conclusion is further supported by the observation that these sulphated polysaccharides are selective inhibitors of a wide range of enveloped DNA and RNA viruses (Baba *et al.*, 1988*c*). Aurintricarboxylic acid, in contrast to the other compounds of this class, has been shown to bind specifically to CD4 and to block adsorption of HIV-1 to CD4 positive cells (Schola *et al.*, 1989).

Compounds that inhibit reverse transcriptase

Reverse transcriptase is an essential virion enzyme of all retroviruses. Inhibitors of this function in animal retroviruses were intensively investigated before the discovery of HIV (summarised in De Clercq, 1986). The first drugs to inhibit antiretroviral activity clinically belong to this class (AZT, ddC, ddA, ddI, foscarnet). The nucleoside analogues will be reviewed first, followed by the other compounds in this class.

Nucleoside analogues

The art and science of synthesis of nucleoside analogues have flourished for decades in the occasionally successful search for anticancer and antiviral drugs. Most such syntheses are reported as accomplishments of organic chemistry but fail to result in a therapeutically useful compound. For example, Horwitz *et al.* (Horwitz, Chua & Noel, 1964) and later Lin and Prusoff (Lin & Prusoff, 1978) reported the synthesis of 3′-azido-3′-deoxythymidine (AZT), which failed to find a role in cancer chemotherapy. The synthesis of 2′,3′-dideoxynucleosides (Robins & Robins, 1964; Horwitz *et al.*, 1967) resulted in the generation of reagents critical to the performance of the Sanger DNA sequencing technique (Sanger, Nicklen & Coulson, 1987). Nevertheless with the experience in nucleoside synthesis and the success in identifying effective inhibitors of herpesvirus DNA polymerases, reverse transcriptase was not surprisingly the first viral target for which an effective anti-HIV compound was identified. Dozens of nucleoside analogues have been shown to inhibit HIV *in vitro* and many of these will undergo clinical trial. The first to be identified and reach clinical trial (AZT and the 2′,3′-dideoxynucleosides) will initially be described.

3′-azido,3′-deoxythymidine (AZT, zidovudine, Retrovir^{TM}) The generic designation for AZT is zidovudine (occasionally abbreviated ZDV). This term conveys no information while AZT or azidothymidine indicates the structure of the compound and represents the designation in common usage by patients, physicians and researchers.

In 1985 Mitsuya and colleagues reported that AZT inhibited the replication of HIV in peripheral blood mononuclear cells and several T cell lines at concentrations of 1 μM, approximately one hundredfold less than the toxic concentration for these cells (Mitsuya *et al.*, 1985). Many other groups have confirmed similar levels of efficacy and toxicity in human cells with one major exception. AZT at concentrations as low as 1 to 3 μM appears to be toxic *in vitro* to normal human granulocyte–macrophage progenitor cells (Sommadossi & Carlisle, 1987). These concentrations approach the peak levels attained in serum. In addition, the replication of LAV strain in primary human macrophages appeared to be resistant to AZT and other dideoxynucleosides (Richman, Kornbluth & Carson, 1987) while the macrophage-tropic Ba-L strain was shown to be fully sensitive by this same group and others (Perno *et al.*, 1988).

Unlike the physiological deoxynucleosides which are transported across cell membranes, AZT appears to enter cells by non-facilitated diffusion (Zimmerman, Mahony & Prus, 1987). The lipophilic 3′-azido group imparts this membrane permeability. Cellular thymidine kinase converts AZT into its monophosphate (Furman *et al.*, 1986). Cellular thymidylate kinase converts the monophosphate into the diphosphate which is further converted to the triphosphate (AZT-TP), the active antiviral compound, by other cellular enzymes (Furman *et al.*, 1986). The predominance of AZT monophosphate (AZT-MP) in human cells exposed to AZT (Furman *et al.*, 1986; Balzarini *et al.*, 1988*c*) suggests that thymidylate kinase is the rate limiting enzyme in the synthesis of the active AZT triphosphate (AZT-TP) which represents only about 1% of intracellular drug. AZT-MP presumably inhibits thymidylate kinase resulting in this anabolic block, also resulting in the inhibition of phosphorylation of thymidine in cells treated with AZT (Furman *et al.*, 1986). This impact on intracellular nucleotide pools has consequences on cell toxicity and antagonism or synergy with other antiviral drugs (Baba *et al.*, 1987*b*; Vogt *et al.*, 1987). This block does not appear to occur in murine cells in which AZT appears to be more effective both *in vitro* and *in vivo* (Dahlberg *et al.*, 1987; Balzarini *et al.*, 1988*c*; Ruprecht *et al.*, 1986).

AZT-TP, as an analogue of thymidine triphosphate, inhibits the reverse transcriptase of HIV-1 and other lentiviruses *in vitro* (St Clair *et al.*, 1987; Vrang *et al.*, 1988). Inhibition of host cell DNA polymerases α and β is considerably weaker, presumably accounting for its therapeutic ratio (St Clair *et al.*, 1987). Of interest, inhibition by AZT of the DNA polymerase of many Gram negative bacteria results in antibacterial activity (Elwell *et al.*, 1987). Bacterial resistance readily develops as a consequence of loss of thymidine kinase activity; nevertheless, toxicological studies of AZT that were undertaken in animals would later prove time saving in its development as an antiviral agent. Chain termination has been demonstrated with the *E. coli* DNA polymerase (Elwell *et al.*, 1987; Mitsuya *et al.*, 1987*a*). Studies *in vitro* with reverse transcriptase and the synthetic template, poly (rA)-oligo$(dT)_{12-18}$, have demonstrated chain termination with AZT-TP (St Clair *et al.*, 1987). With the incorporation of AZT-TP, a 3′-OH group is no longer available to form a 3′-5′phosphodiester bond to permit the addition of another nucleotide triphosphate to the elongating DNA chain. The relative importance of chain termination and competition with thymidine triphosphate as the mechanism of action of AZT-TP remain to be determined.

Addition of thymidine reverses both the efficacy and toxicity of AZT (Mitsuya *et al.*, 1985; Sommadossi *et al.*, 1988), indicating the competition of thymidine triphosphate and AZT-TP for viral and host DNA polymerases. The toxicity of AZT for human granulocyte progenitor cells *in vitro* occurs at concentrations of 1 to 5 μM, approximately one hundredfold less than that required to produce toxicity in lymphoid cell lines (Dainiak *et al.*, 1988; Sommadossi & Carlisle, 1987). This hypersensitivity to the effects of AZT in marrow cells is reversed by uridine but not thymidine (Sommadossi *et al.*, 1988). Both the mechanisms and potential applications of this observation await future investigations.

The inhibition of reverse transcriptase prevents the normal production of viral products and cytopathology. This observation has been made with several laboratory strains of HIV-1 in peripheral blood mononuclear cells and in many continuous cell lines utilising cytopathology, the production of reverse transcriptase and antigens, or even T cell functions as endpoints. Despite the variation in assay systems, inhibition is observed at concentrations as low as 0.01 to 0.1 μM (Sommadossi *et al.*, 1988; Mitsuya *et al.*, 1985; Nakashima *et al.*, 1986; Richman, Kornbluth & Carson, 1987; Lyerly, Cohen & Weinhold, 1987; Larder, Darby & Richman, 1989). Clinical isolates from patients never given AZT display little variation in susceptibility to AZT (Larder, Darby & Richman, 1989). Other retroviruses examined to date also are inhibited by AZT *in vitro*. These include HTLV-1 (Matsushita *et al.*, 1987), simian retroviruses (Tsai, Follis & Benveniste, 1989; Mitsuya & Broder, 1988), avian leukosis virus (Olsen *et al.*, 1987), several murine retroviruses (Ostertag *et al.*, 1974; Dahlberg *et al.*, 1987), feline leukaemia virus (Tavares *et al.*, 1987) and equine infectious anemia virus (Dahlberg *et al.*, 1987). No antiviral activity has been observed against a broad range of human viruses from other families with the exception of one of the human herpesviruses, EBV, for which the 50% effective dose appears to be between 1 and 10 μM (Lin *et al.*, 1988).

Although AZT effectively inhibits the replication of cell-free HIV-1, it does not inhibit virus production by chronically infected cells or transmission of infection to uninfected cells by syncytium formation (Smith, Brian & Pagano, 1987; Nakashima *et al.*, 1987; Nakashima *et al.*, 1986). The viral progeny of infected cells that are chronically producing virus *in vitro*, even in the presence of high but non-toxic levels of AZT, remain sensitive to AZT (Smith, Brian & Pagano, 1987).

In preclinical toxicologic studies, the median lethal dose of AZT in mice and rats is greater than 750 mg/kg intravenously and greater than 3000 mg/kg orally (Ayers, 1988). The intravenous administration of up to 150 mg/kg per day of AZT for four weeks in rats and of up to 85 mg/kg per day for two weeks in dogs resulted in no recognised toxicity (Ayers, 1988). Oral administration of 500 mg/kg AZT to beagle dogs produces vomiting and bloody stools. Oral administration of 30 to 100 mg/kg per day of AZT to mice, rats or monkeys will produce a macrocytic anemia over a period of a month or more (Ayers, 1988; Ruprecht *et al.*, 1986). Mutagenicity can be detected at 1000 to 5000 μg/ml in the L5178Y mouse lymphoma cell mutagenicity assay (Ayers, 1988). Chromosomal aberrations in human lymphocytes cultured in 3 μg/ml AZT and transformation of T3 mouse cells in 0.5 μg/ml occur. No genetic toxicology has been documented in the bacterial mutagenicity assay (Ames test), the marrow cytogenetics of rats assay or in teratogenicity studies in rats and rabbits (Ayers, 1988).

The expeditious introduction of AZT into clinical trials and the documentation of efficacy preceded the evaluation of AZT in animal models of retrovirus infection. As a consequence, the clinical experience with AZT is being used as the standard by which animal models are being validated rather than vice versa. A number of animal retroviral models for evaluation of antivirals have been developed and additional models are certain to be developed. AZT administered intraperitoneally or in drinking water from the time of inoculation is effective in treating mice infected with the Rauscher murine leukaemia virus complex as measured by spleen weight, infectious spleen cells, infectivity titres in plasma, or mortality (Ruprecht *et al.*, 1986). The injection of a murine type C retrovirus (designated Cas-Br-E) into embryos produces a progressive neurologic disease in the newborns which is inhibited by administering AZT in the drinking water of the pregnant mothers (Sharpe, Jaenisch & Ruprecht, 1987). This transplacental delivery of drug is encouraging for the potential in treating congenital HIV infection, as well as the treatment of central nervous system disease produced by retroviruses. AZT administered at the time of inoculation can prevent the establishment of infections with feline leukaemia viruses (Tavares *et al.*, 1987), supporting the possible application to post-exposure prophylaxis of health care workers.

2′,3′-dideoxynucleosides Shortly after the documentation of the antiretroviral activity of AZT, Mitsuya and Broder reported that the 2′,3′-dideoxynucleosides, dideoxycytidine (ddC), dideoxyguanosine (ddG), dideoxyadenosine (ddA) and its deaminated metabolite dideoxyinosine (ddI) all inhibited HIV-1 *in vitro* (Mitsuya & Broder, 1986). Of interest, dideoxythymidine (ddT) which is AZT without the 3′-azido group was relatively impotent (Mitsuya & Broder, 1986; Furmanski *et al.*, 1980). These 2′,3′-dideoxynucleosides structurally are the physiologic deoxynucleosides with the 3′-hydroxyl group removed from the sugar moiety. Both ddA and ddI are approximately 10 to 100 fold less potent than AZT but exhibit little toxicity *in vitro* even at concentrations approaching 1000 μg/ml. ddC was selected for the first clinical development for several reasons. It is more potent than AZT *in vitro* with activity demonstrable at 0.01 μM or lower in some assay systems. It also is resistant to deamination (Starnes & Cheng, 1987; Cooney *et al.*, 1986) and phosphorolytic cleavage of base from sugar. It was well absorbed from the GI tract, had uncomplicated pharmacokinetics in pilot animal studies, and appeared non-toxic in animals (Mitsuya & Broder, 1987; Mitsuya *et al.*, 1987).

The cellular metabolism of ddC appears even less complicated than AZT. It enters the cell by the nucleoside transport system and is anabolised to the triphosphate by deoxycytidine kinase and subsequent cellular enzymes in the phosphorylation of deoxycytidine (Cooney *et al.*, 1986; Eriksson *et al.*, 1987; Balzarini *et al.*, 1987; Ullman *et al.*, 1988). A higher proportion of intracellular nucleotide is ddC-TP probably accounting for some of the greater potency when compared to AZT (Starnes & Cheng, 1987; Hao *et al.*, 1988). In contrast to AZT, ddC is poorly phosphorylated by murine cells (Balzarini *et al.*, 1988*c*). The mouse may thus be a poor model to evaluate ddC. As with AZT, the physiological nucleoside (deoxycytidine in the case of ddC) competes with both the efficacy and toxicity of ddC *in vitro* (Balzarini *et al.*, 1987*a*). The addition of thymidine to cells significantly enhances the phosphorylation of ddC, presumably by reducing the intracellular pools of cytidine deoxynucleotides (Balzarini *et al.*, 1987*a*). These observations point out the complex interactions of intracellular nucleotide pools and the potential for synergy or antagonism between combinations of nucleoside analogues which is well documented (Baba *et al.*, 1987*b*; Vogt *et al.*, 1987).

The triphosphate of ddC inhibits reverse transcriptase to confer its antiviral activity (Chen & Oshana, 1987). Chain termination does occur *in vitro* (Mitsuya *et al.*, 1987). The selectivity of ddC is based on the lower susceptibility of human DNA polymerases α and β to ddC-TP (Waqar *et al.*, 1984; Starnes & Cheng, 1987). Of note, mitochondrial DNA polymerase γ was as sensitive as reverse transcriptase (Starnes & Cheng, 1987) perhaps accounting for the neurotoxicity seen with the drug in one study (Frankel, Bredt & Pabo, 1988) but not in another (Simpson *et al.*, 1989).

The generation of active triphosphates from ddA and ddI are much more complicated than for AZT or ddC. Adenosine deaminase which is abundant in serum and cells rapidly deaminates ddA to ddI (Montefiori & Mitchell, 1986; Cooney *et al.*, 1987). Purine nucleoside phosphorylase converts ddI to the monophosphate (Cooney *et al.*, 1987). If deamination of ddA is prevented, ddA can be phosphorylated to ddA-monophosphate predominantly by deoxycytidine kinase as well as by adenosine kinase (Carson *et al.*, 1988; Cooney *et al.*, 1987). ddAMP is then readily deaminated to ddIMP (Carson *et al.*, 1988). The active triphosphate is then generated by a series of enzyme steps that convert ddIMP to ddATP via ddAMP and ddADP (Ahluwalia *et al.*, 1987). These pathways may thus be subject to variability among different cells (Carson *et al.*, 1988) and to complex interactions with other nucleosides. As with AZT-TP and ddC-TP, ddA-TP then inhibits reverse transcriptase of HIV (Chen & Oshana, 1987; Mitsuya *et al.*, 1987), other lentiviruses (Dahlberg *et al.*, 1987) and host cell DNA polymerases (Montefiori & Mitchell, 1986; Waqar *et al.*, 1984).

No toxicity of ddC is recognised in rodents with any single dose that can be administered or with oral drug at 1000 mg/kg three times daily for five days or with peritoneal infusions of 47 mg/kg daily for 1 week. The only toxicity seen was some decrease in bone colony forming units when cultured *in vitro* from mice receiving the infusion. In dogs receiving constant intravenous infusions for five days, soft, blood-tinged stools became apparent when the dose attained was 3.53 mg/kg/hour. In dogs, oral or parenteral doses approaching 100 mg/kg daily for 4 weeks resulted in similar gastrointestinal toxicity and moderate thrombocytopenia and anaemia. It must be remembered that nucleoside anabolism varies significantly among species with important implications of animal models both for efficacy and toxicity. Following the first clinical trials of ddC, high dose oral administration of drug to *Cynomolgus* monkeys resulted in peripheral neuropathy (W. Soo, personal communication).

The pharmacokinetics of ddC in mouse, dog and rhesus monkey indicate some species variation (Kelley *et al.*, 1987). Oral bioavailability is 100% in monkeys and dogs, but 30–50% in mice. The elimination half-life is 2 hours in monkey and dog but 1 hour in mouse. The total body clearance is 9.5 to 12 ml/min/kg in the monkey, 5 to 6 ml/min/kg in the dog and 30 to 50 ml/min/kg in the mouse. Penetration of ddC into the CSF is 2–6% of the plasma level in dogs and monkeys.

Continuous intravenous infusions of ddA in dogs at 31 mg/kg/hour resulted in the steady state plasma levels of 1.5 μg/ml (6 μM) ddA and 25 μg/ml (95 μM) ddI. No toxicity was observed for ten days of infusion at these levels which exceed the antiviral levels by at least an order of magnitude. Higher infusion levels produce hypotension and tachycardia.

ddA is deaminated completely to ddI in minutes in the dog and rat. In the dog the elimination half-life of ddI is 73 minutes and the total body clearance is 4 to 5 ml/min/kg. Orally administered ddA produces gastric irritation and undergoes hydrolysis to base and sugar unless it is administered with a buffer. Reliable oral bioavailability data will thus depend upon formulation; however, 30–50% of administered ddA or ddI appear to reach the circulation as ddI in various buffers. In rats and dogs, high oral doses of ddA yield a nephrotoxic catabolite generated in the gut. Because ddI and ddA appear to be interconverted by cellular metabolism and yield equivalent levels of the active ddA-triphosphate, ddI has been selected between the two for clinical evaluation.

Other nucleosides Dozens of nucleoside analogues have been shown to inhibit the replication of HIV selectively or as the triphosphate to inhibit reverse transcriptase activity *in vitro*. Many of these will be entering clinical trials. Because assay systems for antiviral activity and cell toxicity vary with each investigator, strict comparisons are difficult. Compounds with selective activity against HIV include the unsaturated 2′,3′-didehydro-derivatives of ddC and ddT termed 2′,3′-didehydrodideoxycytosine (ddeC or d4C), and 2′,3′-didehydrodideoxythymidine (ddeT or d4T) (Lin *et al.*, 1987*b*; Matthes *et al.*, 1987; Balzarini *et al.*, 1986*c*; Lin, Schinazi & Prusoff, 1987*c*; Hamamoto *et al.*, 1987; Chu *et al.*, 1988; Mansuri *et al.*, 1989; Hartmann *et al.*, 1988; Ho & Hitchcock, 1989; Balzarini, Herdewijn & De Clercq, 1989). The 2′,3′-didehydropurines are relatively ineffective (Baba *et al.*, 1987*a, c*; Mitsuya, Matsukura & Broder, 1987; Balzarini *et al.*, 1987*c*); however, the carbocyclic equivalent of didehydroguanosine (Carbovir) has a promising selectivity index against HIV-1 *in vitro* and preliminary pharmacokinetics have been reported for the rat (Yeom *et al.*, 1989). The 3′-azido derivatives of deoxyguanosine (AZG) and several pyrimidines, such as azidodeoxyuridine, are promising *in vitro* (Baba *et al.*, 1987*a*; Lin *et al.*, 1987*a*; Lin *et al.*, 1988; Balzarini *et al.*, 1988*a*), as are the 3′fluoro derivatives of ddT and ddG but not ddA or ddC (Matthes *et al.*, 1987; Herdewijn *et al.*, 1987, 1988; Matthew *et al.*, 1988; Balzarini *et al.*, 1988*a, b*). 2,6-diaminopurine, 2′,3′-deoxyriboside and its 3′fluoro and 3′azido derivatives are effective *in vitro* (Balzarini *et al.*, 1987*b*, 1988*b*). Another effective derivative of this purine with an additional 2-amino group is an acyclic derivative which has a phosphonylmethoxyethyl side chain instead of a sugar (PMEDADP) (Pauwels *et al.*, 1988). Another series of compounds with an acyclic component instead of sugar are 9-(4,-hydroxy-1′,2′-butadienyl)adenine (termed adenallene) and its cytosine equivalent (termed cytallene) (Hayashi *et al.*, 1988). 9-(2-phosphonylmethoxyethyl) adenine (PMEA) is the first of these acyclic nucleosides to be considered for clinical development (Balzarini *et al.*, 1989*c*). Analogous to acyclovir and its relatives, these unsaturated acyclic compounds may structurally resemble a cyclic nucleoside and undergo phosphorylation at a site resembling the 5′carbon. This speculation is under investigation. The synthesis of heterodimers of several of the nucleosides as well as of nucleosides with phosphonoformate have been described (Griengl *et al.*, 1988; Lambert *et al.*, 1989; Busso *et al.*, 1988).

Factors such as preclinical animal toxicology and constraints on synthesis of large quantities of compound may dictate which compounds progress to clinical trial. Additional studies, including structural studies of the triphosphates and reverse transcriptase, may help elucidate the critical structure–activity relationships that will help us to understand how certain analogues work and permit the design of additional, more effective and less toxic compounds.

Properties in addition to antiviral activity may be incorporated into drug design. For example, one possible limitation of ddA and ddI are their susceptibility to gastric acid-catalysed hydrolysis of the glycosidic bond. The derivative 2′-fluoro-ddA is more resistant to this acid catalysed hydrolysis; however, most antiretroviral activity is lost (Marquez *et al.*, 1987). Its stereoisomer 2′-fluoro-ara-ddA is resistant to hydrolysis and retains antiretroviral activity comparable to ddA (Marquez *et al.*, 1987).

Foscarnet

Foscarnet (phosphonoformate, PFA) is a pyrophosphate analogue that inhibits directly a broad range of DNA polymerases and to a certain extent influenza virus RNA polymerase (Oberg, 1983). The compound has been most extensively examined for use for herpesvirus infections. Foscarnet was shown to inhibit the reverse transcriptase activity of all animal retroviruses examined with cell-free assays (Sundquist & Oberg, 1979). One of the first retroviruses shown to be inhibited in culture by foscarnet was the sheep lentivirus, visna (Sundquist & Larner, 1979).

The drug inhibits the replication of HIV-1 in several assay systems at concentrations of 100 μM or less, which are readily attainable clinically (Sandstrom *et al.*, 1985; Sarin *et al.*, 1985). The drug is relatively non-toxic *in vitro* at concentrations approaching 1000 μM (Sandstrom *et al.*, 1985; Sarin *et al.*, 1985*b*). Foscarnet inhibits reverse transcriptase at concentrations below 1 μM by a non-competitive mechanism with respect to deoxynucleotide triphosphate substrates (Sandstrom *et al.*, 1985; Sarin *et al.*, 1985*b*; Vrang & Oberg, 1986; Wondrak, Lower & Kurth, 1988). The discrepancy between concentrations necessary to inhibit enzyme in cell-free assays and virus replication in cells is accounted for by the poor entry into cells of this highly charged molecule. In cell-free systems, cellular DNA polymerases α and β are orders of magnitude less sensitive to inhibition by foscarnet than is the reverse transcriptase of HIV-1, supporting the possibility of a good therapeutic ratio (Lin *et al.*, 1988). Several dozen pyrophosphate analogues have been synthesised and examined and none has yet been found more promising *in vitro* than foscarnet (Vrang & Oberg, 1986). With the encouraging results *in vitro* and clinical experience with foscarnet for the treatment of herpesvirus infections, foscarnet was an early candidate to enter clinical trials for infection with HIV.

HPA-23

HPA-23 (heteropolyanion-23) is a complex compound with a molecular weight of 6800, containing a core of tungsten and antimony, that was shown to inhibit the reverse transcriptase of murine oncoviruses and to protect mice from mammary tumors (Chermann, Sinoussi & Jasmin, 1975; Lidereau *et al.*, 1977). HPA-23 competitively inhibits the reverse transcriptase of

HIV-1 at concentrations down to 0.1 μM (Wondrak, Lower & Kurth, 1988; Dormont *et al.*, 1985). Less encouraging is the lack of selectivity of viral over cellular DNA polymerases (Wondrak, Lower & Kurth, 1988). Little data regarding toxicology or pharmacokinetics appear to be available; nevertheless, clinical trials were initiated in 1983 (Rozenbaum *et al.*, 1985).

Rifamycins

In the 1970s, rifamycin compounds were demonstrated to inhibit the reverse transcriptase of animal retroviruses in cell free systems; however, this occurred at concentrations exceeding those attainable clinically (Gurgo, Ray & Green, 1972; Wu & Gallo, 1974). Rifabutin, also known as ansamycin, was shown to inhibit the production of HIV-1 reverse transcriptase *in vitro* at concentrations between 1 and 10 μg/ml in one study (Anand *et al.*, 1986) but to be inactive at these concentrations in another study (Birch *et al.*, 1988). Although these concentrations exceed the peak concentrations readily attainable in serum, this observation and the experience with this drug in patients with AIDS for the therapy of mycobacterial infection prompted its evaluation as an antiretroviral. No selectivity of viral over cellular DNA polymerases was observed in cell-free assays.

Oligonucleotides

Several oligonucleotides of various composition have been synthesised as potential antisense inhibitors of transcription (Zamecnik & Stephenson, 1978; Agris *et al.*, 1986; Zamecnik *et al.*, 1986; Miller & Ts'o, 1987; Lemaitre, Bayard & Lebleu, 1987; Matsukura *et al.*, 1987; Goodchild *et al.*, 1988; Zaia *et al.*, 1988; Agrawal *et al.*, 1988; Sarin *et al.*, 1988; Shibahara *et al.*, 1989). In addition, 2′,5′-oligoadenylates, which are induced by interferon to inhibit cap methylation of viral mRNA, have been shown to inhibit reverse transcriptase in cell free systems (Bertonis *et al.*, 1988). The activity of most of these oligonucleotides does not appear to be related to the specific sequence. Data are suggestive but not conclusive that these compounds compete with the virion RNA for the template-binding site of the reverse transcriptase (Liu & Owens, 1987). Because their size and charge impair entry into cells and because the synthesis of these compounds is difficult they seem unlikely candidates for clinical use. Nevertheless, template-primer binding remains a theoretical target for therapy acting at reverse transcription. Oligonucleotides as inhibitors of transcription are reviewed below.

Other reverse transcriptase inhibitors

A number of compounds, including suramin, dextran sulfate and sakyomicin A have been reported to inhibit reverse transcriptase in cell-free systems and to inhibit the replication of HIV (Tanaka *et al.*, 1986; Jentsch *et al.*, 1987; De Clercq, 1979; Mitsuya *et al.*, 1988). The concentrations required for enzyme inhibition exceed those for antiviral activity and are similar to concentrations that inhibit host cell polymerases. It is also unlikely that these compounds readily cross cell membranes so that reverse transcriptase is unlikely to be their target.

Compounds that inhibit transcription and translation

Viruses parasitise host cell machinery to transcribe their messenger RNA and to translate these messages into polypeptides. Consequently, to target virus-specific functions at these steps in replication is challenging. At least three approaches have been described: ribavirin, synthetic 'antisense' oligonucleotides and interferons. Interferons may act at several sites and their clinical use will be considered with the biological response modifiers.

Ribavirin

Ribavirin(1-β-D-ribofuranosyl-1,2,4-triazole-3-carboxamide, Virazole) is an analogue of guanosine with a broad spectrum of antiviral activity. It may act by several possible mechanisms including interference with the guanylation step required for 5′-capping of viral messenger RNA and inhibition of viral polymerase by ribavirin triphosphate (Smith & Kirkpatrick, 1980; Goswami *et al.*, 1979).

Ribavirin was one of the first drugs shown to inhibit the replication of HIV in cell culture (McCormick *et al.*, 1984). Concentrations of 50 μM or more, which exceed those attainable clinically, are necessary *in vitro* to inhibit HIV. Nevertheless, the documentation of activity against HIV *in vitro* and the successful experience with the treatment of Lassa fever (McCormick *et al.*, 1986) prompted clinical evaluation.

Ribavirin has several effects on host cell metabolism including the perturbation of intracellular nucleotide pools, at least partially by the inhibition of inosine monophosphate dehydrogenase by ribavirin monophosphate (Streeter *et al.*, 1973). As a consequence, the purine deoxynucleotide triphosphates (dGTP and dATP) diminish intracellularly. In addition, ribavirin results in increased intracellular concentrations of thymidine triphosphate which competes with AZT-TP and which inhibits phosphorylation of AZT (Vogt *et al.*, 1987; Fyfe *et al.*, 1988). These perturbations in intracellular nucleotides result in a marked antagonism of the antiretroviral activity of AZT and other dideoxypyrimidines by ribavirin and enhancement of activity of dideoxypurines like ddA and ddI (Vogt *et al.*, 1987; Baba *et al.*, 1987*b*; Balzarini, Herdewijn & De Clercq, 1989).

Synthetic oligonucleotides

In addition to targeting antivirals to virus specific proteins, synthetic oligonucleotides of 14–30 base pairs in length have been investigated as 'antisense' oligonucleotides with the potential to hybridise with sequences of

HIV messenger RNA, thereby interfering with RNA processing. Methylphosphonate, phosphorothioate and other backbone derivatives have been generated to minimise nuclease sensitivity (Zamecnik & Stephenson, 1978; Lemaitre, Bayard & Lebleu, 1987; Agris *et al.*, 1986; Zamecnik *et al.*, 1986; Goodchild *et al.*, 1988; Matsukura *et al.*, 1987; Zaia *et al.*, 1988; Sarin *et al.*, 1988; Shibahara *et al.*, 1989). Antiviral activity has been documented with such compounds; however, these activities are not always sequence specific, suggesting a number of possible mechanisms including template primer-binding to interfere with reverse transcriptase. The sequence non-specific activity is seen only with the inhibition of acute infection *in vitro* (S. Broder, personal communication). Sequence specific activity has been shown at 1 μM with a 28 base phosphorothioate antisense analogue against *rev* (Matsukura *et al.*, 1987) and at 100 μM with the 8 base methyl phosphonate antisense sequence of the *tat* first splice acceptor site (Zaia *et al.*, 1988). This sequence specific activity can be observed both in acutely and chronically infected cells with the 28 base antisense *rev* compound. This theoretically interesting approach still requires much investigation to determine effective sequences, backbone composition and drug delivery methods, as well as feasibility of economic scale-up of production. This approach will continue to be an active area of investigation (Stein & Cohen, 1988).

A variant of this approach would be the delivery of sequence specific RNA enzymes with endoribonuclease activities (Haseloff & Gerlach, 1988). These 'ribozymes' have been identified for plant viroids and could theoretically be applied to any specific transcripts.

Compounds that inhibit post-translational modification

Retroviral polypeptides are translated as polyproteins that are cleaved post-translationally by a viral protease. Many proteins also undergo glycosylation, phosphorylation or myristoylation by host enzymes. These types of post-translational modifications have been approached as targets of antiviral therapy.

Protease inhibitors

The protease of HIV-1 and other retroviruses cleaves viral polypeptides at well defined sites (Nutt *et al.*, 1988; Hansen *et al.*, 1988; Kohl *et al.*, 1988; Kotler *et al.*, 1988; Meek *et al.*, 1989; Darke *et al.*, 1989). Synthetic peptides of as few as six amino acids including pepstatin A can completely inhibit virus specific proteolysis (Lin *et al.*, 1988; Katoh *et al.*, 1987; Schneider & Kent, 1988). Peptides do not permeate cell membranes; however, the availability of cell-free assays of the proteolytic activity of cloned and expressed viral protease should permit the screening of compounds that act on this target and the analysis of structure-activity relationships to design compounds with therapeutic potential.

Glycosylation inhibitors

The inhibition of glycosylation of the envelope glycoproteins could be expected to affect viral budding, release, binding or penetration by fusion, all functions presumably mediated by the *env* products. Glycosylation inhibitors, typically 2-deoxy-D-glucose, have been examined for years with this target in mind (Blough & Ray, 1988). 2-deoxy-D-glucose acts as a substrate for glycosylating enzymes and exerts some inhibition of the expression of HIV glycoproteins *in vitro* (Blough *et al.*, 1986). Castanospermine is an alkaloid isolated from the seeds of an Australian chestnut tree, *Castanospermum australe*, that inhibits α-1,6-glucosidase (Walker *et al.*, 1987). It is relatively toxic in mice (Ruprecht *et al.*, 1989). 1-deoxynojirimycin and N-butyldeoxynojirimycin inhibit this same enzyme (Bollen & Stalmans, 1989) and 1-deoxymannojirimycin inhibits mannosidase 1. Each of these enzymes trim N-linked glycan structures (Gruters *et al.*, 1987; Karpas *et al.*, 1988*a*). These three compounds inhibit syncytium formation by HIV at concentrations above 50 μg/ml for castanospermine and at 1 to 10 μg/ml for the others. At these concentrations glycosylation of envelope products is inhibited to impair their function (Walker *et al.*, 1987; Gruters *et al.*, 1987; Karpas *et al.*, 1988*a*). Whether compounds which inhibit host cell enzymes can effectively inhibit retroviral replication with acceptable toxicity *in vivo* remains to be determined. N-butyl-deoxynojirimycin entered phase 1 clinical trials in early 1989.

Compounds that interact with the virion structure

The virus envelope is derived from the host cell membrane during the budding process. One would thus expect little antigenically or functionally in this phospholipid bilayer to represent a selective target. Nevertheless, several candidates have been proposed as potential antiretroviral agents that presumably would act upon the virus during budding or after release from the host cell.

AL721 is a mixture of neutral glycerides, phosphatidylcholine and phosphatidylethanolamine in a 7:2:1 ratio that can extract cholesterol from cellular membranes (Lyte & Shinitzky, 1985). The published data regarding AL721 as an antiviral compound have been limited to a letter to the editor (Sarin *et al.*, 1985*a*). A concentration of 100 μg/ml reduced the production of viral products by 50% in cultured, infected H9 cells and 1000 μg/ml reduced these products by >90%. These concentrations did not reduce cell numbers, but viability of cells was not assessed. A great deal of enthusiasm has been generated for the preparation of this compound by infected individuals; however, few data are available about either the effects of this compound in infected cell cultures or the pharmacology of AL721. For example, 10 to 15 grams orally are recommended daily (Sarin *et al.*, 1985*a*). It is unclear whether these lipids are absorbed from the gastrointestinal tract to remain associated and maintain their allegedly effective ratio in the circulation.

Amphotericin B and its water-soluble derivative amphotericin B methyl ester are polyene macrolide antifungal antibiotics that act on the basis of a greater affinity for ergosterol, a component of fungal membranes, than for cholesterol in the membranes of mammalian cells and their viruses. Drug concentrations can be found in cell culture that inhibit replication of HIV and sustain cell viability (Schaffner *et al.*, 1986). The amount of reduction observed is always less than 90% in contrast to nucleoside analogues, for example, where reduction can be measured in orders of magnitude.

In summary, compounds that act upon the viral envelope, including lipid solvents and detergents, may physically inactivate HIV. The critical issue for therapy is whether such an agent can be identified that selectively inactivates virus over host cells.

Hypericin and pseudohypericin are aromatic polycyclic diones extracted from the Saint Johnswort plant of the *Hypericum* family (Meruelo, Lavie & Lavie, 1988). These have antiviral activity against Friend murine leukemia virus *in vitro* and *in vivo* and have been proposed for evaluation against HIV (Meruelo, Lavie & Lavie, 1988). Preliminary data suggest that hypericin inactivates cell free virus and has a prolonged plasma half-life in animals, adding to the interest in these compounds. 'Health-food' stores sell preparations of St Johnswort for infusions which may permit endeavours of self-medication. The concentrations of hypericin and pseudohypericin are extremely low in these preparations which also contain other toxic alkaloids.

Compounds with undetermined site of action

Suramin

Suramin is a complex polyanionic dye synthesised in 1920 at Bayer and used for decades for the therapy of African trypanosomiasis and onchocerciasis (Hawking, 1978). Its mechanism of action for these infections has not been elucidated although interaction with numerous proteins and cell function have been observed (Hawking, 1978). In 1979, de Clercq demonstrated that at about 0.1μM the drug inhibited the reverse transcriptases of several murine and avian retroviruses in cell-free assays (De Clercq, 1979). This observation was confirmed for the reverse transcriptase of HIV-1 (Mitsuya *et al.*, 1984; Jentsch *et al.*, 1987). Derivatives of suramin and structurally similar dyes, like Chicago sky blue, Congo red, direct yellow

50, and Evans blue have similar effects (Balzarini *et al.*, 1986*a*, *b*). In 1984, Mitsuya and colleagues demonstrated that suramin, at concentrations of 10 to 50 μg/ml (7–35 μM), protected T cells *in vitro* from the infectivity and cytopathic effects of HIV (Mitsuya *et al.*, 1984). Cell toxicity was not apparent *in vitro* until concentrations of 100 μg/ml were attained, a relatively narrow margin of safety (Mitsuya *et al.*, 1984). Of historic interest this was the first compound reported to inhibit the replication of HIV. The related dyes were shown also to inhibit virus replication (Balzarini *et al.*, 1986*a*, *b*). Although suramin and the related dyes are potent enzyme inhibitors in cell-free assays, action at this site in infected cells has not been proven and a number of other mechanisms are possible (De Clercq, 1987). Regardless of the mechanism, these *in vitro* results led to studies showing effective antiviral effects in murine leukaemia virus models (Ruprecht *et al.*, 1985) and to clinical trials.

Other candidate compounds

Several other compounds have been reported to inhibit the replication of HIV *in vitro*, like penicillamine, avarol and avarone (Sarin *et al.*, 1987; Chandra & Sarin, 1988). Confirmation of the magnitude and selectivity of the effects are needed. Certain plant extracts have been reported to have antiretroviral activity as well. Glycyrrhizin, a structurally defined polycyclic extract of the liquorice root, *Glycyrrhiza radix*, inhibits the replication of HIV *in vitro* approaching concentrations of 1 mg/ml (Ito *et al.*, 1987*b*). Although these effective concentrations are not toxic *in vitro*, the low potency of this large molecule may not be therapeutically practical. Chemical congeners might merit investigation. Extracts of a large number of other Chinese medicinal herbs have been reported to have activity *in vitro* (Chang & Yeung, 1988). Further studies, including purification and characterisation of the active compounds, will be of interest.

Fusidic acid which has been used as an antibacterial agent in Europe and Canada for years was associated in dramatic clinical improvement in a patient with AIDS who was administered the drug for mycobacterial infection (Faber *et al.*, 1987). This prompted an *in vitro* study showing reduced viral replication in the presence of fusidic acid (Faber *et al.*, 1987), promoting much enthusiasm and further clinical trials (Barnes, 1987*b*). When *in vitro* assays of antiviral activity are conducted with controls for cell toxicity, all the antiviral effects can be attributed to cellular toxicity (Richman *et al.*, 1988*b*; Lloyd, Atkinson & Sutton, 1988; Maehlen & Degre, 1989).

Trichosanthin (GLQ223, Compound Q) is a 26 kD protein derived from the tubers of *Trichosanthes kirilowii* or 'Chinese cucumber' (McGrath *et al.*, 1989). The compound inhibits the replication of HIV in several cell lines at concentrations as low as 0.01 μg/ml (McGrath *et al.*, 1989). Phase 1 clinical studies of tricosanthin were launched in mid-1989.

CLINICAL STUDIES

Suramin

Suramin was the first drug shown to inhibit HIV *in vitro* (Mitsuya *et al.*, 1984) and was soon evaluated clinically because it had undergone extensive prior clinical use in humans (Hawking, 1978). With a newly developed high-performance liquid chromatography (HPLC) assay for suramin, the pharmacokinetics of suramin were evaluated in ten patients with Kaposi's sarcoma or AIDS-related complex (Collins *et al.*, 1986). A total intravenous dose of 6.2 grams was administered over a five-week period. The plasma half-life of the drug was about seven weeks. The plasma levels exceeded 100 μg/ml for two months. 99.7% of drug is bound to plasma proteins. Metabolites are not found in plasma (Collins *et al.*, 1986). Almost all administered drug is ultimately excreted in the urine. In four of the patients, the ability to isolate virus appeared to diminish during peak plasma levels of drug (Broder *et al.*, 1985). Serum p24 antigenemia was not assessed. No significant clinical or immunological improvement was recognised. At least half of the subjects each experienced fever, rash, proteinuria, pyuria, and subclinical hepatitis as determined by aminotransferase elevations in serum (Broder *et al.*, 1985).

Additional clinical trials were undertaken which essentially confirmed and extended these clinical observations in over 100 total patients (Levine *et al.*, 1986; Cheson *et al.*, 1987; Schattenkerk *et al.*, 1988). Suramin sodium at 0.5, 1.0 or 1.5 grams per week for 6 weeks was followed by maintenance therapy of 0.5 or 1.0 grams weekly for 3 to 6 months in most cases. As with the preliminary study, virus isolations were unsuccessful from some patients administered suramin, but no clinical or immunological improvement was appreciated. In addition to the toxicity originally reported, a significant minority of patients experienced malaise, nausea, vomiting, neurological symptoms (headache, metallic taste, parathesias and peripheral neuropathy), neutropenia and adrenal insufficiency. A more recent study of ten patients with AIDS and p24 antigenemia, half of whom were administered suramin could demonstrate no reduction in antigenemia (Schattenkerk *et al.*, 1988). With the failure to observe benefit and the frequent occurrence of serious toxicity, suramin is no longer under active clinical investigation. Further clinical trials must probably await identification of compounds of this class of polyanionic dyes with a higher selectivity index.

AZT

AZT was the first drug demonstrated to be clinically effective and licensed for use for the treatment of HIV infection. It was first administered to patients in 1985 within only 6 months of the demonstration of its efficacy *in vitro* (Mitsuya *et al.*, 1985). In a phase 1 dose escalation study in 19 adults with severe HIV infection (Yarchoan *et al.*, 1986), AZT was shown to be rapidly absorbed from the gastrointestinal tract with mean peak serum levels attained at about 1 hour. Dose independent kinetics occurred at doses up to 10 mg/kg every 4 hours. Approximately 60% of ingested drug is bioavailable because of first passage glucuronidation by the liver (Klecker *et al.*, 1987). This glucuronidated metabolite has no apparent toxicity or antiviral activity and like AZT is excreted almost exclusively by the kidneys with a mean half-life of 1 hour. The mean peak serum concentration in an adult administered 200 mg is approximately 2 to 3 μM. Renal clearance of AZT is approximately 350 ml/minute/70 kg and total body clearance is approximately 1900 ml/minute/70 kg at doses up to 5 mg/kg (Blum *et al.*, 1988). In patients with impaired renal function, the half-life of AZT is only slightly prolonged because of rapid hepatic glucuronidation. The glucuronide of AZT (GAZT) does accumulate in patients with reduced creatinine clearance (Pioger *et al.*, 1988). The toxicity of chronically elevated levels of GAZT is unknown. AZT penetrates into the cerebrospinal fluid (CSF) well, with CSF:plasma ratios of 50–100% at least 4 hours after the first dose (Klecker *et al.*, 1987). Semen concentrations of AZT ranged from 1.3 to 20 fold higher than serum when studied in six patients suggesting pH-dependent trapping of AZT which is a weak base ($pK_a = 9.68$) (Henry *et al.*, 1988).

In addition to pharmacokinetic data, the phase 1 study generated anecdotal evidence of some toxicity during the 6-week regimen (headaches and leukopenia) which was manageable. Therapeutic benefits were also suggested in several patients including weight gain, improved sense of well-being, improved CD4 lymphocyte counts, and return of delayed type hypersensitivity skin test reactions. These observations prompted the design of a multicentre, randomised, placebo-controlled, double blind study (Fischl *et al.*, 1987; Richman *et al.*, 1987*a*).

Two hundred and eighty-two patients with either a first episode of *Pneumocystis carinii* pneumonia diagnosed within 120 days or with advanced AIDS-related complex (ARC) were stratified according to CD4 lymphocyte counts and then randomly assigned to receive either placebo or 250 mg of AZT by mouth every 4 hours for 24 weeks. One hundred and forty-five subjects receiving AZT and 137 receiving placebo entered the study between February and June of 1986. The study was terminated on September 18, 1986 by an independent monitoring board because 19 placebo recipients and 1 AZT recipient had died ($p < 0.001$). At that time 27 subjects had completed 24 weeks of study, 152 had completed 16 weeks and the remainder had completed at least 8 weeks. In addition to prolonging survival, AZT therapy reduced the frequency and severity of opportunistic infections, improved body weight, prevented deterioration of Karnofsky performance score, increased peripheral CD4 lymphocyte counts and reversed skin test anergy in many patients. Of note, the rates of opportunistic infections in the AZT and placebo groups did not begin to diverge until after 6 to 8 weeks of therapy. The drug did not reduce the isolation rates of HIV from peripheral blood mononuclear cells; however, viral p24 antigenemia as measured by ELISA was significantly reduced by AZT therapy (Fischl *et al.*, 1987).

This reduction of p24 antigenemia with AZT administration and its rise following withdrawal of drug has been observed by several investigators (Pizzo *et al.*, 1988; Jackson *et al.*, 1988*a*; Spector *et al.*, 1989; De Wolf *et al.*, 1988; Chaisson *et al.*, 1988; Spear *et al.*, 1988). In some patients the level of p24 antigenemia may increase although the patient continues to ingest drug. The mechanism by which this occurs is unknown. No systematic studies correlating dosing regimens and clearance of p24 antigenemia have yet been published. AZT also consistently has cleared p24 antigen from the cerebrospinal fluid from a small number of patients (Kennedy *et al.*, 1987; De Wolf *et al.*, 1988; Spector *et al.*, 1989; De Gans *et al.*, 1988) although the drug has no benefit on virus isolation (Kennedy *et al.*, 1987). AZT also reduces the elevated levels of serum β_2 -microglobulin that occurs in patients with AIDS or ARC (Jacobson *et al.*, 1989).

The administration of AZT has been associated with the anecdotal reports of significant neurological improvement (Fiala *et al.*, 1988; Yarchoan *et al.*, 1987; Pizzo *et al.*, 1988). These observations have been supported by the analysis of the neuropsychological assessments of the subjects participating in the placebo controlled study. AZT partially reversed and delayed further deterioration of cognitive dysfunction, utilising tests of memory and attention for example (Schmitt *et al.*, 1988). No improvement was appreciated in measures of affective symptoms.

Thrombocytopenia may be another specific complication of HIV infection to respond to the administration of AZT. Although AZT is toxic to cells of the erythroid and granulocytic series slight increases were observed in mean platelet counts in the phase II study (Richman *et al.*, 1987). Several case reports have appeared describing significant improvements of thrombocytopenia associated with the administration of AZT (Hymes, Greene & Karpatkin, 1988; Gottlieb, Wolfe & Chafey, 1987; Oksenhendler *et al.*, 1989). A prospective, placebo-controlled, blinded crossover study of ten patients with severe thrombocytopenia documented an impressive impact of AZT on this condition (Swiss Group for Clinical Studies on AIDS, 1988). Neither the mechanism of the thrombocytopenia nor how AZT reverses it have been elucidated.

Although significant clinical benefit has been documented with AZT, serious adverse clinical reactions occur, particularly bone marrow suppression (Richman *et al.*, 1987*a*). In the phase 2, placebo-controlled trial, nausea, myalgia, insomnia, and severe headaches were reported more frequently by recipients of AZT; macrocytosis developed within weeks in most of the AZT group. Anaemia with haemoglobin levels below 7.5 grams per decilitre developed in 24% of AZT recipients and 4% of placebo recipients ($p<0.001$). These patients have an increased requirement for transfusions with packed red blood cells. Macrocytosis developed within weeks in most of the AZT group. Of note, not all patients with macrocytosis developed anaemia and not all episodes of anaemia were macrocytic. Erythropoietin levels are increased in the presence of red cell hypoplasia (Walker *et al.*, 1988). Neutropenia (<500 cells per cubic millimeter) occurred in 15% of AZT recipients, as compared with 2% of placebo recipients ($p<0.001$). Subjects who entered the study with low CD4 lymphocyte counts, low serum vitamin B12 levels, anaemia, or low neutrophil counts were more likely to have haematological side-effects. Prolonged marrow failure has been associated with AZT administration (Streeter *et al.*, 1973). Although marrow toxicity with AZT is more likely the more severe the underlying disease, no absolute predictors of susceptibility to toxic side-effects are available (Richman *et al.*, 1987*a*). Moreover the underlying mechanism of marrow insufficiency in HIV infection remains obscure.

An apparent increased association of haematologic toxicity with acetominophen (paracetamol) administration (Richman *et al.*, 1987*a*) may not hold up with further scrutiny. No other association of drug interactions was appreciated in the phase 2 study (Richman *et al.*, 1987*a*). Prolongation of the serum half-life of AZT due to administration of probenecid (de Miranda *et al.*, 1986), quinine, and sulphamethoxazole have been documented. Ganciclovir (DHPG) a drug often used for CMV infection in patients with AIDS, has significant additive toxicity with AZT on granulocytes to the extent that the two drugs cannot be safely co-administered at full or even half doses (Hochster *et al.*, 1988*a*). This toxicity cannot be accounted for by pharmacokinetic interaction; neither drug affects the excretion of the other (Hochster *et al.*, 1988*b*).

Anecdotal reports of infrequent toxicities, for example, seizures, have also been published but are always difficult to interpret. Seizures occur in patients with AIDS, even in the absence of neuroradiological abnormalities that are suggestive of cerebral toxoplasmosis or malignancy. The question is whether AZT adds an additional, but small, risk for seizures. Although none of the original 282 study patients, 127 of whom received drug for more than one year, were reported with seizure, several case reports of seizures temporally associated with administration of AZT were reported once thousands of patients were receiving drug (Hagler & Frame, 1986; Harris & Caceres, 1988). Nail pigmentation, especially in blacks, occurs occasionally (Furth & Kazakis, 1987). Other anecdotal reports such as post AZT withdrawal meningitis, have not yet been appreciated by others (D. Barry, personal communication). A myositis-like syndrome, characterised by paraesthesias, myalgias, muscle oedema, muscle wasting and elevations of serum lactate dehydrogenase and creatinine phosphokinase levels, has also been associated with prolonged AZT administration (Bessen *et al.*, 1988; Gertner *et al.*, 1989). Myopathy, however, also appears to be a complication of HIV infection independent of AZT therapy (Gonzales *et al.*, 1988; Simpson & Bender, 1988; Bailey *et al.*, 1987; Stern, Gold & DiCarlo, 1987; Lange *et al.*, 1988; Nordstrom *et al.*, 1989). A case of nemaline rod myopathy in a patient with HIV infection has also been reported (Dalakas, Pezeshkpour & Flaherty, 1987). The causal role of AZT in myopathy is thus difficult to assess. The ongoing large placebo controlled studies should provide useful information regarding this association.

AZT has now been administered to over 20000 subjects, with several receiving drug for over three years. Some conclusions can be gleaned from this experience. Survival rates of patients treated with AZT were higher than those that might have been expected from previous experience with similar patients (Richman, Andrews & AZT Collaborative Working Group, 1988). Patients continued to experience episodes of opportunistic infections; however, these infections were either of decreased severity or were more responsive to conventional therapy. Haematologic toxicities continued to be the major laboratory abnormality associated with drug administration; however, new or more frequent toxicity was not observed with more prolonged therapy (Richman, Andrews & AZT Collaborative Working Group, 1988). Progressive bone marrow suppression did not appear to be associated with prolonged administration.

AZT was made available to 4805 patients with a previous diagnosis of *Pneumocystis carinii* pneumonia, through a compassionate plea program (Treatment IND) following the termination of the placebo-controlled trial in late 1986. The analysis of this large experience with this pre-licensure distribution programme confirmed the positive association of survival and pretherapy clinical status as defined by haemoglobin level, functional status as measured by Karnofsky score, or stage of disease as measured by time since first diagnosis of *Pneumocystis* pneumonia (Creagh-Kirk *et al.*, 1988). A similar experience was reported in an open study of 365 French patients (Dournon *et al.*, 1988). These data indicate that delay of AZT administration in patients with AIDS is not in the best interests of the patient and that efforts are needed to document how much benefit derives with administration even earlier in the course of HIV infection. Encouraging experience has also been obtained with AZT by continuous intravenous infusion in children with symptomatic HIV infection (Pizzo *et al.*, 1988).

The recent reports of preliminary results from two large multicentre studies of the National Institutes of Health AIDS Clinical Trials Program will have important implications. Study 002 has shown that patients status-post *Pneumocystis carinii* pneumonia fared at least as well with less toxicity when administered a regimen of 100 mg every four hours after a first month of the original phase II regimen than did patients who were administered a regimen of 250 mg every four hours. Study 016 demonstrated that patients with 200 to 500 CD4 lymphocytes and a symptom of AIDS-related complex who were administered AZT were less likely to experience disease progression and more likely to elevate their CD4 counts than placebo recipients. Study 019 demonstrated that asymptomatic patients with less than 500 CD4 cells fared better with AZT than placebo. Results were equivalent for subjects receiving either 300 mg or 100 mg AZT every 4 hours. Remarkably little toxicity was observed in either study 016 or 019. In neither study was a difference observed between the AZT and placebo groups in patients with more than 500 CD4 lymphocytes. The results of these trials will undergo more complete and careful analysis before additional conclusions can be drawn.

The documentation of the emergence of AZT resistance in patients with AIDS or ARC raises new considerations in the use of the drug and for the design of new drug regimens (Larder, Darby & Richman, 1989). Little variation in AZT sensitivity was seen in isolates from patients never treated with AZT. No reduction in sensitivity was documented during the first 6 months of treatment. After 6 months, most isolates replicate in the presence of higher concentrations of AZT. Many isolates have more than 100-fold increases in resistance and replicate at drug concentrations not attainable with normal doses of drug. Although the emergence of resistance has not yet been demonstrated to be associated with diminished drug efficacy, this is a reasonable concern. The relative likelihood of the emergence of resistance in other patient populations, such as the asymptomatic seropositive subjects, is under investigation. Mutations in the reverse transcriptase gene of the resistant strains appear to account for the new phenotype. Cross resistance to AZdU was documented but not to other nucleosides or other compounds, suggesting that the 5′-azido group is an important structure for the mutant enzyme. Drug combinations to reduce the likelihood of the emergence of resistance are thus a reasonable strategy for the future.

Dideoxycytidine (ddC)

In phase 1 studies, ddC was administered at 0.03 to 0.25 mg/kg intravenously and then orally at 4 or 8 hour intervals (Yarchoan *et al.*, 1988). The peak concentration of ddC was roughly proportional to the administered dose and a peak concentration of 0.5 μM was observed after a 1 hour infusion of 0.06 mg/kg. The plasma half-life was approximately 1–2 hours with renal clearance accounting for drug elimination. Oral bioavailability was 70–80%. Concentrations of drug in the CSF were 9–37% of the concurrent plasma levels 2–3.5 hours after an intravenous infusion (Yarchoan *et al.*, 1988). No evidence of metabolism of ddC has been noted. The reduction of doses with the appreciation of toxicity unfortunately prevented the acquisition of additional pharmacokinetic data, because plasma levels were below the threshold of detection by current assay methods. The phase 1 studies also indicated that ddC decreased circulating p24 antigenemia as does AZT; however, it produced a moderate dermatitis and mucositis in the first few weeks of therapy and more importantly a severe peripheral neuropathy in the second or third month of therapy (Yarchoan *et al.*, 1988). A larger dose ranging study documented that at 0.06 and 0.03 mg/kg orally every 4 hours, a diffuse erythematous rash and aphthous stomatitis occurred in the first weeks on drug which resolved without interruption of therapy (Merigan *et al.*, 1989). Haematologic toxicity was notably rare. Peripheral sensory neuropathy occurred in all patients at these doses, between the 4th and 14th week, and improved slowly in most patients after discontinuation of ddC with a trend towards a greater reduction in patients with ARC than with AIDS. Serum p24 antigen fell significantly (>70%) in most patients during therapy. CD4 cell numbers tended to rise transiently on study; however, no clear difference in skin test positivity or rate of HIV isolation from blood co-culture positivity was seen. At dose levels of 0.01 and 0.005 mg/kg orally every four hours, skin rash and aphthous stomatitis were mild or absent. Peripheral neuropathy, which occurred by week 12–24, in all patients receiving 0.01 mg/kg orally every 4 hours was less severe and resolved more quickly than at higher doses. Significant suppression of serum p24 antigen was seen in many patients receiving 0.01 mg/kg orally every 4 hours.

Although these initial studies indicated that ddC may have a limited role as a single drug for the treatment of HIV infection, several encouraging observations prompted the evaluation of alternating regimens of ddC and AZT (Yarchoan *et al.*, 1988). First, ddC significantly reduced p24 antigenemia, an effect seen previously only with AZT. Second, ddC resulted in little haematologic toxicity. Thus these two dideoxynucleosides with suppressive activity against p24 antigenemia have significant, but non-overlapping, toxicities. These observations form the rationale for protocols with alternating regimens of AZT and ddC with the hope of reducing toxicity and maintaining antiviral benefits.

Dideoxyinosine (ddI)

Phase 1/2 trials of ddI are in progress. Approximately one-third of an oral dose is absorbed from the gastrointestinal tract. Preliminary results suggest a reduction of p24 antigenemia and an elevation of CD4 cells counts at daily doses of 500 mg or less. Little toxicity was appreciated until daily doses of 1 gram or more were administered which were associated with cases of peripheral neuropathy, pancreatitis and seizures (R. Yarchoan & R. Dolin, personal communication). Phase 3 trials comparing ddI to AZT are scheduled to commence in late 1989.

Foscarnet

A moderate amount of pharmacokinetic data and clinical experience have been acquired with the use of foscarnet for herpesvirus infections. More recent studies of foscarnet in patients with AIDS suggest that this drug, like AZT and ddC, may have antiretroviral activity as measured by suppression of p24 antigenemia (Bergadahl *et al.*, 1988; Gaub *et al.*, 1987; Jacobson *et al.*, 1988), but not reduction of ability to isolate virus from peripheral blood cells by culture (Gaub *et al.*, 1987). Foscarnet has been administered at 120 to 180mg per day divided into two or three 1 to 2 hour infusions for up to 4 weeks or as a continuous infusion at 0.15mg/kg/min for up to 21 days. The clinical experience in patients infected with HIV suggests the drug is relatively well tolerated; however, reversible renal dysfunction, thrombophlebitis at the infusion site, headaches and anaemia are associated with administration of foscarnet (Bergadahl *et al.*, 1988; Gaub *et al.*, 1987). Further evaluation of this compound is warranted; however, it is not orally bioavailable, thus limiting its utility because of the requirement to administer the drug intravenously. Foscarnet is poorly soluble, and large doses are required because of its relatively low potency (Sjovall *et al.*, 1988). Systemic therapy thus requires large doses of intravenous solution. Severe adverse interaction with pentamidine have been observed. In addition, foscarnet stimulates the release of parathyroid hormone and the resulting hyperphosphatemia raises concerns about chronic therapy. Additional investigation is needed to define more clearly the clinical role of foscarnet.

HPA-23

Little pharmacokinetic data are available with regard to HPA-23, although open administration to some patients was begun in 1983 (Rozenbaum *et al.*, 1985). Reports of administration in 1985 to 4 patients (Rozenbaum *et al.*, 1985) and in 1986 to 16 patients (Buimovici-Klein *et al.*, 1986) indicated that thrombocytopenia and occasional elevations in serum aminotransferase were associated with drug administration. In a larger multicentre clinical trial in 69 patients with AIDS, the administration of HPA-23 at 1mg/kg intravenously 5 days per week for 8 weeks was well tolerated; however, the 2mg/kg dose was not tolerated in 8 of 14 patients because of thrombocytopenia and elevated aminotransferases (Moskovitz & The HPA-23 Cooperative Study Group, 1988). No changes in immune function or clinical status were appreciated. P24 antigenemia was not assessed. The manufacturer of HPA-23 has apparently suspended further development of this compound.

Rifabutin (Ansamycin)

Consistent with the *in vitro* observations, no effects on viral or immunological endpoints were observed in a dose escalating study of rifabutin in patients with AIDS or AIDS-related complex (Torseth *et al.*, 1989).

Ribavirin

A moderate amount of clinical experience with ribavirin had been accumulated before its consideration as an antiretroviral drug. Nevertheless its use to treat infection with HIV presented new considerations because of the potential for new or different toxicities with the background of HIV infection and because HIV infection requires chronic therapy.

Using a newly available radioimmunoassay, single dose pharmacokinetics with oral or intravenous ribavirin at 600, 1200, and 2400 mg were conducted in asymptomatic men seropositive for HIV (Laskin *et al.*, 1987). Approximately one-half of orally administered drug is absorbed into the circulation. The most striking characteristic of the pharmacokinetics of ribavirin is its accumulation within erythrocytes. This results in a prolonged terminal phase half-life of 36 hours. This means that a steady state serum concentration will take days to attain and weeks to eliminate.

Two large randomised double-blind, placebo-controlled trials have been conducted with 600 and 800 mg daily oral doses of ribavirin, one in patients with lymphadenopathy and one in patients with ARC. These results have generated a great deal of publicity and controversy; however, the data had not been published in a peer reviewed journal by mid-1989. Preliminary reports of these studies and a phase 1 study of ten patients with AIDS (Crumpacker *et al.*, 1987) indicate that little toxicity occurs at these doses. Even with chronic administration, the only consistent toxic effect is a mild compensated anaemia that reverses with discontinuation of drug. Suppression of p24 antigenemia has not been seen (Spector *et al.*, 1989). The critical conclusions about clinical benefit await the published final analyses of the studies and additional studies that are in progress.

Sulphated polysaccharides

A preparation of dextran sulphate has been used in Japan for decades as an oral agent for anticoagulation and lipid reduction. Little data are published regarding pharmacokinetics, efficacy or toxicity. Nevertheless, the absence of appreciated toxicity in Japan, the remarkable selectivity index *in vitro* (Ueno & Kuno, 1987; Ito *et al.*, 1987; Mitsuya *et al.*, 1988; Baba *et al.*, 1988*a*), and the ready availability of inexpensive, well characterised compound has prompted both controlled phase 1 investigations of oral drug and uncontrolled 'street' use. The first phase 1 evaluation in 30 patients with AIDS or ARC indicated little evidence of efficacy or toxicity (Abrams *et al.*, 1989). Subsequent studies have shown that less than 1% of this large, highly charged compound is absorbed from the gastrointestinal tract (P. Leitman, personal communication). A polysulphated polyxylan (HOE/BAY 946) has entered phase 1 clinical trials in Europe (Biesert *et al.*, 1988).

AL721

One open study with eight seropositive subjects with lymphadenopathy who were administered 10 grams of AL721 twice daily for 18 weeks and later 15 grams twice daily could demonstrate no effects on p24 antigenemia, CD4 cell counts, or toxicity (Grieco *et al.*, 1988). A 3 fold reduction was seen in peak reverse transcriptase levels in the supernatants of cultures of some of these patients; however, this parameter has never been shown to quantitate viral load reliably. A second larger study with higher doses is in progress. Not surprisingly, hyperlipidemia does occur with ingestion of this quantity of lipid.

Other drugs

Inconclusive, small, open clinical studies with little encouragement to pursue further studies have also been reported with D-penicillamine (Schulof *et al.*, 1986) and lithium carbonate (Parenti *et al.*, 1988).

Biological response modifiers

Biological response modifiers are the natural products of cells, or compounds that induce or interact with these products, to modulate immune responses, cell proliferation, cell differentiation or cell metabolism. This class of materials is extensive and rapidly growing in number of identified compounds. For example, the macrophage alone has already been shown to secrete more than a hundred molecular species, many of which have multiple activities, or interact with each other, or are also elaborated by other types of cells or have overlapping activities (Nathan, 1987).

This class of compounds is of keen interest to those involved with the therapy of AIDS for several reasons. The interferons have long interested virologists because they induce antiviral effects in cells, as well as mediate metabolic and functional changes in cells of the immune system. Other cytokines are of theoretical interest for the treatment of patients with HIV infection for at least two reasons. Agents like granulocyte macrophage-colony stimulating factor (GM-CSF) and erythropoietin have been considered for the alleviation of haematologic suppression of HIV disease or haematotoxic drug therapy like AZT or ganciclovir. Agents like interleukin-2 have been proposed to help restore deficiencies of immunity.

Most of the experience with interferons in patients with AIDS has been in open trials for the treatment of Kaposi's sarcoma. These trials have utilised recombinant interferon-α or purified interferon-α prepared from lymphoblastoid cells (Groopman *et al.*, 1984; Krown *et al.*, 1983; Volberding, 1984; Rios *et al.*, 1985; De Wit *et al.*, 1988; Lane *et al.*, 1988). Tumour regression is seen in a minority of patients, with a greater likelihood of response in patients with higher CD4 lymphocyte counts. The dose limiting toxicities of interferon (fever, chills, weakness, fatigue, anorexia, headache, myalgias, arthralgias leukopenia and elevations of aminotransferase) occur in a small proportion of patients. It should be remembered that dose limiting toxicity in oncologic practice may occur at much higher doses than would be tolerated for the suppression of chronic HIV infection. A similar uncontrolled experience with interferons β and γ has been generated with five patients with HTLV-1 associated adult T cell leukemia (Tamura *et al.*, 1987).

The data regarding the use of interferon for the treatment of HIV infection, rather than its malignant complications, are much more limited. *In vitro*, interferons-α,β and γ all inhibit the replication of HIV both in T lymphoid cells (Ho *et al.*, 1985; Hartshorn *et al.*, 1987*a*; Nakashima *et al.*, 1986*b*; Yamamoto *et al.*, 1986; Wong *et al.*,´ 1988), and in primary human macrophages (Kornbluth *et al.*, 1989). In addition, recombinant interferon-α synergistically inhibits HIV replication *in vitro* with two drugs that have antiretroviral activity *in vivo*, AZT (Hartshorn *et al.*, 1987*b*) and foscarnet (Hartshorn *et al.*, 1986). A randomised, double-blind, placebo controlled trial of recombinant human interferon alpha 2a in patients with AIDS was initiated before the recognition of AIDS as a retroviral disease (Interferon Alpha Study Group, 1988). Randomisation of 67 subjects to placebo, 3 million units or 36 million units three times weekly for 12 weeks failed to demonstrate significant differences in tolerance, laboratory endpoints (p24, CD4 cell counts) or clinical status. The power of this study to detect a clinically relevant difference among the study arms of this study was inadequate because of the small number of study subjects and their heterogeneity. Several investigators have reported in abstract form that combined administration of AZT and recombinant or cell derived interferon-α for patients with Kaposi's sarcoma result in at least additive marrow toxicity manifested as leukopenia. In open studies, administration of interferon-α does appear to reduce p24 antigenemia (De Wit *et al.*, 1988; Lane *et al.*, 1988; Brook *et al.*, 1988; Orholm *et al.*, 1989). The critical first questions to be answered for HIV infection regard the dose related effects of different interferon preparations on viral parameters, immune functions and toxicity, and secondarily, what effects are seen in combination with AZT.

Ampligen is a synthetic double stranded RNA consisting of one strand of a homopolymer of inosine and a random copolymer of cytosine and uridine in a ratio of 12:1 (Mitchell *et al.*, 1987). This mismatched double stranded RNA [(formula: $rIn \cdot r(C_{12} \cdot U)n$] was designed as a less toxic interferon inducer than poly I·C because, at the sites of U·I mismatch, there are structural changes that render ampligen susceptible to ribonucleases. Ampligen is a complex preparation with a mean length of 600 bases; however, there is a distribution of molecular sizes and most molecules have overlapping unpaired single strands. Ampligen acts as an interferon inducer and may have other activities as well (Herberman & Pinsky, 1985). *In vitro* it inhibits the replication of HIV-1 in a number of cell lines at concentrations of approximately 10 μg/ml while little toxicity is recognised at 50 μg/ml (Montefiori & Mitchell, 1987; Montefiori, Robinson & Mitchell, 1988). It also acts synergistically with AZT (Mitchell *et al.*, 1987).

Some experience has been acquired with the administration of ampligen to animals and humans, primarily for tumour therapy (Carter & O'Malley, 1976; Carter *et al.*, 1985; Brodsky *et al.*, 1985). Intravenous administration of up to 450 mg twice weekly to patients with tumour appeared to be relatively non-toxic (Brodsky *et al.*, 1985). Ampligen was administered to 10 patients infected with HIV ranging in severity from lymphadenopathy to AIDS (Carter *et al.*, 1987). Two hundred to 250 mg intravenously for up to 18 weeks was associated with little toxicity and some claims of clinical and antiviral effects (Carter *et al.*, 1987). Preliminary reports from several additional controlled and systematic studies have indicated that little toxicity or efficacy can be documented.

Interleukin-2 is a T cell derived glycoprotein with many activities including the induction of T lymphocyte proliferation (Smith, 1988), a function critical for the culture systems used to isolate HTLV-1 and HIV (Poiesz *et al.*, 1980; Barre-Sinoussi *et al.*, 1983; Popovic *et al.*, 1984). Deficient lymphocyte functions in assays such as *in vitro* NK function, can be restored by the addition of interleukin-2 (Ciobanu, Kruger & Welte, 1983; Rook *et al.*, 1983). These observations and the ready availability of both purified natural and recombinant interleukin-2 have prompted several open trials in patients with AIDS. At doses up to 2000000 units/m^2 per day intravenously, no clinical, virologic or immunological benefits have been documented (Lane *et al.*, 1984; Mertelsmann *et al.*, 1984; Ernst *et al.*, 1986; Volberding *et al.*, 1987). Influenza-like symptoms and hypotension were observed in many patients receiving 1000000 units/m,^{2} or more (Volberding *et al.*, 1987). Although interleukin-2 therapy alone is unlikely to be useful, it is possible that immunological restoration in conjunction with antiviral drugs may have a role in the treatment of HIV infection.

Both HIV infection and its treatment can be toxic to the bone marrow (Zon & Groopman, 1988; Spivak, Bender & Quinn, 1984; Richman *et al.*, 1987*a*). Cytokines important for the proliferation and differentiation of marrow precursors are potentially useful in overcoming the severe anaemia or leukopenia. Several trials of the administration of recombinant GM-CSF to patients with AIDS have documented dramatic dose-dependent elevations of leukocyte counts, particularly neutrophils, monocytes, and eosinophils, within hours of initiating infusions, but the counts returned to baseline equally quickly (Groopman *et al.*, 1987). Low-grade fever, myalgia, bone pain, phlebitis and flushing occur in some patients. The newly circulating cells appear functionally normal (Baldwin *et al.*, 1988). No effects on erythroid or platelet elements were seen (Groopman *et al.*, 1987).

Trials of erythropoietin, especially for the anaemia associated with AZT administration, are also in progress.

A large number of biological response modifiers and 'immunomodulating' compounds have been proposed for the therapy of HIV infection. Some of these are well-characterised compounds; some have been systematically evaluated in clinical trials. None has yet been shown effective. No efficacy was shown in small, open trials of cyclosporin (Andrieu *et al.*, 1988), levamisole (Surapaneni *et al.*, 1985), thymopentin (Clumeck *et al.*, 1984, 1985), and transplantations of thymic tissue (Dwyer *et al.*, 1987). Investigators administering transfer factor (not further characterised) or cyclosporin were encouraged by their trials but no viral markers were tested (Carey *et al.*, 1987; Andrieu *et al.*, 1988). Lithium carbonate appeared ineffective but toxic (Parenti *et al.*, 1988).

A final class of compounds are those 'immunomodulating' compounds, such as AS 101 (ammonium trichloro[dioxyethylene-0,0′-] tellurite) (Sredni *et al.*, 1987), DTC (diethyldithiocarbamate, Imuthiol), Imreg-1, disulfiram, isoprinosine (Glasky & Gordon, 1987), thymostimulin (TP-1) (Palmisano *et al.*, 1988), and methionine-enkephalin that are represented by many abstracts at large clinical meetings. Critical interpretation of the merits of such compounds requires information about their composition and purity, well-designed systematic studies, scrutiny of the primary data and statistical analysis. The importance of conclusions of benefit (or the lack of it) for a disease with the impact of AIDS requires that this information is available accurately and expeditiously. A report of a controlled trial of DTC claimed clinical benefit but no impact on p24 antigenemia (Lang *et al.*, 1988). This study merits careful analysis of the primary data and confirmatory studies.

To ascribe potential utility to a biological response modifier and then to assess it properly in clinical trials is even more complicated than for antiviral drugs. The experience with interleukin-2, which is a pure and molecularly defined material, documents several of the difficulties that arise between the production of cytokine preparations by recombinant DNA technology and the documentation of even a potential role in the therapy of disease. Cytokines are being identified, cloned, and expressed faster than their biological roles can be documented. Should a function of a cytokine be identified *in vitro*, there is no guarantee that that function is biologically relevant. Most cytokines identified to date have a multiplicity of functions, often different ones on different cells, which may vary significantly depending on the presence of other cytokines and which may act in microenvironments based upon physical relationships of cells (Miyajima *et al.*, 1988). Serum concentrations may be irrelevant to local concentrations. Cytokine actions may be more a function of changing levels or the presence of other cytokines (Belosevic *et al.*, 1988; Miyajima *et al.*, 1988) than the levels generated *in vitro* or in serum. Studies *in vitro* of biological response modifiers may thus be much less predictive of potential clinical effects than is obtained with antiviral drugs. Proper regimens of administration may be very difficult to ascertain. Further investigation including both laboratory based studies and empirical trials will help to elucidate this complex and cloudy picture.

REFERENCES

Abrams, D. I., Kuno, S., Wong, R., Jeffords, K., Nash, M., Molaghan, J. M., Gorter, R. & Ueno, R. (1989). Oral dextran sulfate (UA001) in the treatment of the acquired immunodeficiency syndrome (AIDS) and AIDS-related complex. *Annals of Internal Medicine*, **110**, 183–8.

Agrawal, S., Goodchild, J., Civeira, M. P., Thornton, A. H., Sarin, P. S. & Zamecnik, P. C. (1988). Oligodeoxynucleoside phosphoramidates and phosphorothioates as inhibitors of human immunodeficiency virus. *Proceedings of the National Academy of Sciences, USA*, **85**, 7079–83.

Agris, C. H., Blake, K. R., Miller, P. S., Reddy, M. P. & Ts'o, P. O. (1986). Inhibition of vesicular stomatitis virus protein synthesis and infection by sequence-specific oligodeoxyribonucleoside methylphosphonates. *Biochemistry*, **25**, 6268–75.

Ahluwalia, G., Cooney, D. A., Mitsuya, H., Fridland, A., Flora, K. P., Hao, Z., Dalal, M., Broder, S. & Johns, D. G. (1987). Initial studies on the cellular pharmacology of 2′,3′-dideoxyinosine, an inhibitor of HIV infectivity. *Biochemical Pharmacology*, **36**, 3797–800.

Anand, R., Moore, J., Feorino, P., Curran, J. & Srinivasan, A. (1986). Rifabutine inhibits HTLV-III. *Lancet*, **i**, 97–8.

Andrieu, J. M., Even, P., Venet, A., Tourani, J. M., Stern, M., Lowenstein, W., Audroin, C., Eme, D., Masson, D., Sors, H., Israel-Biet, D. & Beldjord, K. (1988). Effects of Cyclosporin on T-cell subsets in human immunodeficiency virus disease. *Clinical Immunology and Immunopathology*, **146**, 181–98.

Ayers, K. M. (1988). Preclinical toxicology of zidovudine. *American Journal of Medicine*, **85**, 186–8.

Baba, M., Nakajima, M., Schols, D., Pauwels, R., Balzarini, J. & De Clercq, E. (1988*a*). Pentosan polysulfate, a sulfated oligosaccharide, is a potent and selective anti-HIV agent *in vitro*. *Antiviral Research*, **9**, 335–43.

Baba, M., Pauwels, R., Balzarini, J., Arnout, J., Desmyter, J. & De Clercq, E. (1988*b*). Mechanism of inhibitory effect of dextran sulfate and heparin on replication of human immunodeficiency virus *in vitro*. *Proceedings of the National Academy of Sciences, USA*, **85**, 6132–6.

Baba, M., Pauwels, R., Balzarini, J., Herdewijn, P. & De Clercq, E. (1987*a*). Selective inhibition of human immunodeficiency virus (HIV) by 3′-azido-2′, 3′-dideoxyguanosine *in vitro*. *Biochemical and Biophysical Research Communication*, **145**, 1080–6.

Baba, M., Pauwels, R., Balzarini, J., Herdewijn, P., De Clercq, E. & Desmyter, J. (1987*b*). Ribavirin antagonizes inhibitory effects of pyrimidine 2′,3′-Dideoxynucleosides but enhances inhibitory effects of purine 2′,3′-dideoxynucleosides on replication of human immunodeficiency virus *in vitro*. *Antimicrobial Agents and Chemotherapy*, **31**, 1613–17.

Baba, M., Pauwels, R., Herdewijn, P., De Clercq, E., Desmyter, J. & Vandeputte, M. (1987*c*). Both 2′,3′-dideoxythymidine and its 2′,3′-unsaturated derivative (2′,3′-dideoxythymidinene) are potent and selective inhibitors of human immunodeficiency virus replication *in vitro*. *Biochemical and Biophysical Research Communications*, **142**, 128–34.

Baba, M., Snoeck, R., Pauwels, R. & De Clercq, E. (1988*c*). Sulfated polysaccharides are potent and selective inhibitors of various enveloped viruses, including herpes simplex virus, cytomegalovirus, vesicular stomatitis virus, and human immunodeficiency virus. *Antimicrobial Agents and Chemotherapy*, **32**, 1742–5.

Bailey, R. O., Turok, D. I., Jaufmann, B. P. & Singh, J. K. (1987). Myositis and acquired immunodeficiency syndrome. *Human Pathology*, **18**, 749–51.

Baldwin, G. C., Gasson, J. C., Quan, S. G., Fleischmann, J., Weisbart, R., Oette, D., Mitsuyasu, R. T. & Golde, D. W. (1988). Granulocyte-macrophage colony-stimulating factor enhances neutrophil function in acquired immunodeficiency syndrome patients. *Proceedings of the National Academy of Sciences, USA*, **85**, 2763–6.

Balzarini, J., Baba, M., Pauwels, R., Herdewijn, P. & De Clercq, E. (1988*a*). Anti-retrovirus activity of 3′-fluor- and 3′azido-substituted pyrimidine 2′,3′-dideoxynucleoside analogues. *Biochemical Pharmacology*, **37**, 2847–56.

Balzarini, J., Baba, M., Pauwels, R., Herdewijn, P., Wood, S. G., Robbins, M. J. & De Clercq, E. (1988*b*). Potent and selective activity of 3′-azido-2,6-diaminopurine-2′,3′-dideoxyriboside, 3′-fluoro-2,6-diaminopurine-2′,3′-dideoxyriboside, and 3′-fluoro-2′,3′-dideoxyguanosine against human immunodeficiency virus. *Molecular Pharmacology*, **33**, 243–9.

Balzarini, J., Cooney, D. A., Dalal, M., Kang, G. J., Cupp, J. E., De Clercq, E., Broder, S. & Johns, D. G. (1987*a*). 2′,3′-dideoxycytidine: regulation of its metabolism and anti-retroviral potency by natural pyrimidine nucleosides and by inhibitors of pyrimidine nucleotide synthesis. *Molecular Pharmacology*, **32**, 798–806.

Balzarini, J., Herdewijn, P. & De Clercq, E. (1989*a*). Potentiating effect of ribavarin on the antiretrovirus activity of 3′-azido-2,6-diaminopurine-2′,3′-dideoxyriboside *in vitro* and *in vivo*. *Antiviral Research*, **11**, 161–72.

Balzarini, J., Herdewijn, P. & De Clercq, E. (1989*b*). Differential patterns of intracellular metabolism of 2′,3′-didehyro-2′3′-dideoxythymidine and 3′-azido-2′,3′-deoxythymidine, two potent anti-human immunodeficiency virus compounds. *Journal of Biological Chemistry*, **264**, 6127–33.

Balzarini, J., Mitsuya, H., De Clercq, E. & Broder, S. (1986*a*). Comparative inhibitory effects of suramin and other selected compounds on the infectivity and replication of human T-cell lymphotropic virus (HTLV-III)/lymphadenopathy-associated virus (LAV). *International Journal of Cancer*, **37**, 451–7.

Balzarini, J., Mitsuya, H., De Clercq, E. & Broder, S. (1986*b*). Aurintricarboxylic acid and Evans Blue represent two different classes of anionic compounds which selectively inhibit the cytopathogenicity of human T-cell lymphotropic virus type III/lymphadenopathy-associated virus. *Biochemical and Biophysical Research Communications*, **136**, 64–71.

Balzarini, J., Naesens, L., Herdewijn, P., Rosenberg, I., Holy, A., Pauwels, R., Baba, M., Johns, D. A. & De Clercq, E. (1989*c*). Marked *in vivo* antiretrovirus activity of 9-(2-phosphonylmethoxy-ethyl) adenine, a selective anti-human immunodeficiency virus agent. *Proceedings of the National Academy of Sciences*, **86**, 332–6.

Balzarini, J., Pauwels, R., Baba, M., Herdewijn, P., De Clercq, E., Broder, S. & Johns, D. G. (1988*c*). The in vitro and in vivo anti-retrovirus activity, and intracellular metabolism of 3′-azido-2′,3′-dideoxythymidine and 2′,3′-dideoxycytidine are highly dependent on the cell species. *Biochemical Pharmacology*, **37**, 897–903.

Balzarini, J., Pauwels, R., Baba, M., Robins, M. J., Zou, R. M., Herdewijn, P. & De Clercq, E. (1987*b*). The 2′,3′-dideoxyriboside of 2,6-diaminopurine selectively inhibits human immunodeficiency virus (HIV) replication *in vitro*. *Biochemical and Biophysical Research Communication*, **145**, 269–76.

Balzarini, J., Pauwels, R., Herdewijn, P., De Clercq, E., Cooney, D. A., Kang, G. J., Dalal, M., Johns, D. G. & Broder, S. (1986*c*). Potent and selective anti-HTLV-III/LAV activity of 2′,3′-dideoxycytidinene, the 2′,3′-unsaturated derivative of 2′,3′-dideoxycytidine. *Biochemical and Biophysical Research Communications*, **140**, 735–42.

Balzarini, J., Robins, M. J., Zou, R. M., Herdewijn, P. & De Clercq, E. (1987*c*). The 2′,3′-dideoxyriboside of 2,6-diaminopurine and its 2′,3′-didehydro derivative inhibit the deamination of 2′,3′-dideoxyadenosine, an inhibitor of human immunodeficiency virus (HIV) replication. *Biochemical and Biophysical Research Communications*, **145**, 277–83.

Barnes, D. M. (1987*a*). Debate of potential AIDS drug. *Science*, **237**, 128–30.

Barnes, D. M. (1987*b*). 'On the shelf' AIDS drug in clinical trial. *Science*, **238**, 276.

Barre-Sinoussi, F., Chermann, J. C., Rey, R., Nugeyre, M. T., Chamaret, S., Gruest, J., Dauguet, C. & Axler-Blin, C. (1983). Isolation of a T-lymphotropic retrovirus from a patient at risk for acquired immune deficiency syndrome (AIDS). *Science*, **220**, 868–71.

Belosevic, M., Davis, C. E., Meltzer, M. S. & Nacy, Ca. (1988). Regulation of activated macrophage antimicrobial activities. *The Journal of Immunology*, **141**, 890–6.

Bergadahl, S., Sonnerborg, A., Larsson, A. & Strannegard, O. (1988). Declining levels of HIV p24 antigen in serum during treatment with foscarnet. *Lancet*, **i**, 1052.

Bertonis, J. M., Meier, W., Johnson, V. A., Costopoulos, D. S., Liu, T., Tizard, R., Walker, B. D., Hirsch, M. S., Schooley, R. T. & Flavell, R. A. (1988). HIV infection is blocked in vitro by recombinant soluble CD4. *Nature, London*, **331**, 76–8.

Bessen, L. J., Greene, J. B., Louie, E., Seitzman, P. & Weinberg, H. (1988). Severe polymyositis-like syndrome associated with zidovudine therapy of AIDS and ARC. *The New England Journal of Medicine*, **318**, 708.

Biesert, L., Suhartono, H., Winkler, I., Meichsner, C., Helsberg, M., Hewlett, G., Klimetzek, V., Molling, K., Schlumberger, H.-D., Schrinner, E., Brede, H. -D. & Rubsamen-Waigmann, H. (1988). Inhibition of HIV and virus replication by polysulphated polyxylan: HOE/BAY 946, a new antiviral compound. *AIDS*, **2**, 449–57.

Birch, C., Tachedjian, G., Lucas, C. R. & Gust, I. (1988). *In vitro* effectiveness of a combination of zidovudine and ansamycin against human immunodeficiency virus. *Journal of Infectious Diseases*, **158**, 895.

Blough, H. A., Pauwels, R., De Clercq, E., Cogniaux, J., Sprecher-Goldberger, S. & Thiry, L. (1986). Glycosylation inhibitors block the expression of LAV/HTLV-III (HIV) glycoproteins. *Biochemical and Biophysical Research Communications*, **141**, 33–8.

Blough, H. A. & Ray, E. K. (1988). Glucose analogues in the chemotherapy of herpesvirus infections. *Phamacology and Therapeutics*, **10**, 669–81.

Blum, M. R., Liao, S. H. T., Good, S. S. & Miranda, P. (1988). Pharmacokinetics and bioavailability of zidovudine in humans. *American Journal of Medicine*, **85**, 189–94.

Bollen, M. & Stalmans, W. (1989). The antiglycogenolytic action of 1-deoxynojirimycin results from a specific inhibition of the alpha-1,6-glucosidase activity of the debranching enzyme. *European Journal of Biochemistry*, **181**, 775–80.

Broder, S., Yarchoan, R., Collins, J. M., Lane, H. C., Markham, P. D., Klecker Jr, K. W., Redfield, R. R., Mitsuya, H., Hoth, D. F., Gelmann, E., Groopman, J. E., Resnick, L., Gallo, R. C., Myers, C. E. & Fauci, A. S. (1985). Effects of suramin on HTLV-III/LAV infection presenting as Kaposi's sarcoma or AIDS-related complex: clinical pharmacology and suppression of virus replication *in vivo*. *Lancet*, **ii**, 627–30.

Brodsky, I., Strayer, D. R., Kruger, L. J. & Carter, W. A. (1985). Clinical studies with ampligen (mismatched double-stranded RNA). *Journal of Biological Response Modifiers*, **4**, 669–75.

Brook, M. G., Gor, D., Forster, S. M., Harris, W., Jeffries, D. J. & Thomas, H. C. (1988). Short communication suppression of HIV p24 antigen and induction of HIV anti-p24 antibody by alpha interferon in patients with chronic hepatitis B. *AIDS*, **2**, 391–3.

Buimovici-Klein, E., Ong, K. R., Lange, M., Englard, A., McKinley, G. F., Reddy, M., Grieco, M. H. & Cooper, L. Z. (1986). Reverse transcriptase activity (RTA) in lymphocyte cultures of AIDS patients treated with HPA-23. *AIDS Research*, **2**, 279–83.

Busso, M., Mian, A. B., Hahn, E. F. & Resnick, L. (1988). Nucleotide dimers suppress HIV expression *in vitro*. *AIDS Research and Human Retroviruses*, **4**, 449.

Capon, D. J., Chamow, S. M., Mordenti, J., Marsters, S. A., Gregory, T., Mitsuya, H., Byrn, R. A., Lucas, C., Wurm, F. M., Groopman, J. E., Broder, S. & Smith, D. H. (1989). Designing CD4 immunoadhesins for AIDS therapy. *Nature, London*, **337**, 525–31.

Carey, J. T., Lederman, M. M., Toossi, Z., Edmonds, K., Hodder, S., Calabrese, L. H., Proffitt, M. R., Johnson, C. E. & Ellner, J. J. (1987). Augmentation of skin test reactivity and lymphocyte blastogenesis in patients with AIDS treated with transfer factor. *The Journal of the American Medical Association*, **257**, 651–5.

Carson, D. A., Haertle, T., Wasson, D. B. & Richman, D. D. (1988). Biochemical genetic analysis of 2′,3′-dideoxyadenosine metabolism in human T lymphocytes. *Biochemical and Biophysical Research Communications*, **151**, 788–93.

Carter, W. A. & O'Malley, J. (1976). An integrated and comparative study of the antiviral effects and other biological properties of the polyinosine polycytidylic acid duplex and its mismatched analogues. *Molecular Pharmacology*, **12**, 440–53.

Carter, W. A., Strayer, D. R., Brodsky, I., Lewin, M., Pellegrino, M. G., Einck, L., Henriques, H. F., Simon, G. L., Parenti, D. M., Scheib, R. G., Schulof, R. S., Montefiori, D. C., Robinson, W. D., Mitchell, W. M., Volsky, D. J., Paul, D., Paxton, H., Meyer III, W. A., Kariko, K., Reichenbach, N., Suhadolnik, R. J. & Gillespie, D. H. (1987). Clinical, immunological, and virological effects of ampligen, a mismatched double-stranded RNA, in patients with AIDS or AIDS-related complex. *Lancet*, **i**, 1286–92.

Carter, W. A., Strayer, D. R., Hubbell, H. R. & Brodsky, I. (1985). Preclinical studies with ampligen (Mismatched double-stranded RNA). *Journal of Biological Response Modifiers*, **4**, 495–502.

Chaisson, R. E., Leuther, M. D., Allain, J. P., Nusinoff-Lehrman, S., Boone, G. S., Feigal, D. & Volberding, P. (1988). Effect of zidovudine on serum human immunodeficiency virus core antigen levels. *Archives of Internal Medicine*, **148**, 2151–3.

Chandra, P. & Sarin, P. S. (1988). Selective inhibition of replication of the AIDS-associated virus HTLV-III/LAV by synthetic D-penicillamine. *Arzneim.-Forsch./Drug Res.*, **36**, 184–6.

Chang, R. S. & Yeung, H. W. (1988). Inhibition of growth of human immunodeficiency virus in vitro by crude extracts of Chinese medicinal herbs. *Antiviral Research*, **9**, 163–76.

Chaudhary, V. K., Mizukami, T., Fuerst, T. R., FitzGerald, D. J., Moss, B., Pastan, I. & Berger, E. A. (1988). Selective killing of HIV-infected cells by recombinant human CD4-pseudomonas exotoxin in hybrid protein. *Nature, London*, **335**, 369–72.

Chen, M. S. & Oshana, S. C. (1987). Inhibition of HIV reverse transcriptase by 2′,3′-dideoxynucleoside triphosphates. *Biochemical Pharmacology*, **36**, 4361–2.

Chermann, J. C., Sinoussi, F. C. & Jasmin, C. (1975). Inhibition of RNA-dependent DNA polymerase of murine oncornaviruses by ammonium-5-tungsto-2-antimoniate. *Biochemical and Biophysical Research Communications*, **65**, 1229–36.

Cheson, B. D., Levine, A., Mildvan, D., Kaplan, L. D., Wolfe, P., Rios, A., Groopman, J. E., Parkash, G., Volberding, P. A., Poiesz, B. J., Gottlieb, M. S., Holden, H., Volsky, D. J., Silver, S. S. & Hawkins, M. J. (1987). Suramin therapy in AIDS and related diseases. Initial report of the US suramin working group. *The Journal of the American Medical Association*, **258**, 1347–51.

Chu, C. K., Schinazi, R. F., Arnold, B. H., Cannon, D. L., Doboszewski, B., Bhadti, V. B. & Gu, Z. (1988). Comparative activity of 2′,3′-saturated and unsaturated pyrimidine and purine nucleosides against human immunodeficiency virus type 1 in peripheral blood mononuclear cells. *Biochemical Pharmacology*, **37**, 3543–8.

Ciobanu, N., Kruger, G. & Welte, K. (1983). Defective T cell response to PHA and mitogenic monoclonal antibodies in male homosexuals with acquired immunodeficiency syndrome and its in vitro corrections by interleukin 2. *Journal of Clinical Immunology*, **3**, 332–40.

Clumeck, N., Cran, S., Van de Perre, P., Mascart-Lemone, F., Duchateau, J. & Bolla, K. (1985). Thymopentin treatment in AIDS and pre-AIDS patients. *Surveys of Immunological Research*, **4:suppl 1**, 58–62.

Clumeck, N., Van de Perre, P., Mascart-Lemone, F., Cran, S., Bolla, K. & Duchateau, J. (1984). Preliminary results on clinical and immunological effects of thymopentin in AIDS. *International Journal of Clinical Pharmacology Research*, **IV**, 459–63.

Collins, J. M., Klecker Jr, K. W., Yarchoan, R., Lane, H. C., Fauci, A. S., Redfield, R. R., Broder, S. & Myers, C. E. (1986). Clinical pharmacokinetics of suramin in patients with HTLV-III/LAV infection. *Journal of Clinical Pharmacology*, **26**, 22–6.

Cooney, D. A., Ahluwalia, G., Mitsuya, H., Fridland, A., Johnson, M., Hao, Z., Dalal, M., Balzarini, J., Broder, S. & Johns, D. G. (1987). Initial studies on the cellular pharmacology of 2′,3′-dideoxyadenosine, an inhibitor of HTLV-III infectivity. *Biochemical Pharmacology*, **36**, 1765–8.

Cooney, D. A., Dalal, M., Mitsuya, H., McMahon, J. B., Nadkarni, M., Balzarini, J., Broder, S. & Johns, D. G. (1986). Initial studies on the cellular pharmacology of 2′,3-dideoxycytidine, an inhibitor of HTLV-III infectivity. *Biochemical Pharmacology*, **35**, 2065–8.

Creagh-Kirk, T., Doi, E., Andrews, S., Nusinoff-Lehrman, S., Tilson, H. & Barry, D. (1988). Survival experience among patients with acquired immunodeficiency syndrome receiving zidovudine therapy. *Journal of the American Medical Association*, **260**, 3009–15.

Crumpacker, C., Heagy, W., Bubley, G., Monroe, J. E., Finberg, R., Hussey,

S., Schnipper, L., Lucey, D., Lee, T. H., McLane, M. F., Essex, M. & Mulder, C. (1987). Ribavirin treatment of the acquired immunodeficiency syndrome (AIDS) and the acquired immunodeficiency syndrome related complex (ARC). *Annals of Internal Medicine*, **107**, 664–74.

Dahlberg, J. E., Mitsuya, H., Blam, S. B., Broder, S. & Aaronson, S. A. (1987). Broad spectrum antiretroviral activity of 2′,3′-dideoxynucleosides. *Proceedings of the National Academy of Sciences, USA*, **84**, 2469–73.

Dainiak, N., Worthington, M., Riordan, M. A., Kreczko, S. & Goldman, L. (1988). 3′-azido-3′-deoxythymidine (AZT) inhibits proliferation *in vitro* of human haematopoietic progenitor cells. *British Journal of Haematology*, **69**, 299–304.

Dalakas, M. C., Pezeshkpour, G. H. & Flaherty, M. (1987). Progressive nemaline (ROD) myopathy associated with HIV infection. *The New England Journal of Medicine*, **317**, 1602–3.

Dalgleish, A. G., Beverly, P. C. L., Clapham, P. R., Crawford, D. H., Greaves, M. F. & Weiss, R. A. (1984). The CD4 (T4) antigen is an essential component of the receptor of the AIDS retrovirus. *Nature, London*, **312**, 763–7.

Dalgleish, A. G., Kennedy, R. C., Chanh, T. C., Habeshaw, J., Maddon, P., Malkovsky, M., Thomson, B., Claphan, P., Axel, R., Webster, A. D. B. & Weiss, R. A. (1988). Therapeutic strategies against HIV based on the CD4 molecule: Monoclonal antibody therapy, soluble CD4 and anti-idiotype vaccines. [Abstract]. *IVth International Conference on AIDS, Stockholm, Sweden*.

Darke, P. L., Leu, C.-T., Davis, L. J., Heimbach, J. C., Diehl, R. E., Hill, W. S., Dison, R. A. F. & Sigal, I. S. (1989). Human immunodeficiency virus protease. *Journal of Biological Chemistry*, **264**, 2307–12.

De Clercq, E. (1979). Suramin: a potent inhibitor of the reverse transcriptase of RNA tumor viruses. *Cancer Letters*, **8**, 9–22.

De Clercq, E. (1986). Chemotherapeutic approaches to the treatment of the acquired immune deficiency syndrome (AIDS). *Journal of Medicinal Chemistry*, **29**, 1561–9.

De Clercq, E. (1987). Suramin in the treatment of AIDS: mechanism of action [published erratum appears in Antiviral Res 1987 Mar; 7(3):185]. *Antiviral Research*, **7**, 1–10.

De Gans, J., Lange, J. M. A., Derix, M. M. A., De Wolf, F., Schattenkerk, J. K. M. E., Danner, S. A., de Visser, B. W. O., Cload, P. & Goudsmit, J. (1988). Decline of HIV antigen levels in cerebrospinal fluid during treatment with low-dose zidovudine. *AIDS*, **2**, 37–40.

de Miranda, P., Good, S. S., Blum, M. R. & et al. (1986). The effect of probenecid on the pharmacokinetic disposition of azidothymidine (AZT). (Abstract). *II International Conference on AIDS, Paris, France*.

De Somer, P., De Clercq, E., Billiau, A., Schonne, E. & Claesen, M. (1968). Antiviral activity of polyacrylic acid polymethacrylic acids. I. Mode of action *in vitro*. *Journal of Virology*, **2**, 878–85.

De Wit, R., Boucher, C. A., Veenhof, K. H., Schattenkerk, J. K. M. E., Bakker, P. J. & Danner, S. A. (1988). Clinical and virological effects of high-dose recombinant interferon-alpha in disseminated AIDS-related Kaposi's Sarcoma. *Lancet*, **ii**, 1214–17.

De Wolf, F., Goudsmit, J., De Gans, J., Coutinho, R. A., Lange, J. M. A., Cload, P., Schellekens, P. T. A., Fiddian, A. P. & Van Der Noordaa, J. (1988). Effect of zidovudine on serum human immunodeficiency virus antigen levels in symptom-free subjects. *Lancet*, **i**, 373–6.

Deen, K. C., McDougal, J. S., Inacker, R., Folena-Wasserman, G., Arthos, J., Rosenberg, J., Maddon, P. J., Axell, R. & Sweet, R. W. (1988). A soluble form of CD4 (T4) protein inhibits AIDS virus infection. *Nature, London*, **331**, 82–4.

Dormont, D., Spire, B., Barre-Sinoussi, F., Montagnier, L. & Chermann, J. C. (1985). Inhibition of RNA-dependent DNA polymerases of AIDS and SAIDS retroviruses by HPA-23 (ammonium-21-tungsto-9-antimoniate). *Annals of the Institute Pasteur/Virology*, **136**, 75–83.

Dournon, E., Rozenbaum, W., Michon, C., Perronne, C., De Truchis, P., Bouvet, E., Levacher, M., Matheron, S., Gharakhanian, S., Girard, P. M., Salmon, D., Leport, C., Dazza, M. C., Regnier, R. & The Claude Bernard Hospital AZT Study Group (1988). Effects of zidovudine in 365 consecutive patients with AIDS or AIDS-related complex. *Lancet*, **i**, 1297–302.

Dwyer, J. M., Wood, C. C., McNamara, J. & Kinder, B. (1987). Transplantation of thymic tissue into patients with AIDS. *Archives of Internal Medicine*, **147**, 513–17.

Elwell, L. P., Ferone, R., Freeman, G. A., Fyfe, J. A., Hill, J. A., Ray, P. H., Richards, C. A., Singer, S. C., Knick, V. B., Rideout, J. L. & Zimmerman, T. P. (1987). Antibacterial activity and mechanism of action of 3′-azido-3′-dideoxythymidine (BW A509U). *Antimicrobial Agents and Chemotherapy*, **31**, 274–80.

Eriksson, B., Vrang, L., Bazin, H., Chattopadhyaya, J. & Oberg, B. (1987). Different patterns of inhibition of avian myeloblastosis virus reverse transcriptase activity by 3′-azido-3′-deoxythymidine 5′-triphosphate and its threo isomer. *Antimicrobial Agents and Chemotherapy*, **31**, 600–4.

Ernst, M., Kern, P., Flad, H. D. & Ulmer, A. J. (1986). Effects of systemic in vivo interleukin-2 (IL-2) reconstitution in patients with acquired immune deficiency syndrome (AIDS) and AIDS-related complex (ARC) on phenotypes and functions of peripheral blood mononuclear cells (PBMC). *Journal of Clinical Immunology*, **6**, 170–81.

Faber, V., Dalgleish, A. G., Newell, A.· & Malkovsky, M. (1987). Inhibition of HIV replication *in vitro* by fusidic acid. *Lancet*, **i**, 827–8.

Fiala, M., Cone, L. A., Cohen, N., Patel, D., Williams, K., Casareale, D., Shapshak, P. & Tourtelotte, W. (1988). Responses of neurologic complications of AIDS to 3′-azido-3′-deoxythymidine and 9-(1,3-dihydroxy-2-propoxymethyl) guanine. I. Clinical features. *Journal of Infectious Diseases*, **2**, 250–6.

Fischl, M. A., Richman, D. D., Grieco, M. H., Gottlieb, M. S., Volberding, P. A., Laskin, O. L., Leedom, J. M., Groopman, J. E., Mildvan, D., Schooley, R. T., Jackson, G. G., Durack, D. T., King, D. & AZT Collaborative Working Group (1987). The efficacy of azidothymidine (AZT) in the treatment of patients with AIDS and AIDS-related complex. A double-blind, placebo-controlled trial. *The New England Journal of Medicine*, **317**, 185–91.

Frankel, A. D., Bredt, D. S. & Pabo, C. O. (1988). Fat protein from human immunodeficiency virus forms a metal-linked dimer. *Science*, **240**, 70–3.

Furman, P. H., Fyfe, J. A., St. Clair, M. H., Weinhold, K., Rideout, J. L., Nusinoff-Lehrman, S., Bolognesi, D. P., Broder, S., Mitsuya, H. & Barry, D. W. (1986). Phosphorylation of 3′-azido-3′-deoxythymidine and selective interaction of the 5′-triphosphate with human immunodeficiency virus reverse transcriptase. *Proceedings of the National Academy of Sciences, USA*, **83**, 8333–7.

Furmanski, P., Bourguignon, G. J., Bolles, C. S., Corombos, J. D. & Das, M. R. (1980). Inhibition by 2′,3′-dideoxythymidine of retroviral infection of mouse and human cells. *Cancer Letters*, **8**, 307–15.

Furth, P. A. & Kazakis, A. M. (1987). Nail pigmentation changes associated with azidothymidine (zidovudine). *Annals of Internal Medicine*, **107**, 350.

Fyfe, J. A., Furman, P., Vogt, M. & Sherman, P. (1988). Mechanism of ribavirin antagonism of anti-HIV activity of azidothymidine. [Abstract]. *IVth International Conference on AIDS, Stockholm, Sweden.*

Gaub, J., Pedersen, C., Poulsen, A. G., Mathiesen, L. R., Ulrich, K., Lindhardt, B., Faber, V., Gerstoft, J., Hofmann, B., Lernestedt, J. O., Nielsen, C. M., Nielsen, J. O. & Platz, P. (1987). The effect of foscarnet (Phosphonoformate) on human immunodeficiency virus isolation, T-cell subsets and lymphocyte function in AIDS patients. *AIDS*, **1**, 27–33.

Gertner, E., Thurn, J. R., Williams, D. N., Simpson, M., Balfour Jr, H. H., Rhams, F. & Henry, K. (1989). Zidovudine-associated myopathy. *American Journal of Medicine*, **86**, 814–18.

Glasky, A. J. & Gordon, J. F. (1987). Isoprinosine (Inosine Pranobex BAN, INPX) in the treatment of AIDS and other acquired immunodeficiencies of clinical importance. *Cancer Detection and Prevention Supplement*, **1**, 597–609.

Gonzales, M. F., Olney, R. K., So, Y. T., Greco, C. M., McQuinn, B. A., Miller, R. G. & DeArmond, S. J. (1988). Subacute structural myopathy associated with human immunodeficency virus infection. *Archives of Neurology*, **45**, 585–7.

Goodchild, J., Agrawal, S., Civeira, M. P., Sarin, P. S., Sun, D. & Zamecnik, P. C. (1988). Inhibition of human immunodeficiency virus replication by antisense oligodeoxynucleotides. *Proceedings of the National Academy of Sciences, USA*, **85**, 5507–11.

Goswami, D. B., Borek, E., Sharma, O. K., Fujitaki, J. & Smith, R. A. (1979). The broad spectrum antiviral agent ribavirin inhibits capping of RNA. *Biochemical and Biophysical Research Communications*, **89**, 830–6.

Gottlieb, M. S., Wolfe, P. R. & Chafey, S. (1987). Case report: Response to AIDS-related thrombocytopenia to intravenous and oral azidothymidine (3′-azido-3′-deoxythymidine). *AIDS Research and Human Retroviruses*, **3**, 109–14.

Grieco, M. H., Lange, M., Buimovici-Klein, E., Mohan Reddy, M., Englard, A., McKinley, G. F., Ong, K. & Metroka, C. (1988). Open study of AL-721 treatment in HIV-infected subjects with generalized lymphadenopathy syndrome: An eight week open trial and follow-up. *Antiviral Research*, **9**, 177–90.

Griengl, H., Hayden, W., Penn, G., De Clercq, E. & Rosenwirth, B. (1988). Phosphonoformate and phosphonoacetate derivatives of 5-substituted 2′-deoxyuridines: synthesis and antiviral activity. *Journal of Medical Chemistry*, **31**, 1831–9.

Groopman, J. E., Gottlieb, M. S., Goodman, J., Mitsuyasu, R. T., Conant, M. A., Prince, H., Fahey, J. L., Derezin, M., Weinstein, W. M., Casavante, C., Rothman, J., Rudnick, S. A. & Volberding, P. A. (1984). Recombinant Alpha-2 Interferon Therapy for Kaposi's Sarcoma Associated with the Acquired Immunodeficiency Syndrome. *Annals of Internal Medicine*, **100**, 671–6.

Groopman, J. E., Mitsuyasu, R. T., DeLeo, M. J., Oette, D. H. & Golde, D. W. (1987). Effect of recombinant human granulocyte-macrophage colony-stimulating factor on myelopoiesis in the acquired immunodeficiency syndrome. *The New England Journal of Medicine*, **317**, 593–8.

Gruters, R. A., Neefjes, J. J., Tersmette, M., de Goede, R. E. Y., Tulp, A., Huisman, H. G., Miedema, F. & Ploegh, H. L. (1987). Interference with HIV-induced syncytium formation and viral infectivity by inhibitors of trimming glucosidase. *Nature, London*, **350**, 74–7.

Gurgo, C., Ray, R. & Green, M. (1972). Rifamycin derivatives strongly inhibiting RNA-DNA polymerase (reverse transcriptase) of murine sarcoma viruses. *Journal of the National Cancer Institute*, **49**, 61–79.

Hagler, D. N. & Frame, P. T. (1986). Azidothymidine neurotoxicity. *Lancet*, **ii**, 1392–3.

Hamamoto, Y., Nakashima, H., Matsui, T., Matsuda, A., Ueda, T. & Yamamoto, N. (1987). Inhibitory effect of 2′,3′-didehydro-2′,3′-dideoxynucleosides on infectivity, cytopathic effects, and replication of human immunodeficiency virus. *Antimicrobial Agents and Chemotherapy*, **31**, 907–10.

Hansen, J., Billich, S., Schulze, T., Sukrow, S. & Moelling, K. (1988). Partial purification and substrate analysis of bacterially expressed HIV protease by means of monoclonal antibody. *The EMBO Journal*, **7**, 1785–91.

Hao, Z., Cooney, D. A., Hartman, N. R., Perno, C. F., Fridland, A., De Vico, A. L., Sarngadharan, M. G., Broder, S. & Johns, D. G. (1988). Factors determining the activity of 2′,3′-dideoxynucleosides in suppressing human immunodeficiency virus *in vitro*. *Molecular Pharmacology*, **34**, 431–5.

Harris, P. J. & Caceres, C. (1988). Azidothymidine in the treatment of AIDS. *The New England Journal of Medicine*, **318**, 250.

Hartmann, H., Vogt, M. W., Durno, A. G., Hirsch, M. S., Hunsmann, G. & Eckstein, F. (1988). Enhanced *in vitro* inhibition of HIV-1 replication by 3′-fluoro-3′-deoxythymidine compared to several other nucleoside analogs. *AIDS Research and Human Retroviruses*, **4**, 457.

Hartshorn, K. L., Neumeyer, D., Vogt, M. W., Schooley, R. T. & Hirsch, M. S. (1987*a*). Activity of interferons alpha, beta, and gamma against human immunodeficiency virus replication *in vitro*. *AIDS Research and Human Retroviruses*, **3**, 125–33.

Hartshorn, K. L., Sandstrom, E. F., Neumeyer, D., Paradis, T. J., Chou, T. C., Schooley, R. T. & Hirsch, M. S. (1986). Synergistic inhibition of human T-cell lymphotropic virus type III replication *in vitro* by phosphonoformate and recombinant alpha-A interferon. *Antimicrobial Agents and Chemotherapy*, **30**, 189–91.

Hartshorn, K. L., Vogt, M. W., Chou, T. C., Blumberg, R. S., Byington, R., Schooley, R. T. & Hirsch, M. S. (1987*b*). Synergistic inhibition of human immunodeficiency virus *in vitro* by azidothymidine and recombinant alpha A interferon. *Antimicrobial Agents and Chemotherapy*, **31**, 168–72.

Haseloff, J. & Gerlach, W. L. (1988). Simple RNA enzymes with new and highly specific endoribonuclease activities. *Nature, London*, **334**, 585–91.

Hawking, F. (1978). Suramin: With special reference to onchocerciasis. In: *Advances in Pharmacology and Chemotherapy*, 15th edn, Garatini, S., Hawking, F., Goldin, A. & Kopin, I. J. eds, 289–322. New York: Academic Press.

Hayashi, S., Phadtare, S., Zemlicka, J., Matsukura, M., Mitsuya, H. & Broder, S. (1988). Adenallene and cytallene: Acyclic nucleoside analogues that inhibit replication and cytopathic effect of human immunodeficiency virus *in vitro*. *Proceedings of the National Academy of Sciences, USA*, **85**, 6127–31.

Henry, K., Chinnock, B. J., Quinn, R. P., Fletcher, C. V., Pharm, D., de Miranda, P. & Balfour, H. H. (1988). Concurrent zidovudine levels in semen and serum determined by radioimmunoassay in patients with AIDS or AIDS-related complex. *The Journal of the American Medical Association*, **259**, 3023–6.

Herberman, R. B. & Pinsky, C. M. (1985). Polyribonucleotides for cancer therapy: summary and recommendations for further research. *Journal of Biological Response Modifiers*, **4**, 680–4.

Herdewijn, P., Balzarini, J., Baba, M., Pauwels, R., Van Aerschot, A., Janssen, G. & De Clercq, E. (1988). Synthesis and anti-HIV activity of different sugar-modified pyrimidine and purine nucleosides. *Journal of Medicinal Chemistry*, **31**, 2040–8.

Herdewijn, P., Pauwels, R., Baba, M., Balzarini, J. & De Clercq, E. (1987). Synthesis and anti-HIV activity of various 2′- and 3′-substituted 2′,3′-dideoxyadenosines: a structure–activity analysis. *Journal of Medicinal Chemistry*, **30**, 2131–7.

Ho, D. D., Hartshorn, K. L., Rota, T. R., Andrews, C. A., Kaplan, J. C., Schooley, R. T. & Hirsch, M. S. (1985). Recombinant human interferon alfa-A suppresses HTLV-III replication *in vitro*. *Lancet*, **i**, 602–4.

Ho, H.-T. & Hitchcock, M. J. M. (1989). Cellular pharmacology of 2′,3′-dideoxy-2′,3′-didehydrothymidine, a nucleoside analog active against human immunodeficiency virus. *Antimicrobial Agents and Chemotherapy*, **33**, 844–9.

Hochster, H., Dieterich, D., Laverty, M., Reichman, R., Plank, C., Pettinelli, C., Spector, S. & Richman, D. D. (1988). Toxicity of coadministered AZT and ganciclovir (DHPG) for CMV infection (AIDS Treatment Evaluation Unit (004)). [Abstract]. *IVth International Conference on AIDS, Stockholm, Sweden*.

Hochster, H., Liebes, L., Connor, J., Reichman, R. & Richman, D. D. (1988). Pharmacokinetics of combined AZT and DHPG (Ganciclovir). Results of a multicenter phase I study (ACTG 004). [Abstract]. *28th ICAAC, Los Angeles, CA, October 24*.

Horwitz, J. P., Chua, J. & Noel, M. (1964). Nucleosides. V. The monomesylates of 1-(2′-Deoxy-B-D-lyxofuranosyl) thymine. *Journal of Organic Chemistry*, **29**, 2076–8.

Horwitz, J. P., Chua, J., Noel, M. & Donatti, J. T. (1967). Nucleosides. XI. 2′,3′-dideoxycytidine. *Journal of Organic Chemistry*, **32**, 817–18.

Hussey, R. E., Richardson, N. E., Kowalski, M., Brown, N. R., Chang, H. C., Siliciano, R. F., Dorfman, T., Walker, B., Sodroski, J. & Reinherz, E. L. (1988). A soluble CD4 protein selectively inhibits HIV replication and syncyticum formation. *Nature, London*, **331**, 78–81.

Hymes, K. B., Greene, J. B. & Karpatkin, S. (1988). The effect of azidothymidine on HIV-related thrombocytopenia. *The New England Journal of Medicine*, **318**, 516–17.

Interferon Alpha Study Group (1988). A randomized placebo-controlled trial of recombinant human interferon alpha 2a in patients with AIDS. *Journal of Acquired Immune Deficiency Syndromes*, **1**, 111–18.

Ito, M., Baba, M., Sato, A., Pauwels, R., De Clercq, E. & Shigeta, S. (1987*a*). Inhibitory effect of dextran sulfate and heparin on the replication of human immunodeficiency virus (HIV) *in vitro*. *Antiviral Research*, **7**, 361–7.

Ito, M., Nakashima, H., Baba, M., Pauwels, R., De Clercq, E., Shigeta, S. & Yamamoto, N. (1987*b*). Inhibitory effect of glycyrrhizin on the in vitro infectivity and cytopathic activity of the human immunodeficiency virus [HIV (HTLV-III/LAV)]. *Antiviral Research*, **7**, 127–37.

Jackson, G. G., Paul, D. A., Falk, L. A., Rubenis, M., Despotes, J. C., Mack, D., Knigge, M. & Emeson, E. E. (1988*a*). Human immunodeficiency virus (HIV) antigenemia (p24) in the acquired immunodeficiency syndrome (AIDS) and the effect of treatment with zidovudine (AZT). *Annals of Internal Medicine*, **108**, 175–80.

Jackson, G. G., Rubenis, M., Knigge, M., Perkins, J. T., Paul, D. A., Despotes, J. C. & Spencer, P. (1988*b*). Passive immunoneutralization of human immunodeficiency virus in patients with advanced AIDS. *Lancet*, **ii**, 647–52.

Jacobson, M. A., Abrams, D. I., Volberding, P. A., Bacchetti, P., Wilber, J., Chaisson, R. D., Crowe, S., Howard, W. & Moss, A. (1989). Serum beta2-microglobulin decreases in patients with AIDS or ARC treated with Azidothymidine. *Journal of Infectious Diseases*, **159**, 1029–36.

Jacobson, M. A., Crowe, S., Levy, J., Aweeka, F., Gambertoglio, J., McManus, N. & Mills, J. (1988). Effect of foscarnet therapy on infection with human immunodeficiency virus in patients with AIDS. *Journal of Infectious Diseases*, **158**, 862–5.

U.C.W. ABERYSTWYTH LIBRARY

Jameson, B. A., Rao, P. E., Kong, L. I., Hahn, B. H., Shaw, G. M., Hood, L. E. & Kent, S. B. H. (1988). Location and chemical synthesis of a binding site for HIV-1 on the CD4 protein. *Science*, **240**, 1335–9.

Jentsch, K. D., Hunsmann, G., Hartmann, H. & Nickel, P. (1987). Inhibition of human immunodeficiency virus type 1 reverse transcriptase by suramin-related compounds. *Journal of General Virology*, **68**, 2183–92.

Johnson, V. A., Barlow, M. A., Chou, T.-C., Fisher, Ra., Walker, B. D., Hirsch, M. S. & Schooley, R. T. (1989). Synergistic inhibition of human immunodeficiency virus type 1 (HIV-1) replication in vitro by recombinant soluble CD4 and 3′-azido-3′-deoxythymidine. *Journal of Infectious Diseases*, **159**, 837–44.

Karpas, A., Fleet, G. W., Dwek, R. A., Petursson, S., Namgoong, S. K., Ramsden, N. G., Jacob, G. S. & Rademacher, T. W. (1988*a*). Aminosugar derivatives as potential anti-human immunodeficiency virus agents. *Proceedings of the National Academy of Sciences, USA*, **85**, 9229–33.

Karpas, A., Hill, F., Youle, M., Cullen, V., Gray, J., Byron, N., Hayhoe, F., Tenant-Flowers, M., Howard, L., Gilgen, D., Oates, J. K., Hawkins, D. & Gazzard, B. (1988*b*). Effects of passive immunization in patients with the acquired immunodeficiency syndrome-related complex and acquired immunodeficiency syndrome. *Proceedings of the National Academy of Sciences, USA*, **85**, 9234–7.

Katoh, I., Yasunaga, T., Ikawa, Y. & Yoshinaka, Y. (1987). Inhibition of retroviral protease activity by an aspartyl proteinase inhibitor. *Nature, London*, **329**, 654–6.

Kelley, J. A., Litterst, C. L., Roth, J. S., Vistica, D. T., Poplack, D. G., Cooney, D. A., Nadkarni, M., Balis, F. M., Broder, S. & Johns, D. G. (1987). The disposition and metabolism of 2′,3′-dideoxycytidine, and *in vitro* inhibitor of human T-lymphotropic virus type III infectivity, in mice and monkeys. *Drug Metabolism and Disposition*, **15**, 595–601.

Kennedy, C. J., Teschke, R. S., Hesselink, J., Berger, J., Fischl, M., Richman, D. D. & et al. (1987). Clinical evaluation of the central nervous system in HIV infected patients on azidothymidine (AZT) (Abstract). *III International Conference on AIDS, Washington, D.C., June.*

Klatzmann, D., Champagne, E., Chamaret, S., Gruest, J., Guetard, D., Hercend, T., Gluckman, J.-C. & Montagnier, L. (1984). T-lymphocyte T4 molecule behaves as the receptor for human retroviruses LAV. *Nature, London*, **312**, 767–8.

Klecker Jr, K. W., Collins, J. M., Yarchoan, R., Thomas, R., Jenkins, J. F., Broder, S. & Myers, C. E. (1987). Plasma and cerebrospinal fluid pharmacokinetics of 3′-azido-3′-deoxythymidine: a novel pyrimidine analog with potential application for the treatment of patients with AIDS and related diseases. *Clinical Pharmacology and Therapeutics*, **41**, 407–12.

Kohl, N. E., Emini, E. A., Schleif, W. A., Davis, L. J., Heimbach, J. C., Dixon, R. A. F., Scolnick, E. M. & Sigal, I. S. (1988). Active human immunodeficiency virus protease is required for viral infectivity. *Proceedings of the National Academy of Sciences, USA*, **85**, 4686–90.

Kornbluth, R. S., Oh, P. S., Munis, J. R., Cleveland, P. H. & Richman, D. D. (1989). Interferons and bacterial lipopolysaccharide protect macrophages from productive infection by HIV *in vitro*. *Journal of Experimental Medicine*, **169**, 1137–51.

Kotler, M., Katz, R. A., Danho, W., Leis, J. & Skalka, A. M. (1988). Synthetic peptides as substrates and inhibitors of a retroviral protease. *Proceedings of the National Academy of Sciences, USA*, **85**, 4185–9.

Krown, S. D., Real, F. X., Cunningham-Rundles, S., Myskowski, P. L., Koziner, B., Fein, S., Mittelman, A., Oettgen, H. F. & Safai, B. (1983). Preliminary observations on the effect of recombinant leukocyte A interferon in homosexual men with Kaposi's Sarcoma. *The New England Journal of Medicine*, **308**, 1071–6.

Lambert, R. W., Martin, J. A., Thomas, G. J., Duncan, I. B., Hall, M. J. & Heimer, E. P. (1989). Synthesis and antiviral activity of phosphonoacetic and phosphonoformic acid esters of 5-bromo-2′-deoxyuridine and related pyrimidine nucleosides and acyclonucleosides. *Journal of Medicinal Chemistry*, **32**, 367–74.

Landau, N. R., Warton, M. & Littman, D. R. (1988). The envelope glycoprotein of the human immunodeficiency virus binds to the immunoglobulin-like domain of CD4. *Nature, London*, **334**, 159–62.

Lane, H. C., Feinberg, J., Davey, V., Deyton, L., Baseler, M., Manischewitz, J., Masur, H., Kovacs, J. A., Herpin, B., Walker, R., Metcalf, J. A., Salzman, N., Quinnan, G. & Fauci, A. S. (1988). Anti-retroviral effects of interferon-alpha in AIDS-associated Kaposi's Sarcoma. *Lancet*, **ii**, 1218–22.

Lane, H. C., Siegel, J. P., Rook, A. H., Masur, H., Gelmann, E. P., Quinman, G. V. & Fauci, A. S. (1984). Use of interleukin-2 in patients with acquired immunodeficiency syndrome. *Journal of Biological Response Modifiers*, **3**, 512–16.

Lang, J. M., Trepo, C., Kirstetter, M., Herviou, L., Retornaz, G., Renoux, G., Musset, M., Touraine, J. L., Choutet, P., Falkernrodt, A., Livrozet, J. M., Remoux, M., Caraus, J. & AIDS-Imuthiol French Study Group (1988). Randomised, double-blind, placebo-controlled trial of ditiocarb sodium (Imuthiol) in human immunodeficiency virus infection. *Lancet*, **ii**, 702–5.

Lange, D. J., Britton, C. B., Younger, D. S. & Hays, A. P. (1988). The neuromuscular manifestations of human immunodeficiency virus infections. *Archives of Neurology*, **45**, 1084–8.

Larder, B. A., Darby, G. & Richman, D. D. (1989). HIV with reduced sensitivity to zidovudine isolated during prolonged therapy. *Science*, **243**, 1731–4.

Laskin, O. C., Longstreth, J. A., Hart, C. C., Scavuzzo, D., Kalman, C. M., Connor, J. D. & Roberts, R. B. (1987). Ribavirin disposition in high risk patients for acquired immunodeficiency syndrome. *Clinical Pharmacology and Therapeutics*, **41**, 546–55.

Lemaitre, M., Bayard, B. & Lebleu, B. (1987). Specific antiviral activity of a poly(L-lysine)-conjugated oligodeoxyribonucleotide sequence complementary to vesicular stomatitis virus N protein mRNA initiation site. *Proceedings of the National Academy of Sciences, USA*, **84**, 648–52.

Levine, A. M., Gill, P. S., Cohen, J., Hawkins, J. G., Formenti, S. C., Aguilar, S., Meyer, P. R., Krailo, M., Parker, J. & Rasheed, S. (1986). Suramin antiviral therapy in the acquired immunodeficiency syndrome: clinical, immunological and virologic results. *Annals of Internal Medicine*, **105**, 32–7.

Lidereau, R., Bouchet, C., Barre-Sinoussi, F., Saracino, R. & Chermann, J. C. (1977). Inhibition of spontaneously occurring mammary tumors in the mouse by tungstoantimoniate. *Current Chemotherapy*, **2**, 1323–5.

Lifson, J. D., Hwang, K. M., Nara, P. L., Fraser, B., Padgett, M., Dunlop, N. M. & Eiden, L. E. (1988). Synthetic CD4 peptide derivatives that inhibit HIV infection and cytopathicity. *Science*, **241**, 712–16.

Lin, J. C., Zhang, Z. X., Smith, M. C., Biron, K. & Pagano, J. S. (1988). Anti-human immunodeficiency virus agent 3′-azido-3′-deoxythymidine inhibits replication of Epstein–Barr virus. *Antimicrobial Agents and Chemotherapy*, **32**, 265–7.

Lin, T. S., Chen, M. S., McLaren, C., Gao, Y. S., Ghazzouli, I. & Prusoff, W. H. (1987*a*). Synthesis and antiviral activity of various 3′-azido, 3′-amino, 2′,3′-unsaturated, and 2′,3′-dideoxy analogues of pyrimidine deoxyribonucleosides against retroviruses. *Journal of Medicinal Chemistry*, **30**, 440–4.

Lin, T. S., Guo, J. Y., Schinazi, R. F., Chu, C. K., Xiang, J. N. & Prusoff, W. H. (1988). Synthesis and antiviral activity of various 3′-azido analogues of pyrimidine deoxyribonucleosides against human immunodeficiency virus (HIV-1, HTLV-III/LAV). *Journal of Medicinal Chemistry*, **31**, 336–40.

Lin, T. S. & Prusoff, W. H. (1978). Synthesis and biological activity of several amino analogues of thymidine. *Journal of Medicinal Chemistry*, **21**, 109–12.

Lin, T. S., Schinazi, R. F., Chem, M. S., Kinney-Thomas, E. & Prusoff, W. H. (1987*b*). Antiviral activity of 2′,3′-dideoxycytidin-2′-ene (2′,3′-dideoxy-2′,3′-didehydrocytidine) against human immunodeficiency virus *in vitro*. *Biochemical Pharmacology*, **36**, 311–16.

Lin, T. S., Schinazi, R. F. & Prusoff, W. H. (1987*c*). Potent and selective *in vitro* activity of 3′-deoxythymidin-2′ene (3′-deoxy-2′,3′-didehydrothymidine) against human immunodeficiency virus. *Biochemical Pharmacology*, **36**, 2713–18.

Liu, D. K. & Owens, G. F. (1987). Inhibition of viral reverse transcriptase by 2′,5′-oligoadenylates. *Biochemical and Biophysical Research Communications*, **145**, 291–7.

Lloyd, G., Atkinson, T. & Sutton, P. M. (1988). Effect of bile salts and of fusidic acid on HIV-1 infection of cultured cells. *Lancet*, **i**, 1418–21.

Lyerly, H. K., Cohen, O. J. & Weinhold, K. J. (1987). Transmission of HIV by antigen presenting cells during T-cell activation: prevention by 3′-azido-3′-deoxythymidine. *AIDS Research and Human Retroviruses*, **3**, 87–94.

Lyte, M. & Shinitzky, M. (1985). A special lipid mixture for membrane fluidization. *Biochimica et Biophysica Acta*, **812**, 133–8.

Maehlen, T. & Degre, M. (1989). Lack of activity of fusidic acid against human immunodeficiency virus in monocytes. *Antimicrobial Agents and Chemotherapy*, **33**, 680–3.

Mansuri, M. M., Starrett Jr, J. E., Ghazzouli, I., Hitchcock, M. J. M., Sterzycki, R. Z., Brankovan, V., Lin, T.-S., August, E. M., Prusoff, W. H., Sommadossi, J.-P. & Martin, J. C. (1989). 1-(2,3-Dideoxy-beta-dela-glycero-pent-2-enofuranosyl)thymine. A highly potent and selective anti-HIV agent. *Journal of Medicinal Chemistry*, **32**, 461–6.

Marquez, V. E., Tseng, C. K.-H., Kelley, J. A., Mitsuya, H., Broder, S., Roth, J. S. & Driscoll, J. S. (1987). 2′-,3′-dideoxy-2′-fluoro-ARA-A. An acid-stable purine nucleoside active against human immunodeficiency virus (HIV). *Biochemical Pharmacology*, **36**, 2719–22.

Matsukura, M., Shinozuka, K., Zon, G., Mitsuya, H., Reitz, M., Cohen, J. S. & Broder, S. (1987). Phosphorothioate analogs of oligodeoxynucleotides: Inhibitors of replication and cytopathic effects of human immunodeficiency virus. *Proceedings of the National Academy of Sciences, USA*, **84**, 7706–10.

Matsushita, S., Mitsuya, H., Reitz, M. S. & Broder, S. (1987). Pharmacological inhibition of *in vitro* infectivity of human T lymphotropic virus type 1. *The Journal of Clinical Investigation*, **80**, 394–400.

Matthes, E., Lehmann, Ch., Scholz, D., von Janta-Lipinski, M., Gaetner, K., Rosenthal, H. A. & Langen, P. (1987). Inhibition of HIV-associated reverse transcriptase by sugar-modified derivatives of thymidine 5′-triphosphate in comparison to cellular DNA polymerases alpha and beta. *Biochemical and Biophysical Research Communications*, **1**, 78–85.

Matthew, E., Lehmann, C., Scholz, D., Rosenthal, H. A. & Langen, P. (1988). Phosphorylation, anti-HIV activity and cytotoxicity of 3′-fluorothymidine. *Biochemical and Biophysical Research Communications*, **153**, 825–31.

McCormick, J. B., Getchell, J. P., Mitchell, S. W. & Hicks, D. R. (1984). Ribavirin suppresses replication of lymphadenopathy-associated virus in cultures of human adult T lymphocytes. *Lancet*, **ii**, 1367–9.

McCormick, J. B., King, I. J., Webb, P. A., Scribner, C. L., Craven, R. B., Johnson, K. M., Elliott, L. H. & Belmont-Williams, R. (1986). Lassa Fever: Effective therapy with ribavirin. *The New England Journal of Medicine*, **314**, 20–6.

McDougal, J. S., Kennedy, M. S., Sligh, J. M., Cort, S. P., Mawle, A. & Nicholson, J. K. A. (1986). Binding of HTLV-III/LAV to T4+ T cells by a complex of the 110K viral protein and the T4 molecule. *Science*, **231**, 382–5.

McGrath, M. S., Hwang, K. M., Caldwell, S. E., Gaston, I., Luk, K.-C., Wu, P., Ng, V. L., Crowe, S., Daniels, J., Marsh, J., Deinhart, T., Lekas, P. V., Vennari, J. C., Yeung, H.-W. & Lifson, J. D. (1989). GLQ223: an inhibitor of human immunodeficiency virus replication in acutely and chronically infected cells of lymphocyte and mononuclear phagocyte lineage. *Proceedings of the National Academy of Sciences, USA*, **86**, 2844–8.

Meek, T. D., Dayton, B. D., Metcalf, B. W., Dreyer, G. B., Strickler, J. E., Gorniak, J. G., Rosenberg, M., Moore, M. L., Magaard, V. W. & Debouck, C. (1989). Human immunodeficiency virus 1 protease expressed in *Escherichia coli* behaves as a dimeric aspartic protease. *Proceedings of the National Academy of Sciences, USA*, **86**, 1841–5.

Merigan, T. C., Skowron, G., Bozzette, S., Richman, D. D., Uttamchandani, R., Fischl, M. C., Schooley, R., Hirsch, M., Soo, W., Pettinelli, C., Schaumberg, H. & Study Group of the AIDS Clinical Trials Group (1989). Circulating p24 antigen levels and responses to dideoxycytidine in human immunodeficiency virus (HIV) infection. *Annals of Internal Medicine*, **110**, 189–94.

Mertelsmann, R., Welta, K., Sternberg, C., O'Reilly, R., Moore, M. A., Clarkson, B. D. & Oettgen, H. F. (1984). Treatment of immunodeficiency with interleukin-2: initial exploration. *Journal of Biological Response Modifiers*, **3**, 483–90.

Meruelo, D., Lavie, F. & Lavie, D. (1988). Therapeutic agents with dramatic antiretroviral activity and little toxicity at effective doses: Aromatic polycyclic diones hypericin and pseudohypericin. *Proceedings of the National Academy of Sciences, USA*, **85**, 5230–4.

Miller, P. S. & Ts'o, P. O. P. (1987). A new approach to chemotherapy based on molecular biology and nucleic acid chemistry: Matagen (masking tape for gene expression). *Anti-Cancer Design*, **2**, 117–28.

Mitchell, W. M., Montefiori, D. C., Robinson Jr, W. E., Strayer, D. R. & Carter, W. A. (1987). Mismatched double-stranded RNA (ampligen) reduces concentration of zidovudine (azidothymidine) required for *in-vitro* inhibition of human immunodeficiency virus. *Lancet*, **i**, 890–2.

Mitsuya, H. & Broder, S. (1986). Inhibition of the in vitro infectivity and cytopathic effect of human T-lymphotropic virus type III/lymphadenopathy-associated virus (HTLV-III/LAV) by 2′,3′-dideoxynucleosides. *Proceedings of the National Academy of Sciences, USA*, **83**, 1911–15.

Mitsuya, H. & Broder, S. (1987). Strategies for antiviral therapy in AIDS. *Nature, London*, **325**, 773–8.

Mitsuya, H. & Broder, S. (1988). Inhibition of infectivity and replication of HIV-2 and SIV in helper T-cells by 2′,3′-dideoxynucleosides *in vitro*. *AIDS Research and Human Retroviruses*, **4**, 107.

Mitsuya, H., Jarrett, R. F., Matsukura, M., Di Marzo Veronese, F., De Vico, A. L., Sarngadharan, M. G., Johns, D. G., Reitz, M. S. & Broder, S. (1987*a*). Long-term inhibition of human T-lymphotropic virus type III/lymphadenopathy-associated virus (human immunodeficiency virus) DNA synthesis and RNA expression in T cells protected by 2′,3′-dideoxynucleosides *in vitro*. *Proceedings of the National Academy of Sciences, USA*, **84**, 2033–7.

Mitsuya, H., Looney, D. J., Kuno, S., Ueno, R., Wong-Staal, W. & Broder, S. (1988). Dextran sulfate suppression of viruses in the HIV family: Inhibition of virion binding to CD4+ cells. *Science*, **240**, 646–8.

Mitsuya, H., Matsukura, M. & Broder, S. (1987*b*). Rapid *in vitro* systems for assessing activity of agents against HTLV-III/LAV. In: *AIDS: Modern concepts and therapeutic challenges*, Broder, S. *ed*., pp. 303–33. New York: Marcel Dekker, Inc.

Mitsuya, H., Popovic, M., Yarchoan, R., Matsushita, S., Gallo, R. C. & Broder, S. (1984). Suramin protection of T cells *in vitro* against infectivity and cytopathic effect of HTLV-III. *Science*, **226**, 172–4.

Mitsuya, H., Weinhold, K. J., Furman, P. A., St. Clair, M. H., Nusinoff-Lehrman, S., Gallo, R. C., Bolognesi, D., Barry, D. W. & Broder, S. (1985). 3′-azido-3′-deoxythymidine (BW A509U): An antiviral agent that inhibits the infectivity and cytopathic effect of human T-lymphotropic virus type III/lymphadenopathy-associated virus *in vitro*. *Proceedings of the National Academy of Sciences, USA*, **82**, 7096–100.

Miyajima, A., Miyatake, S., Schreurs, J., De Vries, J., Arai, N., Yokota, T. & Arai, K. I. (1988). Coordinate regulation of immune and inflammatory responses by T cell-derived lymphokines. *FASEB Journal*, **2**, 2462–73.

Montefiori, D. C. & Mitchell, W. M. (1986). Infection of the HTLV-II-bearing T-cell line C3 with HTLV-III/LAV is highly permissive and lytic. *Virology*, **155**, 726–31.

Montefiori, D. C. & Mitchell, W. M. (1987). Antiviral activity of mismatched double-stranded RNA against human immunodeficiency virus *in vitro*. *Proceedings of the National Academy of Sciences, USA*, **84**, 2985–9.

Montefiori, D. C., Robinson Jr, W. E. & Mitchell, W. M. (1988). Mismatched dsRNA (ampligen) induces protection against genomic variants of the human immunodeficiency virus type 1 (HIV-1) in a multiplicity of target cells. *Antiviral Research*, **9**, 47–56.

Moskovitz, B. L. & The HPA-23 Cooperative Study Group (1988). Clinical trial of tolerance of HPA-23 in patients with acquired immune deficiency syndrome. *Antimicrobial Agents and Chemotherapy*, **32**, 1300–3.

Nakashima, H., Kido, Y., Kobayashi, N., Motoki, Y., Neushul, M. & Yamamoto, N. (1987*a*). Purification and characterization of an avian myeloblastosis and human immunodeficiency virus reverse transcriptase inhibitor, sulfated polysaccharides extracted from sea algae. *Antimicrobial Agents and Chemotherapy*, **31**, 1524–8.

Nakashima, H., Matsui, T., Harada, S., Kobayashi, N., Matsuda, A., Ueda, T. & Yamamoto, N. (1986*a*). Inhibition of replication and cytopathic effect of human T cell lymphotropic virus type III/lymphadenopathy-associated virus by 3′-azido-3′-deoxythymidine *in vitro*. *Antimicrobial Agents and Chemotherapy*, **30**, 933–7.

Nakashima, H., Tochikura, T., Kobayashi, N., Matsude, A., Ueda, T. & Yamamoto, N. (1987*b*). Effect of 3′-azido-2′,3′-dideoxythymidine (AZT) and neutralizing antibody on human immunodeficiency virus (HIV)-induced cytopathic effects: implication of giant cell formation for the spread of virus *in vivo*. *Virology*, **159**, 169–73.

Nakashima, H., Yoshida, T., Harada, S. & Yamamoto, N.(1986*b*). Recombinant human interferon gamma suppresses HTLV-III replication *in vitro*. *International Journal of Cancer*, **38**, 433–6.

Nathan, C. F. (1987). Secretory products of macrophages. *The Journal of Clinical Investigation*, **79**, 319–26.

Nordstrom, D. M., Petropolis, A. A., Giorno, R., Gates, R. H. & Reddy, V. B. (1989). Inflammatory myopathy and acquired immunodeficiency syndrome. *Arthritis and Rheumatism*, **32**, 475–9.

Nutt, R. F., Brady, S. F., Darke, P. L., Ciccarone, T. M., Colton, C. D., Nutt, E. M., Rodkey, J. A., Bennett, C. D., Waxman, L. H., Sigal, I. S., Anderson, P. S. & Veber, D. F. (1988). Chemical synthesis and enzymatic activity of a 99-residue peptide with a sequence proposed for the human immunodeficiency virus protease. *Proceedings of the National Academy of Sciences, USA*, **85**, 7129–33.

Oberg, B. (1983). Antiviral effects of phosphonoformate (PFA, Foscarnet Sodium). *Pharmacology and Therapeutics*, **19**, 387–415.

Oksenhendler, E., Bierling, P., Ferchal, F., Clauvel, J.-P. & Seligmann, M. (1989). Zidovudine for thrombocytopenic purpura related to human immunodeficiency virus (HIV) infection. *Annals of Internal Medicine*, **110**, 365–8.

Olsen, J. C., Furman, P., Fyfe, J. A. & Swanstrom, R. (1987). 3′-azido-3′-deoxythymidine inhibits the replication of avian leukosis virus. *Journal of Virology*, **61**, 2800–6.

Orholm, M., Pedersen, C., Mathiesen, L., Dowd, P. & Nielsen, J. O. (1989). Suppression of p24 antigen in sera from HIV-infected individuals with low-dose alpha interferon and zidovudine: a pilot study. *AIDS*, **3**, 97–100.

Ostertag, W., Roesler, G., Krieg, C. J., Kind, J., Cole, T., Crozier, T., Gaedicke, G., Steinheider, G., Kluge, N. & Dube, S. (1974). Induction of endogenous virus and of thymidine kinase by bromodeoxyuridine in cell cultures transformed by Friend virus. *Proceedings of the National Academy of Sciences, USA*, **71**, 4980–5.

Palmisano, L., Chisesi, T., Galli, M., Gritti, F. M., Jelasi, G., Lazzarin, A., Mezzarome, I., Moroni, M., Raise, E., Vaglia, A. & Aiuti, F. (1988). Thymostimulin treatment in AIDS-related complex. *Clinical Immunology and Immunopathology*, **47**, 253–61.

Parenti, D. M., Simon, G. L., Scheib, R. G., Meyer III, W. A., Sztein, M. B., Paxton, H., DiGioia, R. A. & Schulof, R. S. (1988). Effect of lithium carbonate in HIV-infected patients with immune dysfunction. *Journal of Acquired Immune Deficiency Syndromes*, **1**, 119–24.

Pauwels, R., Balzarini, J., Schols, D., Baba, M., Desmyter, J., Rosenberg, I., Holy, A. & De Clercq, E. (1988). Phosphonylmethoxyethyl purine derivatives, a new class of anti-human immunodeficiency virus agents. *Antimicrobial Agents and Chemotherapy*, **32**, 1025–30.

Perno, C.-F., Yarchoan, R., Cooney, D. A., Hartman, N. R., Gartner, S., Popovic, M., Hao, Z., Gerrard, T. L., Wilson, Y. A., Johns, D. G. & Broder, S. (1988). Inhibition of human immunodeficiency virus (HIV-1/HTLV-IIIBa-L) replication in fresh and cultured human peripheral blood monocytes/macrophages by azidothymidine and related 2′,3′-dideoxynucleosides. *Journal of Experimental Medicine*, **168**, 1111–25.

Pert, C. B., Hill, J. M., Ruff, M. R., Berman, R. M., Robey, W. G., Arthur, L. O., Ruscetti, F. W. & Farrar, W. L. (1986). Octapeptides deduced from the neuropeptide receptor like pattern of antigen T4 in brain potently inhibit human immunodeficiency virus receptor binding and T-cell infectivity. *Proceedings of the National Academy of Sciences, USA*, **83**, 9254–8.

Pert, C. B., Ruff, M. R., Ruscetti, F., Farrar, W. L. & Hill, J. M. (1988). HIV receptor in brain and deduced peptides that block viral infectivity. In *Psychologi-*

cal, Neuropsychiatric, and Substance Abuse Aspects of AIDS, Bridge, T. P., Mirsky, A. F. & Goodwin, F. K. eds pp. 73–83. New York: Raven Press.

Pioger, J. C., Taburet, A. M., Fillastre, J. P. & Singlas, E. (1988). Pharmacokinetics of zidovudine (AZT) and its glucuronide in healthy volunteers and in uremic patients. A. *28th ICAAC, Los Angeles, CA, October 24*.

Pizzo, P. A., Eddy, J., Falloon, J., Balis, F. M., Murphy, R. F., Moss, H., Wolters, P., Brouwers, P., Jarosinski, P., Rubin, M., Broder, S., Yarchoan, R., Brunetti, A., Maha, M., Nusinoff-Lehrman, S. & Poplack, D. G. (1988). Effect of continuous intravenous infusion of zidovudine (AZT) in children with symptomatic HIV infection. *The New England Journal of Medicine*, **319**, 889–96.

Poiesz, B. J., Ruscetti, F. W., Mier, J. W., Woods, A. M. & Gallo, R. C. (1980). Detection and isolation of type-C retrovirus particles from fresh and cultured lymphocytes of a patient with cutaneous T-cell lymphoma. *Proceedings of the National Academy of Sciences, USA*, **77**, 7415–19.

Popovic, M., Sarngadharan, M. G., Read, E. & Gallo, R. C. (1984). Detection, isolation and continuous production of cytopathic retroviruses (HTLV-III) from patients with AIDS and pre-AIDS. *Science*, **224**, 497–500.

Richardson, N. E., Brown, N. R., Hussey, R. E., Vaid, A., Matthews, T. J., Bolognesi, D. P. & Reinherz, E. L. (1988). Binding site for human immunodeficiency virus coat protein gp120 is located in the NH2-terminal region of T4 (CD4) and requires the intact variable-region-like domain. *Proceedings of the National Academy of Sciences, USA*, **85**, 6102–6.

Richman, D. D., Andrews, J. & AZT Collaborative Working Group (1988*a*). Results of continued monitoring of participants in the placebo-controlled trial of zidovudine for serious human immunodeficiency virus infection. *American Journal of Medicine*, **85**, 208–13.

Richman, D. D., Fischl, M. A., Grieco, M. H., Gottlieb, M. S., Volberding, P. A., Laskin, O. L., Leedom, J. M., Groopman, J. E., Mildvan, D., Hirsch, M. S., Jackson, G. G., Durack, D. T., Nusinoff-Lehrman, S. & AZT Collaborative Working Group (1987*a*). The toxicity of azidothymidine (AZT) in the treatment of patients with AIDS and AIDS-related complex. A double-blind, placebo-controlled trial. *The New England Journal of Medicine*, **317**, 192–7.

Richman, D. D., Kornbluth, R. S. & Carson, D. A. (1987). Failure of dideoxynucleosides to inhibit human immunodeficiency virus replication in cultured human macrophages. *Journal of Experimental Medicine*, **166**, 1144–9.

Richman, D. D., Mitsuya, H., Broder, S. & Hostetler, K. Y. (1988*b*). Fusidic acid, HIV, and host cell toxicity. *Lancet*, **i**, 1051–2.

Rios, A., Mansell, P. W. A., Newell, G. R., Reuben, J. M., Hersh, E. M. & Gutterman, J. U. (1985). Treatment of acquired immunodeficiency syndrome-related Kaposis' sarcoma with lymphoblastoid interferon. *Journal of Clinical Oncology*, **3**, 506–12.

Robins, M. J. & Robins, R. K. (1964). The synthesis of 2′,3′-dideoxyadenosine from 2′-deoxyadenosine. *Journal of the American Chemistry Society*, **86**, 3585–6.

Rook, A. H., Masur, H., Lane, H. C., Frederick, W., Kasahara, T., Macher, A. M., Djeu, J. Y., Manischewitz, J. F., Jackson, L., Fauci, A. S. & Quinnan Jr, G. V. (1983). Interleukin-2 enhances the depressed natural killer and cytomegalovirus-specific cytotoxic activities of lymphocytes from patients with the acquired immune deficiency syndrome. *The Journal of Clinical Investigation*, **72**, 398–403.

Rozenbaum, W., Dormont, D., Spire, B., Vilmer, E., Gentilini, M., Griscelli, C., Montagnier, L., Barre-Sinoussi, F. & Chermann, J. C. (1985). Antimoniotungstate (HPA 23) treatment of three patients with AIDS and one with prodrome [letter]. *Lancet*, **i**, 450–1.

Ruprecht, R. M., Mullaney, S., Anderson, J. & Bronson, R. (1989). In vivo analysis of castanospermine, a candidate antiretroviral agent. *Journal of Acquired Immune Deficiency Syndromes*, **2**, 149–57.

Ruprecht, R. M., O'Brien, L. G., Rossoni, L. D. & Nusinoff-Lehrman, S. (1986). Suppression of mouse viremia and retroviral disease by 3′-azido-3′-deoxythymidine. *Nature, London*, **323**, 467–9.

Ruprecht, R. M., Rossoni, L. D., Haseltine, W. A. & Broder, S. (1985). Suppression of retroviral propagation and disease by suramin in murine systems. *Proceedings of the National Academy of Sciences, USA*, **82**, 7733–7.

Sandstrom, E. G., Kaplan, J. C., Byington, R. E. & Hirsch, M. S. (1985). Inhibition of human T-cell lymphotropic virus type III *in vitro* by phosphonoformate. *Lancet*, **i**, 1480–2.

Sanger, F., Nicklen, S. & Coulson, A. R. (1987). DNA sequencing with chain-terminating inhibitors. *Proceedings of the National Academy of Sciences, USA*, **74**, 5463–7.

Sarin, P. S., Agrawal, S., Civeira, M. P., Goodchild, J., Ikeuchi, T. & Zamecnik, P. C. (1988). Inhibition of acquired immunodeficiency syndrome virus by oligodeoxynucleoside methylphosphonates. *Proceedings of the National Academy of Sciences, USA*, **85**, 7448–51.

Sarin, P. S., Gallo, R. C., Scheer, D. I., Crews, F. & Lippa, A. S. (1985*a*). Effects of a novel compound (AL 721) on HTLV-III infectivity *in vitro*. *The New England Journal of Medicine*, **313**, 1289–90.

Sarin, P. S., Sun, D., Thorton, A. & Muller, W. E. G. (1987). Inhibition of replication of the etiologic agent of acquired immune deficiency syndrome (human T-lymphotropic retrovirus/lymphadenopathy-associated virus) by avarol and avarone. *Journal of the National Cancer Institute*, **78**, 663–6.

Sarin, P. S., Taguchi, Y., Sun, D., Thornton, A., Gallo, R. C. & Oberg, B. (1985*b*). Inhibition of HTLV-III/LAV replication by foscarnet. *Biochemical and Biophysical Research Communications*, **34**, 4075–9.

Schaffner, C. P., Plescia, O. J., Pontani, D. & et al. (1986). Anti-viral activity of amphotericin B methyl ester: inhibition of HTLV-III replication in cell culture. *Biochemical Pharmacology*, **35**, 4110–13.

Schattenkerk, J. K. M. E., Danner, S. A., Paul, D. A., van Boxtel, C. J., Miedema, F., Schellekens, P. T. A. & Goudsmit, J. (1988). Persistence of human immunodeficiency virus antigenemia in patients with the acquired immunodeficiency syndrome treated with a reverse transcriptase inhibitor, suramin. *Archives of Internal Medicine*, **148**, 209–11.

Schmitt, F. A., Bigley, J. W., McKinnis, R., Logue, P. E., Evans, R. W., Drucker, J. L. & The AZT Collaborative Working Group (1988). Neuropsychological outcome of zidovudine (AZT) treatment of patients with AIDS and AIDS-related complex. *The New England Journal of Medicine*, **319**, 1573–8.

Schneider, J. & Kent, S. B. H. (1988). Enzymatic activity of a synthetic 99 residue protein corresponding to the putative HIV-1 protease. *Cell*, **54**, 363–8.

Schola, D., Baba, M., Pauwels, R., Desmyter, J. & De Clercq, E. (1989). Specific interaction of aurintricarboxylic acid with the human immunodeficiency virus/CD4 cell receptor. *Proceedings of the National Academy of Sciences, USA*, **86**, 3322–6.

Schulof, R. S., Scheib, R. G., Parenti, D. M., Simon, G. L., DiGioia, R. A., Paxton, H. M., Sztein, M. B., Chandra, P., Courtless, J. W., Taguchi, Y. T.,

Sun, D. K., Goldstein, A. L. & Sarin, P. S. (1986). Treatment of HTLV-III/LAV-infected patients with D-penicillamine. *Arzneim.-Forsch./Drug Res.*, **36**, 1531–4.
Sharpe, A. H., Jaenisch, R. & Ruprecht, R. M. (1987). Retroviruses and mouse embryos: A rapid model for neurovirulence and transplacental antiviral therapy. *Science*, **236**, 1671–4.
Shibahara, S., Mukai, S., Morisawa, H., Nakashima, H., Kobayashi, S. & Yamamoto, N. (1989). Inhibition of human immunodeficiency virus (HIV-1) replication by synthetic oligo-RNA derivatives. *Nucleic Acids Research*, **17**, 239–52.
Simpson, D. M. & Bender, A. N. (1988). Human immunodeficiency virus-associated myopathy: analysis of 11 patients. *Annals of Neurology*, **24**, 79–84.
Simpson, M. V., Chin, C. D., Keilbaugh, S. A., Lin, T.-S. & Prusoff, W. H. (1989). Studies on the inhibition of mitochondrial DNA replication by 3′-azido-3′-deoxythymidine and other dieoxynucleoside analogs which inhibit HIV-1 replication. *Biochemical Pharmacology*, **36**, 1033–6.
Sjovall, J., Karlsson, A., Ogenstad, S., Sandstrom, E. & Saarimaki, M. (1988). Pharmacokinetics and absorption of foscarnet after intravenous and oral administration to patients with human immunodeficiency virus. *Clinical Pharmacology and Therapeutics*, **44**, 65–73.
Smith, D. H., Byrn, R. A., Marster, S. A., Gregory, T., Groopman, J. E. & Capon, D. J. (1987). Blocking of HIV-1 infectivity by a soluble, secreted form of the CD4 antigen. *Science*, **238**, 1704–7.
Smith, K. A. (1988). Interleukin-2: Inception, impact and implications. *Science*, **240**, 1169–76.
Smith, M. S., Brian, E. L. & Pagano, J. S. (1987). Resumption of virus production after human immunodeficiency virus infection of T lymphocytes in the presence of azidothymidine. *Journal of Virology*, **61**, 3769–73.
Smith, R. A. & Kirkpatrick, W. (1980). *RIBAVIRIN: A Broad Spectrum Antiviral Agent*, New York: Academic Press.
Sommadossi, J. P. & Carlisle, R. (1987). Toxicity of 3′-azido-3′-deoxythymidine and 9-(1,3-dihydroxy-2-propoxymethyl)guanine for normal human hematopoietic progenitor cells *in vitro*. *Antimicrobial Agents and Chemotherapy*, **31**, 452–4.
Sommadossi, J. P., Carlisle, R., Schinazi, R. F. & Zhou, Z. (1988). Uridine reverses the toxicity of 3′-azido-3′-deoxythymidine in normal human granulocyte-macrophage progenitor cells *in vitro* without impairment of antiretroviral activity. *Antimicrobial Agents and Chemotherapy*, **32**, 997–1001.
Spear, J. M., Benson, C. A., Pottage Jr, J. C., Paul, D. A., Landay, A. L. & Kessler, H. A. (1988). Rapid rebound of serum human immunodeficiency virus antigen after discontinuing zidovudine therapy. *Journal of Infectious Diseases*, **158**, 1132–3.
Spector, S. A., Kennedy, C., McCutchan, J. A., Bozzette, S. A., Straube, R. C., Connor, J. D. & Richman, D. D. (1989). The antiviral effect of zidovudine and ribavirin in clinical trials and the use of p24 antigen levels as a virologic marker. *Journal of Infectious Diseases*, **159**, 822–9.
Spivak, J. L., Bender, B. S. & Quinn, T. C. (1984). Hematologic abnormalities in the acquired immunodeficiency syndrome. *American Journal of Medicine*, **77**, 224–8.
Sredni, B., Caspi, R. R., Klein, A., Kalechman, Y., Danziger, Y., Ben Ya'akov, M., Tamari, T., Shalit, F. & Albeck, M. (1987). A new immunomodulating compound (AS-101) with potential therapeutic application. *Nature, London*, **330**, 173–6.
St. Clair, M. H., Richards, C. A., Spector, T., Weinhold, K. J., Miller, W. H., Langlois, A. J. & Furman, P. A. (1987). 3′-azido-3′-deoxythymidine triphosphate

as an inhibitor and substrate of purified human immunodeficiency virus reverse transcriptase. *Antimicrobial Agents and Chemotherapy*, **31**, 1972–7.

Starnes, M. C. & Cheng, Y. C. (1987). Cellular metabolism of 2′,3′-dideoxycytidine, a compound active against human immunodeficiency virus *in vitro*. *Journal of Biological Chemistry*, **262**, 988–91.

Stein, C. A. & Cohen, J. S. (1988). Oligodeoxynucleotides as inhibitors of gene expression: a review. *Cancer Research*, **48**, 2659–68.

Stern, R., Gold, J. & DiCarlo, E. F. (1987). Myopathy complicating the acquired immune deficiency syndrome. *Muscle and Nerve*, **10**, 318–22.

Streeter, D. F., Witkowski, J. T., Khare, G. P., Sidwell, R. W., Bauer, R. J., Robins, R. K. & Simon, L. N. (1973). Mechanism of action of 1-beta-delta-ribofuranosyl-1,2,4-triazole-3-carboxamide (Virazole), a new broad-spectrum antiviral agent. *Proceedings of the National Academy of Sciences, USA*, **20**, 1174–8.

Sundquist, B. & Larner, E. (1979). Phosphonoformate inhibition of visna virus replication. *Journal of Virology*, **30**, 847–51.

Sundquist, B. & Oberg, B. (1979). Phosphonoformate inhibits reverse transcriptase. *Journal of General Virology*, **45**, 273–81.

Surapaneni, N., Raghunathan, R., Beall, G. N., Daniel, K. & Mundy, T. M. (1985). Levamisole, immunostimulation, and the acquired immunodeficiency syndrome (Letter). *Annals of Internal Medicine*, **102**, 137.

Swiss Group for Clinical Studies on AIDS (1988). Zidovudine for the treatment of HIV-associated thrombocytopenia: a placebo-controlled prospective study. *Annals of Internal Medicine*, **109**, 718–21.

Takemoto, K. K. & Spicer, S. S. (1965). Effects of natural and synthetic sulfated polysaccharides on viruses and cells. *Annals of the New York Academy of Sciences*, **130**, 365–73.

Tamura, K., Makino, S., Araki, Y., Imamura, T. & Seita, M. (1987). Recombinant interferon beta and gamma in the treatment of adult T-cell leukemia. *Cancer*, **59**, 1059–62.

Tanaka, N., Okabe, T., Take, Y., Inougy, Y., Nakamura, S., Nakashima, H. & Yamamoto, N. (1986). Inhibition by sakyomicin A of avian myeloblastosis virus reverse transcriptase and proliferation of AIDS-associated virus (HTLV-III/LAV). *Japanese Journal of Cancer Research*, **77**, 324–6.

Tavares, L., Roneker, C., Johnston, K., Nusinoff-Lehrman, S. & de Noronha, F. (1987). 3′-azido-3′-deoxythymidine in feline leukemia virus-infected cats: a model for therapy and prophylaxis of AIDS. *Cancer Research*, **47**, 3190–4.

Tochikura, T. S., Nakashima, H., Tanabe, A. & Yamamoto, N. (1988). Human immunodeficiency virus (HIV)-induced cell fusion: Quantification and its application for the simple and rapid screening of anti-HIV substances *in vitro*. *Virology*, **164**, 542–6.

Torseth, J., Bhatia, G., Harkonen, S., Skinner, M., Robinson, W. S., Blaschke, T. F. & Merigan, T. C. (1989). Evaluation of the antiviral effect of rifabutin in AIDS-related complex. *Journal of Infectious Diseases*, **159**, 1115–18.

Toyoshima, K. & Vogt, P. K. (1969). Enhancement and inhibition of avian sarcoma viruses by polycations and polyanions. *Virology*, **384**, 414–26.

Traunecker, A., Luke, W. & Karjalainen, K. (1988). Soluble CD4 molecules neutralize human immunodeficiency virus type 1. *Nature, London*, **331**, 84–6.

Tsai, C.-C., Follis, K. E. & Benveniste, R. E. (1989). Antiviral effects of 3′-azido-3′-deoxythymidine, 2′,3′-dideoxycytidine, and 2′,3′-dideoxydenosine against simian acquired immunodeficiency syndrome-associated type D retrovirus *in vitro*. *AIDS Research and Human Retroviruses*, **4**, 359–68.

Ueno, R. & Kuno, S. (1987). Dextran sulfate, a potent anti-HIV agent in vitro having synergism with zidovudine (Letter). *Lancet*, **i**, 1379.

Ullman, B., Coons, T., Rockwell, S. & McCartan, K. (1988). Genetic analysis of 2′,3′-dideoxycytidine incorporation into cultured human T lymphoblasts. *Journal of Biological Chemistry*, **263**, 12391–6.

Vogt, M. W., Hartshorn, K. L., Furman, P. A., Chou, T. C., Fyfe, J. A., Coleman, L. A., Crumpacker, C., Schooley, R. T. & Hirsch, M. S. (1987). Ribavirin antagonizes the effect of azidothymidine on HIV replication. *Science*, **235**, 1376–9.

Volberding, P. (1984). Therapy of Kaposi's Sarcoma in AIDS. *Seminars in Oncology*, **11**, 60–7.

Volberding, P., Moody, D. J., Beardslee, D., Bradley, E. C. & Wofsy, C. B. (1987). Therapy of acquired immune deficiency syndrome with recombinant interleukin-2. *AIDS Research and Human Retroviruses*, **3**, 115–24.

Vrang, L. & Oberg, B. (1986). PPi analogs as inhibitors of human T-lymphotropic virus type III reverse transcriptase. *Antimicrobial Agents and Chemotherapy*, **29**, 867–72.

Vrang, L., Oberg, B., Lower, J. & Kurth, R. (1988). Reverse transcriptases from human immunodeficiency virus type 1 (HIV-1), HIV-2, and simian immunodeficiency virus (SIV mac) are susceptible to inhibition by foscarnet and 3′-azido-3′-deoxythymidine triphosphate. *Antimicrobial Agents and Chemotherapy*, **32**, 1733–4.

Walker, B. D., Kowalski, M., Goh, W. C., Kozarsky, K., Krieger, M., Rosen, C., Rohrschneider, L., Haseltine, W. A. & Sodroski, J. (1987). Inhibition of human immunodeficiency virus syncytium formation and virus replication by castanospermine. *Proceedings of the National Academy of Sciences, USA*, **84**, 8120–4.

Walker, R. E., Parker, R. I., Kovacs, J. A., Masur, H., Lane, H. C., Carleton, S., Kirk, E., Gralnick, H. R. & Fauci, A. S. (1988). Anemia and erythropoiesis in patients with the acquired immunodeficiency syndrome (AIDS) and Kaposi Sarcoma tested with zidovudine. *Annals of Internal Medicine*, **108**, 372–6.

Waqar, M. A., Evans, M. J., Manly, K. F., Hughes, R. G. & Huberman, J. A. (1984). Effects of 2′,3′-dideoxynucleosides on mammalian cells and viruses. *Journal of Cell Physiology*, **121**, 402–8.

Wondrak, E. M., Lower, J. & Kurth, R. (1988). Inhibition of HIV-1 RNA-dependent DNA polymerase and cellular DNA polymerases alpha, beta and lambda by phosphonoformic acid and other drugs. *Journal of Antimicrobial Chemotherapy*, **21**, 151–61.

Wong, G. H. W., Krowka, J. F., Stites, D. P. & Goeddel, D. V. (1988). In vitro anti-human immunodeficiency virus activities of tumor necrosis factor-alpha and interferon-lambda. *The Journal of Immunology*, **140**, 120–4.

Wu, A. M. & Gallo, R. C. (1974). Interaction between murine type-C virus RNA-directed DNA polymerase and rifamycin derivatives. *Biochimica et Biophysica Acta*, **240**, 419–36.

Yamamoto, J. K., Barre-Sinoussi, F., Bolton, V., Pedersen, N. C. & Gardner, M. B. (1986). Human alpha- and beta-interferon but not gamma- suppress the *in vitro* replication of LAV, HTLV-III, and ARV-2. *Journal of Interferon Research*, **6**, 143–52.

Yarchoan, R., Berg, G., Brouwers, P., Fischl, M. A., Spitzer, A. R., Wichman, A., Grafman, J., Thomas, R. V., Safai, B., Brunetti, A., Perno, C. F., Schmidt, P. J., Larson, S. M., Myers, C. E. & Broder, S. (1987). Response of human-immunodeficiency-virus-associated neurological disease to 3′-azido-3′-deoxythymidine. *Lancet*, **i**, 132–5.

Yarchoan, R., Klecker, R. W., Weinhold, K. J., Markham, P. D., Lyerly, H.

K., Durack, D. T., Gelmann, E., Nusinoff-Lehrman, S., Blum, R. M., Barry, D. W., Shearer, G. M., Fischl, M. A., Mitsuya, H., Gallo, R. C., Collins, J. M., Bolognesi, D. P., Myers, C. E. & Broder, S. (1986). Administration of 3′-azido-3′-deoxythymidine, an inhibitor of HTLV-III/LAV replication, to patients with AIDS or AIDS-related complex. *Lancet*, **i**, 575–80.

Yarchoan, R., Thomas, R. V., Allain, J. P., McAtee, N., Dubinsky, R., Mitsuya, H., Lawley, T. J., Safai, B., Myers, C. E., Perno, C.-F., Klecker, R. W., Wills, R. J., Fischl, M. A., McNeely, M. C., Pluda, J. M., Leuther, M., Collins, J. M. & Broder, S. (1988). Phase studies of 2′,3′-dideoxycytidine in severe human immunodeficiency virus infection as a single agent and alternating with zidovudine (AZT). *Lancet*, **i**, 76–81.

Yeom, Y.-H., Remmel, R. P., Huang, S.-H., Hua, M., Vince, R. & Zimmerman, C. L. (1989). Pharmacokinetics and bioavailability of carbovir, a carbocyclic nucleoside active against human immunodeficiency virus, in rats. *Antimicrobial Agents and Chemotherapy*, **33**, 171–5.

Yoshida, O., Nakashima, H., Yoshida, T., Kaneko, Y., Yamamoto, I., Matsuzaki, K., Uryu, T. & Yamamoto, N. (1988). Sulfation of the immunomodulating polysaccharide lentinan: a novel strategy for antivirals to human immunodeficiency virus (HIV). *Biochemical Pharmacology*, **37**, 2887–981.

Zaia, J. A., Rossi, J. J., Murakawa, G. J., Spallone, P. A., Stephens, D. A., Kaplan, B. E., Eritja, R., Wallace, R. B. & Cantin, E. M. (1988). Inhibition of human immunodeficiency virus by using an oligonucleoside methylphosphonate targeted to the tat-3 gene. *Journal of Virology*, **62**, 3914–17.

Zamecnik, P. C., Goodchild, J., Taguchi, Y. & Sarin, P. S. (1986). Inhibition of replication and expression of human T-cell lymphotropic virus type III in cultured cells by exogenous synthetic oligonucleotides complementary to viral RNA. *Proceedings of the National Academy of Sciences, USA*, **83**, 4143–6.

Zamecnik, P. C. & Stephenson, M. L. (1978). Inhibition of Rous sarcoma virus replication and cell transformation by a specific oligodeoxynucleotide. *Proceedings of the National Academy of Sciences, USA*, **75**, 280–4.

Zimmerman, T. P., Mahony, W. B. & Prus, K. L. (1987). 3′-azido-3′-deoxythymidine. *Journal of Biological Chemistry*, **262**, 5748–54.

Zon, L. I. & Groopman, J. E. (1988). Hematologic manifestations of the human immune deficiency virus (HIV). *Seminars in Hematology*, **25**, 208–18.

HERPESVIRUSES IN THE IMMUNOCOMPROMISED HOST

RICHARD J. WHITLEY, STUART GOLDSMITH, & JOHN GNANN

Departments of Medicine, Pediatrics and Microbiology University of Alabama at Birmingham, Birmingham, Alabama 35294, USA

INTRODUCTION

Over the past decade significant progress has been made in understanding the natural history, pathogenesis, and chemotherapy of herpesvirus infections of humans. The successful application of organ and bone marrow transplantation, as documented by prolonged patient survival, has become well established (Sutherland, Fryd & Morrow, 1983; Sutherland, Fryd & Strand, 1984). While such procedures have unequivocally led to improved quality and duration of life for individuals receiving transplanted organs and bone marrow, infections, particularly those associated with herpesviruses, are a continuing source of complication. The herpesviruses are the most common causes of viral infections in individuals undergoing transplantation. Similarly, the use of aggressive chemotherapy regimens in the management of individuals with malignancies has been associated with both chronic and frequently recurrent herpesvirus infections. The recent appearance of human immunodeficiency virus (HIV) infections as a cause of the acquired immunodeficiency syndrome (AIDS) has resulted in the documentation of a plethora of new manifestations of herpesvirus infections in these severely immunocompromised individuals. Such infections are associated with both increased morbidity and mortality.

The treatment of herpesvirus infections in humans with impaired host defences exemplifies successes in clinical care. The enhanced severity of normally benign viral infections is characteristic of the immunocompromised host whose defence mechanisms cannot control progressive infection. Herpes simplex virus (HSV), cytomegalovirus (CMV) and varicella-zoster virus (VZV) infections are the most common causes of viral diseases in these high risk individuals. The incidence, natural history, and clinical manifestations of these infections as well as preventive and therapeutic strategies for immunocompromised individuals have been well documented (Saral, Burns & Prentice, 1984; Richman, 1989; Whitley, 1989*a*, *b*). As with other infectious diseases, the effectiveness of antiviral therapy is ultimately dependent upon host defenses. The unique propensity of immunocompromised patients to develop symptomatic herpesvirus infections deserves special note. The ability of each of the herpesviruses to establish latent infection provides the opportunity for reactivated infection at a high frequency. In renal transplant recipients, for example, 40% to 70% of patients seropositive for HSV will reactivate latent infection within one year of transplantation; 80% to 100% of individuals will reactivate latent CMV and between 5% and 35% will develop herpes zoster (Pass, Long & Whitley, 1978; Betts, 1984; Whitley, 1989*a*). The recent documentation of isolates of HSV, VZV and CMV that are resistant to standard antiviral therapies will introduce new problems in the care of individuals who require repeated treatment.

This chapter will emphasise therapeutic approaches to opportunistic herpesvirus infections in the immunocompromised host and the utility of treatment in altering the natural history of disease. The treatment of human immunodeficiency virus infection and the management of hepatitis B infections in the immunocompromised host are considered elsewhere in this monograph.

General considerations

Several factors should be considered in the treatment of viral infections of the immunocompromised host. First, it has been difficult to provide clinically beneficial antiviral therapy without also affecting replication of normal host cells and thus introducing potential toxicity. Viruses are obligate intracellular parasites that utilise many of the biochemical pathways of normal host cell metabolism. Therefore, the development of antiviral agents that act independently of host cell processes is an essential component of future therapeutic approaches. Second, the success of antiviral therapy is dependent upon early diagnosis. Early diagnosis can be exceedingly difficult for some viral infections because of the lack of specificity of clinical findings, even in the immunocompromised host. Obviously, cutaneous infections such as herpes zoster or genital herpes are usually diagnosed clinically whereas CMV pneumonia and cytomegaloviraemia in the organ transplant recipient are prime examples of diseases in which early diagnosis is difficult. However, effective antiviral therapy is not possible without rapid, sensitive, specific, and practical means of diagnosing viral disease. Third, since some of the syndromes caused by viruses are relatively benign and self limited, even in the immunocompromised host, the therapeutic index (or ratio of efficacy to toxicity), must be high in order for a therapy to be clinically acceptable. An exception to this guideline is cytomegalovirus chorioretinitis in HIV-infected individuals where preservation of sight is a high priority.

The rapid proliferation of applied molecular biology has helped solve two of these problems. By identifying unique enzymes required for viral replication, it has become possible to distinguish the events of viral replication from those of the host cell replication. Thus, these virus-specific pathways can serve as targets for carefully designed antiviral agents. Examples of virus-specific enzymes include thymidine kinase of HSV and reverse transcriptase of HIV. Furthermore, several sensitive and specific diagnostic methods for viral infection have been made possible by recombinant DNA technology. By utilising monoclonal antibodies, DNA hybridisation or, more recently, polymerase chain reaction, early diagnosis is becoming more feasible.

CYTOMEGALOVIRUS INFECTIONS

General considerations and epidemiology

Cytomegalovirus is a ubiquitous agent in humans. Man is thought to be the only reservoir for human CMV infection; transmission usually occurs by direct or indirect person-to-person contact (Alford & Britt, 1989). Transmission of CMV occurs year-round and can either be vertical (i.e. from mother to baby) or horizontal (i.e. person to person). Identified sources of virus transmission include blood, saliva, urine, breast milk, and genital secretions (Hayes *et al.*, 1972; Reynolds, Stagno & Hosty, 1973; Lang & Kummer, 1975; Stagno, Reynolds & Pass, 1980; Alford & Britt, 1989). An important route of horizontal transmission among young children is infected saliva deposited on toys in the day-care environment (Pass, August & Dworsky, 1982). Rates of acquisition of infection appear to relate to socioeconomic status and housing conditions. In less-developed societies, where crowded living conditions are the rule, CMV seroprevalence is higher than in more developed countries (Alford, Stagno & Reynolds, 1981; Gold & Nankervis, 1982; Pass, August & Dworsky, 1982; Alford & Britt, 1989).

The epidemiology of CMV infections is best understood by an analysis of the age when infection is most frequently acquired. The rate of congenital infection is approximately 1% in developed countries with a range of 0.5% to 2.2% (Montgomery, Youngblood & Medearis, 1972; Stagno, Pass & Dworsky, 1982). In addition, many children (as high as 60%) become infected during the first six months of life due to contact with infected maternal genital secretions at the time of birth (natal acquisition), acquisition from maternal breast milk, or crowded living conditions (Hayes *et al.*, 1972; Reynolds *et al.*, 1973; Stagno *et al.*, 1982). Acquisition of infection later in life can occur in the day-care environment or in the school age population (Pass *et al.*, 1982). Other routes of CMV acquisition include blood transfusions or receipt of an organ transplant from an infected donor. By adulthood, 60% to 100% of individuals, depending on geographic area and socioeconomic status, have been infected by this virus.

Cytomegalovirus diseases in the immunocompromised host are encountered most frequently in individuals undergoing solid organ or bone marrow transplantation and patients with AIDS. Cytomegaloviral infection often results from reactivation of latent infection as opposed to acquisition of new infection; although individuals with AIDS can excrete different CMVs at one time (Spector, Hirata & Newman, 1984). This latter observation would imply acquisition of CMV strains from multiple partners upon sexual contact. Two other major routes of infection have been documented. Blood or blood products and the transplanted organ or bone marrow are all potential sources for CMV infection (Betts, 1982; Ho, 1982). Transfusion with multiple units of blood, particularly blood that has not been depleted of leukocytes, is associated with an increased risk. Primary infection of a seronegative individual, reactivation of latent virus in a seropositive individual, or exogenous reinfection of a seropositive individual can occur as consequences of blood transfusion. It is often not possible to differentiate between reactivation and reinfection and so the term recurrent is used to indicate active CMV infection in a seropositive individual. Prospective studies of the seroprevalence of cytomegalovirus infection following transfusions indicate that infection varies between 5% and 67% with an average incidence of approximately 20%. Among seronegative individuals, primary infection is associated with symptoms in approximately one-third of cases. This figure is greatly increased in patients who are immunocompromised and undergoing transplantation, as noted below (Ho, 1982).

The donor organ or bone marrow is another common source of CMV infection in the transplant recipient (Betts, Freeman & Douglas, 1975; Ho, Suwansirikul, & Dowling, 1975; Peterson, Balfour & Marker, 1980; Meyers, Flurnoy, & Thomas, 1986; Chou, 1987). Primary infection (which occurs in about 50% of patients) and recurrent infection (which occurs in about 85% of renal transplant patients) are predictable occurrences rather than unusual complications. Primary and recurrent CMV infections have been reported at similar frequencies for cardiac and bone marrow transplant recipients. Overall, between 70% and 100% of renal transplant recipients will experience CMV infection (either primary or recurrent) following transplantation (Lopez, Simmons, & Mauer, 1974; Simmons, Lopez, Balfour, 1974; Fiala, Payne & Berne, 1975; Peterson, Balfour & Marker, 1980; Glenn, 1981; Marker, Howard & Simmons, 1981). Clinical symptoms appear in virtually all patients with primary infection but in only 40% of those which are recurrent (Betts, 1982).

Life-threatening CMV infections occur at highest frequency following bone marrow transplantation or in individuals with HIV infection (Meyers, Spencer & Watts, 1975; Meyers *et al.*, 1986; Jacobson & Mills, 1988; Alford, & Britt, 1989). Following bone marrow transplantation, the development of CMV infection accounts for 5% to 35% higher mortality, depending upon the underlying disease, its therapy, site of infection, and the method of diagnosis (Saral *et al.*, 1984; Meyers *et al.*, 1986).

Clinical manifestations of disease

The clinical manifestations of CMV infections in the immunocompromised host are broad (Simmons *et al.*, 1974; Peterson *et al.*, 1980; Saral *et al.*, 1984; Meyers *et al.*, 1986; Jacobson & Mills, 1988; Whitley, 1989*b*). Symptoms are more likely to occur and to be serious following primary infection in the immunosuppressed host than in individuals experiencing reactivated infection. Importantly, the newer therapeutic regimens designed to prevent graft rejection, specifically the utilisation of OKT3 human monoclonal antibodies, has been associated with severe clinical disease even with reactivation of CMV. Infection caused by CMV following renal transplantation is usually associated with fever and, often, leukopenia and thrombocytopenia. A similar syndrome has been identified in heart transplant recipients and individuals with HIV infection (Gorensek *et al.*, 1988; Jacobson & Mills, 1988). Development of pneumonia is more common in bone marrow transplant recipients and in those with AIDS than in renal transplant recipients (Meyers *et al.*, 1975; Jacobson & Mills, 1988) and is becoming an increasingly important problem in heart transplant recipients (Dummer *et al.*, 1985). The depression of cell-mediated immune responses which occur either as a consequence of CMV infection or with therapeutic immunosuppressive regimens can be associated with other opportunistic infections, particularly *Pneumocystis carinii* pneumonia. Hepatic and/or gastrointestinal involvement with CMV may occur in as many as 20% of renal transplant recipients. In addition, both unilateral and bilateral CMV chorioretinitis have been identified in organ transplant recipients, and have been the most common opportunistic ocular infection in individuals with HIV infection (Jacobson & Mills, 1988). Other clinical manifestations of CMV infection that have been reported for transplant recipients include glomerulopathy, splenomegaly, arthralgia, myalgia, skin rashes, atypical lymphocytosis, and neuropathy.

Therapeutic strategies

Prevention

While not the primary focus of this discussion, four strategies have been studied for prevention of CMV infections in high-risk individuals. These include: vaccination with a live attenuated vaccine, (Plotkin, Smiley & Friedman, 1984; Balfour, Welo & Sachs, 1985) passive administration of CMV immunoglobulin (Stiehm, 1987), prevention of infection by selection of donors and blood products (matching seronegative donors and recipients), and antiviral chemo-prophylaxis. Antiviral prophylaxis has been accomplished either with CMV immune plasma, hyperimmune globulin, intravenous gamma globulin, and, more recently, acyclovir. Overall, these studies have reduced the incidence of infection by approximately 16%.

Similarly, CMV pneumonia is decreased by a similar percentage in bone marrow transplant recipients. Interferon has also been utilised to prevent the appearance of CMV infections in high risk patient populations, but is of less documented clinical value.

A prospective double-blind placebo controlled randomised study has been completed in renal transplant recipients which documented that the prophylactic administration of 200 mg of acyclovir five times a day for six months post-transplant decreased the frequency and severity of CMV infections in high risk individuals (Balfour *et al.*, 1989). Furthermore, it appeared that frequency of rejection was decreased in the treated patients as well.

Therapy

Therapy of cytomegalovirus infections in high risk organ transplant recipients has been attempted for numerous years with little evidence of success. Early studies with such nucleoside analogues as cytosine arabinoside and vidarabine failed to indicate a beneficial effect for the treatment of acute CMV infections (Dworsky *et al.*, 1984). Therapeutic trials with vidarabine have been reported in transplant recipients with CMV chorioretinitis. At least one published study suggested transient benefit (Pollard, Egbert & Gallagher, 1980). More recently, the administration of acyclovir at high dosages decreased the frequency and time to onset of CMV infections in renal transplant recipients. This observation is somewhat surprising, predicated on the poor *in vitro* sensitivity of CMV to acyclovir, but has been verified in bone marrow recipients with intravenous treatment, as well (Meyers *et al.*, 1988).

Therapy of CMV pneumonia in bone marrow transplant recipients has been particularly frustrating (Meyers *et al.*, 1988; Meyers, 1988). When therapy was administered as a single agent, namely vidarabine, acyclovir, or interferon, or with various combinations of these drugs, it was not of value. Only 8 of 58 patients survived treatment with combinations of vidarabine, acyclovir, and/or interferons.

Ganciclovir (9-di-hydroxyethoxymethyl guanine; DHPG) has been evaluated in extensive studies of CMV infection (specifically chorioretinitis, colitis or pneumonitis) in HIV-infected patients with life and sight-threatening diseases (Buhles *et al.*, 1988). Reports of efficacy have varied from favourable to unfavourable; however, proper controls have never been utilised (Whitley, 1987). Recent information suggesting a trend toward benefit has led to licensure of ganciclovir by the Food and Drug Administration for therapy of severe life and sight-threatening CMV infections. Dosages of 7.5 to 15 mg/kg/day for 14 to 21 days in the treatment of CMV chorioretinitis ameliorate progression and stabilise vision, at least for transient periods. Bone marrow toxicity may offset the potential value of this therapeutic approach for long-term treatment. The most important side-effects of ganciclovir, based on both animal and human studies, are neutropenia and thrombocytopenia. Neutropenia occurs in approximately 24% to 38% of all patients, and most commonly in those individuals with HIV infection (Buhles *et al.*, 1988; Matthews & Boehme, 1988). Neutropenia is almost always reversible with dosage reduction but some patients have been reported to have irreversible leukopenia. Thrombocytopenia occurs in 6% to 19% of patients. Agents with significant myelotoxicity, such as antimetabolites or alkylating agents, cannot be used concomitantly with ganciclovir. Azidothymidine (AZT) is also contraindicated in patients who receive ganciclovir, resulting in difficult management decisions for many AIDS patients. Numerous other side-effects related to ganciclovir include nausea, vomiting, dizziness, and headache which are not usually clinically significant (Food and Drug Administration, 1989). Ganciclovir also has significant gonadal toxicity and is a potent inhibitor of spermatogenesis, at least in animal models. In addition, ganciclovir causes an increased incidence of tumours in male mice, a finding of unknown significance. As an agent affecting DNA synthesis, ganciclovir has carcinogenic potential (Matthews & Boehme, 1988; Food and Drug Administration, 1989).

The most significant application of ganciclovir therapy in CMV infections appear to be for individuals suffering from chorioretinitis. Reports suggesting efficacy of ganciclovir were first published in 1985 and have continued to appear (Bach, Bagwe & Knapp, 1985; Collaborative DHPG Treatment Study Group, 1986; Felsenstein, D'Amico & Hirsch, 1986; Laskin, Stahl-Bayliss & Kalman, 1987). In one of these preliminary studies, intravenous ganciclovir was reported to heal retinal lesions and eliminate both viraemia and viral shedding in two patients with CMV retinitis. Ganciclovir has now been used in thousands of AIDS patients with CMV retinitis, most receiving ganciclovir on a 'compassionate plea' basis, thereby necessitating comparisons to historical controls (Buhles *et al.*, 1988). In one summary, of 125 patients, 84% had either improvement or stabilisation of their retinal lesions based on fundoscopic examinations as compared to only 2 of 20 historical controls. Induction therapy was usually at a dosage of 5 mg/kg twice daily for 10 to 14 days. The efficacy of maintenance therapy was also evaluated by sub-analyses of 61 cases. Median time to relapse for 20 patients receiving no maintenance therapy was 47 days. If this maintenance dosage was increased to 25 to 35 mg/kg per week, there appeared to be a significantly longer median time to relapse of 105 days. Other studies have reported similar outcomes (Matthews & Boehme, 1988). It has been noted that viraemia was cleared in approximately 7 to 10 days. Visual acuity usually stabilised at pre-treatment levels but rarely improved dramatically. Studies to date, therefore, have demonstrated a high probability of efficacy of ganciclovir therapy in AIDS patients with CMV retinitis.

Relapse occurs quickly in the absence of maintenance therapy but will develop eventually. The significance of bone marrow toxicity must be taken into consideration as well since 30% to 40% of these patients will develop neutropenia.

Other uses of ganciclovir in HIV-infected patients have focused on treatment of gastrointestinal tract disease. It has been suggested that as many as 85% of patients will benefit from ganciclovir therapy of CMV colitis (Masur, Lane & Palestine, 1986).

The value of the treatment of CMV infections of the lung is not as well established as with other forms of disease (Shepp, Dandliker & deMiranda, 1985; Reed, Dandliker & Meyers, 1986; Crumpacker *et al.*, 1988; Keay *et al.*, 1988; Sydman, 1988). As noted previously CMV pneumonia occurs in approximately 17% of bone marrow transplant recipients and, when untreated, has a mortality of 85%. In the first series of patients to receive ganciclovir following bone marrow transplantation, it was noted that patients improved from a virological standpoint, namely clearance of virus from blood and urine as well as respiratory secretions, but mortality remained high (Shepp, Dandliker & deMiranda, 1985). Other studies performed at different bone marrow transplant centres have reported more encouraging outcomes (Emanuel *et al.*, 1988; Reed *et al.*, 1988). More recently, ganciclovir combined with intravenous gamma globulin has been suggested to be of value for bone marrow transplant recipients with cytomegalovirus pneumonia (Reed *et al.*, 1988). Survival appears to be increased with this therapy from 5% to 10% with prior treatment or placebo controls to between 55% and 70%, depending upon the series (Emanuel *et al.*, 1988; Reed *et al.*, 1988). Thus, ganciclovir has become an accepted therapy on organ transplant services, either administered alone or with immune globulin.

Ganciclovir therapy of CMV infections in organ transplant recipients is being pursued aggressively, particularly in those individuals undergoing heart or bone marrow transplantation. It has been reported that mortality can be decreased for individuals with proven CMV pneumonia undergoing renal transplantation. For heart transplant recipients, it was noted in one study that approximately 80% of 14 patients had improvement on ganciclovir (Keay *et al.*, 1985). When death occurred it appeared to be related to acute rejection and not to CMV pneumonia. Nevertheless, relapse of CMV disease was documented in four of the 11 patients who improved during ganciclovir therapy.

The indications for maintenance ganciclovir therapy as well as the optimum dose and duration of maintenance therapy are all unresolved issues. Of patients with CMV retinitis, all had at least one episode of relapse of retinitis. Relapse is much less frequent with extraocular infections in the organ transplant population. The risk of infection or relapse due to specific immunosuppressants is also unknown.

Viral resistance to cytomegalovirus

Cytomegalovirus isolates resistant to ganciclovir *in vitro* have been identified (Crumpacker, 1989; Erice *et al.*, 1989). Clinically resistant CMV isolates have been reported more recently (Crumpacker, 1989) despite prolonged treatment at appropriate dosages with ganciclovir.

To summarise, ganciclovir is the most widely tested compound for the treatment of CMV infections. There is strong support (although based on uncontrolled trials) for clinical benefit in individuals with HIV infection, particularly those suffering from CMV retinitis and, perhaps, gastrointestinal infection. While ganciclovir appears to be beneficial in other immunocompromised patients with CMV retinitis as well, benefit remains to be established for other indications. Benefit is suggested but has been less dramatic for CMV pneumonia in AIDS patients and organ transplant recipients, but has nonetheless become the standard treatment. This apparent benefit, as compared to no treatment, makes further evaluation of ganciclovir in placebo controlled trials impossible. As a consequence, the United States Food and Drug Administration recently licensed this compound. Nevertheless, there are many key issues requiring further study to determine the optimum use of ganciclovir, including the precise indications, dose, and duration of acute and maintenance therapy, the prophylactic use of ganciclovir, the usefulness of other modalities combined with ganciclovir (e.g. immunoglobulin therapy).

HERPES SIMPLEX VIRUS

General considerations and epidemiology

Infections caused by HSV have been recognised since ancient times. Although many routes of infection have been suggested, transmission most often occurs in association with intimate, personal contact (Whitley, 1989). Virus must come in contact with mucosal surfaces or abraded skin for infection to be initiated. Following viral replication at the site of infection, the nucleocapsid is transported intra-axonally to the ganglia where latency is established. Infection with HSV type 1, generally limited to the oropharynx, can be transmitted by respiratory droplets or through direct contact of a susceptible individual with infected secretions (such as infectious virus contained in labial vesicular fluid). Acquisition often occurs during childhood. Infection with HSV 1 in the young child is usually asymptomatic, but when illness occurs, is usually manifest as gingivostomatitis. Primary infection in young adults has been associated with pharyngitis and a mononucleosis-like syndrome. Like other herpesvirus infections, seroprevalence studies have demonstrated that acquisition of HSV 1 infection is related to socioeconomic factors. Antibodies, indicative of past infection, are found early in life among individuals of lower socioeconomic groups, presumably the consequence of crowded living conditions which would provide a greater opportunity for direct contact between susceptible and infected individuals (Smith, Peutherer & MacCallum, 1967; Nahmias, Josey & Naib, 1970; Wentworth & Alexander, 1971; Whitley, 1989). As many as 75% to 90% of individuals from lower socioeconomic populations develop antibodies by the end of the first decade of life. In comparison, among the middle and upper socioeconomic groups only 30% to 40% are seropositive by the second decade of life. A change in seroprevalence rates of HSV 1 has been recognised, reflected by the acquisition of infection later in life during early adult years.

Because infections with HSV 2 are usually acquired through sexual contact, antibodies to this virus are rarely found until the age of onset of sexual activity. As with HSV 1 infections, the risk of acquisition of HSV 2 appears related to socioeconomic factors (Stavraky, Rawls & Chiavetta, 1983). Recent serologic studies using type-specific antibody determinations indicate that as many as 35% of middle class women receiving care through a variety of health delivery circumstances have antibodies to HSV 2 (Nahmias, Keyserling & Bain, 1985; Johnson *et al.*, 1989).

The epidemiology of HSV infections in the immunocompromised host reflects prior infection which, with appropriate immune suppression, results in reactivation and clinical disease. It is common for patients who have no history of either primary or recurrent HSV infections to experience recurrent infection following immunosuppressive therapy (Pass *et al.*, 1978).

Clinical manifestations

Clinical illness caused by infection with HSV occurs in the period immediately following the onset or development of immunosuppression. In studies of patients undergoing cardiac, renal or bone marrow transplantation, excretion of HSV persisted for prolonged periods of time, particularly in comparison to the normal host (Naraqui, Jackson & Jonasson, 1977; Meyers, Flournoy & Thomas, 1980; Saral, Burns & Prentice, 1984; Whitley, Levin & Barton, 1984). The average duration of excretion of virus in the untreated patient has been reported to be as long as three weeks. Overall, evidence of clinical disease becomes apparent in approximately 50% to 80% of patients while viral excretion only can be documented in between 18% and 50% of patients. Both viral excretion and clinical evidence of disease was present in between nearly 60% of patients undergoing organ or bone marrow transplantation (Naraqui *et al.*, 1977; Meyers *et al.*, 1980; Whitley *et al.*, 1984). It is important to recognise that nearly one-third of the immunocompromised patients who develop reactivation of HSV remain asymptomatic (Saral *et al.*, 1984). As such, these individuals still represent a source for transmission of virus to susceptible patients or relatives. Under such circumstances, health care providers who fail to wear gloves while providing oropharyngeal care can become infected and may develop a herpetic whitlow.

Clinical disease, which develops in nearly two-thirds of immunosuppressed individuals, involves mainly mucous membranes, including oropharyngeal and genital areas. Lesions heal more slowly than in the normal host. Specifically, organ transplant recipients have a time to total healing which can be as long as six weeks in contrast to 7 to 10 days for the normal host. Furthermore, evidence of contiguous spread to adjacent skin sites or even mucosal spread is not uncommon. For example, the development of HSV oesophagitis is a complication in these high-risk patient populations.

Disseminated HSV infection with multi-organ involvement is less common than that encountered with CMV in both HIV infected, cancer, and organ transplant patients. Nevertheless, in severely immunocompromised individuals, cutaneous dissemination may occur and lead to multisystem disease, involving the lung, central nervous system, adrenals, liver and kidneys (Whitley *et al.*, 1984). Prior to the availability of antiviral therapy, approximately 3% of bone marrow transplant recipients died of HSV pneumonitis (Ramsey, Fife & Hackman, 1982).

Therapeutic strategies

Prevention

The deployment of acyclovir, both intravenously and orally, has led to suppression of reactivation of HSV infection at a time critical for regeneration of host immune responses. Because of the extremely high incidence of HSV infections in immunocompromised patients, as well as the increased morbidity and mortality, prophylaxis of these infections is an important strategy of patient management. When acyclovir was administered intravenously for 18 days after bone marrow transplant, reactivation of latent infection was completely prevented in one study (Saral *et al.*, 1981). Subsequent studies of orally administered acyclovir at low dosages indicate supression of herpetic recurrences in a variety of patient populations (Wade *et al.*, 1982; Gluckman, Lotsberg & Devergie, 1983; Hann, Prentice & Blacklock, 1983; Saral, Ambinder & Burns, 1983). At many transplant centres acyclovir is now the standard suppressive treatment. Acyclovir prophylaxis for at least a year, and in some cases indefinitely, is common.

Therapy

While numerous nucleoside analogues have been utilised for therapy of HSV infections in the immunocompromised host, the greatest success has been achieved with acyclovir. Vidarabine has been shown useful in the management of HSV encephalitis, neonatal herpes, and varicella-zoster virus infections in the immunocompromised host. However, the activity of vidarabine in the management of HSV infections in the immunocompromised host is significantly less striking.

Numerous studies have unequivocally established acyclovir as the treatment of choice for HSV infection in these high risk patients. Intravenous, topical and orally administered acyclovir will lead to acceleration in the events of cutaneous healing (Mitchell, Gentry & Boen, 1981; Meyers, 1982; Straus, Smith, & Braickman, 1982). Duration of virus shedding can be decreased from nearly 17 days to approximately 3 days by acyclovir therapy. Similarly, the duration of pain is decreased from 13 to approximately 9 days. Healing is accelerated by approximately 7 days in treated patients as compared to placebo recipients. Striking consistency exists in outcome irrespective of the therapeutic formulation (Mitchell *et al.*, 1981; Meyers, 1982; Ramsey *et al.*, 1982; Straus *et al.*, 1982; Whitley *et al.*, 1984). At the present time, acyclovir should be considered the treatment of choice for the management of HSV infections in the immunocompromised host. The current intravenous dosage is 5–10 mg/kg administered over a period of one hour every 8 hours for varying durations of time (four to eight weeks). Alternatively, oral therapy can be administered at 200 mg three to five times a day over similar time frames. One major difference between these two routes of therapy is that oral acyclovir leads to significantly lower plasma levels than that associated with administration of intravenous medication.

While toxicity with acyclovir is relatively insignificant, it should be recognised that the drug is cleared by the kidney. Nephrotoxicity has been documented in selected patient populations, manifested by renal tubular necrosis. In addition, neurotoxicity at high doses has also been encountered (Wade & Meyers, 1982). The potential for this complication should increase caution regarding the administration of this medication to patients with impaired renal function. An additional concern is the appearance of HSV isolates which are resistant to acyclovir. Resistance to acyclovir can develop through mutations in two genes, the viral thymidine kinase and DNA polymerase genes (Fields, Darby & Wildy, 1980; Coen, Schaffer & Furnam, 1982; Larder & Darby, 1984). Alterations in viral thymidine kinase expression result in either a selection of thymidine kinase deficient mutants or a selection of mutants possessing a thymidine kinase which is unable to phosphorylate acyclovir (altered substrate specificity). Recently, clinical isolates of HSV resistant to acyclovir have been identified in individuals with HIV infection (Erlich *et al.*, 1989). Selected isolates from this group have also exhibited increased neurovirulence and pathogenicity. Until recently, such resistance has been rare and all virus mutants isolated had reduced virulence. This finding raises the concern that therapy of HSV infections in individuals with HIV infection may no longer be routinely amenable to treatment with acyclovir. As such, the development of new antivirals with alternative mechanisms of action will be required.

VARICELLA-ZOSTER VIRUS

General introduction and epidemiology

Varicella-zoster virus is responsible for two distinct clinical syndromes, the primary infection chicken-pox and its recurrent form herpes zoster, or shingles. Reviews of the natural history and epidemiology of VZV infections are numerous (Gelb, 1985; Whitley, 1986; Straus, Ostrove, & Inchauspe, 1988). Humans are the only known reservoir. Chicken-pox is highly contagious with an attack rate of at least 90% among susceptible seronegative individuals (Hope-Simpson, 1952). Both sexes and individuals of all races are infected equally. The virus is endemic in the population at large; however, it becomes epidemic among susceptible individuals during seasonal periods, usually late winter and early spring in the temperate zone. Children between the ages of 5 and 9 are most commonly affected and account for 50% of all cases. Most other cases occur in children between the ages of 1 and 4 and 10 and 14 (Hope-Simpson, 1952; Gelb, 1985). Over the age of 15 years, approximately 10% of the population of the United States is susceptible to infection. Thus, most immunocompromised individuals, particularly organ transplant recipients, are seropositive to VZV and, therefore, only at risk for recurrent infection. The obvious exception to this is the child who suffers from acute lymphoblastic leukaemia and is at risk for chicken-pox.

Herpes zoster, on the other hand, is a sporadic disease caused by reactivation of latent VZV from the dorsal root ganglia. Zoster is a disease which occurs at all ages but is most frequently seen among the elderly. Most patients who develop herpes zoster have no history of exposure to other individuals with VZV infection in the recent past. The incidence of disease peaks between five and ten cases per 1000 persons for individuals in the sixth to eighth decades of life (Hope-Simpson, 1965). Approximately 2% of herpes zoster patients who do not receive immunosuppressive therapy will later develop a second episode of infection.

Clinical manifestations

VZV disease in the immunocompromised host is significantly prolonged (as was the case with HSV infection) when compared to that encountered in the normal host. For the immunocompromised individual who develops chicken-pox, the clinical course of disease is protracted with prolonged vesicle formation and a delay in the progression of events to pustulation and scabbing. Lesions may often have an associated haemorrhagic base. The duration of disease is usually extended by threefold, at the very least. The propensity for developing visceral complications as a consequence of either primary or recurrent VZV infection is increased. Complications of chicken-pox in the immunocompromised host occur in as many as 25%

to 50% of individuals (Dolin, Reichman & Mazur, 1978; Merigan *et al.*, 1978; Arvin *et al.*, 1982; Whitley, Hilty & Haynes, 1982; Whitley, Soong & Dolin, 1982). These complications include pneumonitis, encephalitis, gastrointestinal disease, bleeding diatheses, hepatitis, and pancreatitis. The overall mortality from chicken-pox in the immunocompromised host is as high as 17% when no therapy is administered (Whitley, 1985).

Herpes zoster occurs in a dermatomal distribution with a vesicular rash. The onset of disease is often heralded by dermatomal pain for 24–48 hours prior to the appearance of a maculopapular or vesicular rash. Subsequently, the evolution of lesions is similar to that of chicken-pox. During herpes zoster in the normal host, the period of active viral replication is between 3 and 5 days with total healing occurring over 14 to 21 days. Two important components of disease are acute neuritis, a nearly uniform event in all infected patients, and post-herpetic neuralgia which can persist in approximately 25% of patients six months after disease onset. As noted for chicken-pox, the course of the disease in the immunocompromised host is severe due to the extension and duration of cutaneous lesions within a dermatome and the risk of cutaneous and visceral dissemination. Cutaneous dissemination occurs in 25% to 50% of patients, depending upon the extent of underlying immunosuppression (Merigan *et al.*, 1978; Arvin *et al.*, 1982; Whitley *et al.*, 1982). Specifically, for individuals with Hodgkin's disease the frequency of cutaneous dissemination when treated is 50%, compared to individuals who receive high dose steroids (for rheumatoid arthritis or nephrotic syndrome) who experience dissemination at a rate of approximately 15% to 20%. Furthermore, as many as 15% of patients will develop visceral organ involvement and 8% to 10% of patients will die as a consequence of disseminated VZV infection in the absence of antiviral therapy. Complications in these patients are similar to those encountered in the immunocompromised host with chicken-pox, namely meningoencephalitis, zoster ophthalmicus, hepatitis, pneumonitis, and motor paralysis, among others.

Therapeutic strategies

Prevention

Though not the subject of this discussion, it should be recognised that two preventive strategies are currently utilised or under development. The first is the routine administration of varicella-zoster immune globulin to exposed individuals at high risk who have not previously developed chicken-pox (Ross, 1962). Studies of zoster immune globulin under these circumstances indicate that infection can at least be attenuated if not prevented (Gershon, Steinberg & Brunell, 1974; Balfour, Groth & McCullough, 1977; Groth, McCullough & Marker, 1978; Zaia, Levin & Wright, 1978).

Second, significant effort has been devoted to the development of a safe and efficacious VZV vaccine (Takahashi, Otsuka & Okuno, 1974;

Takahashi, Okuno & Otsuke, 1975; Asano, Nakayama & Yazaki, 1977; Izawa, Ihara & Hattori, 1977; Ha, Baba & Ikeda, 1980; Gershon, Steinberg & Gelb, 1984; Weibal, Neff & Kuter, 1984). While originally targeted for immunocompromised children at risk for chicken-pox, vaccination may eventually be extended to include the normal host.

Treatment

Numerous therapeutic strategies have been developed for the management of VZV infections in the immunocompromised host. These infections occur sporadically after transplant or after the development of HIV infection, but occur at higher frequency and with increased risk so that targeting these patients for therapy becomes extremely important. Three therapeutic approaches have generated the most interest in controlled therapeutic trials at this time; interferon, vidarabine, and acyclovir therapy.

Leukocyte interferon derived from the Finnish blood bank by Dr Kari Cantell has been utilised in studies at Stanford University in the United States (Merigan *et al.*, 1978; Arvin *et al.*, 1982). These data can be summarised as indicating that leukocyte interferon employed under such circumstances is of value in accelerating the events of cutaneous healing and decreasing complications but not without evidence of adverse clinical and laboratory effects. Specifically, nearly 50% of patients experience fever, malaise, myalgia and flu-like symptoms after receiving one or two doses of leukocyte interferon. The availability of recombinant DNA interferon has allowed for the continuation of such studies in high risk organ transplant recipients. One such trial has suggested the potential value of an alpha interferon preparation for management of herpes zoster in high risk immunocompromised individuals. Patients entered into these therapeutic studies were not selected specifically as individuals undergoing organ transplantation.

Vidarabine has been extensively evaluated as a therapy for VZV infections in the immunocompromised host in controlled clinical trials since the mid-1970s. Data on the therapy of both chicken-pox and herpes zoster in immunocompromised patient populations (but again, not selected for organ transplant recipients solely) indicate the value of vidarabine for accelerating events of cutaneous healing, decreasing progressive cutaneous dissemination, and decreasing visceral complications. Mortality from both chicken-pox and herpes zoster was decreased to 0% in two controlled studies (Whitley *et al.*, 1982*a*; 1982*b*). Similarly, visceral complications from herpes zoster were reduced from 18% to 3%, and from 50% to 8% in patients with chicken-pox. These data provided a basis by which vidarabine became a licensed drug for VZV infections in the immunocompromised host.

More recently, acyclovir has been proven of value in placebo-controlled trials as well as in comparison to vidarabine. A small placebo-controlled trial suggested that acyclovir had value for therapy of chicken-pox in the immunocompromised host. Larger placebo controlled trials have not been reported; nevertheless, acyclovir appears to be used preferentially to vidarabine for this disease (Balfour, Bean & Laskin, 1983). One study reported no significant differences between acyclovir and vidarabine in the treatment of varicella; however, neurotoxicity has been encountered in another treatment trial with vidarabine when utilised in children with acute lymphoblastic leukaemia (Feldman *et al.*, 1986).

Acyclovir has also been demonstrated to be useful for therapy of herpes zoster in immunocompromised patients. An important study was published recently which indicated that in bone marrow transplant recipients, acyclovir is the treatment of choice for cutaneous disease when compared to vidarabine (Shepp, Dandliker & Meyers, 1987). The administration of acyclovir accelerated the events of cutaneous healing, decreased pain and prevented progressive cutaneous dissemination. This was the first study which unequivocally established the clinical value of acyclovir over vidarabine for the management of this infection.

Effective orally administered antiviral therapy for either chicken-pox or herpes zoster in the organ and bone marrow transplant recipient is not currently available. While some physicians have favoured the use of oral acyclovir, it has not been demonstrated that adequate plasma levels can be achieved by this route. Thus, at the present time, this approach should be discouraged. It is likely that newer antiviral medications administered orally will have improved bio-availability such that ambulatory therapy approaches can be developed for the majority of patients. An example of such a drug with good oral bio-availability is bromovinyl arabinosyl uracil. After initial encouraging studies performed in Japan, clinical pharmacologic and pharmacodynamic studies have been initiated in humans in order to determine dosages which could be used in controlled investigations. The preliminary experience with this compound is encouraging and sets the stage for future controlled studies.

EPSTEIN–BARR VIRUS

General considerations and epidemiology

Epstein–Barr virus was discovered fortuitously during the course of attempts to define the aetiology of lymphoma affecting children in certain parts of East Africa. As with other herpesviruses, EBV has the ability to establish latency and recur. However, in distinction to other herpesviruses, EBV does not replicate in continuous cell lines and cause cytopathic effect; its lymphotropic nature results in transformation of cell lines leading to immortalisation of lymphocytes. Infection caused by EBV has not been well characterised in the immunocompromised host.

The assessment of seroprevalence to EBV by an indirect immunofluorescence assay to viral capsid and nuclear antigens is a primary epidemiologic tool for studying prior infection. As with other herpesviruses, infection usually is acquired early in life as a consequence of person-to-person contact. Epstein–Barr virus infections are acquired throughout childhood and early adulthood. The prevalence of antibody to virus in young children differs considerably according to geographic locale. In developed countries, seroprevalence at school age is only 30% to 40% while in underdeveloped and tropical areas seroprevalence can be as high as 90% (Miller, 1985).

Transmission of EBV infection is primarily by saliva and oropharyngeal secretions. Blood transfusion has been incriminated as a route of infection as well (Gerber *et al.*, 1969; Evans, 1982). However, the demonstration of transmission by virus isolation is difficult and requires transformation of naive cord blood lymphocytes by the purportedly infectious oropharyngeal secretions. Thus documentation of infection by culture is not routinely available in diagnostic virology laboratories.

Clinical manifestations

Two clinical syndromes have been described which are manifestations of EBV infection. The classic clinical syndrome is infectious mononucleosis, a disease primarily seen in the normal host. However, in individuals with X-linked lymphoproliferative disease, clinical mononucleosis can be fatal (Hanto, Gajl-Peczalska & Frizzera, ·1983; Hanto, Frizzera & Gajl-Peczalska, 1985).

More importantly, EBV can be associated with the appearance of solid tumours in organ and bone marrow transplant recipients (Hanto *et al.*, 1983). These individuals often present years after transplantation. Sites of involvement have included gastrointestinal lymphomas and central nervous system and lung tumours. The frequency of occurrence of these clinical syndromes is undefined at the present time. However, with progressive increases in immunosuppressive therapeutic regimens, it is likely that they will continue to be recognised. One group of these patients has EBV induced B-cell proliferation that may be responsive to acyclovir in the early stages of disease when B-cell proliferation is polyclonal. If the B-cell growth progresses to involve only a monoclonal cell line, then, acyclovir has no effect on tumour growth. Similarly, no effect of acyclovir has been shown in the X-linked lymphoproliferative syndrome, in which young males may develop lymphoma after an episode of aggressive infectious mononucleosis.

Therapeutic strategies

At the present time, prophylactic and therapeutic approaches for EBV virus have not been effective, with the exception of polyclonal B-cell proliferation with tumour. While numerous investigators have tried nucleoside analogues and immunotherapy regimens, none has yet been proven efficacious. Ganciclovir, a compound used in the therapy of life and sight-threatening CMV infection, does have activity against EBV (Colby *et al.*, 1980); however, it has not been evaluated in extensive field studies.

HUMAN HERPESVIRUS 6

General considerations and epidemiology

The first description of human herpesvirus 6 was reported by Salahuddin *et al* (1986). The virus was fortuitously found in cells obtained from individuals with HIV infection. The seroprevalence of HHV 6 is extremely high, with infection apparently acquired early in life. Diseases associated with HHV 6 have not yet been defined.

NEW THERAPEUTIC APPROACHES

Foscarnet, a pyrophosphate analogue of phosphonoformic acid has potent *in vitro* and *in vivo* activity against all the herpesviruses. However, unacceptable toxicity was demonstrated with the parent compound phosphonoformic acid. Nevertheless, foscarnet is less toxic. These drugs are thought to inhibit DNA polymerase of all human herpesviruses by blocking the pyrophosphate binding site thus inhibiting the formation of the 3′–5′ phosphodiester bond between primer and substrate and preventing chain elongation. Foscarnet acts directly on the viral DNA polymerase and does not require a virus-specific thymidine kinase for activation. Recently, Foscarnet has been used successfully to treat severe mucocutaneous infection due to an acyclovir-resistant HSV 2 infection in an individual with HIV infection. Furthermore, foscarnet has attractive activity as a potential inhibitor of CMV infection in individuals with HIV infection. It remains to be determined whether or not this compound will have true therapeutic utility.

Newer antivirals, the phosphonate series such as HPMPC and PMEA have been demonstrated to have significant activity against CMV *in vitro*. The mechanism of action of these compounds is thought to be similar to that for the acyclic guanine derivatives (such as acyclovir) with inhibition of viral DNA polymerase. To date, the evaluations of HPMPC would indicate that it has the most significant activity against CMV of all available compounds.

ACKNOWLEDGEMENTS

Those studies performed by the authors and reported here were initiated and supported under a contract (NO1-AI-62554) with the Development and Applications Branch of the National Institute of Allergy and Infectious Diseases (NIAID), by grants from the General Clinical Research Center Program (RR-032) and the State of Alabama.

REFERENCES

Alford, C. A. & Britt, W. J., (1989). Cytomegalovirus: clinical, pathogenetic and therapeutic features. In *Virology*, 2nd edn. B. N. Fields, D. M. Knipe, R. Chanock, M. Hirsch, J. Melnick, T. Monath, & B. Roizman, eds., New York: Raven Press. In press.

Alford, C. A., Stagno, S. & Reynolds, D. W. (1981). Congenital cytomegalovirus infection. In *The Human Herpesviruses: An Interdisciplinary Perspective*, A. Nahmias, W. Dowdle & R. Schinazi, eds., pp. 159–71. New York: Elsevier.

Arvin, A. M., Kushner, J. H., Feldman, S., Baehner, R. L., Hammond, D. & Merigan, T. C. (1982). Human leukocyte interferon for the treatment of varicella in children with cancer. *New England Journal of Medicine*, **306**, 761–5.

Asano, Y., Nakayama, H. & Yazaki, T. *et al.* (1977). Protection against varicella in family contacts by immediate inoculation with live varicella vaccine. *Pediatrics*, **59**, 3–7.

Bach, M. C., Bagwe, S. P. & Knapp, N. P. *et al.* (1985). 9-(1,3-dihydroxy-2-ethoxymethyl) guanine for cytomegalovirus infections in patients with the acquired immunodeficiency syndrome. *Annals of Internal Medicine*, **103**, 381–2.

Balfour, H. H., Chace, B. A., Stapleton, J. T. & Simmons, R. L. *et al.* (1989). A randomized, placebo-controlled trial of oral acyclovir for the prevention of cytomegalovirus disease in recipients of renal allografts. *New England Journal of Medicine*, **320**, 1381–7.

Balfour, H. H., Bean, B. & Laskin, O. L. *et al.* (1983) Acyclovir halts progression of herpes zoster in immunocompromised patients. *New England Journal of Medicine*, **308**, 1448–53.

Balfour, H. H., Groth, K. E. & McCullough, J. *et al.* (1977). Prevention or modification of varicella using zoster immune plasma. *American Journal of Diseases of Children*, **131**, 693–6.

Balfour, H. H., Jr, Welo, P. K. & Sachs, G. W. (1985). Cytomegalovirus: Vaccine trial in 400 renal transplant candidates. *Transplantation Procedings*, **17**, 81.

Betts, R. F. (1982). Cytomegalovirus infections in transplant patients. *Progress in Medical Virology*, **28**, 44–64.

Betts, R. F. (1984). The relationship of epidemiology and treatment factors to infection and allograft survival in renal transplantation. In *CMV: Pathogenesis and Prevention of Human Infection*, S. A. Plotkin, S. Michelson, J. S. Pagano & F. Rapp, eds., pp. 87–99. New York: Alan R. Liss, Inc.

Betts, R. F. Freeman, R. B. & Douglas, R. G., *et al.* (1975). Transmission of cytomegalovirus infection with renal allograft. *Kidney International*, **8**, 387–94.

Buhles, W. C., Mastre, B. J., Tinker, A. J. & Strand, V. *et al.* (1988). Ganciclovir treatment of life- or sight-threatening cytomegalovirus infection experience in 314 immunocompromised patients. *Reviews of Infectious Diseases*, **10**, Suppl. 3, 495–503.

Chou, S. (1987). Cytomegalovirus infection and reinfection transmitted by heart transplantation. *Journal of Infectious Diseases*, **155**, 1054–6.

Coen, D. M., Schaffer, P. A. & Furnam, P. A. *et al.* (1982). Biochemical and genetic analysis of acyclovir-resistant mutants of herpes simplex virus type 1. *American Journal of Medicine*, **73(1A)**, 351–60.

Colby, B. M., Shaw, J. E., Elion, G. B. & Pagano, J. S. (1980). Effect of acyclovir [9-(2-hydroxyethoxymethyl) guanine] on Epstein–Barr DNA replication. *Journal of Virology*, **34**, 560–8.

Collaborative DHPG Treatment Study Group. (1986). Treatment of serious cytomegalovirus infections with 9-(1,3-dihydroxy-2-proproxymethyl) guanine in patients with AIDS and other immunodeficiencies. *New England Journal of Medicine*, **314**, 801–5.

Crumpacker, C., Marlowe, S., Zhang, J. L., Abrams, S. & Watkins, P. (1988). Treatment of cytomegalovirus pneumonia. *Reviews of Infectious Diseases*, **10**, 538–46.

Dolin, R., Reichman, R. C. & Mazur, M. H. *et al.* (1978). Herpes zoster-varicella infections in immunosuppressed patients. *Annals of Internal Medicine*, **89**, 375–88.

Dummer, J. S., White, L. T., Ho, M., Griffith, B. P., Hardesty, R. L. & Bahson, H. T. (1985). Morbidity of cytomegalovirus infection in recipients of heart or heart-lung transplants who received cyclosporine. *Journal of Infectious Diseases*, **152**, 1182–91.

Dworsky, M., Pass, R. F., Stagno, S. & Whitley, R. J. (1984). Therapeutic approaches to the control of cytomegalovirus infections. In *CMV: Pathogenesis and Prevention of Human Infection*, S. A. Plotkin, S. Michelson, J. S. Pagano, F. Rapp, eds, pp. 345–52. New York: Alan R. Liss, Inc.

Emanuel, D., Cunningham, S., Jules-Elysee, K. & Brockstein, J. A. *et al.* (1988). Cytomegalovirus pneumonia after bone marrow transplantation successfully treated with the combination of ganciclovir and high-dose intravenous immune globulin. *Annals of Internal Medicine*, **109**, 777–82.

Erice, A., Chou, S., Biron, K. K., Stanat, S. C., Balfour, H. H. & Jordan, M. C. (1989). Progressive disease due to ganciclovir-resistant cytomegalovirus in immunocompromised patients. *New England Journal of Medicine*, **320**, 289–92.

Erlich, K. S., Mills, J., Chatis, P., Merts, G. J., Busch, D. F., Follansbee, S. E., Grant, R. M. & Crumpacker, C. S. (1989). Acyclovir-resistant herpes simplex virus infections in patients with the acquired immunodeficiency syndrome. *New England Journal of Medicine*, **320**, 293–6.

Evans, A. S. (1982). The transmission of EB viral infections. In *Viral Infections in Oral Medicine*, J. Hooks & G. Jordan, eds., pp. 211–25, Holland, Amsterdam, Elsevier/North.

Feldman, S., Robertson, P. K., Lott, L. & Thornton, D. (1986). Neurotoxicity due to adenine arabinoside therapy during varicella-zoster virus infections in immunocompromised children. *Journal of Infectious Diseases*, **154**, 889–93.

Felsenstein, D., D'Amico, D. J. & Hirsch, M. S. *et al.* (1986). Treatment of cytomegalovirus retinitis with 9[2-hydroxy-1-(hydroxymethyl) ethoxymethyl] guanine. *Archives of Ophthalmology*, **104**, 1788–93.

Fiala, M., Payne, J. E. & Berne, T. V., *et al.* (1975). Epidemiology of cytomegalovirus infection after transplantation and immunosuppression. *Journal of Infectious Diseases*, **132**, 421–33.

Fields, H. J., Darby, G. & Wildy, P. (1980). Isolation and characterization of acyclovir-resistant mutants of herpes simplex virus. *Journal of General Virology*, **49**, 115–24.

Food and Drug Administration (1989). Ganciclovir (Cytovene) Package Insert.

Gelb, L. (1985). Varicella-zoster virus. In *Virology*, B. N. Fields, ed., pp. 591–627, New York: Raven Press.

Gerber, P., Walsh, J. N., Rosenblum, E. N. & Purcell, R. H. (1969). Association of EB-virus infection with the post-perfusion syndrome. *Lancet*, **i**, 593–6.

Gershon, A. A., Steinberg, S. & Brunell, P. A. (1974). Zoster immune globulin: A further assessment. *New England Journal of Medicine*, **290**, 243–5.

Gershon, A. A., Steinberg, S. P. & Gelb, L. *et al.* (1984). Live attenuated varicella vaccine: Efficacy for children with leukemia in remission. *The Journal of the American Medical Association*, **252**, 355–62.

Glenn, J. (1981). Cytomegalovirus infections following renal transplantation. *Review of Infectious Diseases*, **3**, 1151–78.

Gluckman, E., Lotsberg, J. & Devergie, A. *et al.* (1983). Prophylaxis of herpes infections after bone marrow transplantation by oral acyclovir. *Lancet*, **ii**, 706–8.

Gold, E. & Nankervis, G. A. (1982). Cytomegalovirus. In *Viral Infections of Humans: Epidemiology and Control*. A. S. Evans, ed., pp. 167–86. New York: Plenum.

Gorensek, M. J., Stewart, R. W., Keys, T. F. & McHenry, M. C. *et al.* (1988). A multivariate analysis of the risk of cytomegalovirus infection in heart transplant recipients. *The Journal of Infectious Diseases*, **157**, 515–22.

Groth, K. E., McCullough, J. & Marker, S. C. *et al.* (1978). Evaluation of zoster immune plasma: Treatment of cutaneous disseminated zoster in immunocompromised patients. *The Journal of the American Medical Association*, **239**, 1877–9.

Ha, K., Baba, K. & Ikeda, T. *et al.* (1980). Application of live varicella vaccine to children with acute leukemia or other malignancies without suspension of anticancer therapy. *Pediatrics*, **65**, 346–50.

Hann, J. M., Prentice, H. G. & Blacklock, H. A. *et al.* (1983). Acyclovir prophylaxis against herpes virus infections in severely immunocompromised patients: randomized double-blind trial. *British Medical Journal*, **287**, 384–8.

Hanto, D. W., Frizzera, G. & Gajl-Peczalska, K. J. *et al.* (1985). Epstein–Barr virus, immunodeficiency, and B-cell lymphoproliferation. *Transplantation*, **39**, 461–72.

Hanto, D. W., Gajl-Peczalska, K. J. & Frizzera, G. *et al.* (1983). Epstein–Barr virus (EBV) induced polyclonal and monoclonal B-cell lymphoproliferative diseases occurring after renal transplantation. *Annals of Surgery*, **198**, 356–69.

Hayes, K., Danks, D. M., Gibas, H. & Jack, I. (1972). Cytomegalovirus in human milk. *New England Journal of Medicine*, **287**, 117–8.

Ho, M. (1982). Epidemiology of cytomegalovirus infection in man. In *Current Topics in Infectious Disease*. W. B. Greenough, and T. C. Merigan, eds., vol. 4, pp. 79–104, New York: Plenum.

Ho, M., Suwansirikul, S. & Dowling, J. N., *et al.* (1975). The transplanted kidney as a source of cytomegalovirus infections. *New England Journal of Medicine*, **293**, 1109–12.

Hope-Simpson, R. E. (1952). Infectiousness of communicable diseases in the household (measles, chickenpox, and mumps). *Lancet*, **ii**, 549–54.

Hope-Simpson, R. E. (1965). The nature of herpes zoster: A long-term study and a new hypothesis. *Proceedings of the Royal Society of Medicine*, **58**, 9–20.

Izawa, T., Ihara, T. & Hattori, A. *et al.* (1977). Application of a live varicella vaccine in children with acute leukemia or other malignant disease. *Pediatrics*, **60**, 805–9.

Jacobson, M. A. & Mills, J. (1988). Serious cytomegalovirus disease in the acquired immunodeficiency syndrome (AIDS): clinical finding, diagnosis, and treatment. *Annals of Internal Medicine*, **108**, 585–94.

Johnson, R. E., Nahmias, A. J., Madger, L. S. & Brooks, C. A. *et al.* (1989). Seroepidemiologic survey of the prevalence of herpes simplex virus type 2 infection in the United States. *New England Journal of Medicine*, **321**, 7–12.

Keay, S., Peterson, E., Icenogle, T., Zeluff, B. J. & Samo, T. *et al.* (1988). Ganciclovir treatment of serious cytomegalovirus infection in heart and heart–lung transplant recipients. *Reviews of Infectious Diseases*, **10**, 563–71.

Lang, D. J. & Kummer, J. F., (1975). Cytomegalovirus in semen: Observations in selected populations. *Journal of Infectious Diseases*, **132**, 472–3.

Larder, B. A. & Darby, D. (1984). Virus drug-resistance: Mechanisms and consequences. *Antiviral Research*, **4**, 1–42.

Laskin, O. L., Stahl-Bayliss, C. M. & Kalman, C. M. *et al.* (1987). Use of ganciclovir to treat serious cytomegalovirus infections in patients with AIDS. *Journal of Infectious Diseases*, **155**, 323–7.

Lopez, C., Simmons, R. L. & Mauer, S. M. *et al.* (1974). Association of renal allograft rejection with virus infections. *American Journal of Medicine*, **56**, 280–9.

Marker, S. C., Howard, R. J. & Simmons, R. L. *et al.* 1981). Cytomegalovirus infection: A quantitative prospective study of 320 consecutive renal transplants. *Surgery*, **89**, 660–71.

Masur, H., Lane, H. C. & Palestine, A. *et al.* (1986). Effect of 9-(1,3-dihydroxy-2-propoxymethyl) quanine on serious cytomegalovirus disease in eight immunosuppressed homosexual men. *Annals of Internal Medicine*, **104**, 41–4.

Matthews, T. & Boehme, R. (1988). Antiviral Activity and Mechanism of Action of Ganciclovir. *Reviews of Infectious Diseases*, **10**, 490–4.

Merigan, T. C., Rand, K. H., Pollard, R. B., Abdallah, P. S., Jordan, G. W. & Freid, R. P. (1978). Human leukocyte interferon for the treatment of herpes zoster in patients with cancer. *New England Journal of Medicine*, **298**, 981–7.

Meyers, J. D. (1982). Intravenous acyclovir to treat mucocutaneous herpes simplex virus infection after marrow transplantation: a double-blind trial. *Annals of Internal Medicine*, **96**, 265–9.

Meyers, J. D. (1988). Management of cytomegalovirus infection. *The American Journal of Medicine*, **85**, 102–6.

Meyers, J. D., Flournoy, N. & Thomas, E. D. (1980). Infection with herpes simplex virus and cell-mediated immunity after marrow transplant. *Journal of Infectious Diseases*, **142**, 338–46.

Meyers, J. D., Flournoy, N. & Thomas, E. D. (1986). Risk factors for cytomegalovirus infection after human marrow transplantation. *Journal of Infectious Diseases*, **153**, 478–88.

Meyers, J. D., Spencer, H. C. & Watts, J. C. *et al.* (1975). Cytomegalovirus pneumonia after human bone marrow transplantation. *Annals of Internal Medicine*, **82**, 181–8.

Meyers, J. E., Reed, E. C., Shepp, D. H., Thornquist, M. & Dandliker, P. S. *et al.* (1988). Acyclovir for prevention of cytomegalovirus infection and disease after allogeneic marrow transplantation. *New England Journal of Medicine*, **318**, 70–5.

Miller, G. (1985). Epstein–Barr virus. In *Virology*, B. N. Fields, ed., pp. 563–96, New York: Raven Press.

Mitchell, C. D., Gentry, S. R. & Boen, J. R. *et al.* (1981). Acyclovir therapy of mucocutaneous herpes simplex infections in immunocompromised patients. *Lancet*, **i**, 1389–92.

Montgomery, R., Youngblood, L. & Medearis, D. N. (1972). Recovery of cytomegalovirus from the cervix in pregnancy. *Pediatrics*, **49**, 524–31.

Nahmias, A. J., Josey, W. E. & Naib, Z. M. *et al.* (1970). Antibodies to herpesvirus hominis types 1 and 2 in humans. *American Journal of Epidemiology*, **91**, 539–46.

Nahmias, A. J., Keyserling, H. & Bain, R. *et al.* (1985). Prevalence of herpes simplex virus (HSV) type-specific antibodies in a USA prepaid group medical practice population. Abstract, Sixth International Meeting of the International Society for STD Research, London, England.

Naraqui, S., Jackson, G. G. & Jonasson, O. *et al.* (1977). Prospective study of prevalence, incidence, and source of herpesvirus infections in patients with renal allografts. *Journal of Infectious Diseases*, **136**, 531–40.

Pass, R. F., August, A. M. & Dworsky, M. E. *et al.* (1982). Risk of day care acquisition of cytomegalovirus. *New England Journal of Medicine*, **307**, 477–9.

Pass, R. F., Long, W. K. & Whitley, R. J., *et al.* (1978). Productive infection with cytomegalovirus and herpes simplex virus in renal transplant recipients: Role of source of kidney. *Journal of Infectious Diseases*, **137**, 556–63.

Peterson, P. K., Balfour, H. H. & Marker, S. C. *et al.* (1980). Cytomegalovirus disease in renal allograft recipients: A prospective study of the clinical features, risk factors and impact on renal transplantation. *Medicine*, **59**, 283–300.

Plotkin, S. A., Smiley, M. L. & Friedman, H. M. *et al.* (1984). Towne-vaccine-induced prevention of cytomegalovirus disease after renal transplants. *Lancet*, **i**, 528–30.

Pollard, R. B., Egbert, P. R. & Gallagher, J. G. *et al.* (1980). Cytomegalovirus retinitis in immunosuppressed hosts. I. Natural history and effects of treatment and adenine arabinoside. *Annals of Internal Medicine*, **93**, 655–64.

Ramsey, P. G., Fife, K. H. & Hackman, R. C. *et al.* (1982). Herpes simplex virus pneumonia: Clinical, virologic, and pathologic features in 20 patients. *Annals of Internal Medicine*, **97**, 813–20.

Reed, E. C., Bowden, R. A., Dandliker, P. S., Lilleby, K. E. & Meyers, J. D. (1988). Treatment of cytomegalovirus pneumonia with ganciclovir and intravenous cytomegalovirus immunoglobulin in patients with bone marrow transplants. *Annals of Internal Medicine*, **109**, 783–8.

Reed, E. C., Dandliker, P. S. & Meyers, J. D. (1986). Treatment of cytomegalovirus pneumonia with 9-[2-hydroxy-1-(hydroxymethyl) ethoxymethyl] guanine and high-dose corticosteroids. *Annals of Internal Medicine*, **105**, 214–15.

Reynolds, D. W., Stagno, S. & Hosty, T. S. *et al.* (1973). Maternal cytomegalovirus excretion and perinatal infection. *New England Journal of Medicine*, **289**, 1–5.

Richman, D. (1989). Human Immunodeficiency Virus Infection. In *Antiviral Agents and Human Viral Infections*, G. Galasso, T. Merigan & R. J. Whitley, eds., New York: Raven Press. In press.

Ross, A. H. (1962). Modification of chickenpox in family contacts by administration of gamma globulin. *New England Journal of Medicine*, **267**, 369–76.

Salahuddin, S. Z., Albashi, D. V., Markham, P. D. *et al.* (1986). Isolation of a new virus, HBLV, in patients with lymphoproliferative disorders. *Science*, **234**, 596–601.

Saral, R., Ambinder, R. F. & Burns, W. H. *et al.* (1983). Acyclovir prophylaxis against herpes simplex virus infection in patients with leukemia: a randomized, double-blind, placebo controlled study. *Annals of Internal Medicine*, **99**, 773–6.

Saral, R., Burns, W. H. & Laskin, O. L. *et al.* (1981). Acyclovir prophylaxis of herpes simplex virus infections: a randomized double-blind controlled trial in bone marrow transplant recipients. *New England Journal of Medicine*, **305**, 63–7.

Saral, R., Burns, W. H. & Prentice, H. G. (1984). Herpes virus infections: clinical manifestations and therapeutic strategies in immunocompromised patients. *Clinics in Haematology*, **13**, 645–57.

Shepp, D., Dandliker, P. S. & Meyers, J. D. (1987). Treatment of varicella-zoster virus in severely immunocompromised patients: a randomized comparison of acyclovir and vidarabine. *New England Journal of Medicine*, **314**, 208–12.

Shepp, D. H., Dandliker, P. S. & deMiranda, P. *et al.* (1985). Activity of 9-[2-hydroxy-1-(hydroxymethyl) ethoxymethyl] guanine in the treatment of cytomegalovirus pneumonia. *Annals of Internal Medicine*, **103**, 368–73.

Simmons, R. L., Lopez, C. & Balfour, H. H. *et al.* (1974). Cytomegalovirus: Clinical virological correlations in renal transplant recipients. *Annals of Surgery*, **180**, 623–34.

Smith, I. W., Peutherer, J. F. & MacCallum, F. O. (1967). The incidence of herpesvirus hominis antibodies in the population. *Journal of Hygiene (Camb)*, **65**, 395–408.

Spector, S. A., Hirata, K. K. & Neuman, T. R. (1984). Identification of multiple cytomegalovirus strains infecting homosexual men with acquired immunodeficiency syndrome. *Journal of Infectious Diseases*, **150**, 953–6.

Stagno, S., Pass, R. F. & Dworsky, M. I. *et al.* (1982). Maternal Cytomegalovirus infection and perinatal transmission. In *Clinical Obstetrics and Gynecology*. G. E. Knox, ed., pp. 563–76. Philadelphia, Lippincott.

Stagno, S., Reynolds, D. W. & Pass, R. F. *et al.* (1980). Breast milk and the risk of cytomegalovirus infection. *New England Journal of Medicine*, **302**, 1073–5.

Stavraky, K. M., Rawls, W. E. & Chiavetta, J. *et al.* (1983). Sexual and socioeconomic factors affecting the risk of past infections with herpes simplex virus type 2. *American Journal of Epidemiology*, **118**, 109–21.

Stiehm, E. R. (1987). Intravenous immunoglobulin. *Annals of Internal Medicine*, **107**, 367–82.

Straus, S. E., Ostrove, J. M. & Inchauspe, G. *et al.* (1988). Varicella-zoster virus infections: Biology, natural history, treatment and prevention. *Annals of Internal Medicine*, **108**, 221–37.

Straus, S. E., Smith, H. A. & Braickman *et al.* (1982). Acyclovir for chronic mucocutaneous herpes simplex virus infection in immunosuppressed patients. *Annals of Internal Medicine*, **96**, 270–6.

Sutherland, D. E. R., Fryd, D. S. & Morrow, C. V. *et al.* (1983). Kidney transplantation in diabetics at the University of Minnesota: An analysis of the results by era. *Transplantation Proceedings*, **15**, 1110.

Sutherland, D. E. R., Fryd, D. S. & Strand, M. *et al.* (1984). Results of renal transplantation in diabetics at the University of Minnesota since 1979, including a comparison of outcome in diabetic and nondiabetic recipients randomized to cyclosporine versus azathioprine for immunosuppression. *Transplantation Proceedings*, **16**, 629–32.

Sydman, D. R. (1988). Ganciclovir therapy for cytomegalovirus disease associated with renal transplants. *Reviews of Infectious Diseases*, **10**, 554–60.

Takahashi, M., Okuno, Y. & Otsuka, T. *et al.* (1975). Development of a live attenuated varicella vaccine. *Biken Journal*, **18**, 25–33.

Takahashi, M., Otsuka, T. & Okuno, Y. *et al.* (1974). Live vaccine used to prevent the spread of varicella in children in hospitals. *Lancet*, **ii**, 1288–90.

Wade, J. C. & Meyers, J. D. (1982). Neurologic symptoms associated with parenteral acyclovir treatment after marrow transplantation. *Annals of Internal Medicine*, **98**, 921–5.

Wade, J. C., Newton, B., Flournoy, N. & Meyers, J. D. (1982). Oral acyclovir prophylaxis of herpes simplex virus infections after marrow transplant. Abstracts of the 22nd Interscience Conference of Antimicrobial Agents and Chemotherapy. Washington, DC, pp. 184.

Weibel, R. E., Neff, B. J. & Kuter, B. J. *et al.* (1984). Live attenuated varicella virus vaccine: Efficacy trial in healthy children. *New England Journal of Medicine*, **310**, 1409–15.

Wentworth, B. B. & Alexander, E. R. (1971). Seroepidemiology of infections due to members of the herpesvirus group. *American Journal of Epidemiology*, **94**, 496–507.

Whitley, R. J. (1985). Chickenpox in the immunocompromised host. In *Advances in Therapy Against Herpesvirus Infections in Immunocompromised Hosts*. H. H. Balfour, ed., pp. 25–30, New York: Park Row.

Whitley, R. J. (1986). Varicella-zoster virus infections. In *Harrison's Principles of Internal Medicine*, E. Braunwald, K. J. Isselbecher, R. G. Petersdorf, J. D. Wilson, J. B. Martin & A. S. Fauchi, eds., pp. 689–92, New York: McGraw Hill Publishing.

Whitley, R. J. (1987). Ganciclovir – have we established clinical value in the treatment of CMV infections? *Annals of Internal Medicine*. Editorial.

Whitley, R. J. (1989*a*). Herpes simplex virus. In *Virology*, 2nd Edition, B. N. Fields, D. M. Knipe, R. Chanock, M. Hirsch, J. Melnick, T. Monath & B. Roizman, eds., New York: Raven Press. In press.

Whitley, R. J. (1989*b*). Herpesvirus infections of organ transplant recipients. In *Medical Aspects of Transplantation*, R. G. Luke, J. C. Curtis, eds., Philadelphia: W. B. Saunders Company. In press.

Whitley, R. J., Hilty, J. & Haynes, R. *et al.* (1982*a*). Vidarabine therapy of varicella in immunosuppressed patients. *Journal of Pediatrics*, **101**, 125–31.

Whitley, R. J., Levin, M. J. & Barton, N. *et al.* (1984). Infections caused by herpes simplex virus in the immunocompromised host: Natural history and topical acyclovir therapy. *Journal of Infectious Diseases*, **150**, 323–33.

Whitley, R. J., Soong, S. J. & Dolin, R. *et al.* (1982*b*). Early vidarabine therapy to control the complications of herpes zoster in immunosuppressed patients. *New England Journal of Medicine*, **307**, 971–5.

Zaia, J. A., Levin, M. J. & Wright, G. G. *et al.* (1978). A practical method for preparation of varicella-zoster immune globulin. *Journal of Infectious Diseases*, **137**, 601–4.

NEW DEVELOPMENTS IN ANTIVIRAL THERAPY

JANET M. CAMERON

Department of Virology, Glaxo Group Research, Greenford Road, Greenford, Middlesex UB6 0HE, UK

In this review I will consider several new antiviral compounds that are under development as well as new developments in the search for antiviral drugs. The compounds described in detail are confined to those which are at the experimental or preclinical stages of development and not those already in clinical trials. The main human disease for which antiviral therapies are being developed and for which compounds are presently in clinical trials, are those caused by the herpes viruses, various respiratory viruses (e.g. influenza A and B virus, rhinovirus, respiratory syncytial virus) and the human retroviruses, especially human immunodeficiency virus. There is also a need to develop effective therapy against hepatitis B virus and papillomavirus.

Recently, much of the intellectual and financial effort in discovering novel antiviral agents has been focussed on the search for drugs effective in the treatment of HIV infected individuals. In many cases, this work has proceeded at the expense of other antiviral research, although there is still a strong interest in the development of novel anti-herpes drugs. Thus, over the past few years, the new compounds to emerge as possible development candidates have been mainly anti-HIV agents, with a few anti-herpes compounds. This account therefore concentrates on compounds under investigation as therapy for diseases caused by herpes viruses (HSV, VZV, CMV and EBV), or human immunodeficiency virus (HIV) infections, although some of the approaches taken to devise such therapies are also applicable to respiratory viruses. However, the prospects for developing antiviral agents active against hepatitis B virus and the papillomaviruses will also be discussed.

At present, the only antiviral compounds approved by the United States Food and Drug Administration for clinical use are adenine arabinoside (AraA), iododeoxyuridine (IDU), trifluorothymidine (TFT), and acyclovir (ACV) for various herpes virus infections, ribavirin for severe respiratory syncytial virus infections in infants and children, amantadine for influenza A virus infections and azidothymidine (AZT) for HIV infections (Fig. 1). With the exception of amantadine all these compounds are nucleoside analogues which, as triphosphates, act by inhibiting the viral DNA or RNA polymerase. Of these compounds ACV and AZT are the most successful

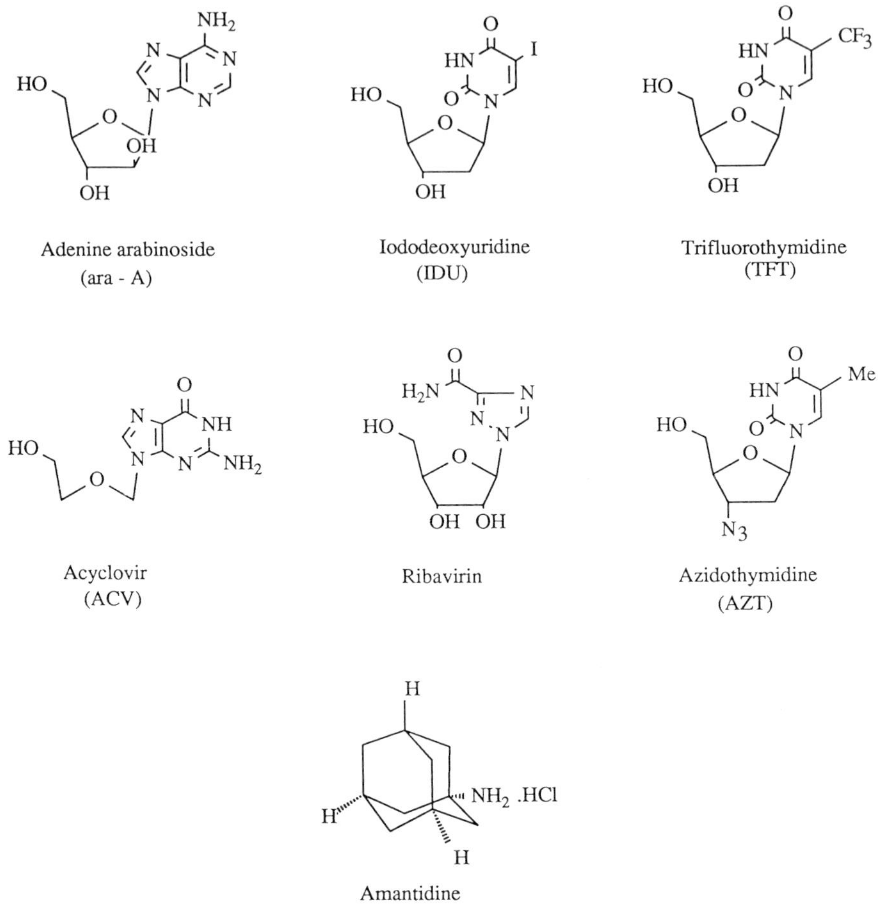

Fig. 1. Antivirals approved by FDA for clinical use.

in both clinical and commercial terms. With ACV licensed for the treatment of several diseases caused by HSV-1, HSV-2 or VZV, the clinical reasons to develop further therapies for these conditions are discussed.

It is clear that AZT, although giving benefit to particular groups of AIDS patients, is not a sufficient therapy for these patients or HIV positive individuals, and further drugs, either singly or in combination, are required.

Success with nucleosides as antiviral agents has generated the synthesis of numerous analogues, directed against one of a range of DNA or RNA viruses. Several of the more promising nucleoside antivirals are considered in this chapter.

One other feature of the success of nucleosides such as ACV and AZT has been to establish the virally encoded polymerases as good targets for the development of antiviral drugs. Work in many laboratories has identified other viral proteins (enzymes, receptor binding proteins or control proteins) which may also have potential as targets for antiviral chemotherapy. These targets are only beginning to be exploited in a rational way. The detailed biochemical and structural study of such proteins will allow the design of a new generation of inhibitors which should complement the traditional screening approach of drug discovery.

Progress in these areas and towards more experimental therapies, such as the use of antisense oligonucleotides is also described.

HERPES VIRUSES

The initial successes in antiviral therapy have been made in the development of treatment for herpes virus infections, more specifically those caused by HSV-1, HSV-2 and VZV. Acyclovir, despite its poor oral absorption, is the most successful of these therapies and has been clinically useful especially in the treatment of primary genital herpes infections, herpes simplex encephalitis, neonatal herpes simplex infections, and herpes simplex virus and varicella zoster virus infections in the immunocompromised host (reviewed by Fiddian, 1988). Despite these advances, oral or topical acyclovir treatment for recurrent oral and genital HSV infections gives only marginal benefit. The treatment of herpes zoster infection in the normal host is not ideal (poor oral absorption and bioavailability) (Fiddian, 1988) and there is also a need for improved therapy for herpes encephalitis (Whitley, 1988*a*). With the increase in numbers of immunocompromised patients, therapy is required to control not only recurrent HSV and VZV infections, but also those caused by CMV and EBV. There is no effective therapy at present for either CMV or EBV, although ganciclovir (DHPG) is being developed for CMV infections in the immunocompromised (reviewed by Whitley, 1988*b*) (Fig. 2). Lastly, none of these therapies affects the virus in the latent state.

Thus, effective anti-herpes agents are sought for treatment of the various diseases described above. Also, any compound which also prevents recurrence, shortens the course of the recurrent infection, reduces viral shedding or is better absorbed orally and so would allow for less frequent daily dosing, would have a clinical advantage over acyclovir.

It is improbable that a single drug will be effective against all diseases caused by the human herpes viruses and so the future will see the development of a number of effective therapies.

Until recently, attempts to find a successor to acyclovir for the treatment of HSV infections have concentrated on the synthesis of analogues of acyclovir and other nucleosides, analogues of phosphonoformic acid (PFA) or on the random screening of compounds to detect inhibitors of HSV replication *in vitro*. Thus, rationally designed compounds are directed towards the HSV DNA polymerase either as inhibitors which act directly on that enzyme or indirectly after specific phosphorylation by the viral thymidine kinase.

An alternative approach to antiviral therapy is to define further target enzymes essential for herpes virus replication. The complete sequence of the HSV-1 genome is now known (McGeogh *et al.*, 1986; McGeogh *et al.*, 1988) and sequence data suggest that HSV-1 encodes at least 70 polypeptides, although few have been assigned biochemical functions. However, the ability to clone, overexpress and mutate genes has helped in the identification of the function of several HSV proteins (Zhu and Weller, 1988; Goldstein & Weller, 1988*a*; Olivo, Nelson & Challberg, 1989; Mullaney, Moss & McGeogh, 1989; Smith & Smith, 1989; Worrad & Caradonna, 1988). Reports indicate that seven early polypeptides are essential for HSV DNA replication (Challberg, 1986; Olivo, Nelson & Challberg, 1988; McGeogh *et al.*, 1988; Wu *et al.*, 1988). Only one of these, the DNA polymerase, has been studied in any detail, whilst the others have poorly defined functions. One (UL9) binds to the origin of replication whilst three others (UL5, UL8 and UL52) may act as a complex with helicase/primase activity (Crute *et al.*, 1989). These proteins may provide further targets for the design of antiherpes drugs.

More is known of the early proteins involved in nucleoside and nucleotide metabolism, e.g. thymidine kinase and ribonucleotide reductase. Such enzymes appear to be non-essential in rapidly dividing cells *in vitro* but animal studies suggest that both of these enzymes are essential for virulence *in vivo* (see below). As such, they are attractive targets for viral chemotherapy and several groups report work on the development of inhibitors of the HSV thymidine kinase or ribonucleotide reductase as potential antivirals. One further target is thymidylate synthetase. Unlike HSV-1 or 2, VZV appears to encode a novel thymidylate synthetase activity (Thompson *et al.*, 1987). Its specificity and its requirement for VZV replication are as yet unclear but it may prove to be useful in the development of VZV therapies.

Nucleoside analogues

BRL 39123 (Fig. 2)

Since the discovery of acyclovir, many analogues have been prepared in attempts to increase the potency of this drug, extend its use to infections by other viruses or to increase its bioavailability. Harnden (1989) reviews one such series – the 9-[4-hydroxy 3-(hydroxymethyl)but-1-yl] purines – which are candidates for clinical trials. The guanine derivative (BRL 39123) has been reported to have activity comparable to acyclovir against laboratory strains and clinical isolates of HSV-1, HSV-2 and VZV in tissue culture. Thus, in plaque reduction assays with HSV-1, HSV-2 and VZV the mean 50% inhibiting dose (ID_{50}) concentrations were for BRL 39123 1.6, 5.9

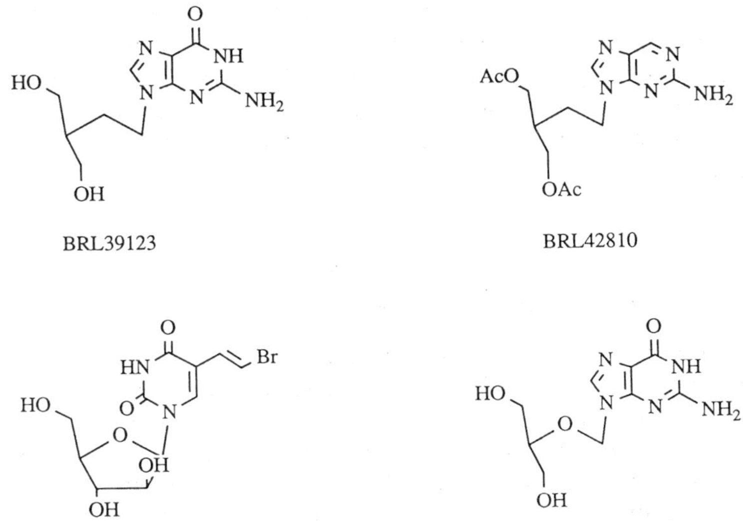

BRL39123 BRL42810

BVaraU DHPG

Fig. 2. Inhibitors of Herpes viruses.

and 12.0 μM respectively and for acyclovir 0.8, 2.4 and 15.0 μM respectively. This compound also was as active as, or more active than, acyclovir against HSV in a number of topical and systemic animal models. However, like acyclovir, the compound was poorly absorbed orally and so attempts were made to synthesise prodrugs of BRL 39123 with improved gastrointestinal absorption after oral dosing. Of these, the ester of the 2-aminopurine, BRL 42810 (9-[4-acetyl-3(acetoxymethyl)but-1-yl]-2-aminopurine) was rapidly absorbed after oral dosing and was efficiently converted to BRL 39123 *in vivo*. (Deacetylation was demonstrated *in vitro* by incubation with liver homogenates, and oxidation via xanthine oxidase leading to metabolism of this compound to BRL 39123) (Harnden, 1989). Clinical trials against HSV infections are planned using intravenous and topical formulations of BRL 39123 and oral formulations of BRL 42810.

BVaraU (Fig. 2)

There is a need for an effective treatment for recurrent VZV infections (shingles) in the normal host, as well as improved treatment for this recurrent infection in the immunocompromised (especially AIDS and transplant patients). One series of nucleosides with potent activity *in vitro* against a number of clinical isolates of VZV, is the 5-halogenovinyl-arabinosyluracil analogues (Machida, 1986). Of these compounds, the most potent is 1-β-D-arabinofuranosyl-E-5-(2-bromovinyl) uracil with an ID_{50} in a VZV plaque reduction assay of 0.0006–0.0025 μM and a therapeutic index* of >3,000,000. As with other compounds of this class, it is also active against HSV-1, but shows little effect against HSV-2 *in vitro*. It is now in further development for the treatment of recurrent VZV infections and of HSV-1 infections. Its excellent activity against VZV makes it an attractive antiviral candidate.

Other targets

Inhibitors of HSV thymidine kinase (Fig. 3)

Acyclovir and many other nucleoside analogues active against herpes viruses are substrates of the viral thymidine kinase (TK) and are selectively phosphorylated by the viral enzyme. The HSV TK is not essential for virus replication in dividing tissue culture cells but replication of TK^- mutants is compromised in serum-starved cells (Jamieson, Gentry & Subak-Sharpe, 1974). However, HSV TK is necessary for virulence *in vivo* (Field & Wildy, 1978). There is evidence that TK^- virus may not efficiently establish latent infections (Field & Darby, 1980; Tenser, Miller & Rapp, 1979) and more recently it has been suggested that such mutants may establish latency in mouse trigeminal ganglia but that the virus may not reactivate (Coen *et al.*, 1989). Thus a specific, irreversible inhibitor of viral TK may provide a novel approach to therapy.

Several groups have described nucleoside analogues which are inhibitors of HSV-1 and HSV-2 thymidine kinases. Sim *et al.* (1988) described a series of 5′ substituted thymidine analogues, and Focher *et al.* (1988) a series of N^2-phenyldeoxyguanosine nucleosides as being specific inhibitors of the HSV enzymes. The best of these compounds inhibited the viral enzyme in the μM range. Further thymidine derivatives have been described by Martin *et al.* (1988) and guanine derivatives by Ashton *et al.* (1988) which are several orders of magnitude more potent. This last series of compounds also acted *in vitro* as shown by their ability to antagonise the action of ganciclovir. So far there is no information on the activity *in vivo* of any of these compounds. As HSV-TK is also known to be involved in the establishment and/or reactivation of latent virus, these compounds may not only be of use in the treatment of a primary or recurrent infection, but may also inhibit or limit the reactivation of the latent virus. Detailed animal studies are required to investigate these possibilities.

Inhibitors of ribonucleotide reductase

Several of the herpes viruses (HSV-1, HSV-2, EBV, VZV and PRV) have

* $$\text{Therapeutic Index (TI)} = \frac{\text{Toxic dose}}{\text{Effective dose}}$$

ID_{50} (μM)

a) 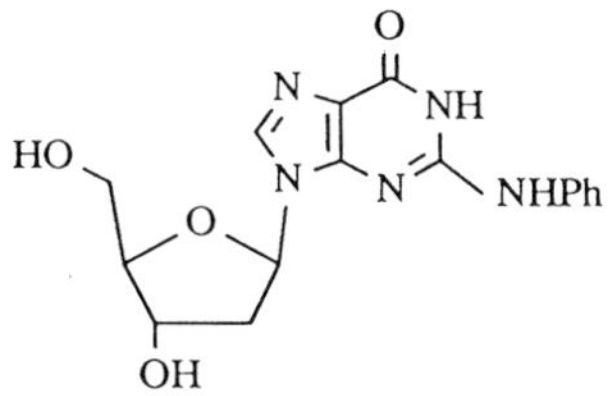 HSV1-TK 0.3

b) 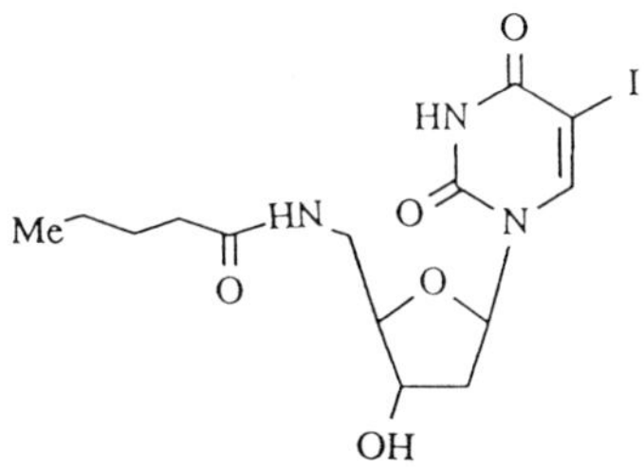 HSV1-TK 38(+/-15), HSV2-TK 2(+/-0.4)

c) 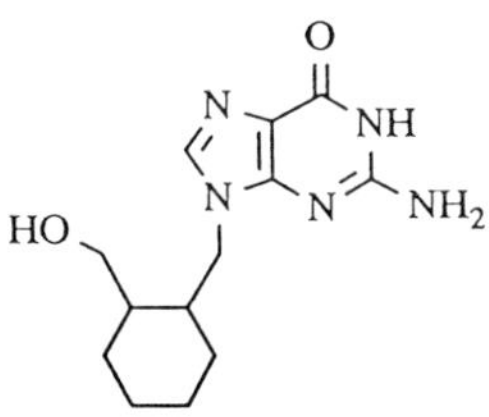 HSV1-TK 0.07

a) from Focher *et al.*,1988

b) from Sim *et al.*,1988

c) from Ashton *et al.*,1988

Fig. 3. Examples of inhibitors of HSV thymidine kinase.

been shown to induce a novel ribonucleotide reductase activity *in vitro* (Cohen, 1972; Cohen, Factor & Ponce de Leon, 1974; Henry *et al.*, 1978; Spector *et al.*, 1987; Lankinen, Gräslund & Thelander, 1982) and work, especially on the HSV system, suggests that this enzyme is virally coded (Dutia, 1983). Like many of the other reductases studied, the HSV enzyme consists of a tight complex of two pairs of subunits (the large, RR1, and small, RR2, subunits) (Bacchetti, Evelegh & Muirhead, 1986; Frame, Marsden & Dutia, 1985; Ingemarson & Lankinen, 1987). Nutter, Gill & Cheng (1985) suggest, from experiments with hydroxyurea *in vitro*, that HSV-2 ribonucleotide reductase is not essential in proliferating cells. Also, experiments *in vitro* using mutants have shown that, whilst the enzyme (like the HSV-TK) is not essential for virus replication in dividing cells, viral replication is impaired in non-dividing cells (Goldstein & Weller, 1988*b*). It appears that the cellular enzyme, if active, can support HSV replication. Further studies with other mutants *in vivo* suggest the enzyme is essential for virulence (Cameron *et al.*, 1988). Therefore, this enzyme, encoded by a number of the human herpes viruses, has potential as a target for broad spectrum herpes virus chemotherapy.

Two approaches have been taken in the use of ribonucleotide reductase as a target for anti-herpes chemotherapy. The first is based on the synergistic effect of various compounds with acyclovir and the other on a nonapeptide inhibitor.

The ribonucleotide reductase of the various herpes viruses so far studied differ from their eukaryotic (or indeed bacterial) counterparts in that they are not under tight allosteric control (Averett *et al.*, 1983; Averett, Furman & Spector, 1984; Spector *et al.*, 1987; Lankinen *et al.*, 1982). Physiological concentrations of deoxy or ribonucleoside triphosphates neither inhibit nor stimulate the enzyme's activity. As a result of this, high concentrations of deoxyribonucleoside triphosphates necessary for viral DNA replication are synthesised in cells infected by these herpes viruses. However, as acyclovir triphosphate (ACV-TP) and dGTP compete for binding to the viral DNA polymerase, high levels of dGTP can antagonise the action of acyclovir (Elion *et al.*, 1977; Furman, Lambe & Nelson, 1982). An inhibitor of the viral ribonucleotide reductase would be expected to decrease the deoxyribonucleoside triphosphate pools (including those of dGTP) and might prevent this competition (Spector *et al.*, 1989). One series of compounds known to inhibit eukaryotic and viral reductases are the 2-acetyl thiosemicarbazones, and Shipman *et al.* (1981) have reported that several of these compounds are selective inhibitors of viral replication *in vitro*. Subsequently, Shipman *et al.* (1986) described several of these compounds as having some effect *in vivo* in the topical treatment of cutaneous HSV-1 guinea pig infections. One compound, A723U (2-acetylpyridine 4-(2-morpholinoethyl)thiosemicarbazone) was tested *in vitro* in combination with acyclovir to see whether an inhibitor of HSV ribonucleotide reductase would reduce the accumulation of dGTP and subsequently reduce the competition between dGTP and ACV-TP (Spector *et al.*, 1985). It was found that these two compounds had a synergistic effect *in vitro* and biochemical studies showed that the dGTP pools were indeed reduced in the presence of A723U. Subsequently, a further analogue, A1110U, (2-acetylpyridine 5-[(dimethylamino)thiocarbosyl]thiocarbonohydrazone) was found to be more potent than A723U (Spector *et al.*, 1989). This compound inhibited HSV-1, HSV-2, and VZV reductases (ID_{50} ~1 μM) and potentiated the action of ACV *in vitro*. It also inhibited human ribonucleotide reductase but at concentrations 10 times higher than those required to inhibit the herpes enzymes. (The two compounds were not synergistic as regards toxicity to uninfected cells.) As with A723U, A1110U reduced dGTP pools but also elevated ACV-TP pools (by so far unknown mechanism(s)).

Such synergy indicates a possible combination therapy for HSV and VZV infections. Preliminary studies on topically infected nude mice treated with topical or oral ACV and topical A1110U suggest that combination therapy with ACV and A1110U (but not A723U) was synergistic (Ellis, Lobe & Spector, 1986). These observations may provide useful leads in the design of novel therapy using an inhibitor of ribonucleotide reductase and ACV or indeed other anti-herpes nucleoside analogues.

Although combination therapies may be effective, treatment with a single therapeutic agent is always preferable. Also, these thiosemicarbazones are active against the human ribonucleotide reductase which may result in toxicity to dividing cells *in vivo*. Thus, a specific inhibitor of the viral enzyme is sought.

One such compound, a nonapeptide, may represent a lead for future development of a novel anti-herpes drug, although it should be remembered that peptide inhibitors of enzymes pose additional problems such as cell penetration, stability and oral absorption. However, the long-term aim is a small molecular weight non-peptide analogue of the original compound. The peptide Y A G A V V N D L, which represents the carboxy terminus of the RR2 subunit of HSV ribonucleotide reductase, has been shown to inhibit specifically ribonucleotide reductase of HSV-1 and HSV-2 (Cohen *et al.*, 1986; Dutia *et al.*, 1986) and also the VZV and EBV enzymes (Gaudreau *et al.*, 1987). McClements *et al.* (1988) have shown that this peptide acts by dissociating the two subunits of the viral ribonucleotide reductase and work by Paradis *et al.* (1988) gives support for the hypothesis that the peptide competes with RR2 for a site on the large subunit-RR1. (However, this site on RR1 has not yet been described.)

The nonapeptide has provided a starting point for the synthesis of peptide and non-peptide analogues with the aim of obtaining a more potent small molecular weight inhibitor. In all cases, initial studies have concentrated on identifying the minimum sequence required for activity, investigating the role of key amino acids in the core sequence and studying the effect of introducing conformational constraints into the sequence.

Work by several groups has shown that the minimum sequence to inhibit the HSV-1 and HSV-2 enzymes is V V N D L (Gaudreau *et al.*, 1987) and substitutions within this sequence suggests that each amino acid of this core sequence is important. Although these compounds do not inhibit viral replication *in vitro*, probably due to problems with cell penetration and/or intracellular stability, they constitute significant leads. The next step is to attempt to remove the peptide bonds from the optimal sequence. As these peptides inhibit HSV-1, HSV-2, VZV and EBV ribonucleotide reductase they may lead to the identification of broad spectrum anti-herpes agents.

HUMAN IMMUNODEFICIENCY VIRUS

Anti-HIV drug design is directed towards the discovery of a novel agent(s) active against HIV which can be developed as a treatment for AIDS, AIDS related complex (ARC) and also possibly the treatment of HIV seropositive individuals. To be considered for development and clinical studies, any HIV drug should have broad spectrum activity against a range of HIV isolates grown in a number of primary and established lymphocyte and monocyte cell lines. As well as being able to cross the blood/brain barrier, it should ideally, but not essentially, be well absorbed orally and have fewer or less severe toxic side effects than other drugs on the market or in clinical trials (e.g. avoid the bone marrow toxicity of AZT or the neuropathy seen with dideoxycytidine (ddC)).

With our present knowledge of the virus, it is unlikely that any drug will completely clear the virus from an individual (especially from the integrated state) and so patients may expect to be on life-long therapy. Obviously any long-term toxicity of a drug will pose problems.

Therefore, drugs may also be expected to be given in combination or as alternating therapies in order to avoid this long-term toxicity (Hirsch & Kaplan, 1987). It is therefore desirable that a range of drugs be developed to combat HIV infection and ideally these drugs should be targeted against several of the different stages of the viral replication cycle.

Since the first description of HIV as the causative agent of AIDS and related conditions (Barré-Sinoussi *et al.*, 1983; Popovic *et al.*, 1984) numerous laboratories have contributed to an understanding of the molecular biology of this virus. From this work several potential targets for anti-HIV drugs have been defined (Mitsuya & Broder, 1987; Hirsch & Kaplan, 1987). Thus the gp120/CD4 interaction, the reverse transcriptase and protease, and the two *trans*-acting regulatory proteins, *tat* and *rev*, have all been shown to be essential for HIV replication *in vitro*. Further enzyme activities of the *pol* gene product, RNase H and endonuclease, may also prove to be useful targets (Hansen, Shulze & Moelling, 1987; Hansen *et al.*, 1988*b*; Lightfoot *et al.*, 1986). Other possible targets include the glycosylation of the viral glycoprotein and the subsequent trimming of these groups by glucosidases and mannosidases. However, these processes are catalysed by host enzymes and it may be difficult to incorporate specificity and lack of long-term toxicity into inhibition of these processes.

As regards antiviral targets, as with other viruses, the enzyme involved with genome replication – reverse transcriptase – first attracted the most attention. Several inhibitors of the reverse transcriptase of other retroviruses had previously been described and they, as well as numerous nucleo-

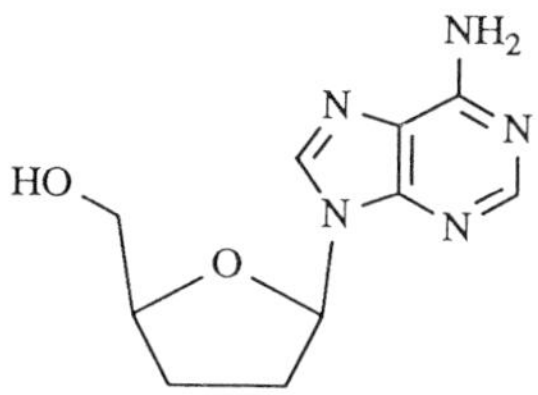

2′,3′-Dideoxy-adenosine
(ddA)

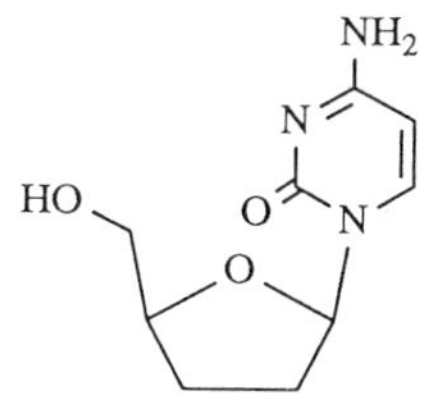

2′,3′-dideoxy-cytidine
(ddC)

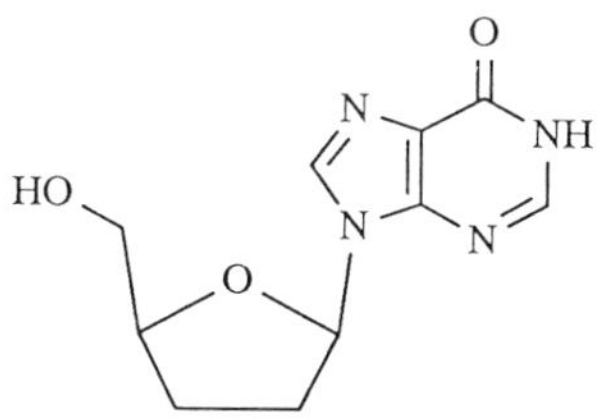

2′,3′-dideoxyinosine
(ddI)

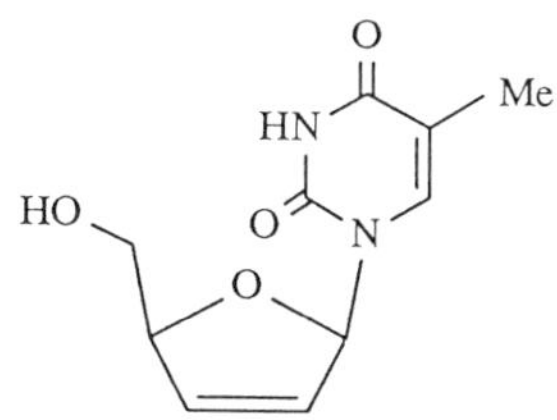

2′,3′dideoxydidehydro-thymidine
(d4T)

Fig. 4(a) Inhibitors of Human Immunodeficiency Virus.

sides, were tested *in vitro* for anti-HIV activity. From such work, AZT has been progressed to the clinic and may be followed by a number of other nucleoside analogues which are presently undergoing clinical trials. These and related compounds have either been shown to, or are thought to, act as inhibitors of the viral reverse transcriptase after they have been phosphorylated by host kinases to the 5′ triphosphate.

Many of these nucleoside analogues are based on the dideoxynucleoside or dideoxydidehydronucleoside series (reviewed by de Clercq, 1989). Of these, ddC, ddA, ddI and d4T are undergoing clinical trials (Fig. 4(a)), but others in initial studies have therapeutic indices comparable to AZT, ddC or ddA (de Clercq, 1989). They may therefore be sufficiently potent to merit further study as they may have pharmacological or toxicological advantages over other members of the series and, considering the need for multiple therapies, may be useful when given in combination.

In the next section several further classes of nucleoside with activity against HIV *in vitro* are described, some of which are expected to enter clinical trials in the next year. They represent different classes of nucleoside from the 2′3′dideoxy and 2′3′dideoxydidehydronucleoside series, and are the carbocyclic nucleosides, e.g. carbovir, acyclic nucleoside derivatives, e.g. PMEA, the oxytanocin analogues, some of which are carbocyclic, and a novel series of nucleosides with substitutions at the 3′ position of the pentose ring. Also, progress with finding inhibitors of the HIV protease, *tat*, and gp120 will be discussed, as well as attempts to find inhibitors of gp120 glycosylation. Lastly, the use of oligodeoxynucleosides as antiviral therapy will be discussed.

Nucleoside inhibitors of HIV

Carbocyclic nucleosides–carbovir and analogues (Fig. 4 (b))

Carbocyclic nucleosides have previously been shown to have antiviral activity against viruses of several groups (Marquez & Lim, 1986). These nucleosides are structurally analogous to their 'sugar' equivalent, except that a methylene group replaces the oxygen atom of the sugar ring. Their advantage over the sugar analogues is that they lack the labile glycosidic bond and so are stable to the action of phosphorylases and hydrolases, but in many cases retain their affinity for a number of enzymes involved in nucleoside and nucleotide metabolisms (Marquez & Lim, 1986). Thus, not only are they more stable *in vivo*, and therefore longer acting, but they may not yield potentially toxic bases or other metabolites.

Recently, Vince *et al.* (1988) described a novel carbocyclic nucleoside which had potent selective activity against HIV *in vitro* in the same range as AZT and ddC. Depending on the cell line, carbovir inhibited HIV ($HTLVIII_B$)-induced cytopathic effect with an ID_{50} of 0.6–0.8 μM (AZT–ID_{50} 0.19–0.38 μM, ddC–ID_{50} 0.52–1.37 μM). Toxicity experiments suggested that carbovir has a similar level of toxicity to several lymphocyte cell lines as AZT and ddC (50% cytotoxic dose (CD_{50}) for carbovir 140–250 μM; AZT 100–700 μM; ddC 50–200 μM). This compound, carbocyclic 2′,3′didehydro-2′-3′-dideoxyguanosine (carbovir, NSC 614846) was the first carbocyclic nucleoside reported to have activity against HIV and, since the initial report, a number of analogues of this nucleoside have been synthesised. In structure–activity relationship studies, Vince *et al.* (1988) found that replacement of the guanine base by adenine decreased selectivity by a factor of 10 and activity was also lost if the purine ring was substituted by pyrimidine. Saturation of the carbocyclic ring either abolished or greatly decreased activity and none of the analogues with substitutions at the 2′ or 3′ position of the carbocyclic moiety (even an azido group) were active. Thus, in general, it seems that either a 2′,3′dideoxy sugar or a 2′,3′-unsaturated carbocyclic ring is necessary for activity against HIV.

Mode of action studies indicate that carbovir acts, at least in part, at the level of the triphosphate, the triphosphate being a potent inhibitor of the HIV-1 reverse transcriptase (White *et al.*, 1989). Thus, using polyCdG as template and dGTP as substrate, carbovir triphosphate inhibited HIV ($HTLVIII_B$) reverse transcriptase with an ID_{50} of 0.03 μM compared with an ID_{50} of 0.03 μM for AZT triphosphate (where polyAdT is template

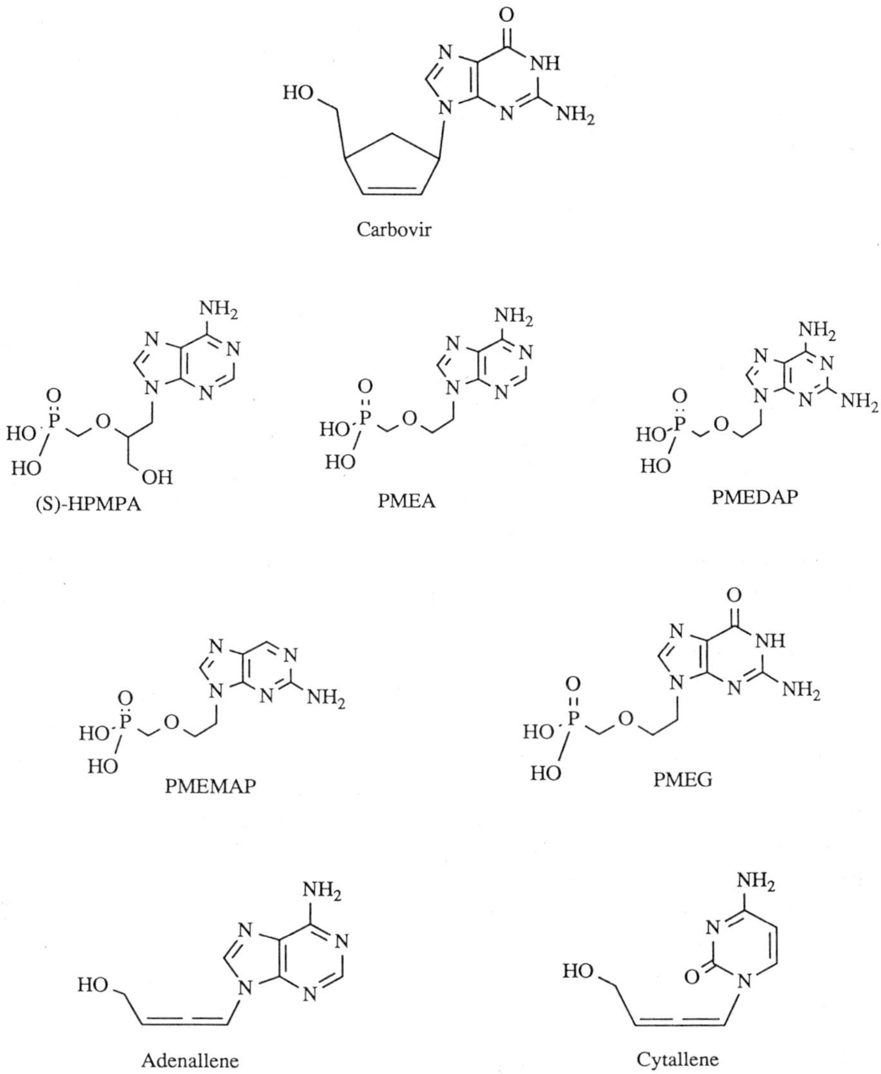

Fig. 4(b)

and dTTP substrate). Carbovir triphosphate is selective for the HIV reverse transcriptase, having little effect on human DNA polymerases α, β or γ (at 100 μM), unlike AZT-TP which had an inhibitory effect on the β and γ DNA polymerases at 16 μM and 11 μM respectively.

PMEA and analogues (Fig. 4(b))

Previously, de Clercq *et al.* (1986) described a series of acyclic adenine derivatives as having activity against a range of DNA viruses (e.g. herpes viruses, adenoviruses, iridoviruses and poxviruses) and that the most potent, (S)-HPMPA, ((S)-9-(3-hydroxy-2-phosphonylmethoxypropyl)-adenine) and its analogue PMEA (9-(2-phosphonylmethoxyethyl)adenine) also inhibited transformation of murine fibroblast cells by the Moloney murine sarcoma virus. PMEA, (S)-HPMPA and several further phosponomethoxyalkyl derivatives have now been studied in more detail for their activity against HIV *in vitro* (Pauwels *et al.*, 1988); in all cases, they were compared with ddC. PMEA, PMEDAP, PMEMAP and the guanosine analogue PMEG inhibited HIV (HTLV III_B) replication in MT4 cells with ID_{50}s of 2, 1, 45 and 3.8 μM respectively (compared to 0.47 μM for ddC). Substitution of the base with hypoxanthine, uracil, cytosine or thymine abolished activity (ID_{50}s >125 μM). PMEA and PMEDAP were also active (at 1.5 μM and 13 μM) against HIV (HTLV III_B) in ATH8 cells. In all cases cytotoxicity was only seen at concentration >18 times the antiviral dose. The mechanism of action of these compounds remains unclear, but as previously shown for (S)-HPMPA they may need to be further phosphorylated to the active form (Votruba *et al.*, 1987). Their therapeutic potential deserves further study.

Adenallene and Cytallene (Fig. 4(b))

A further class of acyclic nucleoside analogues, lacking an oxacyclopentane moiety has been reported by Hayashi *et al.* (1988) to have some activity against HIV. The best of these, the adenine derivatives, 9-(4′-hydroxyl-1′,2′-butadienyl) adenine (adenallene), and the cytosine derivatives, 1-(4′-hydroxyl-1′,2′-butadienyl) cytosine (cytallene), protected ATH8 and TM11 cells from the cytopathic effect of HIV-1, and ATH8 cells from the cytopathic effect of HIV-2. Cytallene protected ATH8 cells at 0.5–1 μM (in the same concentration range as ddC), whereas adenallene was an order of magnitude less active. However, some breakthrough of virus replication was observed after 14 days of culture. Several guanine, hypoxanthine and thymine derivatives of cytallene had little activity in these systems.

Oxetanocin analogues (Fig. 4(c))

Hoshino *et al.* (1987) first reported the activity of oxetanocin A (9-(2-deoxy-2-hydroxymethyl-β-D-erythro-oxetanosyl)adenine) against HIV (strains HTLV-III_B and GUN-1) in MT4 cells. This natural product which contains oxetanosyl-N-glycoside as a sugar moiety with adenine as a base was active at concentrations at least 10 times lower than the cytotoxic dose. Seki *et al.* (1989*a*) extended these studies to a series of oxetanocin analogues. The 2-amino adenine, guanine, hypoxanthine and xanthine analogues of oxetanocin A were tested for their ability to protect MT-4 cells from the cytopathic effect of HIV (HTLV-III_B) and compared with 2′3′-dideoxy-adenosine in the same system. Oxetanocin A was the most active (ID_{50} =

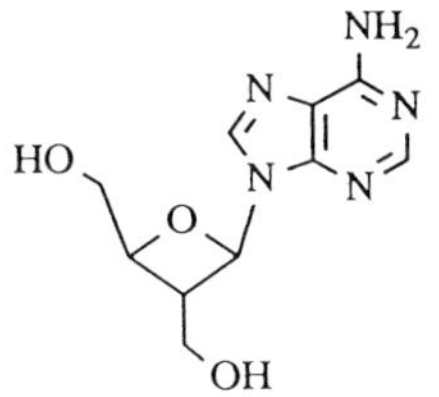

Oxetanocin A

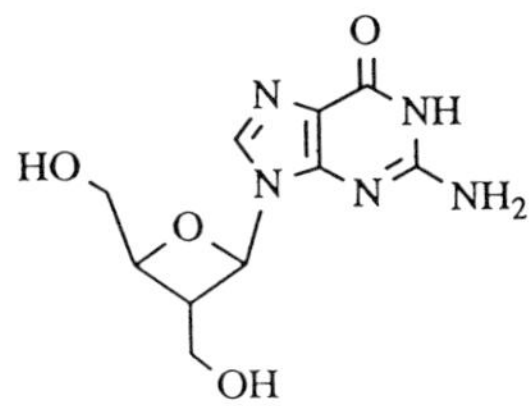

Oxetanocin G

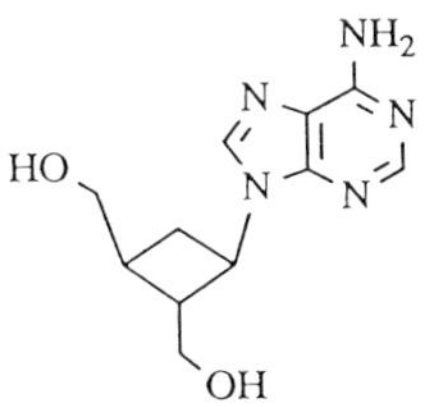

Carbocyclic oxetanocin A

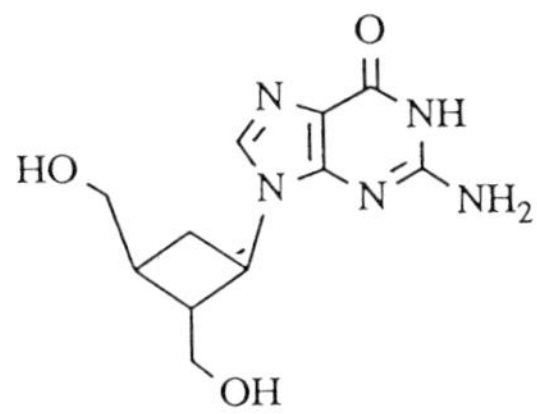

Carbocyclic oxetanocin G

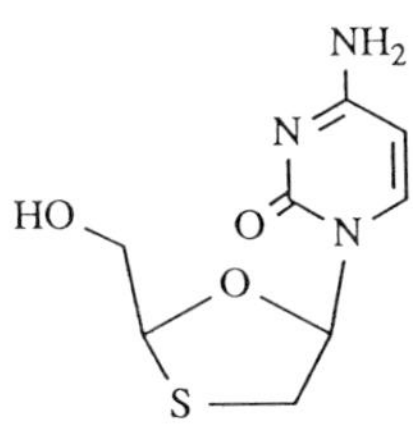

NGPB-21(BCH-189)

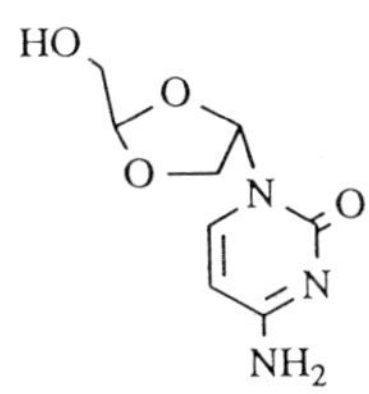

DDIII 30A (BCH-203)

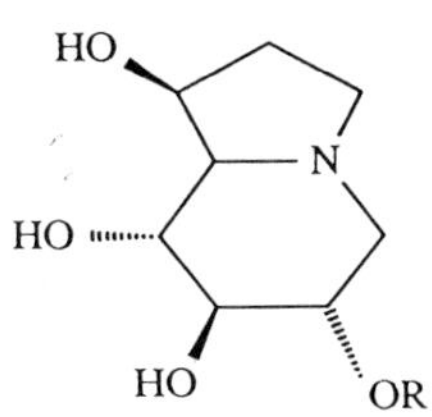

R = H Castanospermine

R = $CH_3CH_2CH_2$–CO–

6-*o*-butyryl-castanospermine

Fig. 4(c).

5,6 μM, CD_{50} = 64 μM). However, of the others, the hypoxanthine analogue had the highest therapeutic index (ID_{50} = 8,7 μM, CD_{50} = 1300 μM, TI = 150). Several further derivatives of oxetanocin A, with substitutions in the oxetanosyl-N-glycoside ring, are also more active, and less cytotoxic than the parent compound (Seki *et al.*, 1989*b*). Several of these were active with ID_{50}s less than 1 μM. However, they also exhibited some cytotoxicity. Work with the triphosphate of oxetanocin A suggested this compound inhibits purified HIV reverse transcriptase. Recently, Hayashi *et al.* (1989) described the carbocyclic derivatives of oxetanocin A (A69463) and oxetanocin G (A69992). Both compounds were found to be capable of protecting T-cells against infectivity and the cytopathic effect of HIV-1 and HIV-2. In monocyte/macrophage cultures exposed to HIV-1, production of p24 protein was almost completely suppressed at 0.5 μM of either carbocyclic oxetanocin G or A. However, both compounds showed considerable toxicity in T-cells at 100 μM. The activity of the G version was readily reversed by 2′deoxyguanosine, but neither 2′deoxyadenosine nor 2′deoxycytidine had any effect on the action of the A analogues.

This series of compounds requires further study in a range of cell lines and also comparison with AZT and other compounds in development. The cytotoxicity of some of the oxetanocin derivatives may preclude them from further clinical development.

Other nucleosides (Fig. 4(c))

Wainberg *et al.* (1989) and Belleau *et al.* (1989) described a novel class of cytosine analogues in which the 3′ carbon of the pentose ring was replaced by either sulphur or oxygen. Of these, NGPB-21 (BCH-189) had the best selectivity index, quoted by these workers to be in the same range as AZT. A further cytidine analogue (DDIII30A) (BCH-203), with the sugar-base linkage in the unnatural α-configuration and with an oxygen at the 3′ position of the pentose ring, was also reported to have a favourable therapeutic index *in vitro*.

The mode of action of these compounds is not known, but again is thought to be at the level of the reverse transcriptase.

HIV protease

The *pol* gene of HIV encodes three proteins; a protease at the 5' end, reverse transcriptase and an endonuclease. The *pol* coding region overlaps *gag* and is expressed as a *gag–pol* fusion protein after a translational frameshifting event within the *gag–pol* overlapping region. The *gag* and *pol* gene products are subsequently processed out of the *gag* and *gag–pol* precursors by the action of the HIV encoded protease, which is itself released by autocatalytic cleavage from the *gag – pol* fusion protein (Debouck *et al.*, 1988; Mous, Hemier & Le Grice, 1988; Hansen *et al.*, 1988*a*). Retroviral proteases are essential for viral maturation (e.g. Moloney murine leukaemia virus (Katoh *et al.*, 1985; Crawford & Goff, 1985)) and using protease deficient mutants this has also been shown for HIV (Kohl *et al.*, 1988; Peng *et al.*, 1989). In both cases, the mutants produced non-infectious or immature particles. This enzyme is therefore a suitable target of anti-HIV therapy and may lead to the design of novel inhibitors.

It has been shown in studies using pepstatin A and in mutational studies that the HIV protease is an aspartyl protease (Schneider & Kent, 1988; Von der Helm *et al.*, 1989; Richards *et al.*, 1989; Loeb *et al.*, 1989) which probably acts as a dimer (Meek *et al.*, 1989; Navia *et al.*, 1989). Several groups have reported the expression of HIV protease in bacterial cells (Hansen *et al.*, 1988*a*; Debouck *et al.*, 1988) and also the total chemical synthesis of this enzyme (Schneider & Kent, 1988). The enzyme has been purified, crystallised, and preliminary structural studies attempted (Navia *et al.*, 1989; Wlodawer *et al.*, 1989) which should contribute to the rational design of protease inhibitors.

Various substrates have been used to characterise the enzyme. Such substrates include bacterially expressed *gag* (Debouck *et al.*, 1988; Graves *et al.*, 1988), shortened versions of *gag–pol* which will not produce protease by autocatalytic cleavage but which retain the protease cleavage site (Hensen *et al.*, 1988*a*). Small synthetic peptides which mimic the putative cleavage sites within *gag* and *pol* (Billich *et al.*, 1988; Moore *et al.*, 1989) are also substrates. Small peptides might be expected to adopt a number of conformations of which some are likely to represent the structure of the cleavage site. As such they seem appropriate substrates. Peptides representing the *gag* p17/p24 and *gag* p24/p7 junctions, the N terminus of protease and the protease/reverse transcriptase junction have been used in such studies (Billich *et al.*, 1988; Darke *et al.*, 1988).

Several analyses have shown that the heptapeptide SQNYPIV (representing the *gag* p17/p24 cleavage site) was the minimum sequence cleaved (Darke *et al.*, 1988). In addition, alteration of the cleavage site (in some cases) resulted in inhibition of the protease. Substitution of amino acid residues with isosteres or statine analogues also resulted in enzyme inhibition (Billich *et al.*, 1988; Moore *et al.*, 1988).

The HIV protease is potentially a good target for the design of novel anti-HIV agents and such inhibitors would obviously be of use in combination with the nucleoside inhibitors described earlier. Apart from random screening for inhibitors using one of the small peptides as substrate, the small peptide inhibitors described above may provide the start for a chemical programme. It is probable, however, that such peptides require modification to non-peptide analogues to overcome problems such as cell entry.

There are reports of possible peptide inhibitors of this HIV enzyme. Marchetto, Nisato & Chermann (1989) described the peptide ZYVStaASta OMe (SR 41476) as inhibiting HIV replication at 1–10 μM (as measured by syncytia formation and inhibition of reverse transcriptase activity). Processing of p55 to p24 was also inhibited. As this peptide inhibited the protease of an avian retrovirus it is possible that it also inhibits HIV protease. No data on specificity of action (i.e. its action against non-viral aspartyl proteases) was given. However, it is encouraging that this peptide appeared to inhibit HIV replication *in vitro*.

Inhibitors of CD4/gp120 interactions

The CD4 molecule, a cell surface protein of human helper T lymphocytes has been shown to interact with the outer envelope glycoprotein – gp120 – of HIV and acts as the receptor for this virus (Dalgleish *et al.*, 1984; Maddon *et al.*, 1986). This represents one of the first steps in the replication cycle of HIV and an antagonist of this interaction may lead to effective anti-HIV therapy.

Several groups have shown that soluble CD4 (sCD4) can effectively block infectivity of the virus (Deen *et al.*, 1988; Fisher *et al.*, 1988; Hussey *et al.*, 1988; Smith *et al.*, 1987; Traunecker, Lüke & Karjalainen, 1988) and clinical trials with sCD4 are already underway.

The CD4 molecule, a 55K protein, consists of four tandem immunoglobulin-like VJ regions, a transmembrane domain and a cytoplasmic segment (Maddon *et al.*, 1987; Clark *et al.*, 1987). Studies with truncated forms of CD4 suggest that the V domain is essential for gp120 recognition (Traunecker *et al.*, 1988; Berger, Fuerst & Moss, 1988). Analysis of truncated forms of CD4 and with substitution mutations show that the determinants for high affinity gp120 binding lie within the first 106–113 amino acids of the V1 region of CD4 (Arthos *et al.*, 1989; Chao *et al.*, 1989) and that within this domain there is a single binding site defined by a region of 12 amino acids (Arthos *et al.*, 1989). It appears that inhibition of HIV infection and subsequent virally mediated cell fusion by sCD4 possibly depends on an association with this binding site. Analysis of the envelope glycoproteins of influenza A virus (Weis *et al.*, 1988) and of the receptor binding protein of rhinovirus (Rossman *et al.*, 1985) indicate that recognition of the cellular receptor by these proteins may depend on a sequence found in a 'canyon' or receptor binding pocket. The same may be true for gp120.

As Rossman *et al.* (1985) suggest, this restriction on the size and the position of the receptor binding domain may protect this essential site from antibody recognition, allowing the virus to evade the immune response. Further understanding of the three dimensional structure of the HIV binding domain will certainly encourage the design of new small molecular weight inhibitors of the CD4/gp120 interactions (Rossman, 1988). Such an approach is being used in the design of anti-rhinovirus drugs (Badger *et al.*, 1988; Colonno *et al.*, 1988) and has also been suggested for the development of small molecular weight inhibitors of influenza A virus (Pritchett *et al.*, 1987; Weis *et al.*, 1988).

Inhibitors of glycosylation

The interaction between the outer membrane glycoprotein gp120 of HIV and the CD4 receptor depends, in part, on the correct processing of this glycoprotein. Modification of the glycosylation pattern of gp120 by inhibitors of the host α-glucosidase-1, has been shown to inhibit HIV replication as measured by syncytia formation (Sunkara *et al.*, 1987; Walker *et al.*, 1987). Castanospermine (1,6,7,8-tetrahydroxyoctahydroindolizine), a plant alkaloid extracted from *Castanospermum australe*, inhibits the α-glucosidase-1 of the endoplasmic reticulum, preventing removal of glucose residues during the normal processing of glycoproteins (Saul *et al.*, 1983, 1985) (Fig. 4(c)). This compound totally inhibited HIV syncytia formation (at ~500 μM) (Walker *et al.*, 1987), possibly by the inhibition of processing of the gp160 envelope precursor protein which in turn may result in decreased cell surface expression of the mature gp120 outer envelope protein. Not only is syncytia formation inhibited by castanospermine, but evidence suggests that there is a reduction in infectivity of progeny virus and a reduction of cell to cell spread of virus in castanospermine treated HIV infected cultures. Both these processes are dependent on mature, functional gp120.

Further analogues of castanospermine have been studied (Sunkara *et al.*, 1989). One analogue, the 6-*o*-butyryl derivative of castanospermine had over 10 times the potency of castanospermine against glucosidase-1, and 40 times the antiviral activity (Fig. 4(c)). Castanospermine had an ID_{50} of 34 μM, and the 6-*o*-butyryl derivative an ID_{50} of 0.58 μM against HIV (GB8) in JM cells (there was no cytotoxicity at 750 μM). This latter compound has been chosen for further development. However, it should be remembered that the action of this compound depends on the inhibition of a host enzyme(s) and it may be difficult to avoid long term toxicity. If further studies are promising, it may at least have a use in combination with other compounds.

Tat

The HIV *tat* function has been shown to be essential for the replication of the virus (Dayton *et al.*, 1986; Fisher *et al.*, 1986) and, as such, is a possible target for anti-HIV therapy. However, as yet the exact biochemical function of the *tat* protein is not fully understood and makes the identification of specific inhibitors of this function difficult. Various groups have reported stable human cell lines expressing functional HIV-1 *tat* (Rosen *et al.*, 1986; Cullen, 1986) or a *tat*-inducible LTR-linked reporter gene (in this case, chloramphenicol acetyltransferase–CAT) (Wright *et al.*, 1986). However, neither of these systems is ideal for the detection of inhibitors of the *tat* function. Recently, Bacheler *et al.* (1989) described the construction of a stable HeLa-derived cell line expressing high levels of β-galactosidase under the control of the *trans*-acting HIV-1 *tat* gene product, and the *cis*-acting HIV-1 LTR. This cell line should form the basis for a screen to detect compounds which inhibit the *tat* transactivating function and which may ultimately inhibit HIV-1 replication.

This is one of the first assays to be described which is suitable for the screening of large numbers of compounds for activity against a transacting function. Similar whole cell assays could also be developed for other transactivating functions, e.g. those of the herpes viruses.

Antisense oligodeoxynucleotides

The use of antisense oligodeoxynucleotides as inhibitors of virus replication *in vitro* was first shown for Rous sarcoma virus (Zamecnik & Stephenson, 1978; Stephenson & Zamecnik, 1978). Subsequently, similar studies on other viruses including vesicular stomatitis virus, HSV-1, and influenza A virus have shown that oligodeoxynucleotide targeted to strategic regions of the genome inhibit these viruses (Lemaitre, Bayard & Lebleu, 1987; Smith *et al.*, 1986; Zerial, Thuong & Hélène, 1987). It is assumed that these antisense sequences act by inhibiting the expression of various genes by binding to and blocking translation of mRNAs. Binding would depend on many factors including the secondary structure of the mRNA, the presence of proteins associated with mRNA and the action of nucleases. In most cases, the oligodeoxynucleotide was present in the tissue culture fluid and, as the uptake and stability of these molecules are expected to pose problems in the further development of these compounds as drugs, various modified oligodeoxynucleotides have also been investigated. These include methylphosphonate analogues (Smith *et al.*, 1986) as well as those with terminal blocking group (Zemecnik & Stephenson, 1978), those modified by the addition of polylysine (Lemaitre *et al.*, 1987) or intercalating agents (Zerial *et al.*, 1987). With these last compounds, the interacting agent stabilised the complex formed by the oligodeoxynucleotides with its complementary sequence, and also facilitated uptake of the compound into cells.

Zamecnik *et al* (1986) described the inhibition of HIV replication in H9 cells by a series of oligodeoxynucleotides complementary to mRNA donor or acceptor splice sites. Experiments with random oligodeoxynucleotides suggested that this inhibition was specific. Goodchild *et al* (1988) extended this work and described several oligodeoxynucleotides complementary to a number of sequences throughout the HIV genome. Using syncytia formation as a measure of HIV (HTLV-III_B or HTLV-III_{CC}) replication in MOLT-3 cells, sequences complementary to sites at or near the 5′ ends of HIV-RNA and to certain splice acceptor sites were inhibitory ($ID_{50}=0.6$–$3.0\,\mu M$). Activity was greatest amongst the longer (20mer) sequences.

Matsukuru *et al.* (1987) described work with methylphosphonate and phosphorothioate analogues of a series of oligodeoxynucleotides complementary to regions of the *rev* gene. Both classes of analogue were expected to be nuclease-resistant whilst the methylphosphonates may have increased hydrophobicity and thus allow easier passage through the cell membrane. The phosphorothioates were also shown to hybridise more efficiently to complementary DNA than the methylphosphonate analogues. Several of these antisense phosphorothioate oligodeoxynucleotides inhibited HIV ($HTLV\text{-}III_B$) replication in ATH8 cells as measured by cell viability (ID_{50} = 5–10 μM) whereas the methylphosphonate derivative had little effect. It was shown that the phosphorothioates were nuclease-resistant and entered cells, and that the antiviral effect was related to base composition and length of the oligodeoxynucleotides. Surprisingly, a 28mer phosphorothioate oligodeoxycytidine inhibited HIV replication, as described above, at 1 μM. Subsequent work showed that this compound did not inhibit expression of *gag* protein, rather it possibly acted at the level of proviral DNA synthesis.

There are also many problems (e.g. stability, oral availability) associated with this approach to antiviral therapy, and at the moment it is difficult to envisage the development of a viable drug based on these compounds.

HEPATITIS B VIRUS

Hepatitis B virus is a causative agent of acute and chronic hepatitis, conditions which affect over 200 million people. Most of those affected are from the Far East and transmission is probably from mother to child. In developed countries, infection occurs later in life by parenteral or sexual transmission. The virus is the major cause of chronic liver disease, up to 90% of neonatally infected individuals becoming chronic carriers, whilst up to 10% of those infected in late childhood/early adulthood develop the chronic carrier state.

Of these chronic carriers, many will develop some form of often progressive liver disease including cirrhosis and hepatocellular carcinoma. These chronic carriers continue to secrete infectious virus into their serum and virus is also found in semen, saliva and urine. These individuals are thus infectious to close contacts.

The main aims of therapy for this chronic state are inhibition of HBV replication, and thus the reduction of infectivity, as well as the control of hepatocyte necrosis. Antiviral therapy is therefore directed towards patients with actively replicating virus, whilst those with only integrated virus are unlikely to benefit. So far adenine arabinoside, adenine arabinoside monophosphate, acyclovir or different interferons have been used in clinical trials and found to be relatively ineffective and toxic (reviewed by Lever, 1988). Testing for anti-HBV agents has been hampered by the lack of *in vitro* test systems: HBV replicates in humans and chimpanzee liver but not in tissue culture.

Various animal viruses – duck, woodchuck, and ground squirrel hepatitis viruses (Mason, Seal & Summers, 1980; Summers, Smolec & Snyder, 1978; Marion *et al.*, 1980) – have proved to be useful models for the study of the human virus and for the testing of potential antiviral compounds

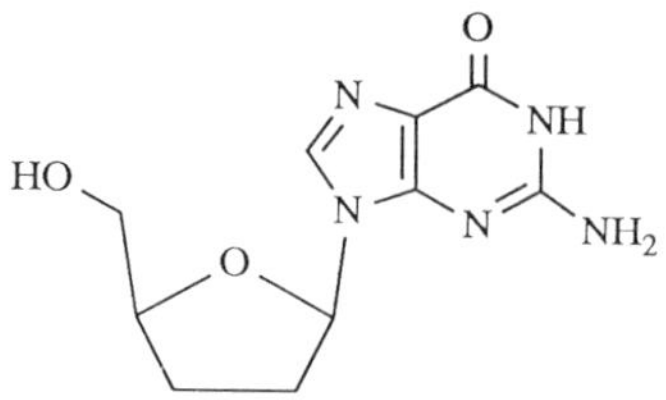

2′,3′-dideoxyguanosine

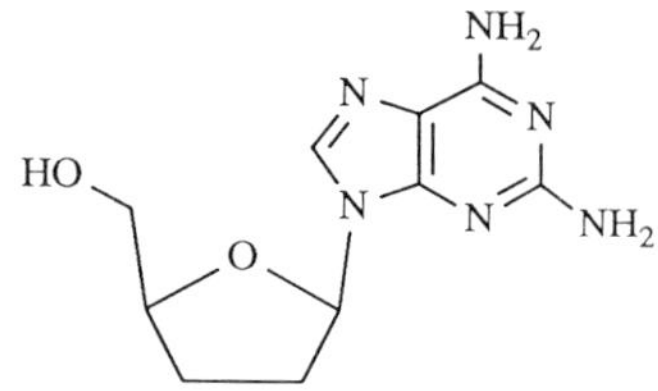

2,6-diaminopurine, 2′3′-dideoxyriboside

Fig. 5. Inhibitors of Hepatitis B virus.

(Offensperger *et al.*, 1988). The model most frequently used is that of duck HBV in duck hepatocytes. Using this system, Suzuki *et al.* (1988) report the action of various 2′,3′ dideoxypurines and pyrimidines. The purine analogues inhibited viral replication, whereas the pyrimidine derivatives were not active. Of the purines ddG and ddDAPR (2,6-diaminopurine 2′,3′-dideoxyriboside) were the most effective (ddDAPR is deaminated by adenosine deaminase to ddG) (Fig. 5). ddDAPR was also used to treat ducklings persistently infected with DHBV. 10 mg/kg given intramuscularly twice daily for two weeks cleared the virus from the duck sera but the virus reappeared, albeit at low levels, after therapy stopped. Despite this, the dideoxypurines may prove a useful start in the development of a new therapy for chronic HBV and it would be interesting to see them tested in the HB611 *in vitro* system described below.

Ueda *et al.* (1989) described this *in vitro* system which may have potential for the screening of anti-hepatitis B agents. The cell line (HB611) was derived from a human hepatoblastoma line by integration of tandemly-arranged HBV genomes. Subsequent analysis indicated that, in these cells, viral transcription occurred with the subsequent production of several of the viral proteins. The cells continuously produce HBV DNA which can be examined by Southern blot analysis. Potential inhibitors of HBV DNA replication can therefore be tested in this system and Ueda *et al.* (1989) report the effect of a number of nucleosides (acyclovir, AraA, AraC, oxetanocin A, ddA, ddC and AZT) on HBV DNA synthesis and compared the effects on cellular DNA synthesis. Acyclovir and ddC showed some specificity in this assay. This system needs further verification but may have some potential for the testing of inhibitors of HBV DNA synthesis.

HUMAN PAPILLOMAVIRUS

Human papillomaviruses, of which there are up to 57 types (reviewed by McCance, 1986; Reichmann & Strike, 1989) are associated with a number of lesions on squamous epithelial surfaces as well as the skin. As with animal papillomaviruses they have been associated with various neoplasias.

The HPV have been linked to genital, oral and laryngeal cancers as well as more benign conditions. HPV-1 is associated with planar warts, HPV-2 with common hand warts, and HPV-6 and 11 with benign genital and respiratory tract condyloma acuminatum, whereas HPV-16 and 18 are associated with genital tract neoplasias.

The human papillomaviruses have not been studied using standard virological techniques as they have not yet been grown successfully in tissue culture. They have a tropism for epithelial cells and the replicative cycle of the virus is synchronised to the differentiation of the epithelium. Although viral DNA is found in the lower epithelial cell layers, infectious virus is only found in the top layers. This dose-dependence of the virus replicative cycle on epithelial cell differentiation has made it difficult to propagate the virus *in vitro*. Recently, it has been possible to clone HPV DNA from tissues, to identify different types of HPV and to begin to study the organisation and function of the HPV genome.

The lack of good tissue culture system for the HPV have made the development of antiviral agents difficult. Therapy of common hand warts or genital warts (condyloma acuminatum) has relied on either cryotherapy, surgery or podophyllin preparations, or the use of agents such as 5′fluorouracil, tri- or biacetic acids or interferons. Apart from the cryotherapy and podophyllin treatment of common warts, these therapies have not been particularly successful especially for the treatment of genital HPV infections (Reichmann & Strike, 1989). There is therefore a need for an effective treatment of genital HPV infections as well as common hand warts and in order to develop compounds specifically for these conditions, rather than using antivirals developed for use against other viruses, good testing systems must be developed. There have been attempts to devise such assays. Using partially differentiated human keratocytes, La Porta & Taichinan (1982) claimed that they could detect replication of HPV-1 DNA in these cells although no virus particles were found. It also appears that the DNA is lost after a few passages (Burnett & Gallimore, 1983). Watts *et al.* (1984) report the transformation of mouse C127 cells by pBR322/HPV-1 and pBR322/HPV-5 but this system has not been further developed. Other systems are based on transformation of mouse C127 cells by bovine papilloma virus (Dvoretzky *et al.*, 1980).

Rational design of HPV inhibitors therefore awaits further characterisation of the gene products and good assay systems. These problems are being addressed.

CONCLUSIONS

Over the past decade, nucleoside analogues have provided the most potent and specific antiviral agents. However, the growing understanding of viral enzymes, receptors and control proteins coupled with novel whole cell assays such as those described for TAT or for measuring hepatitis

B virus DNA replication should lead to the design of inhibitors of virally encoded functions and the detection of specific inhibition from the screening of large numbers of compounds.

At present, the most promising pre-clinical compounds are still nucleosides, but the next decade should see the emergence of novel classes of compounds which can be given either singly or in combination.

REFERENCES

Arthos, J., Deen, K. C., Chaikin, M. A., Fornwald, J. A., Sathe, G., Sattentau, Q. J., Clapham, P. R., Weiss, R. A., McDougal, J. S., Pietropaulo, C., Axel, R., Truneh, A., Maddon, P. J. & Sweet, R. W. (1989). Identification of the residues of human CD4 critical for binding of HIV. *Cell*, **57**, 469–81.

Ashton, W. T., Meurer, L. C., Tolman, R. L., Karkas, J. D., Liou, R., Perry, H. C., Czelusniak, S. M. & Klein, R. J. (1988). A potent selective non-substrate inhibitor of HSV-1 thymidine kinase: (±)-9-[[(Z)-2(hydroxymethyl)cyclohexyl] methyl] guanine and related compounds. *8th International round table on nucleosides, nucleotides and their biological applications. Perdido Beach, Alabama, USA, October 1988.*

Averett, D. A., Furman, P. A. & Spector, T. (1984). Ribonucleotide reductase of herpes simplex virus type 2 resembles that of herpes simplex virus type 1. *Journal of Virology*, **52**, 981–3.

Averett, D. A., Lubbers, C., Elion, G. B. & Spector, T. (1983). Ribonucleotide reductase induced by herpes simplex type 1: characterisation of a distinct enzyme. *Journal of Biological Chemistry*, **258**, 9831–8.

Bacchetti, S., Evelegh, M. J. & Muirhead, B. (1986). Identification and separation of the two subunits of the herpes simplex virus ribonucleotide reductase. *Journal of Virology*, **57**, 1177–81.

Bacheler, L. T., Strehl, L. L., Neubauer, R. H., Petteway Jr, S. R. & Ferguson, B. Q. (1989). Stable indicator cell lines exhibiting HIV-1 *tat* function. *AIDS Research and Human Retroviruses*, **5**, 275–8.

Badger, J., Minor, I., Kremer, M. J., Oliveira, M. A., Smith, T. J., Griffith, J. P., Guerin, D. M. A., Krishnaswamy, S., Luo, M., Rossman, M. G., McKinley, M. A., Diana, G. D., Dutko, F. J., Fancher, M., Ruechert, R. R. & Heinz, B. (1988). Structural analysis of a series of antiviral agents complexed with human rhinovirus 14. *Proceedings of the National Academy of Sciences, USA*, **85**, 3304–8.

Barré-Sinoussi, F., Chermann, J-C., Rey, F., Nugeyre, M. J., Chamaret, S., Gruest, J., Dauguet, C., Axel-Blin, G., Vézinet-Brun, F., Rouzioux, C., Rozenbaum, W. & Montagnier, L. (1983). Isolation of a T-lymphotrophic retrovirus from a patient at risk for acquired immunodeficiency syndrome (AIDS), *Science*, **220**, 868–71.

Belleau, B., Dixit, D., Nguyen-Ba, N. & Krus, J. L. (1989). Design and activity of a novel class of nucleoside analogues effective against HIV-1. *Abstract from the 5th International AIDS Conference, Montréal*, p. 515.

Berger, E. A., Fuerst, T. R. & Moss, B. A. (1988). Soluble recombinant polypeptide comprising the amino terminal half of the extracellular region of the CD4 molecule contains an active binding site for human immunodeficiency virus. *Proceedings of the National Academy of Sciences, USA*, **85**, 2357–61.

Billich, S., Knoop, M-T., Hansen, J., Strop, P., Sedlacek, J., Mertz, R. & Moelling, K. (1988). Synthetic peptides as substrates and inhibitors of human immunodeficiency viruses-1 protease. *Journal of Biological Chemistry*, **263**, 17905–8.

Burnett, T. S. & Gallimore, P. H. (1983). Establishment of a human keratinocyte cell line carrying complete human papillomavirus type 1 genomes: Lack of vegetative viral DNA synthesis upon keratinization. *Journal of General Virology*, **64**, 1509–20.

Cameron, J. M., McDougall, I., Marsden, H. S., Preston, V. G., Ryan, D. M. & Subak-Sharpe, J. H. (1988). Ribonucleotide reductase encoded by herpes simplex virus is a determinant of the pathogenicity of the virus in mice and a valid antiviral target. *Journal of General Virology*, **69**, 2607–12.

Challberg, M. D. (1986). A method for identifying the viral genes required for herpes DNA replication. *Proceedings of the National Academy of Sciences, USA*, **83**, 9094–98.

Chao, B. H., Costopoulos, D. S., Curiel, T., Bertonis, J. M., Chisholm, P., Williams, C., Schooley, R. T., Rosa, J. J., Fisher, R. A. & Maraganore, J. M. (1989). A 113-amino acid fragment of CD4 produced in *Escherichia coli* blocks human immunodeficiency virus-induced cell fusion. *Journal of Biological Chemistry*, **264**, 5812–17.

Clark, S. J., Jeffries, W. A., Barclay, A. N., Gagnon, J. & Williams, A. F. (1987). Peptide and nucleoside sequences of rat CD4 (W3/25) antigen: evidence for derivation from a structure with four immunoglobulin-related domains. *Proceedings of the National Academy of Sciences, USA*, **84**, 1649–53.

Coen, D. M., Kosz-Vnenchak, M., Jacobson, J. G., Leib, D. A., Bogard, C. L., Schaffer, P. A., Tyler, K. L. & Knipe, D. M. (1989). Thymidine kinase-negative herpes simplex virus mutants establish latency in mouse trigeminal ganglia but do not reactivate. *Proceedings of the National Academy of Sciences, USA*, **86**, 4736–40.

Cohen, E. A., Gaudreau, P., Brazeau, P. & Langelier, Y. (1986). Specific inhibition of herpesvirus ribonucleotide reductase by a nonapeptide derived from the carboxy terminus of subunit 2. *Nature, London*, **321**, 441–3.

Cohen, G. H. (1972). Ribonucleotide reductase activity of synchronised KB cells infected with herpes simplex virus. *Journal of Virology*, **9**, 408–18.

Cohen, G. H., Factor, M. N. & Ponce de Leon, M. (1974). Inhibition of herpes simplex type 2 replication by thymidine. *Journal of Virology*, **14**, 20–5.

Colonno, R. J., Condra, J. H., Mizutani, S., Callohan, P. L., Davies, M-E. & Murcko, M. A. (1988). Evidence for the direct involvement of the rhinovirus canyon in receptor binding. *Proceedings of the National Academy of Sciences, USA*, **85**, 5449–53.

Crawford, S. & Goff, S. P. (1985). A deletion in the 5′ part of the *pol* gene of Moloney murine leukaemia virus blocks proteolytic processing of the *gag* and *pol* polyproteins. *Journal of Virology*,. **53**, 899–907.

Crute, J. J., Tsurumi, T., Zhu, L., Weller, S. K., Olivo, P. D., Challberg, M. D., Mocarski, E. S., & Lehman, I. R., (1989). Herpes simplex virus 1 helicase – primase: a complex of three herpes-encoded gene products. *Proceedings of the National Academy of Sciences, USA*, **86**, 2186–9.

Cullen, B. R. (1986). Transactivation of human immunodeficiency virus occurs via a bimodal mechanism. *Cell*, **46**, 973–82.

Dalgleish, A. G., Beverly, P. C. L., Clapham, P. R., Crawford, D. H., Greaves, M. F. & Weiss, R. A. (1984). The CD4 (T4) antigen is an essential component of the receptor for the AIDS retrovirus. *Nature, London*, **312**, 763–6.

Darke, P. L., Nutt, R. F., Brady, S. F., Garsky, V. M., Ciccarone, T. M., Leu, C-T., Lumma, P. K., Freidinger, R. M., Veber, D. F. & Sigal, I. S. (1988). HIV-1 protease specificity of peptide cleavage is sufficient for processing of gag and pol polyproteins. *Biochemical and Biophysical Research Communications*, **156**, 297–303.

Dayton, A. I., Sodroski, J. G., Rosen, C. A., Goh, W. C. & Haseltine, W. A., (1986). The transactivator gene of the human T cell lymphorophic virus type 3 is required for replication. *Cell*, **44**, 941–7.

Debouck, C., Gorniak, J. G., Strickler, J. E., Meek, T. D., Metcalf, B. W. & Rosenberg, M. (1988). Human immunodeficiency virus protease expressed in *Escherichia coli* exhibits autoprocessing and specific maturation of the gag precursor. *Proceedings of the National Academy of Sciences, USA*, **84**, 8903–6.

De Clercq, E., Holý, A., Rosenberg, I., Sakuma, T., Balzarini, J. & Maudgal, P. C. (1986). A novel selective broad spectrum anti-DNA virus agent. *Nature, London*, **323**, 464–7.

De Clercq, E. (1989). Potential drugs for the treatment of AIDS. *Journal of Antimicrobial Chemotherapy*, **23** (supplement A), 35–46.

Deen, K. C., McDougal, J. S., Inacker, R., Folena-Wasserman, G., Arthos, J., Rosenberg, J., Maddon, P. J., Axel, R. & Sweet, R. W. (1988). A soluble form of CD4 (T4) protein inhibits AIDS virus infection. *Nature, London*, **331**, 81–4.

Dutia, B. M., (1983). Ribonucleotide reductase induced by herpes simplex virus has a virus specified constituent. *Journal of General Virology*, **64**, 513–21.

Dutia, B. M., Frame, M. C., Subak-Sharpe, J. H., Clark, W. N., Marsden, H. S. (1986). Specific inhibition of herpes virus ribonucleotide reductase by synthetic peptides. *Nature, London*, **321**, 439–41.

Dvoretsky, I., Shober, R., Chattopadhyay, S. K. & Lowy, D. R. (1980). A quantitative *in vitro* focus assay for bovine papilloma virus. *Virology*, **103**, 369–75.

Elion, G. B., Furman, P. A., Fyfe, J. A., De Miranda, P., Beauchamp, L. & Schaeffer, H. J. (1977). Selectivity of the action of an anti-herpetic agent 9-(2-hydroxyethoxymethyl) guanine. *Proceedings of the National Academy of Sciences, USA*, **74**, 5716–20.

Ellis, M. N., Lobe, D. C., Spector, T. (1986). Combination antiviral chemotherapy for treatment of HSV-1 infection in athymic mice. *26th Interscience Conference on Antimicrobial Agents and Chemotherapy, New Orleans, USA*, September 28th–October 1st, 1986, abstract number 625.

Fiddian, A. P. (1988). Herpes group viruses. In: *Frontiers of Infectious Diseases: New Antiviral Strategies*, Norrby, S. R., ed., pp. 72–90. Edinburgh. Churchill Livingstone.

Field, H. J. & Darby, G. (1980). Pathogenicity of mice of strains of herpes virus which are resistant to acyclovir *in vitro* and *in vivo*. *Antimicrobial Agents and Chemotherapy*, **17**, 209–16.

Field, H. J. & Wildy, P. (1978). The pathogenicity of thymidine kinase-deficient mutants of herpes simplex virus in mice. *Journal of Hygiene*, **81**, 261–77.

Fisher, A. G., Feinberg, M. B., Josephs, S. F., Harper, M. E., Marselle, L. M., Reyes, G., Gonda, M. A., Aldovini, A., Debouck, C., Gallo, R. C. & Wong-Staal, F. (1986). The transactivator gene of HTLV-3 is essential for virus replication. *Nature, London*, **320**, 367–71.

Fisher, R. A., Bertonis, J. M., Mier, W., Johnson, V. A., Costopoulos, D. S., Liu, T., Tizaid, T., Walker, B. D., Hirsh, M. S., Schooley, R. T. & Flavell, R. A. (1988). HIV infection is blocked *in vitro* by recombinant soluble CD4. *Nature, London*, **331**, 76–9.

Focher, G., Hildebrand, C., Freese, S., Ciarrocchi, G., Noonan, T., Sangalli, S., Brown, N., Spadari, S. & Wright, G. (1988). N^2-phenyldeoxyguanosine: A novel selective inhibitor of herpes simplex virus thymidine kinase. *Journal of Medicinal Chemistry*, **31**, 1496–500.

Frame, M. C., Marsden, H. S. & Dutia, B. M. (1985). The ribonucleotide reductase induced by herpes simplex virus type 1 involves minimally a complex of two polypeptides (136K and 38K). *Journal of General Virology*, **66**, 1581–7.

Furman, P. A., Lambe, C. U. & Nelson, D. J. (1982). Effect of acyclovir on the deoxyribonucleoside triphosphate pool levels in Vero cells infected with HSV-1. *American Journal of Medicine*, **73 (1A)**, 14–17.

Gaudreau, P., Michaud, J., Cohen, E. A., Langelier, Y. & Brazeau, P. (1987). Structure–activity studies on synthetic peptides inhibiting herpes simplex virus ribonucleotide reductase. *Journal of Biological Chemistry*, **262**, 12413–16.

Goldstein, D. J. & Weller, S. K. (1988a). An ICP6Δ: *lacZ* insertional mutagen is used to demonstrate that the UL52 gene of herpes simplex virus type 1 is required for virus growth and DNA synthesis. *Journal of Virology*, **62**, 2970–7.

Goldstein, D. J. & Weller, S. K. (1988b). Herpes simplex virus type 1 induced ribonucleotide reductase activity is dispensible for virus growth and DNA synthesis: isolation and characterisation of an ICP6 *lacZ* insertion mutant. *Journal of Virology*, **62**, 196–205.

Goodchild, J., Agrawel, S., Civeira, M. P., Sarin, P. S., Sun, D. & Zamecnik, P. C. (1988). Inhibition of human immunodeficiency virus replication by antisense oligdeoxyonucleotides. *Proceedings of the National Academy of Sciences, USA*, **85**, 5507–11.

Graves, M. C., Lim, J. J., Heimer, E. P. & Kramer, R. A. (1988). An 11-kDa form of human immunodeficiency virus protease expressed in *Escherichia coli* is sufficient for enzymatic activity. *Proceedings of the National Academy of Sciences, USA*, **85**, 2449–53.

Hansen, J., Billich, S., Schulze, T., Sukrow, S. & Moelling, K. (1988a). Partial purification and substrate analysis of bacterially expressed HIV protease by means of monoclonal antibody. *The EMBO Journal*, **7**, 1785–91.

Hansen, J., Schulze, T. & Moelling, K. (1987). RNaseH activity was associated with bacterially expressed reverse transcriptase of human T-cell lymphotrophic virus III/lymphadenopathy associated virus. *Journal of Biological Chemistry*, **262**, 12393–6.

Hansen, J., Schulze, T., Mellert, W. & Moelling, K. (1988b). Identification and characterisation of HIV-specific RNaseH by monoclonal antibody. *The EMBO Journal*, **7**, 239–43.

Harnden, M. R. (1989). Development of 9-[4-hydroxy-3-(hydroxymethyl)but-1-yl] purines as potential therapeutic agents for the treatment of human herpes viruses. *Drugs of the Future*, **14**, 347–58.

Hayashi, S., Norbeck, D. W., Plattner, J., Broder, S. & Mitsuya, H. (1989). Carbocyclic oxetanocin analogues that inhibit *de novo* replication of human immunodeficiency virus in T-cells and monocytes/macrophages *in vitro*. *Abstracts of the 5th International Conference on AIDS. Montréal*, p. 564.

Hayashi, S., Phadtare, S., Zemlicka, J., Matsukuru, M., Mitsuya, H. & Broder, S. (1988). Adenallene and cytallene: Acyclic nucleoside analogues that inhibit replication and cytopathic effect of human immunodeficiency virus *in vitro*. *Proceedings of the National Academy of Sciences, USA*, **85**, 6127–31.

Henry, B. E., Glaser, R., Hewetson, J. & O'Callaghan, D. J. (1978). Expression of altered ribonucleotide reductase activity associated with the replication of Epstein Barr Virus. *Virology*, **89**, 262–71.

Hirsch, M. S. & Kaplan, J. C. (1987). Treatment of human immunodeficiency virus infections. *Antimicrobial Agents and Chemotherapy*. **31**, 839–43.

Hoshino, H., Shimizu, N., Shimada, N., Takita, T. & Takenuchi, T. (1987). Inhibition of infectivity of human immunodeficiency virus by oxetanocin. *Journal of Antibiotics*, **40**, 1077–8.

Hussey, R. E., Richardson, N. E., Kowalski, M., Brown, N. T., Chang, H-C., Siliciano, R. F., Dorfman, T., Walker, B., Sodroski, J. & Reinherz, E. L. (1988). A soluble CD4 protein selectivity inhibits HIV replication and syncytium formation. *Nature, London*, **331**, 78–81.

Ingemarson, R. & Lankinen, H. (1987). The herpes simplex virus type 1 ribonucleotide reductase is a tight complex of the type composed of 40K and 140K proteins of which the latter shows multiple forms due to proteolysis. *Journal of Virology*, **156**, 417–22.

Jamieson, A. T., Gentry, G. A. & Subak-Sharpe, J. H. (1974). Induction of both thymidine and deoxycytidine kinase activity by herpes simplex virus. *Journal of General Virology*, **24**, 465–80.

Katoh, I., Yoshinaka, Y., Rein, A., Shibuya, M., Otaka, T. & Oraszlan, S. (1985). Murine leukaemia virus maturation: protease region is required for conversion from 'immature' to 'mature' core form and for virus infectivity. *Virology*, **145**, 280–92.

Kohl, N. E., Eimin, E. A., Schleif, W. A., Davis, L. J., Heimbach, J. C., Dixon, R. A. F., Scolnick, E. M. & Sigal, I. S. (1988). Active human immunodeficiency virus protease is required for viral infectivity. *Proceedings of the National Academy of Sciences, USA*, **85**, 4686–90.

Lankinen, H., Gräsland, A. & Thelander, L. (1982). Induction of a new ribonucleotide reductase after infection of mouse L cells by pseudorabies virus. *Journal of Virology*, **41**, 893–900.

La Porta, R. F. & Taichinan, L. B. (1982). Human papillomaviral DNA replicates as a stable episome in cultured human keratinocytes. *Proceedings of the National Academy of Sciences, USA*, **79**, 3393–7.

Lemaitre, M., Bayard, B. & Lebleu, B. (1987). Specific antiviral activity of a poly(L-lysine)-conjugated oligodeoxyribonucleotide sequence complementary to vesicular stomatitis virus N protein mRNA initiation site. *Proceedings of the National Academy of Sciences, USA*, **84**, 648–52.

Lever, A. M. L. (1988). Treatment of the chronic hepatitis B virus carrier state. *Journal of Infection*, **16**, 221–9.

Lightfoot, M. M., Coligan, J. E., Folks, T. M., Fauci, A. S., Martin, M. A. & Venkatesan, S. (1986). Structural characterisation of reverse transcriptase and endonuclease polypeptides of the acquired immunodeficiency syndrome virus. *Journal of Virology*, **60**, 771–5.

Loeb, D. D., Hutchinson III, C. A., Edgell, M. H., Farmerie, W. G. & Swanstrom, R. (1989). Mutational analysis of human immunodeficiency virus type 1 protease suggests functional homology with aspartic proteinases. *Journal of Virology*, **63**, 111–21.

McCance, D. J. (1986). Human papillomaviruses and cancer. *Biochemica et Biophysica Acta*, **823**, 195–205.

McClements, W., Yamanaka, G., Garsky, V., Perry, H., Baccehetti, S., Colonno, R. & Stein, R. B. (1988). Oligopeptides inhibit the ribonucleotide reductase of herpes simplex viruses by causing subunit separation. *Virology*, **162**, 270–3.

McGeogh, D. J., Dolan, A., Donald, S. & Brauer, D. H. K. (1986). Complete DNA sequence of the short repeat region in the genome of herpes simplex virus type 1. *Nucleic Acids Research*, **14**, 1727–45.

McGeogh, D. J., Dalrymple, M. A., Davison, A. J., Dolan, A., Frame, M. C., McNab, D., Perry, L. J., Scott, J. E. & Taylor, P. (1988). The complete DNA sequence of the long unique region in the genome of herpes simplex virus type 1. *Journal of General Virology*, **69**, 1531–74.

Machida, H. (1986). Comparison of susceptibilities of varicella zoster virus and herpes simplex virus to nucleoside analogues. *Antimicrobial Agents and Chemotherapy*, **29**, 524–6.

Maddon, P. J., Dalgleish, A. G., McDougal, J. S., Clapham, P. R., Weiss, R. A. & Axel, R. (1986). The T4 gene encodes the AIDS virus receptor and is expressed in the immune system and the brain. *Cell*, **47**, 333–48.

Maddon, P. J., Moleneaux, S. M., Maddon, D. E., Zimmerman, K. A., Godfrey, M., Alt, F. W., Chess, L. & Axel, R. (1987). Structure and expression of the human and mouse T4 genes. *Proceedings of the National Academy of Sciences, USA*, **84**, 9155–9.

Marchetto, S., Nisato, D. & Chermann, J-C. (1989). Inhibition of the maturation of HIV structural proteins by a renin peptide inhibitor. *Abstracts of the 5th International Conference on AIDS. Montréal*, p. 627.

Marion, P. L., Oshiro, L. S., Regnery, D. C., Scullard, G. H. & Robinson, W. S. (1980). A virus in Beechey ground squirrels that is related to hepatitis B virus of humans. *Proceedings of the National Academy of Sciences, USA*, **77**, 2941–5.

Marquez, V. E. & Lim, M-I. (1986). Carbocyclic nucleosides. *Medicinal Research Reviews*, **6**, 1–40.

Martin, J. A., Duncan, I. B., Hall, M. J., Lambert, R. W., Thomas, G. J. & Kai-In, P. W. (1988). Inhibitors of herpes simplex virus thymidine kinase: synthesis and biological properties. *Abstracts of the Second International Conference of Antiviral Research, Williamsburg, West Virginia, USA*, April 1988. *Antiviral Research*, **9**, 84.

Mason, W. S., Seal, G. & Summers, J. (1980). Virus of Pekin ducks with structural and biological relatedness to human hepatitis B virus. *Journal of Virology*, **36**, 829–36.

Matsukuru, M., Shinozuka, K., Zon, G., Mitsuya, H., Reitz, M., Cohen, J. S. & Broder, S. (1987). Phosphorothioate analogs of oligodeoxynucleotides: Inhibitors of replication and cytopathic effects of human immunodeficiency virus. *Proceedings of the National Academy of Sciences, USA*, **84**, 7706–10.

Meek, T. D., Dayton, B. D., Metcalf, B. W., Dreyer, G. B., Strickler, J. E., Gorniak, J. G., Rosenberg, M., Moore, M. L., Magaard, V. W. & Debouck, C. (1989). Human immunodeficiency virus 1 protease expressed in *Escherichia coli* behaves as a dimeric aspartic protease. *Proceedings of the National Academy of Sciences, USA*, **86**, 1841–5.

Mitsuya, H. & Broder, S. (1987). Strategies for antiviral therapy in AIDS. *Nature, London*, **325**, 773–8.

Moore, M. L., Byron, W. M., Fakhoury, S. A., Magaard, V. W., Huffman, W. F., Dayton, B. D., Meek, T. D., Hyland, L., Dreyer, G. B., Metcalf, B. W., Stickler, J. E., Gorniak, J. G. & Debouck, C. (1989). Peptide substrates and inhibitors of the HIV-1 protease. *Biochemical and Biophysical Research Communications*, **159**, 420–5.

Mous, J., Hemier, E. P. & Le Grice, S. F. J. (1988). Processing protease and reverse transcriptase from human immunodeficiency virus type 1 polyprotein in *Escherichia coli*. *Journal of Virology*, **62**, 1433–6.

Mullaney, J., Moss, H. W. McL. & McGeogh, D. J. (1989). Gene UL2 of herpes simplex virus type 2 encodes a uracil-DNA glycosylase. *Journal of General Virology*, **70**, 449–54.

Navia, M. A., Fitzgerald, P. M. D., McKeever, B. M., Leu, C-T., Heimbach, J. C., Herber, W. K., Sigal, I. S., Darke, P. L. & Springer, J. P. (1989). Three-dimensional structure of aspartyl protease from human immunodeficiency virus HIV-1. *Nature, London*, **337**, 615–20.

Nutter, L. M., Grill, S. P. & Cheng, Y-C. (1985). Can ribonucleotide reductase be considered as an effective target for developing antiherpes simplex virus type II (HSV-2) compounds. *Biochemical Pharmacology*, **34**, 777–80.

Offensperger, W-B., Walter, E., Offensperger, S., Zeschnigk, C., Blum, H. E & Gerok, W. (1988). Duck hepatitis B virus: DNA polymerase and reverse transcriptase activities of replicative complexes isolated from liver and their inhibitions *in vitro*. *Virology*, **164**, 48–54.

Olivo, P. D., Nelson, N. J. & Challberg, M. D. (1988). Herpes simplex virus DNA replication: the UL9 gene encodes an origin-dependent binding protein. *Proceedings of the National Academy of Sciences, USA*, **85**, 5414–8.

Olivo, P. D., Nelson, N. J. & Challberg, M. D. (1989). Herpes simplex virus type 1 gene products required for DNA replication: identification and overexpression. *Journal of Virology*. **63**, 196–204.

Paradis, H., Gaudreau, P., Brazeau, P. & Langelier, Y. (1988). Mechanism of inhibition of herpes simplex virus (HSV) ribonucleotide reductase by a nonapeptide corresponding to the carboxyl terminus of its subunit 2. Specific binding of a photoaffinity analogue, [4'-azido-Phe^6]HSV H2-(6–15), to subunit 1. *Journal of Biological Chemistry*, **263**, 16045–50.

Pauwels, R., Balzarini, J., Schols, D., Baba, M., Besmyter, J., Rosenberg, I., Holý, A. & De Clercq, E. (1988). Phosphorylmethoxyethyl purine derivatives, a new class of anti-human immunodeficiency virus agents. *Antimicrobial Agents and Chemotherapy*, **32**, 1025–30.

Peng, C., Ho, B. K., Chang, T. W. & Chang, N. T. (1989). Role of human immunodeficiency virus type 1 specific protease in core protein maturation and viral infectivity. *Journal of Virology*. **63**, 2530–56.

Popovic, M., Sarngadharan, M. G., Reed, E. & Gallo, R. C. (1984). Detection, isolation and continuous production of cytopathic retroviruses (HTLV-III) from patients with AIDS and pre-AIDS. *Science*, **224**, 497–500.

Pritchett, T. J., Brossmer, R., Rose, U. & Paulson, J. C. (1987). Recognition of monovalent sialosides by influenza virus H3 haemagglutinin. *Virology*, **160**, 502–6.

Reichman, R. C. & Strike, D. G. (1989). Pathogenesis and treatment of human genital papillomavirus infections: a review. *Antiviral Research*, **11**, 109–18.

Richards, A. D., Roberts, R., Dunn, B. M., Graves, M. C. & Kay, J. (1989). Effective blocking of HIV-1 protease activity by characteristic inhibitors of aspartic proteinases. *FEBS Letters*, **247**, 113–17.

Rosen, C. A., Sodroski, J. G., Campbell, K. & Haseltine, W. A. (1986). Construction of recombinant murine retroviruses that express the human T-cell leukaemia virus type II and human T-cell lymphotrophic virus type III *trans*activator genes. *Journal of Virology*, **57**, 379–84.

Rossman, M. G. (1988). Antiviral agents targeted to interact with viral capsid proteins and a possible application to human immunodeficiency virus. *Proceedings of the National Academy of Sciences, USA*, **85**, 4625–7.

Rossman, M. G., Arnold, E., Erickson, J. W., Frankenberger, E. A., Griffith, J. P., Hecht, H-J., Johnson, J. E., Kamer, G., Luo, M., Mosser, A. G., Rueckert, R. R., Sherry, B. & Vriend, G. (1985). Structure of a human common cold virus and functional relationship to other picornaviruses. *Nature, London*, **317**, 145–52.

Saul, R., Chambers, J. P., Molyneux, R. S. & Elbein, A. D. (1983). Castanospermine, a tetrahydroxylated alkaloid that inhibits β-glucosidase and β-glucocerebrosidase. *Archives of Biochemistry and Biophysics*, **221**, 593–7.

Saul, R., Ghidoni, J. J., Molyneux, R. J. & Elbein, A. D. (1985). Castanospermine inhibits α-glucosidase activities and alters glycogen distribution in animals. *Proceedings of the National Academy of Sciences, USA*, **82**, 93–7.

Schneider, J. & Kent, S. B. H. (1988). Enzymatic activity of a synthetic 99 residue protein corresponding to the putative HIV-1 protease. *Cell*, **54**, 363–8.

Sekyi, J-I., Shimada, N., Takahashi, K., Takita, T., Takeuchi, T. & Hoshino, H. (1989a). Inhibition of infectivity of human immunodeficiency virus by a novel nucleoside, oxetanocin and related compounds. *Antimicrobial Agents and Chemotherapy*, **33**, 773–5.

Sekyi, J-I., Takeuchi, T., Shimada, N., Takahashi, K., Takita, T. & Hoshino, H. (1989b). Anti-HIV activities of a novel nucleoside oxetanocin and its derivatives. *Abstracts of the 5th International Conference on AIDS. Montréal*, p. 562.

Shipman, Jr, C., Smith, S. H., Drach, J. C. & Klayman, D. L. (1981). Antiviral activity of 2-acetylpyridine thiosemicarbazones against herpes simplex virus. *Antimicrobial Agents and Chemotherapy*, **19**, 682–5.

Shipman, Jr, C., Smith, S. H., Drach, J. C. & Klayman, D. L. (1986). Thiosemicarbazones of 2-acetylpyridine, 2-acetylquinoline, 1-acylylisoquinoline and related compounds as inhibitors of herpes simplex viruses *in vitro* and in a cutaneous herpes guinea pig model. *Antiviral Research*, **6**, 197–222.

Sim, I. S., Picton, C., Cosstick, R., Jones, A. S., Walker, R. T. & Chamiec, A. J. (1988). Inhibition of the herpes simplex virus thymidine kinase by 5′-substituted thymidine analogues. Comparison of the types 1 and 2 enzymes. *Nucleosides and Nucleotides*, **7**, 129–35.

Smith, C. C., Aurelian, L., Reddy, M. P., Miller, P. S. & Ts'o, P. O. P. (1986). Antiviral effects of an oligo(nucleoside methylphosphonate) complementary to the splice junction of herpes simplex virus type 1 immediate early pre-mRNA's 4 and 5. *Proceedings of the National Academy of Sciences, USA*, **83**, 2787–91.

Smith, D. H., Byrn, R. A., Marsters, S. A., Gregory, T., Groopman, J. E. & Capon, D. J. (1987). Blocking of HIV-1 infectivity by a soluble secreted form of the CD4 antigen. *Science*, **238**, 1704–7.

Smith, R. F. & Smith, T. F. (1989). Identification of new protein kinase-related genes in three herpesviruses, herpes simplex virus, varicella zoster virus and Epstein–Barr Virus. *Journal of Virology*, **63**, 450–5.

Spector, T., Averett, D. R., Nelson, D. J., Lambe, C. U., Morrison, Jr, R. W., St Clair, M. H. & Furman, P. A. (1985). Potentiation of anti-herpetic activity of acyclovir by ribonucleotide reductase inhibition. *Proceedings of the National Academy of Sciences, USA*, **82**, 4254–7.

Spector, T., Harrington, J. A., Morrison, Jr, R. W., Lambe, C. U., Nelson, D. J., Averett, D. R., Biron, K. & Furman, P. A. (1989). 2-acetylpyridine 5-[(dimethylamino)thiocarbonyl]-thiocarbonohydrazone (A1110U), a potent inactivator of ribonucleotide reductases of herpes simplex and varicella zoster viruses and a potentiator of acyclovir. *Proceedings of the National Academy of Sciences, USA*, **86**, 1051–5.

Spector, T., Stonehuerner, J. G., Biron, K. K. & Averett, D. R. (1987). Ribonucleotide reductase induced by varicella zoster virus: characterisation and potentiation of acyclovir by its inhibition. *Biochemical Pharmacology*, **36**, 4341–6.

Stephenson, M. C. & Zamecnik, P. C. (1978). Inhibition of Rous sarcoma viral RNA translation by a specific oligodeoxynucleotide. *Proceedings of the National Academy of Sciences, USA*, **75**, 285–8.

Summers, J., Smolec, J. M. & Snyder, R. (1978). A virus similar to human hepatitis B virus association with hepatitis and hepatoma in woodchucks. *Proceedings of the National Academy of Sciences, USA*, **75**, 4533–7.

Sunkara, P. S., Bowlin, T. L., Liu, P. S. & Sjoerdsma, A. (1987). Antiretroviral activity of castanospermine and deoxynojirimycin, specific inhibitors of glycoprotein processing. *Biochemical and Biophysical Research Communications*, **148**, 206–10.

Sunkara, P. S., Taylor, D. L., Kang, M. S., Bowlin, T. L., Liu, P. S., Thyms, A. S. & Sjoerdsma, A. (1989). Anti-HIV activity of castanospermine analogues. *Lancet*, **i**, 1206.

Suzuki, S., Lee, B., Luo, W., Tovell, D., Robins, M. J. & Tyrrell, D. L. J. (1988). Inhibition of duck hepatitis B virus by purine 2′3′ dideoxynucleosides. *Biochemical and Biophysical Research Communications*, **156**, 1144–51.

Tenser, R. B., Miller, R. L. & Rapp, F. (1979). Trigeminal ganglion infection by thymidine kinase negative mutants of herpes simplex virus. *Science*, **205**, 915–17.

Thompson, R., Honess, R. W., Taylor, L., Morran, J. & Davison, A. J. (1987). Varicella zoster virus specifies a thymidylate synthetase. *Journal of General Virology*, **68**, 1449–55.

Traunecker, A., Lüke, W. & Karjalainen, K. (1988). Soluble CD4 molecules neutralise human immunodeficiency virus type 1. *Nature, London*, **331**, 84–6.

Ueda, K., Tsurimoto, T., Nagahata, T., Chisaka, O. & Matsubara, K. (1989). An *in vitro* system for screening anti-hepatitis B virus drugs. *Virology*, **169**, 213–16.

Vince, R., Hua, M., Brownwell, J., Daluge, S., Lee, F., Shannon, W. M., Lavelle, G. C., Qualls, J., Weislow, O. S., Kiser, R., Canonico, P. G., Shultz, R. H., Narayanan, V. L., Mayo, J. G., Shoemaker, R. H. & Boyd, M. R. (1988). Potent and selective activity of a new carbocyclic nucleoside analogue (carbovir: NSC 614846) against human immunodeficiency virus *in vitro*. *Biochemical and Biophysical Research Communications*, **156**, 1046–53.

Von der Helm, K., Gürtler, L., Eberle, J. & Deinhardt, F. (1989). Inhibition of HIV replication in cell culture by the specific aspartic protease inhibitor pepstatin A. *FEBS Letters*, **247**, 349–52.

Votruba, I., Bernaerts, R., Sakuma, T., De Clercq, E., Merta, A., Rosenberg, I. & Holý, A. (1987). Intracellular phosphorylation of broad spectrum anti-DNA virus agent (S)-9-(3 hydroxy-2-phosphonylmethoxypropyl)adenine and inhibition of viral DNA synthesis. *Molecular Pharmacology*, **32**, 524–9.

Wainberg, M. A., Stern, M. D., Martel, R., Belleau, B. & Soudeyns, H. (1989). Antiv-HIV activity, toxicity and pharmacokinetics of totally novel nucleoside analogs. *Abstract from 5th International AIDS Conference. Montréal*, p. 552.

Walker, B. D., Kowalski, M., Goh, W. C., Kozarsky, K., Krieger, M., Rozen, C., Rohrschneider, L., Haseltine, W. A. & Sodroski, J. (1987). Inhibition of human imunodeficiency virus syncytium formation and virus replication by castanospermine. *Proceedings of the National Academy of Sciences, USA*, **84**, 8120–4.

Watts, S. L., Phelps, W. C., Ostrow, R. S., Zachow, K. R. & Faras, A. J. (1984). Cellular Transformation by human papillomavirus DNA *in vitro*. *Science*, **225**, 634–6.

Weis, W., Brown, J. H., Cusack, S., Paulson, J. C., Skehel, J. J. & Wiley, D. C. (1988). Structure of the influenza virus haemagglutinin complexed with its receptor, sialic acid. *Nature, London*, **333**, 426–31.

White, E. L., Parker, W. B., Macy, L. J., Shaddix, S. C., McCaleb, G., Secrist III, J. A., Vince, R. & Shannon, W. M. (1989). Comparison of the effect of carbovir, AZT and dideoxynucleoside triphosphates on the activity of human immunodeficiency virus reverse transcriptase and selected human polymerases. *Biochemical and Biophysical Research Communications*, **161**, 393–8.

Whitley, R. J. (1988a). Herpes simplex virus infections of the central nervous system. *American Journal of Medicine,* **85** (supplement 2A), 61–7.

Whitley, R. J. (1988b). Ganciclovir – have we established clinical value in the treatment of cytomegalovirus infections? *Annals of Internal Medicine*, **108**, 452–4.

Wlodawer, A., Miller, M., Jaskólski, M., Sathyanarayana, B. K., Baldwin, R., Weber, I. T., Selk, L. M., Clawson, L., Schneider, J. & Kent, S. B. H. (1989). Conserved folding in retrovirus proteases: Crystal structure of a synthetic HIV-1 protease. *Science*, **245**, 616–21.

Worrad, D. M. & Caradonna, S. (1988). Identification of the coding sequence for herpes simplex virus uracil-DNA glcosylase. *Journal of Virology*, **62**, 4774–7.

Wright, C. M., Felber, B. K., Paskalis, H. & Pavlakis, G. N. (1986). Expression and characterisation of the *trans*-activator of HTLV-III/LAV virus. *Science*, **234**, 988–92.

Wu, C. A., Nelson, N. J., McGeogh, D. J. & Challberg, M. D. (1988). Identification of herpes simplex virus type 1 genes required for origin-dependent DNA synthesis. *Journal of Virology*, **62**, 435–43.

Zamecnik, P. C. & Stephenson, M. L. (1978). Inhibition of Rous sarcoma virus replication and cell transformation by a specific oligodeoxynucleotide. *Proceedings of the National Academy of Sciences, USA*, **75**, 280–4.

Zamecnik, P. C., Goodchild, J., Taguchi, Y. & Sarin, P. S. (1986). Inhibition of replication and expression of human T-cell lymphotrophic virus type III in cultured cells by exogenous synthetic oligonucleotides complementary to viral RNA. *Proceedings of the National Academy of Sciences, USA*, **83**, 4143–6.

Zerial, A., Thuong, N. T. & Hélène, C. (1987). Selective inhibitors of the cytopathic effect of type A influenza virus by oligodeoxynucleotides covalently linked to an interacting agent. *Nucleic Acids Research*, **15**, 9909–19.

Zhu, L. & Weller, S. K. (1988). UL5, a protein required for HSV DNA synthesis: genetic analysis, overexpression in *Escherichia coli* and generation of polyclonal antibodies. *Virology*, **166**, 366–78.

U.C.W. ABERYSTWYTH LIBRARY